CONTRIBUTIONS TO THE STRATIGRAPHY OF NEW ENGLAND

The Geological Society of America, Inc.
Memoir 148

Contributions to the Stratigraphy of New England

Edited by
Lincoln R. Page

1976

The Memoir series was originally made possible through the bequest of Richard Alexander Fullerton Penrose, Jr.

Library of Congress Catalog Card Number 76-9220
I.S.B.N. 0-8137-1148-7

Published by
The Geological Society of America, Inc.
3300 Penrose Place
Boulder, Colorado 80301

Printed in the United States of America

Contents

Geological Society of America
Memoir 148

Introduction and Summary

LINCOLN R. PAGE
P.O. Box 171
Melvin Village, New Hampshire 03850

Contributions to the Stratigraphy of New England is an outgrowth of a symposium that I organized at the request of the program committee for the Northeast Section meeting of the Geological Society of America held in Hartford, Connecticut, in 1971. Because the then recently published "Billings volume" (Zen and others, 1968) contained many excellent papers that thoroughly described the stratigraphy of many parts of New England, particularly in Vermont, New Hampshire, and Maine, the symposium emphasized the Massachusetts and Connecticut areas and the more recent work done in Maine (Fig. 1). The intent was to select papers that would complement this published work and present different points of view—not to review the stratigraphy of New England. Pavlides, Dixon, Zen, and Hatch and Stanley, whose papers are included in this volume, had previously published definitive papers of the areas being considered in this symposium. They were given preliminary copies of the pertinent papers and were asked to present prepared discussions pointing out contrary points of view. B. A. Hall, Pollock, and Dolan have contributed a written comment that adds materially to an understanding of the upper part of the stratigraphic section; Goldsmith has done the same for the lower part.

The authors of the symposium papers were asked first to emphasize the internal stratigraphy or lithologic sequences within or between mapped units in their areas; the nature of layering, bedding, and other primary structures such as cross-bedding, graded bedding, and channel fills; the type and source of the protolith; and the environment of deposition. Second, the authors were to compare their lithologic and map units with previously named and published stratigraphic units in nearby areas. Thickness of sections, determined from maps or field measurement; ages based on superposition, fossils, and radiometric methods; lateral and vertical facies changes within the units; nonrepetition of sections; and unconformities were items of special concern because these have important implications with reference to the paleogeography, structure, and geologic history of New England, where most of the nonigneous terranes are underlain by metamorphic rocks. The papers by Brookins and Naylor discuss isotopic ages in relation to the stratigraphy of the metamorphic rocks and their absolute ages.

The individual papers in this volume (Fig. 1) present detailed information at intervals along the New England part of the Appalachian geosyncline and on each of the major north- and northeast-trending structural units in it. These are, from west to east, the Green Mountain anticlinorium, the Connecticut-Gaspe synclinorium, the Bronson Hill-Boundary Mountains anticlinorium, and the Merrimack synclinorium. The papers also point out the presence of some of the major fault zones that truncate these units.

Some of the areas described have a stratigraphy based on fossils, but most of the areas have been so highly metamorphosed that the stratigraphy must be based on superposition determined from relict primary structures and field relationships. In some areas isotopic ages can be used successfully; in others they do not appear to match the field data. The presence of similar rocks in various parts

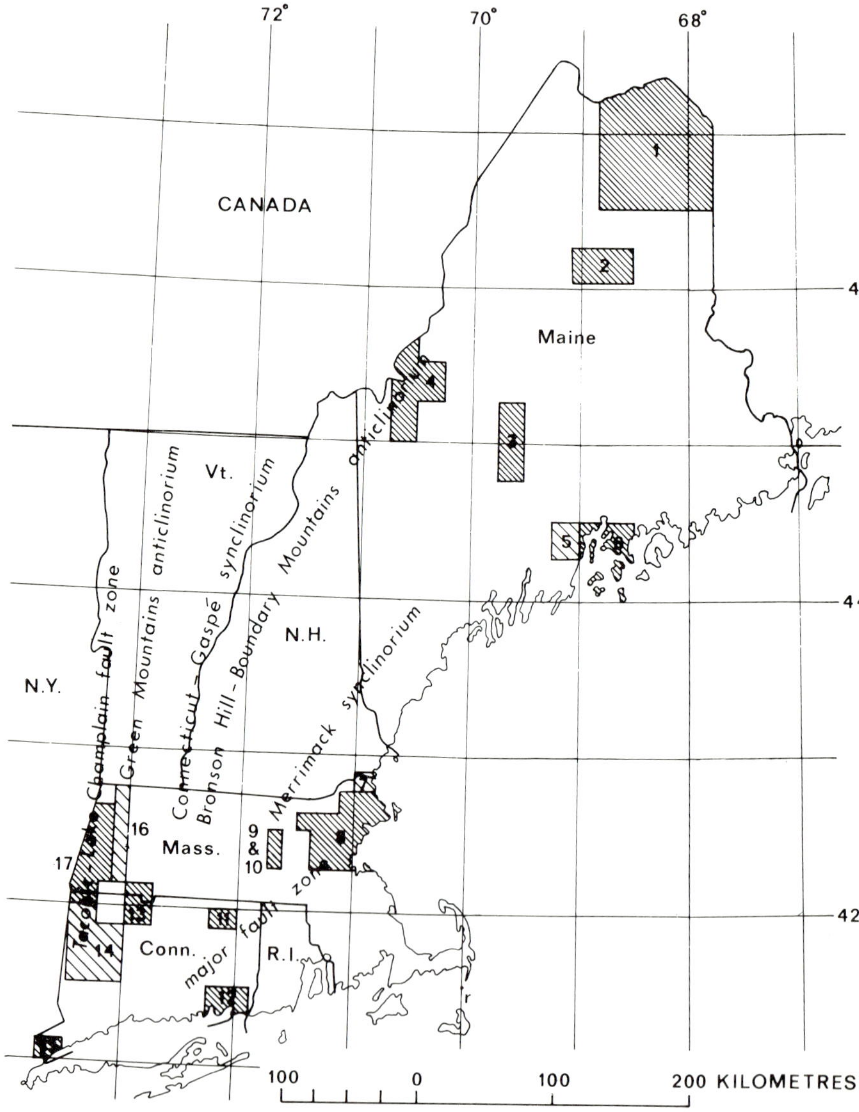

Figure 1. Major structural units of New England and areas described in the papers of this volume. 1, Roy and Mencher; 2, B. A. Hall, Pollock, and Dolan; 3, Ludman; 4, Boudette and Boone; 5, Bickel; 6, Brookins; 7, Shride; 8, Bell and Alvord; 9, Skehan and Abu-moustafa; 10, Peck; 11, Peper and Pease; 12, Goldsmith; 13, Schnabel; 14, Gates and Martin; 15, L. M. Hall; 16, Norton; 17, Ratcliffe and Zartman.

of the section (see Fig. 2, for example) makes reliability of correlations uncertain if they are based on lithologic similarity alone, and for places in New England where lithologic similarity has been used as the primary basis for correlation, the resulting interpretations yield very complex or improbable structures. The presence of minor folds within units is common and suggests the presence of larger folds of similar pattern, although few such folds show in the map patterns of the many areas where the internal stratigraphy of units is well known. This is particularly evident on the east side of the Berkshires in Massachusetts where the rocks of Silurian-Devonian age are isoclinally folded on a broad scale but their lower contact, for many kilometres, is not folded. Isoclinal folds in the older rocks are several orders of magnitude smaller and much less abundant, although they record a similar sense of movement. Skehan and Abu-moustafa discuss places, such as in the Wachusett-Marlborough Tunnel, where it is evident that many of these minor folds, particularly those with folded foliation, are related to regional faulting, not folding, and are postmetamorphic in age. Other folds with axial plane foliation are related to regional folding or to faulting during the Acadian orogeny.

Bell and Alvord, Skehan and Abu-moustafa, Peck, and Peper and Pease describe very thick stratigraphic sections and point out the absence of significant folds superimposed on the Merrimack synclinorium. These and other papers also show the importance of faults, younger than the metamorphism of the Acadian orogeny, that cut the synclinorium and other major regional structures. The width of some of the fault zones and the displacements on them make them truly regional in extent and significant in shaping the eastern edge of the North American continent.

The data in this volume are from many sources, and they are the result of many interpretations. Each paper is independent of the others in terms of the nomenclature used, and the papers may not necessarily be entirely compatible. I have attempted to summarize what I consider the most important contributions of ideas concerning stratigraphy, paleogeography, age, and environment of deposition. On the basis of my fifteen or more years of research on the geology of New England, I have tried to point out the implications of these papers as I see them in the area (see Figs. 2 through 5). Where my views differ with those of the other authors, I have taken pains to present both points of view. I thank all the contributors and those who have critically reviewed the manuscripts for their patience and helpfulness in making this volume possible.

STRATIGRAPHY

The most important contributions to stratigraphy in this volume are the individual stratigraphic sections developed for each area. The detailed succession and thicknesses of lithologic units given for the local areas, together with the basis for determining superposition, should lead to better correlation between areas and thus allow for more accurate structural interpretations. Because of the limitations of petrographic nomenclature, the same rock names reappear in several parts of the section, but the characteristics of bedding or layering differ and thus allow recognition of different units in the field. One of the very real shortcomings of this volume is the variation in the use of terms, such as "granite gneiss," "granitic gneiss," "granulite," "granofels," "gneiss," and "schist." "Granulite," for example, is applied to several very different rocks by different workers, and as a result, correlations such as shown in Figures 2 through 5 must be considered generalizations.

New data on thickness of section, lateral and vertical facies changes, lack of

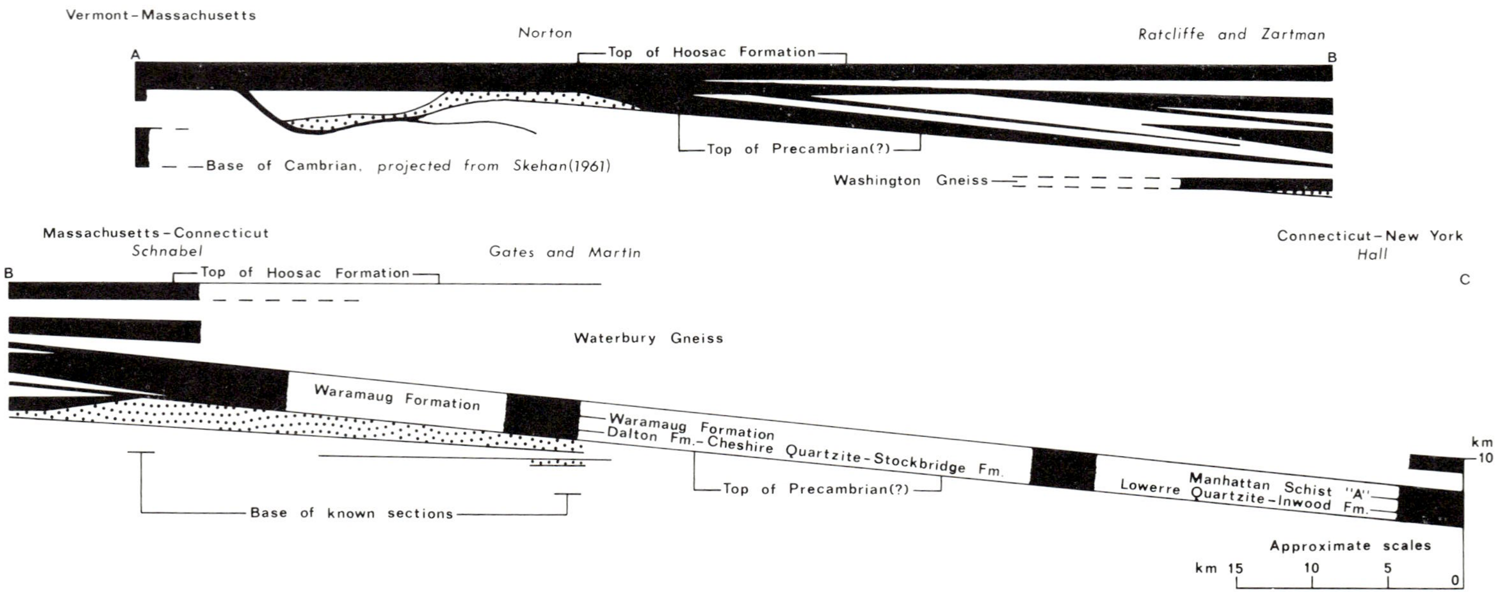

Figure 2. Diagrammatic cross section of the original sedimentary wedge of Paleozoic and Precambrian rocks north to south across western Massachusetts and Connecticut. Datum is the top of the Hoosac Formation (controlled primarily by stratigraphic sections and maps in this volume as indicated in text). Epiclastic rocks—black; felsic and mafic volcaniclastic rocks—dots; felsic volcaniclastic rocks—no pattern.

repetition of major lithologic sequences, and observed unconformities are summarized below from a regional point of view and presented diagrammatically in Figures 2 through 5. These have a direct bearing on the solution of some of the more important regional stratigraphic problems, such as (1) the nature of the Precambrian-Cambrian boundary, (2) the lateral and vertical facies changes along the limestone bank or reef that existed in western New England in Cambrian and Ordovician time, (3) the facies changes within the submarine volcanic rocks that predominate to the southeast of this bank, (4) correlation of rock sequences of Ordovician and older age, (5) distribution and source of rocks of Silurian and Devonian age, and (6) variations in relative and radiometric ages.

The correlation of individual facies within time units has been made in some areas (see papers by Roy and Mencher; B. A. Hall, Pollock, and Dolan; Ludman; and Schnabel), but in many of the metamorphic terranes, lithologies have been grouped and structural interpretations have been made that hide the many variations in composition and thickness of time units. These variations must be distinguished to allow correlation of the various facies. The following discussions attempt to point out some of the more pertinent facts from various detailed correlations relative to these facies changes and to indicate problems and areas for further study.

Thickness of Section

The maximum thickness of the stratigraphic sections described in these papers, correlated on the base of the upper highly aluminous clastic sequence that is of probable Silurian and Devonian age, is more than 38,000 m (Fig. 5); the sections are 12,000 m thick above and 26,000 m thick below the marker horizon. This is believed to be a minimum thickness because all known repetition by folding and faulting has been removed; an additional amount of section is known to have been cut out by faults, igneous bodies, or unconformities, and not all of the Precambrian section is exposed. In general, the entire section thins from southeast to northwest, north of a line from Boston to New Haven; to the south the known section thins southeastward as the result of erosion in pre-Silurian time and nondeposition in Silurian and Devonian time. Part of this widespread land area, along what is now the southeast side of New England, was covered by sediments in late Paleozoic and Mesozoic time. The subaerial volcanic rocks of the Newbury Volcanic Complex and the Castine Volcanics (described by Shride and Brookins, respectively) with their interbedded marine units of Silurian-Devonian age mark the western edge of this high area. This area also could have served as the eastern source of the Devonian delta studied by B. A. Hall, Pollock, and Dolan. The western edge of this land mass is now marked by the major northeast-trending fault zone in southern New England, mentioned by Bell and Alvord, Skehan and Abu-moustafa, Dixon, and Peper and Pease, that abruptly terminates the thicker Paleozoic section on the northwest. The original position of the rocks south of the fault zone is unknown, but the total horizontal displacement on this right-lateral fault, indicated by the fossiliferous rocks, is possibly several hundred kilometres. The contrasts in thickness between the Paleozoic section described by Bickel and discussed by Brookins in Maine also illustrate the effect of this faulting on stratigraphic correlations in nearby areas. Bell and Alvord state that in northeastern Massachusetts about 20,000 m of pre-Silurian rocks are brought up in this fault zone. These are overlain by the Lynn Volcanics that may be correlative with the Newbury Volcanic Complex of latest Silurian or earliest Devonian age.

To the north of this major fault zone, Peck records more than 6,600 m of rocks of presumed Silurian and Devonian age; farther to the southwest, Pease and Peper record a maximum thickness of 16,000 m of rocks of probable Late Ordovician to Devonian age. In the Belfast area, southern Maine, Bickel has about 2,700 m of similar rocks overlying about 9,200 m of pre-Silurain units. To the north and northwest of the Belfast area, thicker sections of each of these main groups are indicated. In the Rangeley area, Boudette and Boone record 7,300 m of pre-Silurian rocks, including at least 3,300 m of Precambrian rocks. The 3,600-m-thick section of rocks of Silurian age described by Ludman thins to 2,100 m in Aroostook County, according to Roy and Mencher. Above this, B. A. Hall, Pollock, and Dolan find at least 4,000 m of rocks of Devonian age in the Matagamon area; total thicknesses elsewhere in Maine exceed 6,000 m.

Lateral Facies Changes

Lateral facies changes are described in detail in the papers by Roy and Mencher; B. A. Hall, Pollock, and Dolan; Ludman; and Schnabel. The data presented by all of the authors, taken together, suggest some lateral facies changes of regional extent. These are diagrammatically shown in Figures 2 through 5.

The oldest rocks in western New England are gneisses, schists, amphibolites, calc-silicate rocks, and quartzites that were derived from arkosic epiclastic and volcaniclastic sedimentary rocks and lava flows deposited in a basin across the trend of older rocks of Precambrian age exposed in the Adirondacks of New York, as indicated by the new magnetic map of Zietz and others (1972). Some epiclastic units—such as the Washington Gneiss, according to Ratcliffe and Zartman—show a decrease in grain size from the northwest to the southeast together with an increase in the thickness of interlayered volcaniclastic units. There also seems to be a decrease in the potassium feldspar and blue quartz content upward and eastward. These features suggest that the western edge of this early depositional basin lies to the west of the study area and that the volcanic source is to the east. Figure 2, representing the eastern edge of the carbonate bank or reef, shows the position of the Washington Gneiss with reference to the rocks of known Cambrian age and indicates that a similar facies change exists in the younger units in southern New England that are above, and in part equivalent to, the miogeosynclinal units (the Dalton Formation, Cheshire Quartzite, and Stockbridge Formation) described by L. M. Hall, Gates and Martin, and Ratcliffe and Zartman.

The miogeosynclinal rocks of western Connecticut (described by Gates and Martin and by L. M. Hall) and of Massachusetts (described by Ratcliffe and Zartman and by Norton) represent, in my opinion, carbonate sediments and beach sands deposited on the eastern edge of an offshore bank, platform, or barrier reef with the main shoreline being in what is now New York. Thicker sections of carbonate rocks to the west represent the main part of the limestone banks or reefs that developed in Cambrian and Ordovician time. These carbonate rocks grade westward into organic-rich pelitic and then clastic units along the shoreline; to the east these bank or reef deposits interfinger and are interlayered with fine-grained clastic and pelitic sedimentary rocks (Hoosac Formation) that grade eastward into submarine volcanic flows and volcaniclastic and graywacke sedimentary rocks (Figs. 2 through 5). The Ordovician carbonate rocks also grade laterally and vertically into pelitic rocks (Walloomsac Formation) that thin and interfinger eastward with rocks of volcanic origin; to the west (back reef), the Ordovician carbonate rocks grade into the clastic and pelitic sedimentary rocks of the New York area. The 3,786-m-deep Shell-St. Armand West No. 1 well, drilled north of Phillipsburg, Quebec, in 1972-73, cut 2,175 m of what appears to be this part of the section below a large fault zone at 1,605-m depth. The same lateral changes east of the reef extend northward through Vermont, New Hampshire, and Maine into Canada. In northern Maine, Ordovician carbonate-rich, fine-grained clastic and pelitic sedimentary rocks are dominant over the volcaniclastic sedimentary units, according to Roy and Mencher.

In Silurian time, coarse beach and littoral deposits formed along what is now the Bronson Hill-Boundary Mountains anticlinorium in New Hampshire, Massachusetts, and Connecticut. These grade eastward into more pelitic units that are interfingered and interlayered with volcanic debris as noted by Skehan and Abu-moustafa and by Peper and Pease. In Maine, on the south side of the Bronson Hill-Boundary Mountains anticlinorium, Boudette and Boone describe the northwestward thinning of a coarse clastic sequence of this age that grades southeastward into the thicker clastic units mapped by Ludman. This coarse clastic sequence with little or no volcanic component was derived from the northwest and represents an offshore, shallow-water environment. Roy and Mencher describe in detail the eastward lateral change of the individual formation units of Ordovician to Devonian age in northern Maine where the volcanic components are sporadic and lenticular but are more abundant in the southern part of this area.

Some of the most interesting lateral facies changes are in the rocks of the Lake Matagamon delta of Devonian age studied by B. A. Hall, Pollock, and Dolan. This deposit records an abrupt change in the depositional history of the basin. The coarser sedimentary deposits of the delta grade westward into widespread pelitic units. In places the pelitic rocks of Devonian age contain sporadic lenses of volcanic debris, perhaps partly submarine as noted by Peper and Pease; the volcanic debris is more abundant along the southeast edge of this depositional basin. Stanley and Hatch do not record appreciable volcanic components in rocks of Silurian-Devonian age in western Massachusetts and Connecticut.

Vertical Facies Changes

All of the papers record local vertical facies changes that suggest differences in the environments of deposition or the source of the sediments. Perhaps the most important are the alternations that record the relative importance of mafic versus felsic volcanic material and a change in the proportions of epiclastic sediments contributed to the depositional basin throughout its history. (Figs. 2 through 5). If the entire stratigraphic sequence described in this volume is considered on a regional basis, there are four general facies groups, as follows:

1. The pre-Ordovician rocks are quartzites, aluminous and potassium-rich gneisses and schists interlayered with amphibolite and plagioclase, biotite, and hornblende gneisses. The oldest rocks in western Massachusetts are conglomeratic quartzites, described by Ratcliffe and Zartman, that contain pebbles of quartz and microcline feldspar; the quartz is commonly blue. In northwestern Maine, Boudette and Boone report that the oldest conglomeratic quartzites contain boulders of previously highly metamorphosed rocks and white vein-quartz pebbles. These are interlayered with potassium-rich quartzwackes. In southeastern New England, quartzites and aluminous schists overlie and are interlayered with microcline-rich, dark and light banded gneisses and amphibolites. In southwestern New England, this older sequence of clean quartzite (derived from beach sands) and aluminous schists is interlayered with, and overlain by, microcline-rich gneisses, in part conglomeratic. Marbles (derived from reef deposits) overlie the quartzite and interfinger with the schists that form the overlying Hoosac and Waramaug Formations. Similar schists interfinger eastward and southward (Fig. 2) with dominantly potassium-rich gneisses (Norton, however, attributes this interfingering to thrusting).

2. The Ordovician rocks (Figs. 3 through 5) are characterized by potassium-poor

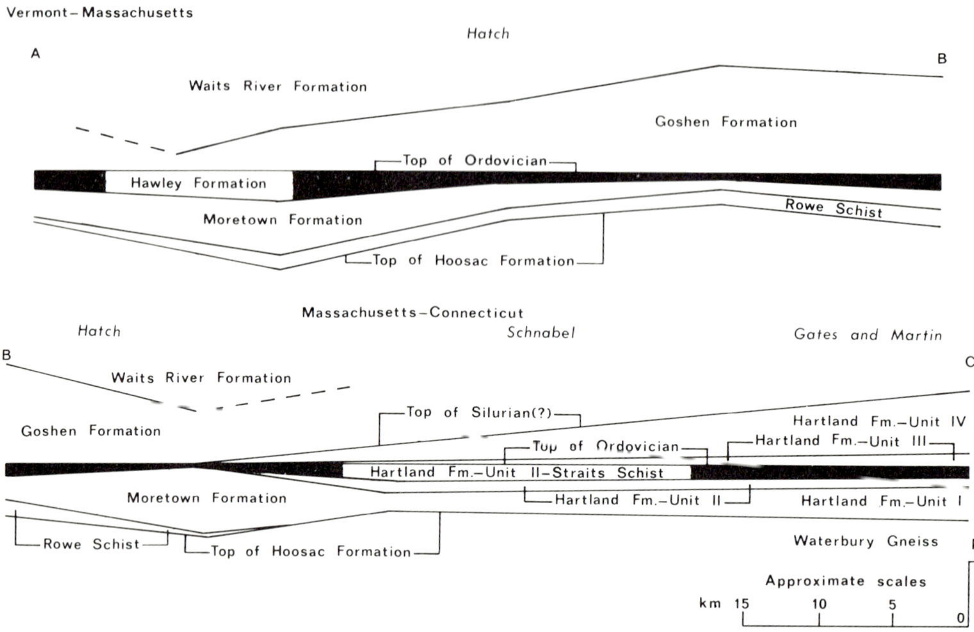

Figure 3. Diagrammatic cross section north to south across west-central Massachusetts and Connecticut showing the original thickness of Ordovician, Silurian, and Devonian rocks. Datum is the unconformity at the top of Ordovician rocks (controlled primarily by stratigraphic sections and maps in this volume as indicated in text).

felsic and mafic gneisses, sulfidic graphite schists, and iron- and manganese-rich carbonate and calcsilicate rocks. The gneisses are dominant in the lower part of this group in eastern and southern New England. These are overlain by, and interfinger with, thin sulfidic schists and gneisses that are relatively low in aluminum. There are at least four carbonate-rich zones within the group of gneisses and schists. Above the sulfidic schists is a pelitic and carbonate-rich zone that underlies, and is interlayered with, porphyroblastic biotite gneisses in east-central Massachusetts and Connecticut, as described by Pease and Peper. There is a general increase in epiclastic and a decrease in volcanic components upward in the Ordovician sequence in the center of the depositional basin.

In the Rangeley area of Maine, Boudette and Boone find that a similar sequence of dominantly metavolcanic rocks at the base of the Ordovician rocks grades upward into one of dominantly siliceous schists. In Aroostook County, Maine, Roy and Mencher find that volcanic components are subordinate and that carbonate rocks become more important in the phyllites and schists.

3. Most of the Silurian rocks described in this volume are quartzites or quartzose schists and gneisses that grade upward through interlayering and lateral facies changes into aluminous and carbonate-rich phyllites and schists. Rocks of this age on the southern end of the Bronson Hill-Boundary Mountains anticlinorium show a similar change from conglomeratic quartzite upward through quartzite to carbonate rocks and aluminous schists, both of which in places are rich in iron, sulfur, and carbon. Volcanic rocks occur sporadically in this part of the section, but they usually are subordinate in the areas described by Skehan and Abu-moustafa, Peper and Pease, Schnabel, and Gates and Martin. They are dominant, however, in the Newburyport area of Massachusetts and the Castine area of Maine, according to Shride and to Brookins.

Roy and Mencher describe in detail the metashale and metacarbonate facies with minor coarser clastic and volcanic components of the Silurian rocks in northern Maine. To the south, Boudette and Boone find metaconglomerates at the base that grade upward and southeastward into the metasandstones and metashales mapped by Ludman. Farther to the southeast, Bickel describes a section that is similar to that recorded by Skehan and Abu-moustafa in southern New England; these rocks perhaps more closely represent the eastern edge of the depositional basin. Bickel, however, designates these rocks as of Cambrian to Ordovician age.

4. Above the quartzose, sulfidic, and aluminous Silurian rocks is a very thick section of aluminous schists and quartzites characterized by graded bedding and lateral and vertical facies changes into iron-carbonate marbles and schists. Such rocks are discussed by Roy and Mencher; B. A. Hall, Pollock, and Dolan; Stanley and Hatch; Schnabel; Peper and Pease; and Peck. In places these rocks are part of deltaic structures (such as described by B. A. Hall, Pollock, and Dolan) where the facies change upward from fine-grained to coarser grained rocks.

Nonrepetition of Section

The lateral and vertical facies changes within and between the major lithologic sequences of the New England area indicate the complexity of the problem of correlation from one part of the area to another on the basis of lithologic similarity alone. Because much of the total section is made up of volcanic or coarse epiclastic material, both the entire sequence as well as the individual units are lenticular, and few units serve as good marker horizons over long distances. Thus, map units are bundles of these lenses. Within these units the lensing produces a pattern commonly mistaken for isoclinal folds; thus, the superposition needs to be determined for each rock type throughout the units. Widespread pelitic carbonate-rich units and flow sequences, therefore, serve as the most reliable horizons upon which to base structural interpretations or time correlations. The detailed sections presented in this volume are conclusive evidence that many of the lithologic units described, although similar in name, are distinct and not a repetition by folding or faulting. In most places, original sedimentary and volcanic structures have been used to determine the sequence of units. In Maine, fossils were used to determine unit sequence by Ludman, by Roy and Mencher, and to some extent by Boudette and Boone; in Connecticut, structural position was used by Gates and Martin, and in other areas facies changes were used. The careful mapping and reporting of Skehan and Abu-moustafa in the Wachusett-Marlborough Tunnel show clearly that the sequences of even thin units and bundles of units are not repeated.

Unconformities

Several unconformities of major importance and others of lesser significance are described in this volume; however, in my opinion, none of the authors describes or conclusively recognizes the profound unconformity that exists at the top of the Precambrian rocks of the Adirondacks in New York and the Canadian Shield in Quebec. This unconformity is marked by sandstone of Cambrian age that overlaps transgressively—and with sharp angularity—intensely metamorphosed rocks of Precambrian age. Where Cambrian fossiliferous rocks are found in southwestern New England, they are interlayered and interfinger with feldspathic gneisses and aluminous quartzites and schists designated as Precambrian by Ratcliffe and Zartman, Norton, Gates and Martin, Schnabel, and L. M. Hall and by D. S. Harwood at the symposium. In my opinion, the rocks below this unconformity are no more intensely metamorphosed than the overlying quartzites, metaconglomerates, marbles, schists, and gneisses. Conglomerates occur locally at this boundary but in most places do not appear to be of regional significance. The conglomerate that shows the greatest discordance with the underlying rocks contains white and black chert pebbles, according to Ratcliffe and Zartman; this feature suggests that the unconformity really formed during the Taconic orogeny. These Precambrian schists for the most part, are highly aluminous like those of the Hoosac Formation which interfingers and is interlayered with the Dalton Formation that is dated by *Olenellus*. The Hoosac Formation, in turn, contains interfingering layers and lenses of gneisses, as described by Norton and by Schnabel, of the same type as those found below the break (Fig. 2). The epiclastic units in both the Hoosac and older rocks are the type of sediment normally derived from deep chemical weathering of a large crystalline land area, such as the Adirondacks and Canadian Shield must have been in Precambrian time. The aeromagnetic maps of Zietz and others (1972) show the broad anomalies of the highly magnetic rocks of the Adirondacks trending nearly east-northeast into the less magnetic north-trending rocks of supposed Precambrian age in the northern Berkshires. Thus it seems probable to me that the Precambrian unconformity recognized in New York either has not been exposed at all in this part of New England or has a far more restricted distribution than has generally been assumed. The conglomerates and unconformities at the base of the sequence that comprises the Dalton Formation, Chesire Quartzite, and Stockbridge Formation are more likely of local significance in that they record the tectonic adjustments that first raised the older volcanic and epiclastic sequence to the positions where it could be transgressed by the sea and be covered by beach deposits (Cheshire Quartzite) and that then depressed the area to form the carbonate bank and reef deposits.

The boulder conglomerate described by Boudette and Boone at the base of the Chain of Lakes massif in northwestern Maine, however, may have formed directly on the Precambrian rocks of the Canadian Shield. This unit contains boulders of metamorphic rocks (sillimanite grade) as well as white vein-quartz pebbles and aluminous matrix.

In eastern Massachusetts, Bell and Alvord record a possible unconformity above an amphibolitic pillow-lava unit (Middlesex Fells Formation) that overlies the Westboro Formation. Above this unconformity are scattered thin gray quartzites. According to Goldsmith, a similar sequence of quartzite above amphibolite occurs near the top of the Mamacoke Formation that overlies the Plainfield Formation in southern Connecticut; this break, therefore, is of regional significance. In Figure 4B I indicate this break as the top of the Cambrian sequence. Bell and Alvord believe that this unconformity may separate the Blackstone Series of Precambrian(?) age from Cambrian(?) rocks (see Fig. 4A); Goldsmith considers the Mamacoke and Plainfield to be Paleozoic. The lithologic sequence above and below this unconformity is also the same as recorded by Ratcliffe and Zartman at the base of their Precambrian section where the Washington Gneiss is overlain by amphibolite derived from mafic flows. If one correlates all of these sequences, then the Westboro Formation and overlying amphibolites, Plainfield Formation and Mamacoke Gneiss, and the Washington Gneiss and amphibolite become equivalent; furthermore, the thin gray quartzites above the amphibolite in the east would be equivalent to the Dalton Formations and Cheshire Quartzite of Cambrian age in the west. A study of aeromagnetic maps (Zietz and others, 1972) supports such a correlation; the microcline gneisses and other rocks below the upper quartzites have similar magnetic properties in all areas. I consider, however, that this break most likely is at the top of the Cambrian rocks as shown diagrammatically in Figure 4B. This conclusion

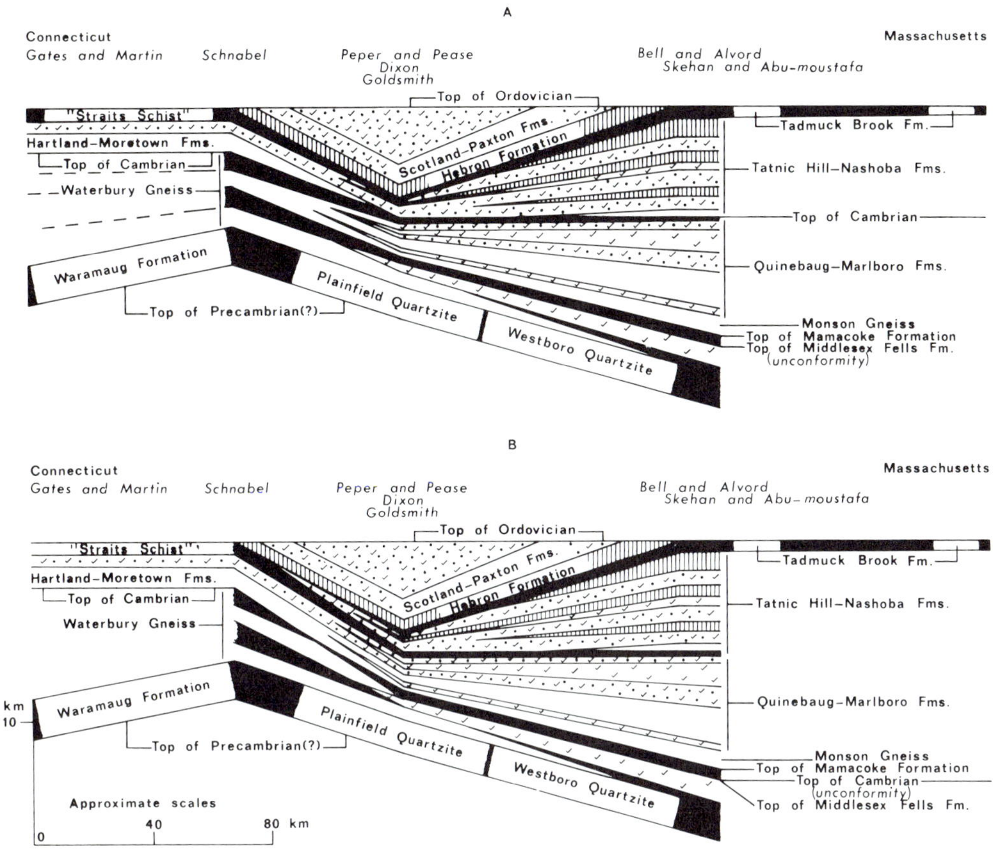

Figure 4. Diagrammatic cross section west to east across Connecticut and Massachusetts showing original thickness of Ordovician and Cambrian rocks. Datum is top of the Ordovician (control as indicated in text). Epiclastic units—black; carbonate-rich zones—lined; mixed epiclastic and volcaniclastic rocks—checks and dots; mafic volcanic flow rock—checks; felsic volcanic rocks—no pattern. (A) Correlation of base of Moretown Formation, Tatnic Hill Formation, and Billerica Schist Member of the Nashoba Formation as the base of Ordovician rocks. (B) Correlation of base of the Moretown Formation, the base of the epiclastic upper part of the Mamacoke Formation, and unconformity at the top of the Middlesex Fells Formation.

is based upon (1) the fact that *Olenellus* is younger than other known Cambrian fossils (Theokritoff, 1968) and that, therefore, there should be an underlying Lower Cambrian section of unfossiliferous rocks in New England, (2) the fact that this unconformity separates highly aluminous and potassium-rich rocks of dominantly epiclastic origin below from rocks of dominantly volcaniclastic origin above (Figs. 3 and 4), (3) the fact that the lowermost members of the Hoosac Formation interfinger with and lie above the Dalton Formation, as noted by Norton, (4) the fact that the Hoosac Formation, of known Cambrian age, is interlayered with microcline-rich gneisses similar to those lower in the section in all of these areas (Fig. 2), (5) the absence of clear-cut evidence of two periods of progressive regional metamorphism, and (6) the presence of mafic volcanic flow rocks in the Cambrian section of Vermont (Skehan, 1961).

Probably the most significant and widespread unconformity recognized in New England is that between rocks of Ordovician and Silurian age—that is, the Taconic unconformity. In northwestern Maine, Boudette and Boone show that it is clearly marked by conglomerates of Silurian age and by an angularity with older rocks, whereas in northern Maine, Roy and Mencher find that these rocks are conformable. The unconformity in northwestern Maine can be traced southward along the Bronson Hill-Boundary Mountains anticlinorium across New Hampshire and Massachusetts into Connecticut. It is marked by a zone of quartz-rich coarse clastic rocks that truncate the older units at a low angle. In Vermont and southward into western Massachusetts and Connecticut, the Ordovician-Silurian boundary is marked by fine-grained quartz-rich schists or carbonaceous phyllites on the east side of the carbonate reef. On the west side, it is possible that the chert-bearing conglomerates at the Dalton Formation horizon and some of the "breccias" at the base of the Everett Formation actually represent the unconformity.

Gates and Martin as well as Schnabel record quartzose schists above the sulfidic schists that have been mapped in the Hawley Formation (Ordovician). These quartzose schists are at the base of the pelitic sequences of probable Silurian age as mapped by Stanley and Hatch farther north and may be equivalent to their Russell Mountain Formation. Stanley and Hatch trace the Silurian and Devonian rocks of Vermont southward into southern Connecticut but do not separate them as to age. These Silurian and Devonian rocks are unique in that where their age is unquestioned, they are only slightly magnetic. In central Connecticut, however, the rocks of Silurian-Devonian age that were mapped by Stanley and Hatch have two magnetic characters: one unit is nonmagnetic, but the other is very magnetic and, in my opinion, is probably the sulfidic schist of Ordovician age referred to above and also is part of the "Straits Schist" of Gates and Martin. These nonmagnetic and magnetic units are exposed in the Waterbury dome and are mapped together by Gates and Martin as "Straits Schist." Here, they are probably separated by an unconformity. They also appear in the domes that arch up the Silurian and Devonian rocks to the north; for example, both a magnetic and a nonmagnetic unit of "Straits" are known in the Granby dome where an unconformity is mapped by Schnabel. In northern Massachusetts and Vermont, metapelitic rocks of the Cram Hill and Hawley Formations of Ordovician age are highly magnetic, and thus, it seems quite possible that the highly magnetic "Straits" unit is also of Ordovician age. Figures 3 and 4 present my view of the possible eastward correlation of this magnetic "Straits" or Cram Hill-Hawley unit. It is truncated where the Taconic unconformity is visible along the east side of the Connecticut River at Bolton Notch in central Connecticut. To the east of the Bronson Hill-Boundary Mountains anticlinorium, near the center of the Merrimack synclinorium, Pease and Peper do not recognize this unconformity; it would appear that the Upper Ordovician schists and Lower Silurian schists are conformable. Skehan and Abu-moustafa, on the other hand, record an angularity below conglomeratic clastic units at this horizon. This may be the same unconformity recorded by Bell and Alvord at the base of the Tadmuck Brook Formation, but I prefer to consider the Tadmuck Brook Formation as Ordovician as shown in Figure 4A, because it is highly magnetic.

Radiometric Ages

Naylor notes that the difference in ages on the two flanks of the Precambrian geosyncline—1,000 to 1,200 m.y. on the west and 500 to 600 m.y. on the east—might be caused by the provenance of the materials in the sedimentary section. That on the west may come from the Grenville and that on the east from volcanic sources (as suggested for the Yonkers Gneiss by L. M. Hall) rather than requiring a partial resetting of the time clock. Naylor points out that "radiometric dating confirms the lithologic and structural correlation of basement rocks of the Green Mountain massif with the 'Grenville' basement of the Adirondacks." Some Rb-Sr muscovite ages, as young as 800 m.y., are attributed to subsequent disturbances; the least disturbed rocks are 1,050 m.y. old. Ratcliffe and Zartman give Th-Pb ages of about 1,000 m.y. for the Washington Gneiss and the Tyringham Gneiss in the Berkshire massif; Long (1969) gave Rb-Sr whole-rock ages as young as 570 m.y. for the Yonkers Gneiss north of New York City; Faul and others (1963) had a Pb-α age of 900 m.y. for the core of the Chester dome and 1,100 m.y. for sample Vt6 on the east flank of the Green Mountain massif. Sample Vt6 is actually from the basal Hoosac Formation according to Naylor. Ratcliffe and Zartman show photographs of rounded zircons in clasts in their 1,000-m.y.-old gneisses; this fact suggests that part of the problem of diverse ages lies in the origin of the rocks sampled.

Naylor quotes the Rb-Sr whole-rock ages of three granitic rocks in eastern Massachusetts: 514 ± 17 m.y. for the Hoppin Hill Granite, 569 ± 4 m.y. for the Northbridge and Milford Granites, and 591 ± 28 m.y. for the Dedham Granodiorite. These are intrusive into rocks that are designated Precambrian but that appear to me to be part of the Cambrian section, except perhaps at Hoppin Hill where the relationships may suggest an unconformity. The ages of about 500 m.y. quoted by Naylor for volcanic rocks in New Brunswick and Nova Scotia are compatible with the relative ages of similar volcanic rocks in southeast New England that are described by Bell and Alvord as being above the Westboro Formation. Also the Rb-Sr whole-rock ages of 650 ± 50 m.y. (Hills and Dasch, 1968, p. 136) for the gneisses of the Stony Creek dome in Connecticut are compatible with the interpretation by Goldsmith that the gneisses below the Plainfield Quartzite are considered Cambrian.

The Massabesic Gneiss in the Manchester area of New Hampshire is another rock of nonintrusive origin that, according to Naylor, yields Rb-Sr whole-rock ages of 600 ± 200 m.y. This age suggests that it is low in the volcanic pile as recorded in northeastern Massachusetts, but the possibility of contamination by detrital material also must be considered.

The anomalously young Rb-Sr whole-rock ages discussed by Naylor and also by Brookins as typical of intermediate and felsic volcanic rock need careful consideration. Most of the volcanic rocks described in New England are marine volcaniclastic rocks which show evidence of considerable sorting and also of mixing of nonvolcanic components. The only rocks in this volcanic pile that are 100 percent volcanic and that might yield true isotopic ages are basaltic pillow lavas, a few rhyolite and andesite flows, perhaps the very finely layered air borne tuffs (commonly confused with mylonite), and the premetamorphic dikes and sills that cut them. By far, most of the volcanic pile is mafic and felsic volcanic debris that in most places is difficult to separate from arkosic and feldspathic weathering debris from older source rocks. Thus, extreme care must be used in sampling to select a rock that gives a "primary" age rather than a mixed age weighted toward the older source rock.

There are few rocks in New England that are of pre-Devonian age and that have physical evidence of original glass components, yet the chemistry of many of the rocks and their textures and structures suggest that many were originally, at least in part, glass. The absence of cross cutting silicified zones, common in devitrified volcaniclastic rocks, suggests that devitrification and zeolitization took place while these were still porous and saturated with saline water. The abnormally high sodium content suggests that many of these devitrified rocks lost potassium and probably rubidium and strontium during devitrification. The lack of potassium feldspar in many of these rocks and its presence in others may well be the result of the interchange of these solutions; an unchanged Rb-Sr ratio under these conditions would be fortuitous. Devitrification probably would have less effect on the abundance and age of zircon, but the volcanic zircon would be mixed with zircon from detrital sources.

Naylor's point that there is a lack of reliable petrographic criteria on which to reject "bad" samples is valid. More critical analysis of the field relations, of the paragenesis of the contained minerals, and of the complete mineralogic and chemical data is needed to make sure that a sample will give the age of deposition. I have seen instances in which samples of cross-bedded arkose of Cambrian age have been taken for dating apparently on the assumption that they were granites cutting the Cambrian formations; there were other instances in which samples of volcaniclastic felsic gneiss with basaltic clasts were treated as if the gneiss were intrusive.

My experience with radiometric ages in New England has been frustrating in terms of having the ages determined in the laboratory match the ages from field observations in terms of the geologic time scale. The radiometric ages, however (with few exceptions), show the same sequence as expected from the field observations but are older or younger than the known paleontologic age. For example, rocks intruded into Silurian rocks, dated by fossils, commonly give an Ordovician age, and as a result, the geologist has to choose what data shall be given priority in working out the stratigraphy of his area. There is no doubt that the determinations made in the laboratory are accurate in so far as the methods of analysis are concerned; what is in doubt is the history of the rocks and their contained minerals and perhaps the numbers on the time scale. There is, I believe, a reasonable doubt as to whether the numbers assigned to the beginning of the Cambrian and the Devonian Periods are compatible with the geologic relationships of Cambrian and Devonian rocks in New England. Perhaps this problem will disappear if more careful petrographic and chemical work is done on the samples used for isotopic studies and as a better understanding is gained of the geochemical factors responsible for consistent but anomalous radiometric ages.

PRECAMBRIAN-CAMBRIAN BOUNDARY

One of the most controversial issues related to the stratigraphy of New England is the nature and position of the Precambrian-Cambrian contact. The Precambrian rocks of the Adirondacks in New York, however, have a different structural trend and topographic expression, a different magnetic pattern, and a different metamorphic history. These facts suggest strongly that the rocks of southwestern New England described as Precambrian in this volume are younger and were deposited in a basin whose shoreline was between the Adirondacks and western Massachusetts and Vermont. This shoreline is now covered by the overlapping, younger, unmetamorphosed Cambrian sandstones and Ordovician rocks. Its trend probably was originally somewhat east of north (probably N30° to 40°E); its present northerly trend is the result of faulting.

The patterns on the aeromagnetic maps (Zietz and others, 1972) show an abrupt truncation of the east-northeast trend of the Precambrian rocks of the Adirondacks against the more northerly trend of units in southwestern New England. This truncation probably represents the western edge of a deep depositional basin, although Harwood and Zietz (1974) believed that the truncation marks a zone of thrusting. There is no doubt that thrusts complicate the relationships, but this abrupt change in the magnetic pattern cannot be related directly to the Precambrian gneisses exposed and mapped in Massachusetts and Vermont because in places these rocks overlie but, for the most part, do not modify the broad, highly magnetic anomalies. The equally highly magnetic rocks east of this area are at the surface, interlayered with weakly magnetic rocks, and are relatively thin elongate sources; thus, the truncation may mark a scarp formed by a large gravity fault on the western edge of the depositional basin. D. S. Harwood at this symposium described more than 3,000 m of nonrepeated section, some 2,000 m more than the estimated depth to the highly magnetic rocks at Copake, New York. To the north, on the Vermont, Massachusetts, and New York state lines, depths of about 3,000 m to the highly magnetic rocks are compatible with Harwood's thicknesses, but Skehan (1961, p. 24) showed about 6,000 m of section not far to the east. Thus, it would appear that this deep part of the basin was filled with the volcanic and clastic rocks of so-called Precambrian age as the basin was dropped down by faulting. The unconformity noted in western New England at the base of the Dalton Formation may record this period of gravity faulting. Carbonate deposition followed on the edge of this basin, owing to the increasing shallowness of the seas. Later movements on this fault could perhaps be correlated with thick sections of coarser clastic units such as the schists and massive microcline gneisses of the Hoosac Formation (see Fig. 2).

The detailed stratigraphy of these rocks of questionable Precambrian age in southern New England given in this volume by Ratcliffe and Zartman, Gates and Martin, Norton, L. M. Hall, and Schnabel indicates two main sources of materials: (1) the saprolitized rocks of Precambrian age to the northwest in the Adirondacks and Canada and (2) a chain of submarine volcanoes to the southeast that extended northeastward across New England. During mapping, the upper limit of these rocks has been placed arbitrarily at the base of the Dalton Formation and Cheshire Quartzite of "*Olenellus* age," but these quartzites interfinger eastward into aluminous and volcaniclastic rocks of the Hoosac Formation.

The history of rock formation in the area seems to be one in which the importance of volcanic sources decreased through time with an accompanying lateral facies change from continentally derived, coarse to fine clastic rocks on the west, to carbonates, and then to aluminous pelites that grade into clastic volcanic rocks on the east. The presence of sedimentary rocks indicating beach and shallow-water environments (Dalton Formation, Cheshire Quartzite, and Stockbridge Formation) above the deeper water marine epiclastic and volcaniclastic sedimentary rocks (Fig. 2) suggests considerable vertical movement just before *Olenellus* time that formed an offshore bank or reef and a deep basin to the southeast.

L. M. Hall, Gates and Martin, Ratcliffe and Zartman, Norton, and Zen take the traditional approach and accept the light and dark banded gneisses of western New England as of Precambrian age and unconformably below the Dalton, Hoosac, and Waramaug Formations. They have made structural interpretations based on this assumption that involve complicated isoclinal folding and thrusting of diverse ages. They present the arguments in favor of their positions and at the same time point out some anomalous features that suggest to me that an alternative explanation for the observed geologic relationships is quite probable.

Schnabel departs from the traditional approach and considers the light-colored gneisses as interbedded with the schists of the Hoosac Formation and therefore of Cambrian age. Gates and Martin consider the interlayering and interfingering of the schists of the Waramaug Formation (Cambrian) with the upper 75 m of the Berkshire Gneiss Complex (considered by them to be of Precambrian age) to be a normal sedimentary relationship. Norton, on the other hand, explains this type of interlayering as tectonic and the result of the thrusting of slivers of Precambrian gneiss into the Hoosac schists. Zen accepts this type of thrusting, in part, but is disturbed by the fact that the Precambrian clearly dips under the Hoosac at North Adams. Ratcliffe and Zartman are concerned that the Dalton conglomerates on Day Mountain contain "unmetamorphosed" white and black chert pebbles (see section above titled Unconformities). The isotopic ages given by Ratcliffe and Zartman and also by Naylor appear to support a Precambrian age for the Washington and Tyringham Gneisses; the Tyringham is considered intrusive into the Washington Gneiss although the radiometric ages are the same.

These inconsistencies coupled with my field observations of sedimentary structures in the gneisses, the lack of structures indicating thrusting at the contacts of some of the schists and gneisses, and an analysis of aeromagnetic maps (Zietz and others, 1972), side-view radar, (U.S. Geol. Survey 1968, 1971), and space photography have prompted the alternative solution given diagrammatically in Figure 2. The stratigraphic columns and map data given in this volume have been plotted in their approximate relative horizontal and vertical positions and referred to the top of the Hoosac Formation that appears as an essentially straight line on maps of the east side of the Berkshires and is assumed to have been a horizontal sedimentary surface. The base of the uppermost unit of the Hoosac, according to Norton, is also parallel to this surface, in part.

Figure 2 shows a wedge of interfingering gneisses and schists, thickening southward, capped by a continuous and uniform schist layer. There appears to be no break in this pattern below the Dalton Formation, and the Washington and Tyringham Gneisses appear to be part of this deltalike or fanlike structure. The Waterbury Gneiss, whose relationship to these older units is poorly known, fits into such a stratigraphic column in the position suggested by Gates and Martin as does the apparent duplication of the Hoosac section mapped by Schnabel. If this section was extended into Woodford, Vermont, it would show the *Olenellus* horizon to be about 30 m above the blue quartz rocks of the Washington Gneiss (J. W. Skehan, 1974, oral commun.).

The interfingerings and interlayerings of these apparently conformable rocks of Precambrian and Cambrian age need explanation on some basis other than faulting, because in most places breccia, fractures, and other evidences of faulting are absent in the metamorphosed rock and no relicts of such structures have been preserved. My determinations of original orientations for these rocks also indicate the absence of folding in many places. Therefore, it seems most probable that the interfingering and interlayering observed is in large part a depositional feature. The light-colored gneisses that interfinger with the lower parts of the Hoosac Formation thicken where the Hoosac thins and would seem to represent coarse-grained epiclastic and volcaniclastic sediments of Cambrian age (Fig. 2). At the base of these units, the conformable metasedimentary rocks such as the Washington Gneiss grade southeastward into the finer grained rocks similar to the Hoosac. It seems reasonable that these are part of the same sedimentary sequence both on stratigraphic grounds and the present isotopic age data. The unconformities recorded in western Massachusetts at the base of the Dalton-Cheshire-Hoosac part of the section thus would be of only local importance; they record the time of emergence and reef building. The angularities noted would thus represent differences in original dip of volcaniclastic and littoral sediments. Theokritoff (1968, p. 9) also suggested that the rocks representing earliest Early Cambrian time are an unfossiliferous section lying conformably below the *Olenellus* zone. If one assumes this is so—as indicated by the stratigraphic sections and conformable sequence of these silica- and alumina-rich rocks and the light-colored gneisses—many of the complexities of structure disappear, and the patterns on the aeromagnetic maps can be related more readily to the observed geology. Specifically, many of the isoclinal folds, especially the early recumbent type, would be explained by lensing, but others are related to the north-trending postmetamorphic (Acadian) thrusts that are responsible for the many minor structures that show mass movement from east to west. These thrusts are superimposed on the major folds of the Acadian orogeny whose axial planes trend north and dip westward, as on the east side of the Berkshires, as well as on the open folds of the Taconic orogeny (see Figs. 3 and 4). In Connecticut the folds of the Acadian orogeny also have been truncated by several large left-lateral northeast-trending faults that give an apparent thickening to this part of the section. These faults are relatable to the northeast-trending segments of Cameron's line described by Gates and Martin.

In eastern Massachusetts, the light and dark banded gneisses and the underlying quartzite in the major northeast-trending fault zone are considered by Bell and Alvord to be Precambrian. According to Skehan and Abu-moustafa, the gneisses may be Ordovician. Goldsmith regards the Plainfield Quartzite, at the base of such gneissic sections in southern Connecticut, as Cambrian. Brookins gives isotopic data for rocks of Precambrian age in south-central Maine; Bickel assigns a Precambrian age to the lowermost unit in the Belfast area, Maine.

Bell and Alvord equate the quartzite sequence in eastern Massachusetts with the Blackstone Formation of Rhode Island and give it a Precambrian age. They also note the presence of an unconformity above the pillow lavas that overlie it. The unconformity suggests a major break in the section, either the Precambrian-Cambrian or Cambrian-Ordovician boundary. This quartzite horizon is interlayered with the lower part of the flows, and they obviously are conformable units. The clastic units above the flows are felsic volcaniclastic rocks with abundant feldspar clasts that are interlayered with fine-grained units (possibly airborne felsic tuffs) containing only minor mafic constituents. Thus, these clastic units record an abrupt termination to the mafic volcanism associated with the quartzite. This change from mafic volcanic flows, associated with quartzite and other quartz-rich rocks, to a sequence of clastic felsic volcanic rocks (Wilmington Gneiss) has been noted in southern Vermont (Skehan, 1961) where the quartz-rich rocks are of Cambrian age. To the south the number of amphibolites of this age that may have been mafic flows decreases. Ratcliffe and Zartman record the same sequence of mafic flows above, or equivalent to, the Washington Gneiss and consider them to be Precambrian. Goldsmith has a quartzite sequence with felsic volcanic units directly above it. In parts of eastern Connecticut, however, I have seen fine-grained amphibolite, possibly mafic flows, directly above the quartzite unit.

A similar sequence of events has taken place in northwestern Maine where the Chain of Lakes massif underlies rocks characterized by mafic flows and epiclastic units. The rocks of the massif are felsic volcaniclastic debris mixed with epiclastic materials such as quartz pebbles, sillimanite nodules, quartzwacke layers, and aluminous schist layers. These rocks dip 20° steeper than the Ordovician rocks and show a difference in metamorphism; such features suggest an unconformity. The absence of a Cambrian section seems strange in view of the distribution of fossiliferous Cambrian rocks in other parts of New England; perhaps these rocks, like the other volcaniclastic rocks to the south that lie below known Cambrian rocks, really should be grouped with the post-*Olenellus* rocks into the Cambrian section, the base of which would be the unconformity on the Adirondack and Canadian Shield rocks.

The similarity of the Cambrian section of Bickel in south-central Maine to the sections in southwestern New England suggests that it is in the *Olenellus* group and that the gneisses of Islesboro are part of an older sequence that needs to be redefined as of Cambrian rather than Precambrian age.

The widespread relationship of mafic flows of marine origin and clastic rocks of continental origin, as well as the indirect evidence from fossils, suggests that a Cambrian (rather than Precambrian) age for this group of rocks would be more appropriate and that the unconformity noted by Bell and Alvord in this volume is either the base of the Ordovician section or a surface of nondeposition of mafic flows. Isotopic ages of these flows might be definitive, but ages determined on these felsic gneisses, which contain epiclastic material, and the quartzites probably would be of questionable value in determining the true age of this unit. At present, because of the lack of fossils, radiometric methods must be used in the correlation of these widely scattered parts of the section; however, refinements in mapping techniques and more careful stratigraphic description of exposed sections could aid greatly in the determination of definitive ages. There are some very distinctive felsic gneisses that need additional study as possible correlative units—for example, the ferrohastingsite-bearing Tyringham Gneiss of Ratcliffe and Zartman, the aegerine gneiss of Goldsmith, ferrohastingsite gneisses in central Massachusetts (Peter Robinson, 1972, oral commun.), and other similar Na-rich and rare-earth-rich gneisses in Rhode Island. These units seem to be of distinctive composition, are widespread, and occupy a specific position in the volcanic sequences in southern New England. A study of rare-earth content of these rocks might prove that it is sufficiently distinctive that these units together could be used as a marker horizon.

In order to clarify the relationships discussed above, Figure 4 has been prepared from lithologic data given in the papers by Gates and Martin, Schnabel, Peper and Pease, Dixon, Goldsmith, Peck, Skehan and Abu-moustafa, and Bell and Alvord. The horizontal spacing of sections is the approximate distance between areas (see Fig. 1); the horizontal plane of reference is the Ordovician-Silurian boundary; the "folding" below this plane (Figs. 4A and 4B), therefore, is a measure of the Taconic orogeny. The quartzite sequences on the east are correlated with the lower part of the Hoosac Formation and the Waramaug Formation as shown in Figure 2, and the thin quartzites above the mafic flow rocks are correlated with the base of the Missisquoi Formation. Figure 4 shows diagrammatically the thickening and thinning of the Cambrian and Precambrian sections described in this volume.

In summary, the main problem posed by the widely separated parts of this older group of gneisses and quartzites in New England is whether they are related more closely to the Cambrian rocks of western New England or to the Precambrian rocks of the Adirondacks and the Canadian Shield—that is, the Grenville and Archean rocks. I prefer to consider these an unfossiliferous pre-*Olenellus* part of the Cambrian section for the following reasons:

1. The rocks of Precambrian age in the Adirondack area of New York and Canada were metamorphosed intensely, highly folded, and deeply weathered and eroded before being covered by deposition of the unmetamorphosed Potsdam Sandstone and other rocks of Paleozoic age. No such major, widespread, angular unconformity has been mapped in New England.

2. East of the Taconic and associated thrust faults, rocks of known Cambrian age interfinger and are interbedded at their base with a thick section of volcaniclastic and epiclastic rocks that are magnetically distinct from the rocks of Precambrian

age to the west but similar to rocks in the Cambrian section (Fig. 2).

3. The epiclastic rocks of Cambrian or older age are highly aluminous fine-grained and quartz-rich coarse-grained units derived from a deeply saprolitized terrane of silicic plutonic rocks, mafic and felsic volcanic rocks, and crystalline metamorphic rocks. One of these epiclastic units, the Washington Gneiss, has a U-Pb zircon age of about 1,000 m.y., according to Ratcliffe and Zartman, that is compatible with an age derived from a saprolitized source of Grenville age and, thus, probably is of earliest Cambrian age.

4. The volcaniclastic and epiclastic rocks of this group in southern and western New England have been regionally metamorphosed only once (though perhaps locally more than once)—that is, during the Acadian orogeny—not twice (as suggested by some authors) as would be expected if these rocks were equivalent to the rocks of Precambrian age in the Adirondacks and Canadian Shield.

5. High-grade metamorphic rock clasts are, however, present in the Chain of Lakes massif and have been remetamorphosed. Part of the Chain of Lakes massif includes rocks with a variety of boulders and pebbles, some containing sillimanite, that were obviously derived from a pre-existing metamorphic terrane together with vein-quartz pebbles in a matrix, which is also of epiclastic origin. These have been recrystallized to the lower grade of the overlying Paleozoic sedimentary rocks. Boudette and Boone believe that the entire massif was metamorphosed twice, once in Precambrian time and then during the Acadian orogeny. I suggest this is a local—not regional—metamorphism, on the basis of my observations of some of these rocks, and not comparable to the metamorphism in the Canadian Shield areas.

6. In eastern Massachusetts and southeastern Connecticut and probably also in the Penobscot Bay area of Maine, the rocks designated as Precambrian do not have two regional metamorphic fabrics, and they appear to be essentially conformable with rocks that appear to be Paleozoic in age. Quartzite is associated with felsic and mafic gneisses in the same position in the section as in western New England and is associated with highly aluminous rocks derived from a saprolitized terrane. Pebbles of similar quartzite in the Pennsylvanian and Pleistocene rocks southeast of the igneous dome in Rhode Island contain fossils of Cambrian age. Their distribution suggests that before Pennsylvanian time, the quartzite of Cambrian age capped the igneous dome and was eroded to form the conglomerates. The quantity and types of quartzite in the conglomerates suggest that they are those designated as Precambrian by Bell and Alvord. The youngest volcanic rocks of Paleozoic age in Rhode Island are similar to those in the Boston area, and they also overlie these older rocks.

7. The Cheshire Quartzite of known Cambrian age appears to correlate on the basis of fossils with the lower part of the fossiliferous group of Bell and Alvord. The top of the Westboro Formation and also the quartzites in the top of the Plainfield Formation as mapped by Goldsmith and Dixon appear to correlate on the basis of mapping. The feldspathic sandstone units of the Dalton Formation and the Washington Gneiss probably are equivalent to the lower part of these formations, and in both areas feldspathic gneisses occur between the epiclastic zones. Midway between these eastern and western sequences are similar rocks in the Pelham dome area of north-central Massachusetts (described by Thompson and others, 1968). Locally above the quartzite sequence, both in eastern and western New England, are limestones that were formed in shallow water; the main limestone banks and reefs were formed offshore from the western edge of the depositional basin that extended from Long Island north into Canada.

ORDOVICIAN SECTION

The Ordovician rocks of southern New England are nonfossiliferous, and consequently, there are many problems involved in correlating them with the fossiliferous rocks in Maine, western New England, and New York. In New Hampshire and Vermont, rocks of known Ordovician age have been traced southward from fossiliferous areas, and a good general sequence has been established along the Bronson Hill-Boundary Mountains and Green Mountains anticlinoriums as far south as Connecticut. Complications arise here as a result of faulting along northeast trends (part of Cameron's line) southwest of the Massachusetts state line and west of the Connecticut River, along north trends to the east of the river (Ammonoosuc thrust?), and along northeast trends (extentions of the Clinton-Newbury fault zone) and east trends (Honey Hill fault) in southeastern Connecticut. An added complication is the fact that the Ordovician rocks have lateral and vertical facies changes within the sedimentary wedge that thickens rapidly eastward. Figure 3 shows diagrammatically the approximate relative thickness of these rocks along strike across Massachusetts where they are exposed in the west limb of the asymmetric syncline that forms the east side of the Berkshires. Figure 3 is based only on the maps and sections in this volume and, thus, is a controlled generalization rather than an accurate representation of all available information. It, however, does show the thickening and thinning of the Ordovician section between the Taconic unconformity and the top of the Cambrian rocks (Fig. 2), the irregularities of the Ordovician sea bottom, and the minor folding that occurred at the end of Cambrian time and also during the Taconic orogeny. The lenticular shape of the Rowe Formation may be either a depositional phenomenon or the result of erosion at the top of the Cambrian section. As shown in Figure 4, which is a representation controlled by the data in this volume of the sedimentary wedge from this area eastward, the stratigraphic section thickens abruptly to the east. Figure 4 also shows that the uppermost part of the Ordovician section is preserved only in a small area east of the Bronson Hill-Boundary Mountains anticlinorium. Here, the top of the Ordovician is preserved in a syncline that was formed during the Taconic orogeny and that also must have been the deepest part of the Silurian depositional basin. These uppermost Ordovician rocks are similar to those described by Bickel as Precambrian in the Belfast area.

Figure 4A has been drawn on the assumption that the Tadmuck Brook Formation of Bell and Alvord is equivalent to the top of the Tatnic Hill Formation (Yantic), of Dixon, the Hawley Formation of Norton and of Stanley and Hatch, and also the magnetic part of the "Straits" of Gates and Martin. This surface appears to be the approximate top of the Middle Ordovician. Another similar zone containing sulfidic rocks, the Billerica Schist Member of the Nashoba Formation recognized by Bell and Alvord, has been correlated with the base of the Tatnic Hill Formation of Dixon and the base of the Rowe and Moretown Formations of Stanley and Hatch. To the south of the major fault zone in Connecticut, the Brimfield Schist (the equivalent of the Tatnic Hill Formation, as mapped) is a very thin sulfidic unit and, although similar lithologically, is probably not the Rowe or Moretown but one of the several thin lenticular sulfidic schist zones in the Ordovician wedge. The Hebron Formation in this area also probably is a carbonate-rich zone that is distinct from the unit labeled Hebron Formation on Figure 4.

Figure 4B is drawn on the assumption that the magnetic "Straits Schists" and base of the Tatnic Hill Formation are correlative.

Whether the correlations of Figure 4A or Figure 4B are correct is not known; the answer depends on more detailed data, but what is very clear is that in the

Ordovician basin, all of the units shown in this wedge are distinct from the sulfidic schists in the area of Brimfield, Massachusetts (the original type locality of the Brimfield Schist). In the Brimfield area, above the Scotland and Hebron Formations described by Dixon and by Skehan and Abu-moustafa, Peper and Pease mapped a continuous section of rocks that have sulfidic schists in the upper unnamed units of probable Silurian age above the Nashoba Formation (equivalent to the Tatnic Hill Formation) or Ordovician age. The older units such as the Monson Gneiss and Partridge Formation have been thrust eastward over the western edge of this younger sequence of sulfidic rocks, and they in turn override the Brimfield Schist-Monson Gneiss sequence and the older "Hebron" at Hebron, Connecticut, on the major fault zone (Fig. 1). Further clarification of the problem of the Brimfield Schist will require additional mapping of lithologic units in central Massachusetts and southern New Hampshire.

The problems of correlation within the Ordovician in the Rangeley area of Maine, like those in southern New England, are related to the interlensing of rock types known to be repetitive in other areas, the presence of facies changes, and the marked thickening of the section southeastward as shown in Figure 5. The Taconic unconformity in this area cuts down to Middle Ordovician in the Boundary Mountains area (Fig. 5), where the section is thin and dominantly fine-grained epiclastic and volcaniclastic material. At the base of the Ordovician section are some mafic flows that are in a stratigraphic position similar to those in southern New England where mafic volcanic flows record the end of quartz-rich clastic deposition of Cambrian or Precambrian age.

One of the problems related to the Ordovician of New England is its widespread distribution north of the northeast-trending major fault zone (Fig. 1) in southern New England northward into Maine and its absence to the south and east of the Rhode Island-Connecticut border. There is little or no evidence for thinning and considerable evidence for thickening eastward. The abruptness with which the Ordovician section disappears on the east and the conformability of the youngest Ordovician with the overlying Silurian rocks in the central part of the basin that strongly suggest that there was differential uplift of the eastern part of the basin that started in the Late Ordovician time perhaps along a gravity fault in about the same position as the major fault zone (Fig. 1) of late Paleozoic age. Such an uplifted area could have been the source of the clastic sediments of probable Late Ordovician age above Dixon's Hebron Formation and could explain the absence of Ordovician rocks in this part of New England. This area did not receive any sediments until latest Silurian and earliest Devonian time and also was above sea level at least for short periods between latest Silurian and Pennsylvanian time. It also could have been the source for the Devonian delta described by B. A. Hall, Pollock, and Dolan. Later, it must have acted as a buttress against which the major northeast-trending thrust faults pushed rocks from the center of the depositional basin.

SILURIAN-DEVONIAN SECTION

The stratigraphic problems of the Silurian and Devonian are largely those relating to the lateral and vertical facies changes, the abrupt thickening of the section toward the middle of the depositional basin, and the presence of both eastern and western shoreline facies. A further complication is the fact that these rocks are intensely folded—apparently internally—into complex masses, especially near

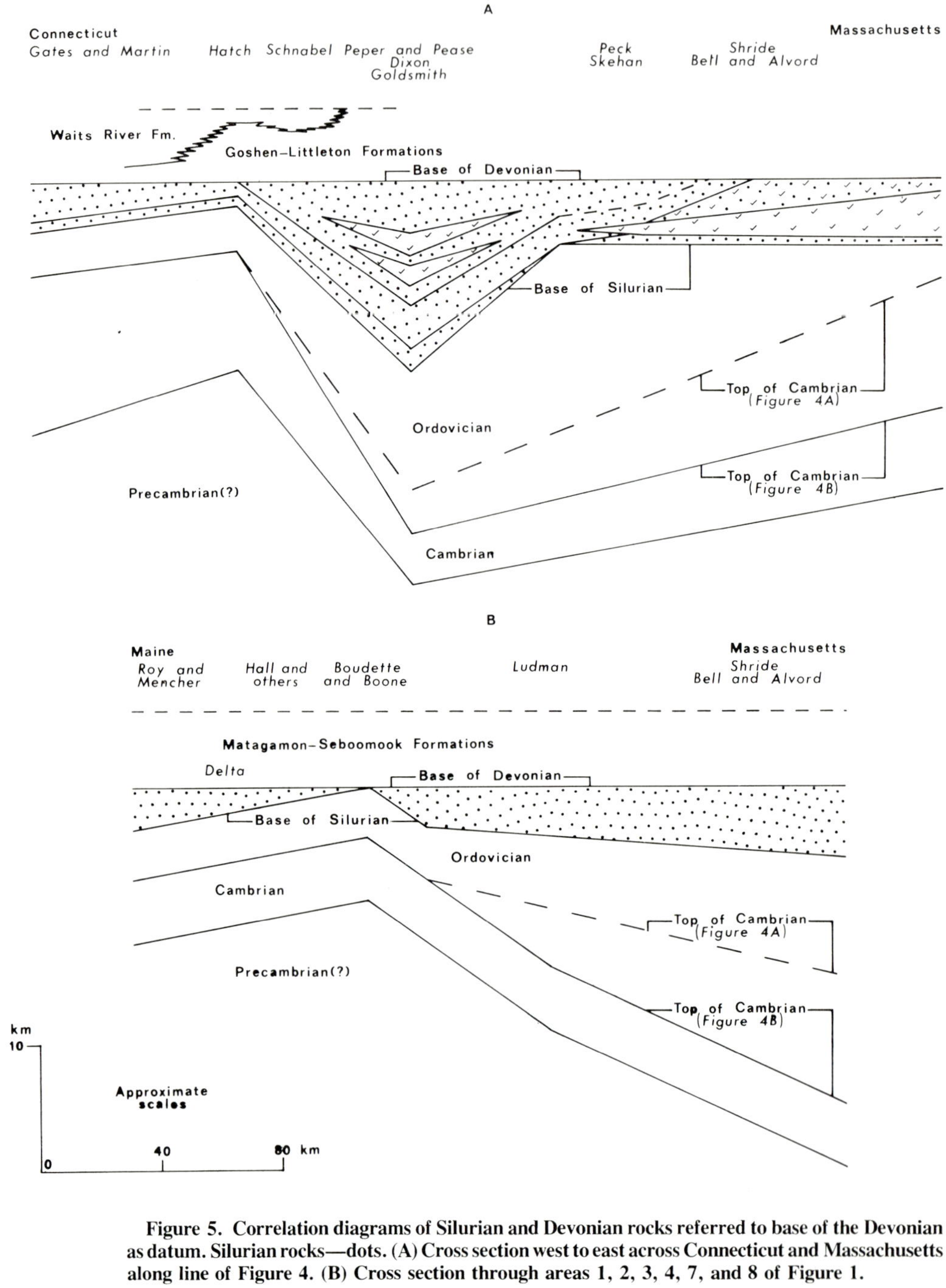

Figure 5. Correlation diagrams of Silurian and Devonian rocks referred to base of the Devonian as datum. Silurian rocks—dots. (A) Cross section west to east across Connecticut and Massachusetts along line of Figure 4. (B) Cross section through areas 1, 2, 3, 4, 7, and 8 of Figure 1.

intrusive rocks. Figure 5 is a generalization, controlled by data in this volume, that depicts the stratigraphic relationships referred to an assumed horizontal plane marking the uppermost Silurian or lowermost Devonian boundary, described by Shride, and the base of obviously graded-bed sequences in southern New England. This boundary is of questionable validity, but appears to be reasonably definite in Maine (Fig. 5) where there is fossil control.

The Silurian section along the east side of the Green Mountains anticlinorium in Vermont has been traced southward into Connecticut (Figs. 3 and 5), where it is a thin- to massive-bedded aluminous mica schist grading southward into more quartzitic rocks. To the east, on the Bronson Hill-Boundary Mountains anticlinorium, it has a littoral facies of conglomeratic quartzite, quartzite, and schist that marks a stillstand of the sea during transgression. These are essentially "shoestring sands" for the most part, but in places, such as in the Rangeley area of Maine, they form part of a widespread deltaic structure. In southern New England they are overlapped by younger finer grained units, such as those along the Green Mountains anticlinorium.

On the east side of this basin, Peck records fine-grained quartzite at the base of his section, and Skehan and Abu-moustafa in the same area have mapped fine-grained quartzite interbedded with aluminous and sulfidic schists unconformably above Ordovician(?) rocks. These quartzites appear to represent a littoral depositional environment and are not recognized by Peper and Pease to the west where the equivalent rocks are sulfidic schists and volcanic rocks. Near Belfast, Maine, Bickel has mapped lenticular and channel-shaped quartzites at this horizon but considers them of Ordovician or Cambrian age. Figure 5 shows a generalization of the thicknesses in adjacent parts of this section. The variations imply that the quartzites at the base of the eastern section formed at a lower sea level and at an earlier time than those on the Bronson Hill-Boundary Mountains anticlinorium. The thickness of rocks between these shoreline facies suggests that the depositional basin in Ordovician time was reactivated and continued to deepen throughout Silurian time with its edges perhaps being marked by large gravity fault zones.

PENNSYLVANIAN-TRIASSIC SECTION

A paper on rocks of Pennsylvanian age in the Worcester, Massachusetts, area was presented by Grew at this symposium, but has been published elsewhere (1973). The presence of plants designated as Pennsylvanian in rocks of this area caused Grew to argue for an unconformity above the rocks of Silurian and Devonian age described by Peck and by Peper and Pease. Such an unconformity is recognized in southeastern Massachusetts and Rhode Island between rocks of Ordovician or older age and fossil plant-bearing rocks of Pennsylvanian age. There is no record of deposition in Pennsylvanian time in Connecticut; the Triassic rocks appear to lie unconformably on rocks of Silurian and Devonian age.

REFERENCES CITED

Faul, Henry, Stern, T. W., Thomas, H. H., and Elmore, P.L.D., 1963, Ages of intrusion and metamorphism in the northern Appalachians: Am. Jour. Sci., v. 261, no. 1, p. 1-19.

Grew, E. S., 1973, Stratigraphy of the Pennsylvanian and pre-Pennsylvanian rocks of the Worcester area, Massachusetts: Am. Jour. Sci., v. 273, p. 113-129.

Harwood, D. S., and Zietz, Isidore, 1974, Configuration of Precambrian rocks in southeastern New York and adjacent New England from aeromagnetic data: Geol. Soc. America Bull., v. 85, p. 181-188.

Hills, F. A., and Dasch, E. J., 1969, Rb-Sr evidence for metamorphic remobilization of the Stony Creek Granite, southeastern Connecticut [abs.]: Geol. Soc. America Spec. Paper 121, p. 136-137.

Long, L. E., 1969, Whole-rock Rb-Sr age of the Yonkers Gneiss, Manhattan prong: Geol. Soc. America Bull., v. 80, p. 2087-2090.

Skehan, J. W., 1961, The Green Mountain anticlinorium in the vicinity of Wilmington and Woodford, Vermont: Vermont Development Dept. Bull. 17, Montpelier, 159 p.

Theokritoff, George, 1968, Cambrian biogeography and biostratigraphy in New England, *in* Zen, E-an, and others, eds., Studies of Appalachian geology: Northern and Maritime: New York, Interscience Pub., p. 9-22.

Thompson, J. B., Jr., Robinson, Peter, Clifford, T. N., and Trask, N. J., Jr., 1968, Nappes and gneiss domes in west-central New England, *in* Zen, E-an, and others, eds., Studies of Appalachian geology: Northern and Maritime: New York, Interscience Pubs., p. 203-218.

U.S. Geological Survey, 1968, Side-looking radar mosaic (APS-94) of the state of Massachusetts, with parts of adjacent states—Uncontrolled (east looking); uncontrolled (west looking): U.S. Geol. Survey Open-File Map, 2 sheets (mosaics), scale 1:500,000.

——1971, Side-looking radar mosaic (APS-94D), Connecticut and Rhode Island, with parts of adjacent states—Uncontrolled (south looking); uncontrolled (north looking): U.S. Geol. Survey Open-File Map, 2 sheets (mosaics), scale 1:500,000.

Zen, E-an, White, W. S., Hadley, J. B., and Thompson, J. B., Jr., 1968, Studies of Appalachian geology: Northern and Maritime: New York, Interscience Pubs., 451 p.

Zietz, Isidore, Gilbert, Francis, and Kirby, J. R., 1972, Northeastern United States regional aeromagnetic maps: U.S. Geol. Survey Open File Rept., 13 sheets, scale 1:250,000.

MANUSCRIPT RECEIVED BY THE SOCIETY MARCH 3, 1975
REVISED MANUSCRIPT RECEIVED JULY 28, 1975
MANUSCRIPT ACCEPTED AUGUST 21, 1975

Printed in U.S.A.

Geological Society of America
Memoir 148

Ordovician and Silurian Stratigraphy of Northeastern Aroostook County, Maine

David C. Roy
Department of Geology and Geophysics
Boston College
Chestnut Hill, Massachusetts 02167

and

Ely Mencher
Department of Earth and Planetary Sciences, CUNY
New York, New York 1003

ABSTRACT

Consideration of stratigraphic sections from five areas of northeastern Maine suggests that marked lateral lithofacies changes occur within the Ordovician and Silurian systems over very short distances. Three regionally extensive Ordovician (Caradocian-Ashgillian) lithofacies are present in the area of this study: (1) a western volcanic-graywacke-slate facies (Winterville Formation), (2) a medial slate-graywacke facies (Madawaska Lake Formation), and (3) an eastern slate-limestone-graywacke facies (lower Carys Mills Formation). The zones of gradual transition between these Ordovician lithofacies are subparallel to the trends of major structural elements. All of the lithofacies appear to have formed in relatively deep water and relatively far from cratonal source areas.

The transition from Ordovician to Silurian is conformable and gradual in the eastern part of the area in contrast to abrupt and unconformable transitions in the central and western portions. Mild Taconian deformation during the latest Ordovician and earliest Silurian created western emergent areas that shed clastic debris into a continuing Silurian sedimentary basin in the east.

In the eastern part of the area, continuous and fossiliferous stratigraphic successions representing almost the entire Silurian are present. The Early Silurian (Llandoverian-early Wenlockian) is represented by three principal lithofacies: (1) a western sandstone-conglomerate-slate facies (Frenchville Formation), (2) a central phyllitic slate-limestone-ironstone facies (New Sweden Formation), and (3) an eastern calcareous mudstone-limestone facies (Spragueville Formation). The Frenchville and New Sweden Formations contain material eroded from exposed Winterville and Madawaska Lake rocks in the Taconian land area(s) to the west. The Lower

Silurian units are overlain conformably by thin-bedded flysch (Jemtland Formation) of late Wenlockian-early Ludlovian age that was also derived for the most part from western source areas. The youngest eastern Silurian unit is the Fogelin Hill Formation which overlies the Jemtland Formation conformably and consists of thinly interlayered red slate, green slate, and fine-grained, laminated, calcareous graywacke.

In the western part of the region, the Lower and Middle Silurian are not present, and Upper Silurian sandstone, conglomerate, limestone, shale, and volcanic rocks unconformably overlie the Winterville Formation. The Upper Silurian sedimentary rocks, derived largely from the Winterville Formation, formed along irregular and possibly narrow shelves around emergent areas of the western Taconian land mass that had become largely submerged by Late Silurian time.

No melange units have been found in the Ordovician and Silurian rocks of northeastern Maine to suggest the presence of a subduction zone as proposed by Bird and Dewey (1970). The Winterville volcanic rocks may be interpreted to be part of a complex island arc with a trench located outside the area to the southeast or northwest.

INTRODUCTION

Numerous recent investigations have shown that northern Maine is uniquely suited for studies of the stratigraphic history of the axial region of the New England part of the Appalachian mountain belt. First, the rocks of the Ordovician, Silurian, and Devonian systems contain several major eugeosynclinal lithofacies that may be traced for several hundred kilometres to the southwest and northeast (for example, Boucot and Thompson, 1963; Naylor and Boucot, 1965; Berry, 1968; Boucot, 1968; Ayrton and others, 1969). Second, the moderate deformation, low metamorphic grade, numerous fossil localities, and well-preserved sedimentary features permit the age and paleoenvironments of the rock units to be determined in a detail uncommon elsewhere in New England. It is our intent to present an outline of the Ordovician and Silurian systems in a large part of northeastern Aroostook County, Maine (Fig. 1); it is becoming increasingly clear that the lithofacies syntheses now possible in this area of Maine may serve as useful models for other more highly metamorphosed parts of the eugeosyncline, particularly in southwestern Maine.

Stratigraphic studies, stimulated largely by the numerous fossil localities in the region, extend back at least 70 yr. These studies have produced a number of stratigraphic syntheses, which are summarized in Figure 2. The present understanding of the stratigraphy of the eastern part of the area of Fig. 1 is based on that provided by Boucot and others (1964), with modifications and additions by Roy (1970a) and Pavlides (1968). Knowledge of the stratigraphy west of Maine Route 11 is the result of work by Boone (1962, 1970), Horodyski (1968), and Roy and Mencher (unpub. data).

The generalized geologic and tectonic map of the region (Fig. 1) shows the distribution of rocks of Ordovician and Silurian ages with respect to two anticlinoria and an intervening synclinorium. In the western anticlinorium, here designated as the Pennington Mountain anticlinorium, rocks of Ordovician age, consisting of two prominent lithofacies, are exposed: a volcanic-rich facies (Winterville Formation) and a slate-graywacke facies (Madawaska Lake Formation). In the Aroostook-Matapedia anticlinorium (Pavlides and others, 1964) to the east, an Ordovician and Lower Silurian limestone-slate-graywacke ("ribbon rock") lithofacies, the Carys Mills Formation of Pavlides (1965, 1968), is the oldest rock unit.

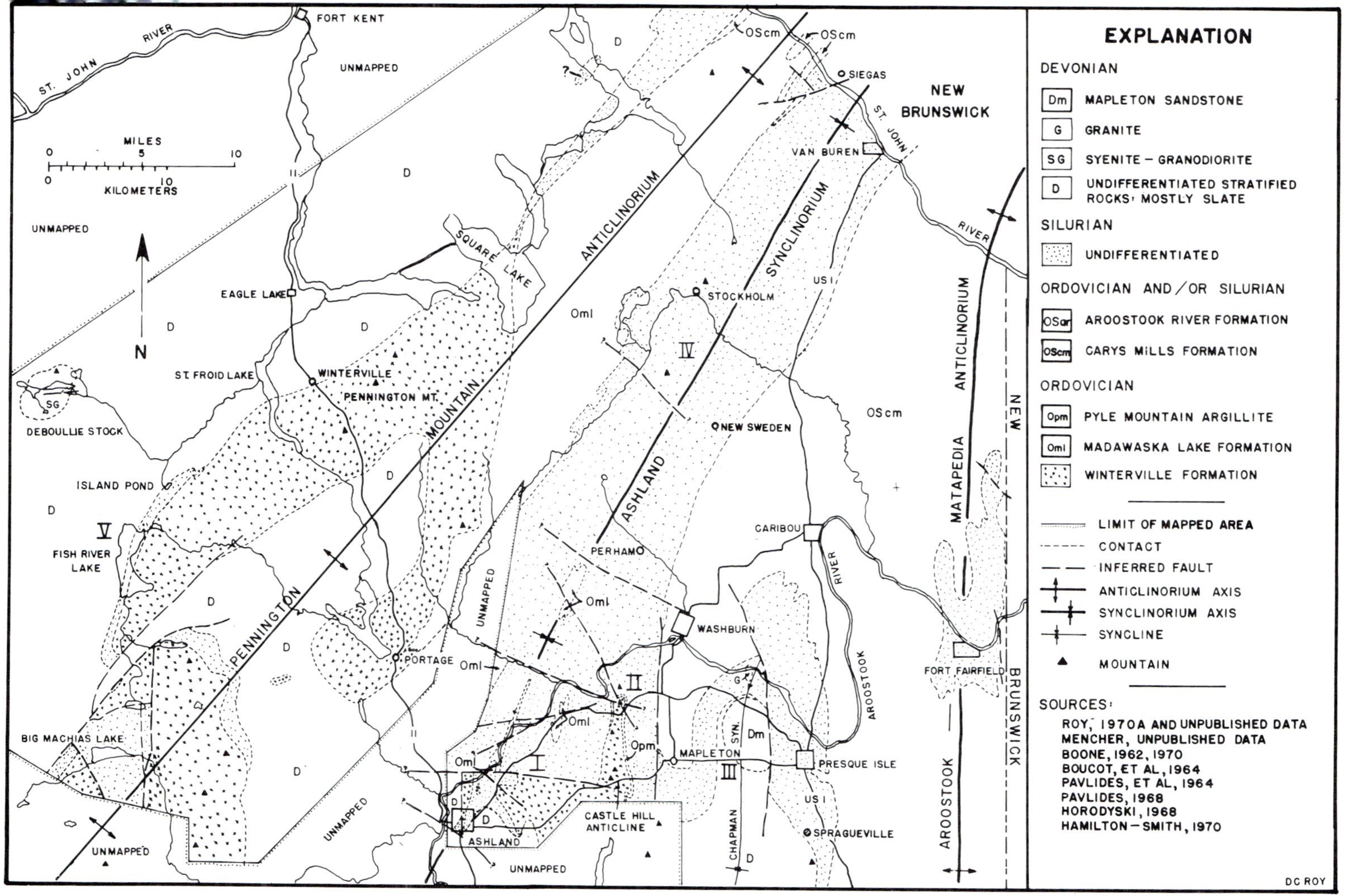

Figure 1. Generalized geologic and tectonic map of a portion of northeastern Aroostook County, Maine. Roman numerals refer to sections described in this paper. Undifferentiated Devonian east of Ashland may contain some Late Silurian strata as shown in Figure 3.

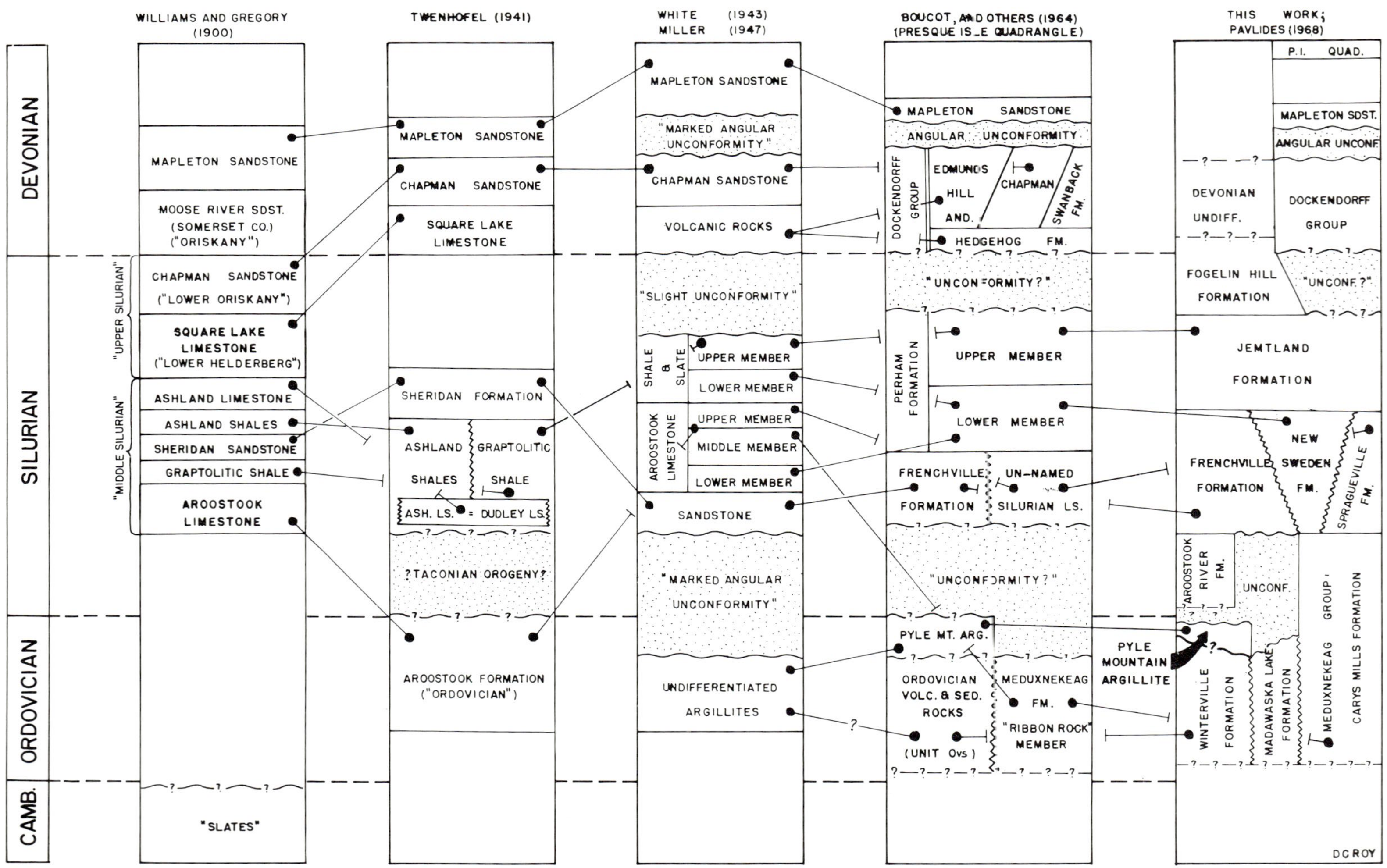

Figure 2. History of stratigraphic nomenclature in the area of Figure 3. Tielines indicate the changes in stratigraphic positions and names of rock units.

Rocks of Silurian and Devonian ages predominate in the Ashland synclinorium, but also are present in synclinal infolds in the anticlinoria. Rocks of Early, Middle, and Late Silurian age are documented by fossil localities in the eastern anticlinorium, the Ashland synclinorium, and along the northwest flank of the Pennington Mountain anticlinorium between Cross Lake and the St. John River. West of Maine Route 11, in the vicinity of Fish River Lake (Fig. 1), only Upper Silurian rocks are found resting unconformably above the volcanic-rich Winterville Formation (Ordovician) and disconformably(?) beneath a slate-dominated Devonian sequence.[1]

A variety of Devonian rocks overlies the Silurian in the eastern part of the region and overlies both Silurian and Ordovician rocks to the west. In the Presque Isle area, Devonian volcanic and coarse-grained clastic rocks are abundant (Dockendorff Group of Boucot and others, 1964), but elsewhere slate forms most of the Devonian.

The rocks of the Ordovician and Silurian systems display rapid lateral and vertical changes in lithofacies. These changes will be illustrated here by considering the stratigraphic sections for five areas shown by Roman numerals in Figures 1 and 3. Areas I through IV are in the Ashland synclinorium and Aroostook-Matapedia anticlinorium where Lower through Upper Silurian rocks are present; area V is at Fish River Lake on the west flank of the Pennington Mountain anticlinorium. In this discussion, four new formations are established, two units are given formational rank and redesignated, and five provisional members of the Frenchville Formation are proposed; formalization of these additions and changes is given in Appendix 1.

ORDOVICIAN AND LOWER SILURIAN LITHOFACIES VARIATIONS IN THE PRESQUE ISLE-ASHLAND REGION

In the region between Ashland and Presque Isle (Fig. 3), rapid lateral changes in lithofacies occur in rocks of both Ordovician and Lower Silurian ages. The nature of the variations is shown by stratigraphic sections from areas I, II, and III; because no complete or even lengthy partial sections are available in any of these areas, the sections are areal composites. Although the sections give the entire Ordovician and Silurian succession in each area, only the Ordovician and Lower Silurian rocks are treated in this discussion; the Upper Silurian units are discussed in connection with sections from areas IV and V where they are better known.

Ashland-Frenchville Area (I)

Madawaska Lake Formation. The oldest rocks in the immediate vicinity of the village of Frenchville are green to green-gray slate with lesser graywacke (Fig. 4). This sequence, the Madawaska Lake Formation (see Appendix 1), is exposed in the central part of small anticlines within the Ashland synclinorium and forms

[1]Boone (1970) published the results of his stratigraphic studies (Boone, 1958) in the region between Fish River Lake and Eagle Lake along the northwestern flank of the Pennington Mountain anticlinorium. He proposed a new formation, the Fish River Lake Formation, which contains fossiliferous rocks of both Late Silurian and Early Devonian age. More recent work by Horodyski (1968) and Roy and Mencher (unpub. data) indicates that the Silurian rocks of the formation may be mapped separately. For the purposes of this discussion, the distribution of these Silurian rocks is shown in Figure 1, and the Devonian strata of the Fish River Lake Formation are included with the undifferentiated Devonian system.

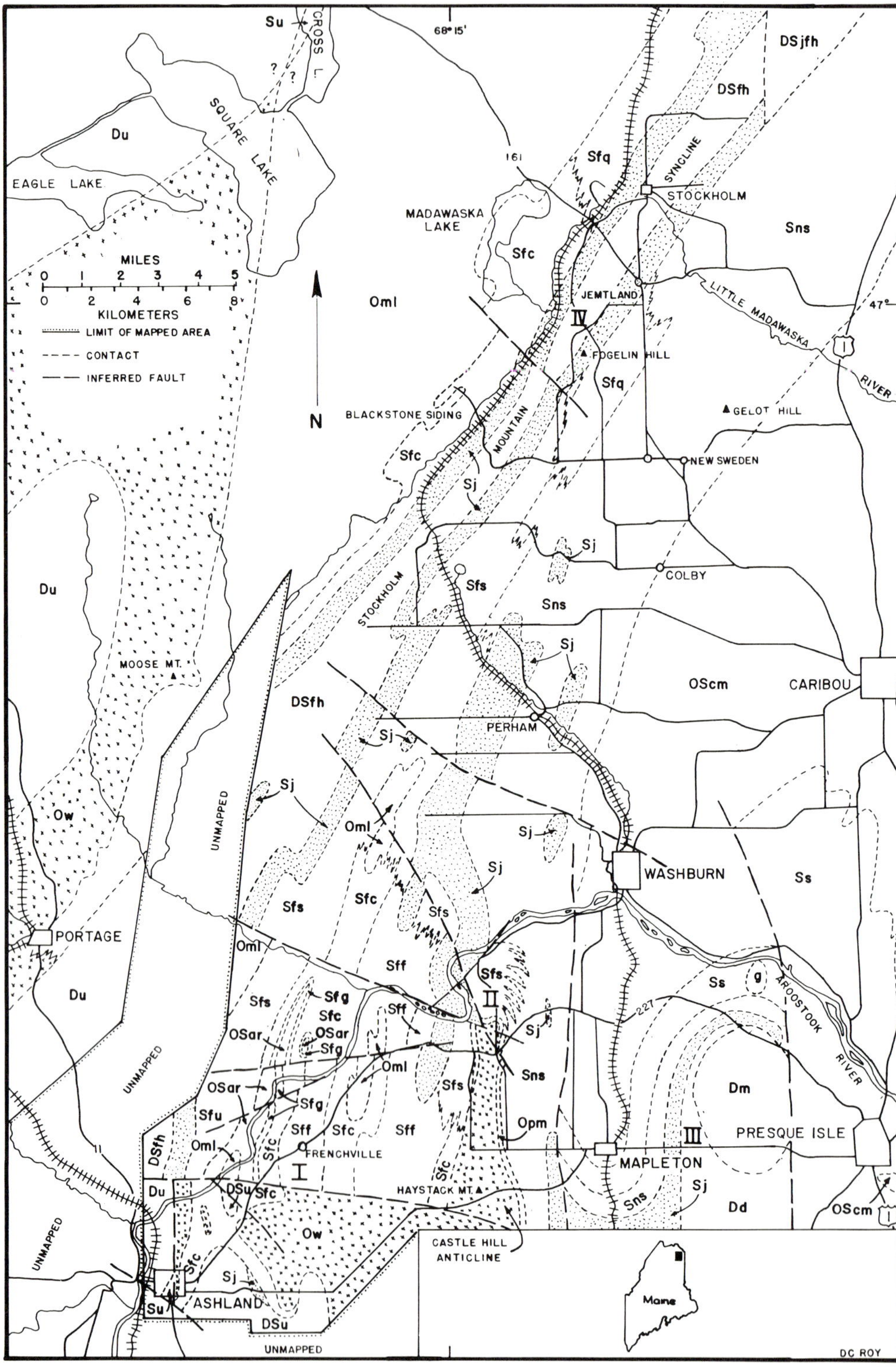

68° 15'
47°
MILES
0 1 2 3 4 5
0 2 4 6 8
KILOMETERS
LIMIT OF MAPPED AREA
CONTACT
INFERRED FAULT
N
CROSS L.
SQUARE LAKE
EAGLE LAKE
MADAWASKA LAKE
STOCKHOLM
SYNCLINE
JEMTLAND
FOGELIN HILL
LITTLE MADAWASKA RIVER
GELOT HILL
BLACKSTONE SIDING
MOUNTAIN
NEW SWEDEN
COLBY
STOCKHOLM
MOOSE MT.
CARIBOU
PERHAM
WASHBURN
UNMAPPED
PORTAGE
AROOSTOOK RIVER
PRESQUE ISLE
MAPLETON
FRENCHVILLE
HAYSTACK MT.
CASTLE HILL ANTICLINE
ASHLAND
Maine
DC ROY
Su
Du
Oml
DSjfh
DSfh
Sfq
Sfc
Sns
Sj
Sfs
OScm
Ow
Ss
Sff
Sfg
OSar
Sfu
Dm
Dd
Opm
DSu
g
161
227
11
1
I
II
III
IV

much of the central terrain of the Pennington Mountain anticlinorium (Figs. 1, 3). Although greenish pelite predominates in the formation, red slate is present locally, especially where the Madawaska Lake is overlain unconformably by the Frenchville Formation. The graywacke beds of the Madawaska Lake commonly show sedimentary features arranged in a sequence consistent with the now classic turbidite model of Bouma (1962) as illustrated by the specimen in Figure 5.

Aroostook River, Formation. The Aroostook River Formation, a succession of gray to green-gray, commonly calcareous slate with thin fine calcareous graywacke beds, overlies the Madawaska Lake Formation in the anticline just west of Frenchville (Fig. 3). Medium- and coarse-grained, generally graded, graywacke beds that range up to 30 cm in thickness are present throughout the sequence, but appear to be slightly more abundant near the conformable upper contact with the Frenchville Formation. The graywacke beds of the Aroostook River Formation display turbidite sedimentary features and typically contain pelite fragments derived from underlying pelite beds.

Though possibly as much as 600 m thick along the axis of the anticline west of Frenchville, the Aroostook River Formation is absent in the small anticline just northeast of Frenchville (Fig. 3); indeed, the formation has not been observed outside of the more western anticline. The contact between the Aroostook River and Madawaska Lake Formations has not been observed, but it is inferred to be conformable or possibly disconformable; the Aroostook River is interpreted to be a local transitional phase between the Madawaska Lake and Frenchville Formations. The Aroostook River Formation may be more extensive west of Frenchville, but is hidden beneath younger rocks.

Frenchville Formation. The Frenchville Formation of Boucot and others (1964) has been subdivided into five provisional members (see Appendix 1); three of these members are in the vertical succession in the vicinity of Frenchville as shown in Figure 4. Conformably overlying the Aroostook River Formation is the 153-m sequence of medium- to thin-bedded lithic graywacke of the Graywacke Member. No pelite beds have been observed in the Graywacke Member; thin conglomerate beds containing abundant pelite clasts occur locally. The Graywacke Member has not been observed beyond the limits of underlying Aroostook River Formation.

The Graywacke Member is overlain by the Conglomerate Member, which consists of 75 to 80 percent polymictic pebble conglomerate, 20 to 25 percent lithic graywacke, and minor pebbly and nonpebbly slate. The conglomerate and graywacke beds contain lithic fragments of a variety of volcanic rocks and chert that clearly are derived from the Ordovician Winterville Formation (Roy, 1970a). The conglomerate beds display little fabric except for the alignment of elongate pebbles parallel to bedding; in many beds the graywacke matrix is in excess of 50 percent. At one locality, apparently graded bedding is present in the member as shown in Figure 6.

Figure 3. Geologic map showing the distributions of Ordovician and Silurian rocks in the Ashland synclinorium and location of sections I through IV described in the text. Stratigraphic units: Ow, Winterville Formation; Oml, Madawaska Lake Formation; Opm, Pyle Mountain Argillite; OSar, Aroostook River Formation; OScm, Carys Mills Formation; Sfu, Frenchville Formation undifferentiated; Sfg, Sfc, Sff, Sfs, Sfq, respectively, the Graywacke, Conglomerate, Feldspathic Sandstone, Sandstone-Slate, and Quartzose Sandstone Members of the Frenchville Formation; Sns, New Sweden Formation; Ss, Spragueville Formation; Sj, Jemtland Formation; DSfh, Fogelin Hill Formation; DSjfh, Jemtland-Fogelin Hill undifferentiated; Su, Silurian undifferentiated; Dd, Dockendorff Group (Presque Isle area); Dsu, Devonian-Silurian undifferentiated (Ashland area); Du, Devonian undifferentiated; Dm, Mapleton Sandstone; g, granite. Sources of information are given in Figure 1.

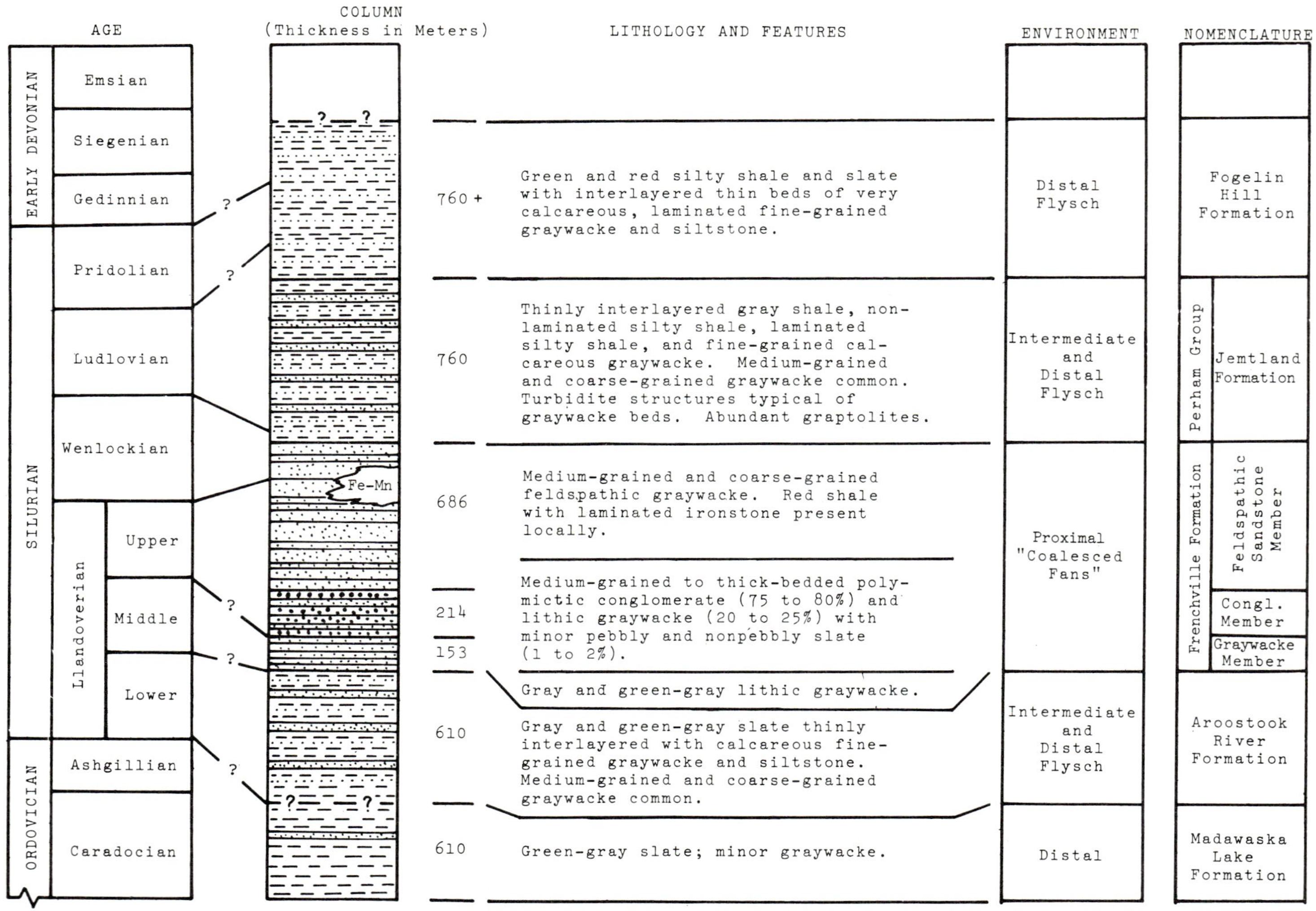

Figure 4. Stratigraphic section in the vicinity of Frenchville, Maine, area I. The lithology is generalized. Thicknesses determined from structure sections.

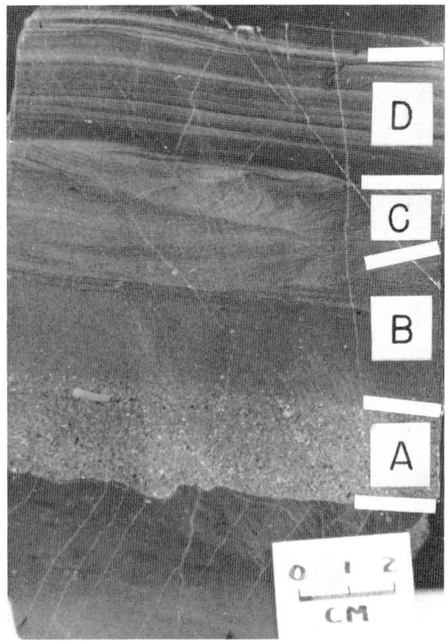

Figure 5. Sawed specimen of a graded graywacke bed from the Madawaska Lake Formation showing the sedimentary structures of intervals A through D as defined by Bouma (1962). Note the erosional contact between the lower parallel-laminated interval (B) and cross-laminated interval (C); we feel an erosional B-C contact is typical of turbidite layers of all ages in the area. D intervals (here consisting of the interlamination of fine siltstone and pelite) are commonly difficult or impossible to distinguish from overlying E intervals, which are in turn typically inseparable from "interturbidite" pelitic beds (Walker, 1967; Roy, 1970b).

Although approximately 200 m thick in the Frenchville area, the Conglomerate Member is locally much thicker, especially where it comprises the entire Frenchville Formation (Fig. 3). The Conglomerate Member interfingers northward and eastward from the Frenchville area with the Sandstone-Slate Member (Appendix 1).

The upper 686 m of the Frenchville Formation in this area is composed of medium- to thick-bedded feldspathic graywacke and arenite which are locally interlayered with lesser red slate and manganiferous ironstone (Fig. 4); these rocks compose the Feldspathic Sandstone Member of the Frenchville Formation. Pebble conglomerates have not been observed in this member, and the sandstones do not contain abundant volcanic and chert rock fragments. Roy (1970a) interprets the sandstones of the Feldspathic Sandstone Member to have been largely derived from a quartz diorite or granodiorite source with much lesser contribution of detritus from the volcanic rocks of the Winterville Formation. The Feldspathic Sandstone Member is restricted to the southeastern portion of the region shown in Figure 3 and, like the Conglomerate Member, interfingers northward and eastward with the Sandstone-Slate Member.

The Frenchville Formation is inferred to be internally conformable in the vicinity of the village of Frenchville, although stratigraphic continuity cannot everywhere be demonstrated owing to incomplete exposure. The lower contact of the Frenchville with the Aroostook River Formation has been observed and is gradational; the upper contact of the Frenchville with the Jemtland Formation (Late Silurian) is not observed in area I, but elsewhere (area IV) can be shown to also be gradational.

Six kilometres south of the location of the column shown in Figure 4, in Ashland, the Conglomerate Member of the Frenchville Formation overlies the Winterville Formation, a lateral equivalent of the Madawaska Lake Formation, with angular unconformity. The Aroostook River Formation and the Graywacke and Feldspathic Sandstone Members of the Frenchville Formation are absent. The Conglomerate Member in overlain conformably by the Jemtland Formation consisting of slate and fine-grained graywacke. The differences between the Ashland succession and

Figure 6. Apparently graded beds in the Conglomerate Member of the Frenchville Formation.

that found in area I just to the north are explained, in part, by rapid lithofacies changes in the Ordovician and southward onlap of the Frenchville and Jemtland lithofacies onto a local source area (Roy, 1970a).

Castle Hill Area (II)

Winterville Formation. A markedly different stratigraphic succession (Fig. 7) is present to the northeast of Frenchville on the west flank and northern end of the Castle Hill anticline (Fig. 3). The oldest rocks exposed in the core of the anticline form the easternmost occurrence of the Winterville Formation (Appendix 1).

The Winterville Formation consists of andesitic, basaltic, keratophyric, rhyolitic, and pyroclastic volcanic rocks with variable interlayered black slate, graywacke, and chert. The formation may be traced by surface exposures to the vicinity of Ashland and is inferred to connect in the subsurface to the large terrains of the formation in the Pennington Mountain anticlinorium (Fig. 1).

The internal stratigraphy of the Winterville Formation is not well known. Regional variations in the relative abundances of the rock types in the Winterville have been observed; for example, black slate is only sparsely exposed in the Castle Hill anticline, but is very abundant in the Portage-Moose Mountain area where the Winterville Formation interfingers with the slate-rich Madawaska Lake Formation to the east (Fig. 3).

Pyle Mountain Argillite. Disconformably(?) overlying the Winterville Formation in the Castle Hill anticline is the Pyle Mountain Argillite of Boucot and others (1964). The Pyle Mountain Argillite is a poorly exposed succession of gray and green-gray slate (locally calcareous) and richly-fossiliferous brown-weathering silty argillite. The Pyle Mountain is limited in distribution to the flanks of the Castle Hill anticline where it has a maximum thickness of approximately 90 m.

Frenchville Formation. The Sandstone-Slate Member of the Frenchville Formation (Appendix 1) overlies the Pyle Mountain Argillite disconformably or with only a slightly angular unconformity on the west flank and northern end of the Castle Hill anticline. The Sandstone-Slate Member consists of lithic and feldspathic graywacke interlayered with approximately equal amounts of slate. Pebbly graywacke and conglomerate are found locally near the base of this sequence; turbidite features

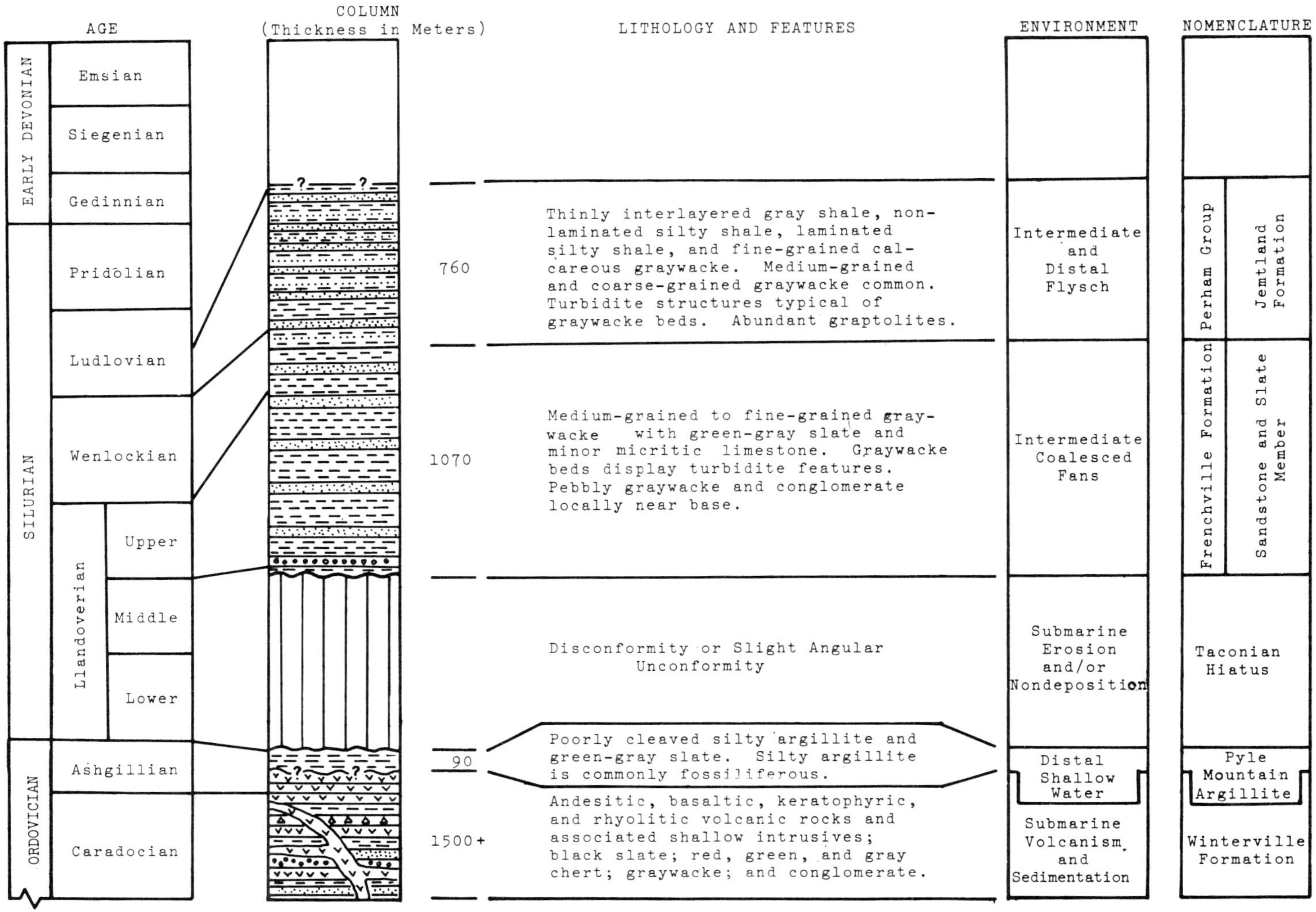

Figure 7. Stratigraphic section at Castle Hill, Maine, area II. The lithology column is generalized. Thicknesses determined from structure sections.

are common in the sandstone beds. Lenticular and extensive thin beds of aphanitic limestone are observed in some exposures. The Sandstone-Slate Member is considered the lateral equivalent of the more western Conglomerate and Feldspathic Sandstone Members on the basis of both age and sandstone composition (Roy, 1970a).

The Sandstone-Slate Member interfingers with the calcareous slates of the New Sweden Formation in the vicinity of the Castle Hill anticline. The New Sweden Formation overlies the Pyle Mountain Argillite along much of the eastern flank of the structure.

Mapleton–Presque Isle Area (III)

Carys Mills Formation. East of the Castle Hill anticline, in area III of Figure 3, the Ordovician and Lower Silurian section is greatly different from those previously described. Thinly interlayered aphanitic limestone and slate with lesser lithic graywacke (Carys Mills Formation of Pavlides, 1962, 1965, 1968) form the oldest rocks of the area (Fig. 8). This lithofacies contains fossils of Caradocian (graptolite zone 13) to early Llandoverian (zones 18 and 19) ages and is therefore a time equivalent of the more westerly Winterville and Madawaska Lake Formations, the Pyle Mountain Argillite, and the Aroostook River Formation. The Carys Mills Formation forms the southwestern portion of an extensive carbonate-rich lithofacies that extends from Houlton, Maine, to the Gaspe Peninsula (Pavlides and others, 1964; Pavlides, 1968).

The internal stratigraphy of the Carys Mills Formation cannot be documented in detail at present. In the Mapleton–Presque Isle area, the "ribbon rock" interbedding of dark gray limestone and slate predominates in the Carys Mills. The limestone beds (typically 1.3 to 15.0 cm thick) are characteristically argillaceous and many are silty and sandy with parallel and cross-lamination. Ankeritic limestone beds generally show a rusty weathering color whereas weathered surfaces of more calcic varieties are light gray.

The distribution of graywacke beds in the Carys Mills Formation is not fully known, but they are abundant in dated exposures of Ordovician age (Carodocian and Ashgillian) and are virtually absent in dated Lower Silurian exposures. The graywacke beds contain abundant volcanic detritus, thus their apparent abundance in the Ordovician phase of the Carys Mills suggests that the graywacke beds were derived from contemporaneous volcanic rocks to the west (Winterville Formation) and possibly also to the east (Tetagouche Group, Davies, 1966).

Spragueville Formation. The Carys Mills Formation is overlain conformably by the Spragueville Formation (Pavlides, 1965), a sequence of thinly interbedded and interlaminated gray silty pelite, light-gray silty limestone, and calcareous siltstone. In the Mapleton–Presque Isle area, exposures of this unit show a general absence of cleavage and have a platy or slabby appearance; this absence of cleavage is probably the result of the open fold style of the Chapman syncline (Boucot and others, 1964; Figs. 1, 3). In the vicinity of Fort Fairfield, Maine, to the east of Presque Isle (Fig. 1), the Spragueville is in tight, isoclinal synclines and is commonly well cleaved (Roy and Mencher, unpub. data).

The Spragueville Formation is the only formation in the region that displays obvious bioturbation. Knots, visible in outcrop surfaces cut perpendicular to bedding (Fig. 9), have been recognized as animal (worm?) borings; these bioturbation features are the "nubbles" of White (1943). Gray aphanitic limestone typical of the underlying Carys Mills Formation is present locally as thin beds within the Spragueville.

New Sweden Formation. Conformably overlying the Spragueville Formation in

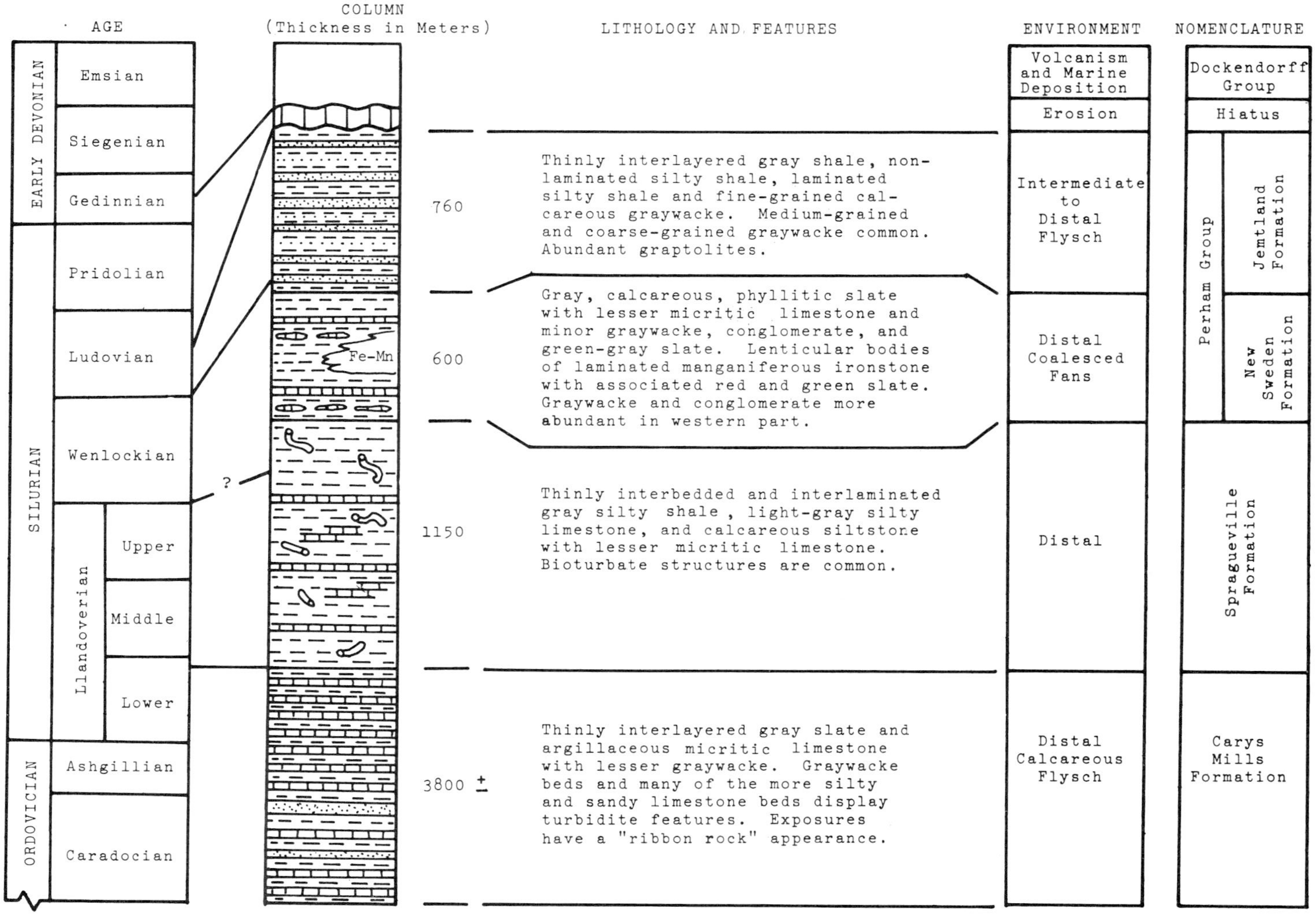

Figure 8. Stratigraphic section in the Mapleton–Presque Isle area, Maine, area III. The lithology is generalized. Thicknesses determined from structure sections.

Figure 9. Pavement outcrop of Spragueville Formation showing the thinly laminated character of the bedding, truncated animal burrows (arrows), and thin micritic limestone bed (behind knife).

this area is the New Sweden Formation of the Perham Group (Appendix 1), formerly the lower member of the Perham Formation of Boucot and others (1964). The New Sweden is composed principally of calcareous, commonly phyllitic, locally laminated slate containing both extensive and lenticular thin beds of aphanitic limestone. Most of the limestone beds are argillaceous; traces of ankerite(?) in some beds produce a rusty weathered surface. Miller (1947) and Pavlides (1962) have described the geochemistry and local stratigraphic setting of apparently lenticular bodies of manganiferous ironstone and associated red and green slates which occur within the New Sweden Formation and its correlative units to the south in Maine near Houlton. The ironstone is at various stratigraphic levels widely spaced throughout the New Sweden outcrop belt from Mapleton to Van Buren, Maine, and into New Brunswick.

From east to west within the Mapleton-Presque Isle area, the Spragueville Formation thins beneath the thickening New Sweden Formation; between Mapleton and the Castle Hill anticline, the Spragueville is absent and the New Sweden rests on the pre-Silurian (Pyle Mountain Argillite) along the eastern flank of the Castle Hill anticline. East of Presque Isle, the New Sweden section pinches out, and only the Spragueville Formation is present. Along the contact between the towns of Washburn and Van Buren (Figs. 1, 3), the New Sweden rests conformably on the Carys Mills (Boucot and others, 1964; Roy 1970a); the Spragueville Formation is absent. All of these relationships, together with a similarity in age, indicate that the New Sweden and Spragueville Formations are lateral equivalents.

UPPER SILURIAN AND LOWER DEVONIAN(?) LITHOFACIES

In areas I, II, and III discussed above, the Lower Silurian (Llandoverian and early Wenlockian) units are overlain conformably by a thin-bedded "flysch-type"

(Van Der Lingen, 1969) sequence, the Jemtland Formation (late Wenlockian-early Ludlovian) that is in turn overlain conformably by the "flysch-like" Fogelin Hill Formation (early Ludlovian-Siegenian(?)). These two formations are best exposed in the Stockholm Mountain syncline, a relatively simple structure that lies along the western margin of the Ashland synclinorium in the vicinity of Jemtland, Maine (area IV, Fig. 3).

Late Silurian rocks of a shelf lithofacies locally rest with angular(?) unconformity on the Ordovician Winterville Formation in the region west of Maine Route 11 (Fig. 1). The limited distribution of these Upper Silurian rocks possibly is in part the result of original patchy sedimentation around emergent islands, but may also be the result of minor uplift and erosion in the latest Silurian and earliest Devonian times (Pavlides and others, 1964).

Stockholm-Jemtland Area (IV)

Pre-Jemtland Rocks. The stratigraphic succession in the Stockholm-Jemtland area (area IV, Figs. 1, 3) is shown in Figure 10. The Madawaska Lake, Carys Mills, and New Sweden Formations are little different in this area from their occurrences discussed above. Within the Frenchville Formation, the Sandstone-slate Member of more southerly areas is replaced by the Quartzose Sandstone Member (Appendix 1) on the east flank of the Stockholm Mountain syncline; the Graywacke and Feldspathic Sandstone members are absent. The Quartzose Sandstone Member contains quartz arenite and graywacke interlayered with lithic and feldspathic graywacke together with variable, but locally abundant slate; quartz-rich sandstone beds are restricted to this member of the Frenchville Formation.

The Conglomerate Member of the Frenchville Formation on the west flank of the Stockholm Mountain syncline rests with angular unconformity on the slates of the Madawaska Lake Formation. To the northeast, the Conglomerate Member intertongues with the Quartzose Sandstone Member, and structural data suggest that the Frenchville-Madawaska Lake contact is a disconformity or a very slightly angular unconformity. The intertonguing of the members of the Frenchville Formation and the transition of the basal Frenchville contact from one of pronounced angularity to one of disconformity are inferred to take place across the syncline from northwest to southeast in this area (Fig. 10).

Jemtland Formation. The thinly bedded Jemtland Formation conformably overlies the Frenchville and New Sweden formations (Fig. 10). This unit, previously called the Upper Member of the Perham Formation by Boucot and others (1964), consists predominantly of four rock types interlayered on a small scale: indistinctly laminated silty shale; non-laminated silty shale; fine-grained shale (slate); and fine-grained graywacke. All four rock types are calcareous. Graptolites are commonly abundant and well preserved in the indistinctly laminated shale. The pelitic rocks are generally moderately to well cleaved in the eastern synclinal occurrences of the formation (Fig. 3); where uncleaved, these rocks usually have a platy bedding fissility.

Interlayered with the above four rock types are beds of medium- to coarse-grained lithic graywacke which compose as much as 50 percent of the section along the western flank of the Stockholm Mountain syncline. All of these coarser graywacke beds are calcareous and display grading and sedimentary structures consistent with Bouma's (1962) turbidite model (Fig. 5). As discussed by Roy (1970b), the relative abundances of both the fine- and coarser-grained graywacke beds decrease in the formation from northwest to southeast across the structural trends.

Locally along both flanks of the Stockholm Mountain syncline and approximately in the middle of the Jemtland section are horizons of devitrified lithic and vitric

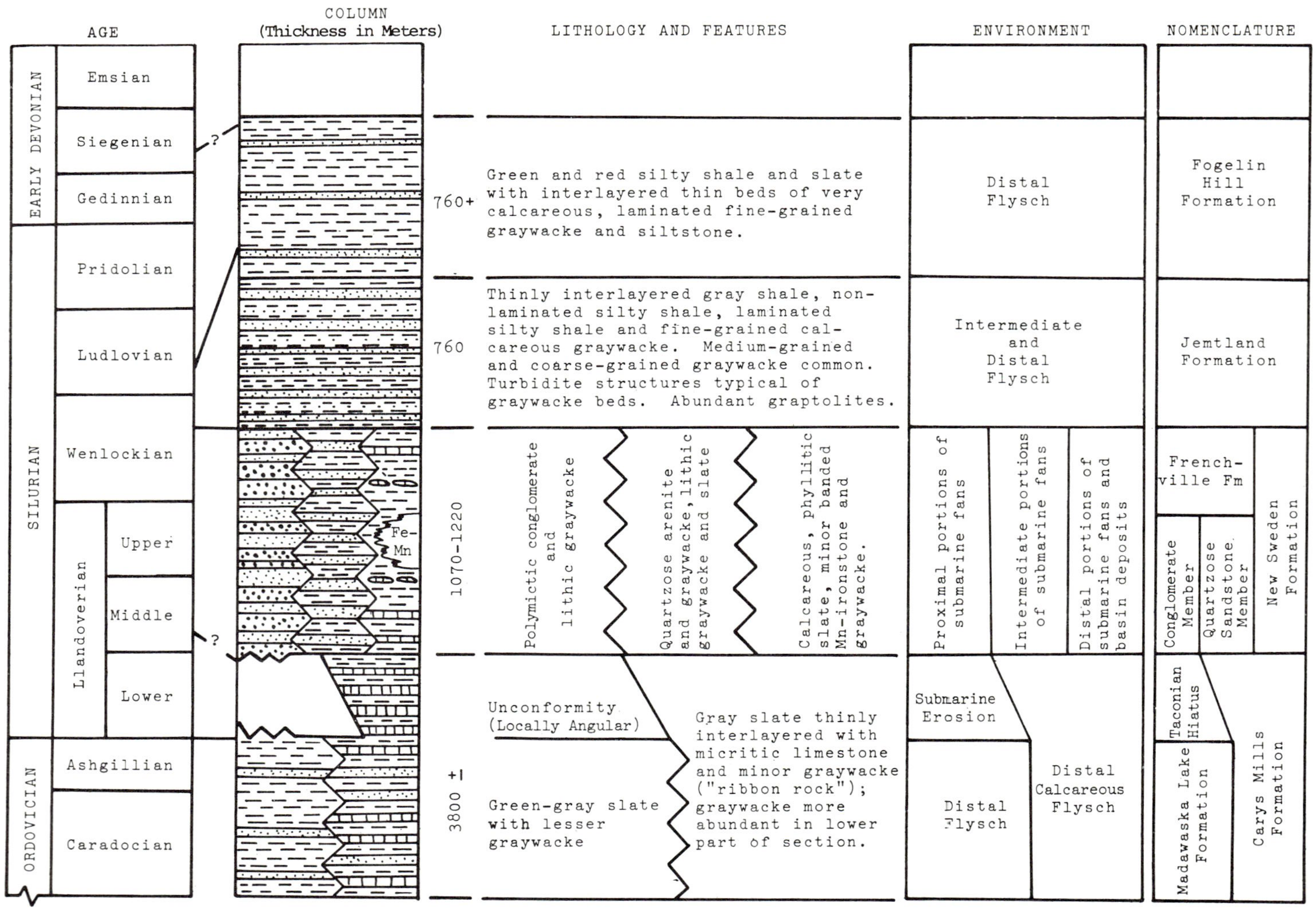

Figure 10. Stratigraphic section in the Stockholm-Jemtland area, Maine, area IV. The lithology is generalized. Thickness determined from structure sections.

aquagene tuff as much as 20 m thick. The tuff sequences are excellent local marker horizons and are probably formed from pyroclastic debris derived from sites of Late Silurian volcanism in the Fish River Lake (Horodyski, 1968), Munsungun Lake (Hall, 1970), and Island Pond (Mencher, unpub. data) areas to the west and southwest.

Micritic limestone similar to that found in the older New Sweden and Carys Mills Formations is present in the Jemtland Formation as a very minor lithology.

Fogelin Hill Formation. The Jemtland Formation is conformably superseded by a sequence of green and red silty shale (slate) with interlayered calcareous and laminated fine graywacke and siltstone. This sequence, the Fogelin Hill Formation (Appendix 1), forms the axial terrain of the Stockholm Mountain syncline and is also sparsely exposed just north of Ashland (Fig. 3). The top of the Fogelin Hill has not been observed.

Fish River Lake Area (V)

The stratigraphy of the Fish River Lake area (area V, Fig. 1) is depicted in Figure 11. The oldest rocks are of the Winterville Formation; these are overlain unconformably by a highly variable upper Silurian succession of volcanic rocks (mostly flow rocks), pebble conglomerate, sandstone, fossiliferous calcareous mudstone, and silty limestone with thin coral biostromes that forms a part of the Fish River Lake Formation of Boone (1970).

The Devonian succession has been studied in reconnaissance fashion by Horodyski (1968), Boone (1970), and Roy and Mencher (unpub. data). The stratigraphic nomenclature for the Silurian-Devonian shown in Figure 11 is that of Boone (1970). There is a strong possibility that the Silurian and Devonian rocks of the Fish River Lake Formation are separated at least locally by an unconformity as suggested by the apparant absence of the earliest Devonian (early Gedinnian) in the Fish River Lake area and the presence of probable disconformities reported by Horodyski (1968) and Hall (1970) between similar rock sequences to the southwest.

SUMMARY OF ORDOVICIAN AND SILURIAN STRATIGRAPHIC RELATIONSHIPS

A summary of the stratigraphic relationships of the Ordovician and Silurian units in the region of Figure 3 is provided by the correlation chart of Figure 12. This chart illustrates the facies relationships in the Presque Isle-Ashland area where stratigraphic changes occur over very short distances and where all of the formations are present; however, the chart also indicates the general east-west and vertical distributions of the units that are applicable to the entire region shown in Figure 3. It therefore is possible to represent the succession of units at any point in the region by a suitably selected vertical line on the chart.

The age resolution of fossil localities in each unit is also indicated on the correlation chart; although most of the localities are from the region, a few from surrounding areas are also plotted for completeness. The fossil localities are discussed in detail by Roy (1970a, Appendix) and are presented here only for the purpose of showing the extent of age control that is presently available.

In the Presque Isle-Ashland area, the easternmost occurrence of the Winterville Formation is in the Castle Hill anticline (Fig. 3). The Winterville rocks in the northern end of the anticline are inferred to interfinger westward with the Madawaska Lake Formation and eastward with the lower part of the Carys Mills Formation

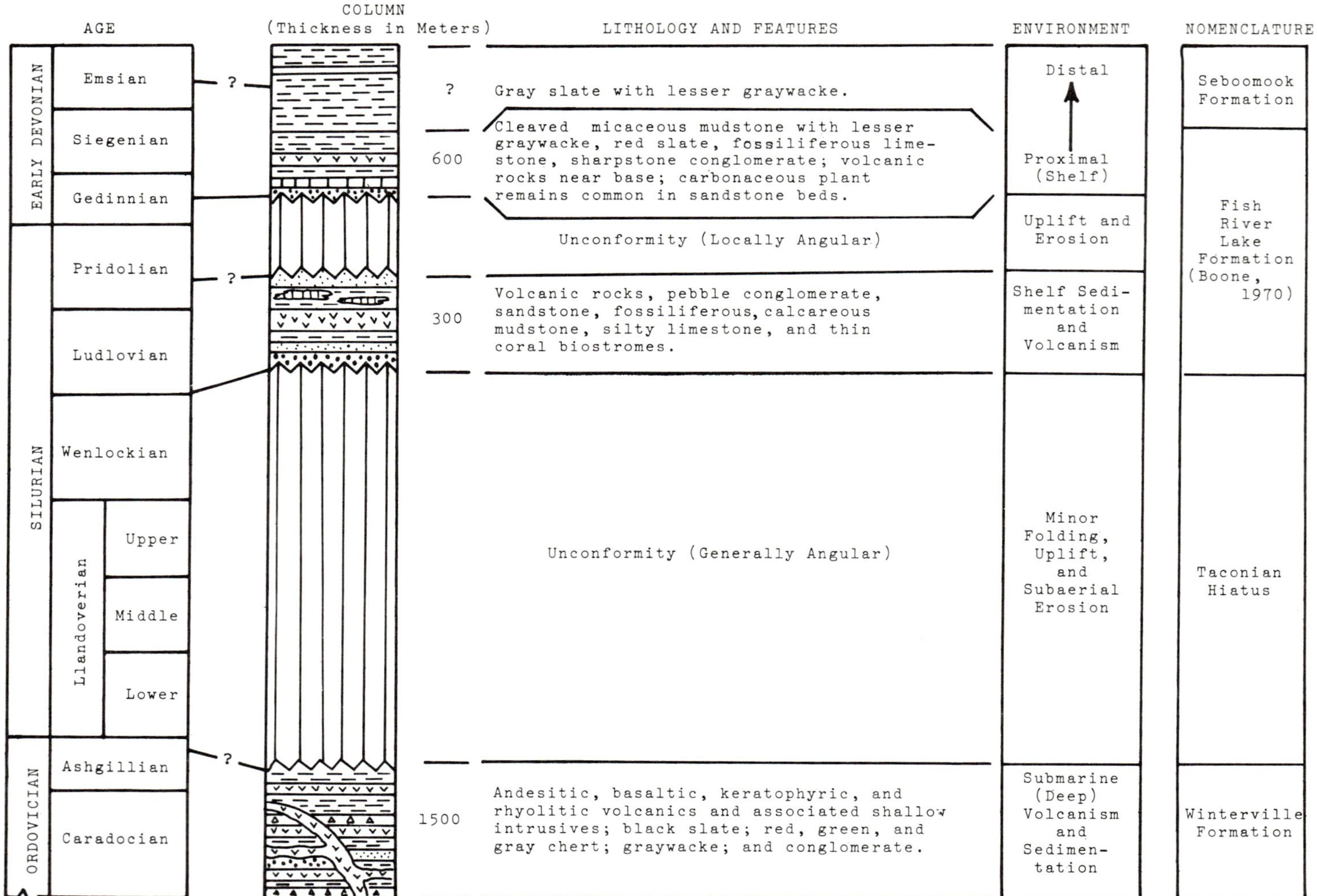

Figure 11. Stratigraphic section in the Fish River Lake area (area V), Maine. The lithology is generalized. Thicknesses determined from structure sections. The Silurian-Devonian nomenclature is that of Boone (1959, 1970), but the stratigraphic column is based on more recent field work by Horodyski (1968) and Roy and Mencher (unpub. data) which has produced more fossil localities in the area.

as shown on the correlation chart (Fig. 12). Along an east-west line from Moose Mountain to Caribou, however, the Madawaska Lake Formation and lower Carys Mills formations are inferred to grade into each other with the zone of transition located in the subsurface beneath the Silurian rocks of the synclinorium.

ENVIRONMENTS OF DEPOSITION

In Figures 4, 7, 8, 10, and 11, a brief statement of the environment of deposition of each of the sequences is given. Justification of these interpretations requires more space than is possible here and will be reserved for subsequent papers on each of the units. In general, the Winterville, Madawaska Lake, and Carys Mills Formations are inferred to have formed in relatively deep water and far from major land areas. The volcanic rocks of the Winterville Formation appear to have formed largely as a result of submarine volcanism because pillowed flow rocks are commonly interlayered with graptolitic shales (slate), radiolarian(?) cherts, and water-laid tuffs; local emergent volcanic islands or shoal areas were probably present to provide the clastic debris of the associated graywacke and conglomerate beds as suggested by Neuman (1967).

The Frenchville Formation is interpreted to represent coalesced submarine fans forming a clastic wedge at the base of a slope along the margin of a Taconian land area, "Highlandcroftia" of Rodgers (1971), to the west and southwest. The slate-rich members of the formation compose the intermediate part of the wedge and give way eastward to the distal slate and lesser limestone succession of the New Sweden Formation. The easternmost and most distal (with respect to Highlandcroftia) upper Llandoverian-lower Wenlockian lithofacies is the bioturbated calcareous mudstone and silty limestone sequence of the Spragueville formation. In the vicinity of Frenchville, the Frenchville Formation is conformably underlain by the Aroostook River Formation that contains graywacke turbidites interlayered with slate and is inferred to be a local transitional flysch from the slates of the Madawaska Lake Formation. Elsewhere, the Frenchville Formation rests unconformably on older rocks.

As the Silurian basin in the eastern part of the area of Figure 1 deepened and became more distal with respect to western emergent terrain, the thinly bedded turbidite sequences of the Jemtland and Fogelin Hill Formations were deposited. Coeval with at least the lower Ludlovian and Pridolian(?) parts of these thin-bedded flysch units, are shallow-water and near-shore sedimentary rocks and interlayered volcanic rocks in the Fish River Lake area. These western Upper Silurian rocks are interpreted as reflecting the presence of shelf areas, possibly narrow, around emergent small remnant islands of a previously (Llandoverian-early Wenlockian) more extensive tectonic land area.

CONCLUSIONS

The general parallelism of lithofacies boundaries and the trends of major tectonic elements in the northern Appalachians has long been recognized (see, for example, Naylor and Boucot, 1965; Berry, 1968; Boucot, 1968; Naylor, 1968; and Ayrton and others, 1969) and this parallelism largely accounts for the general difficulty in cross-structure correlation in regions with appreciable Acadian metamorphism. In northeastern Maine, cross-structure lithofacies changes are well documented and indicate that coeval sequences in adjacent areas can be markedly different.

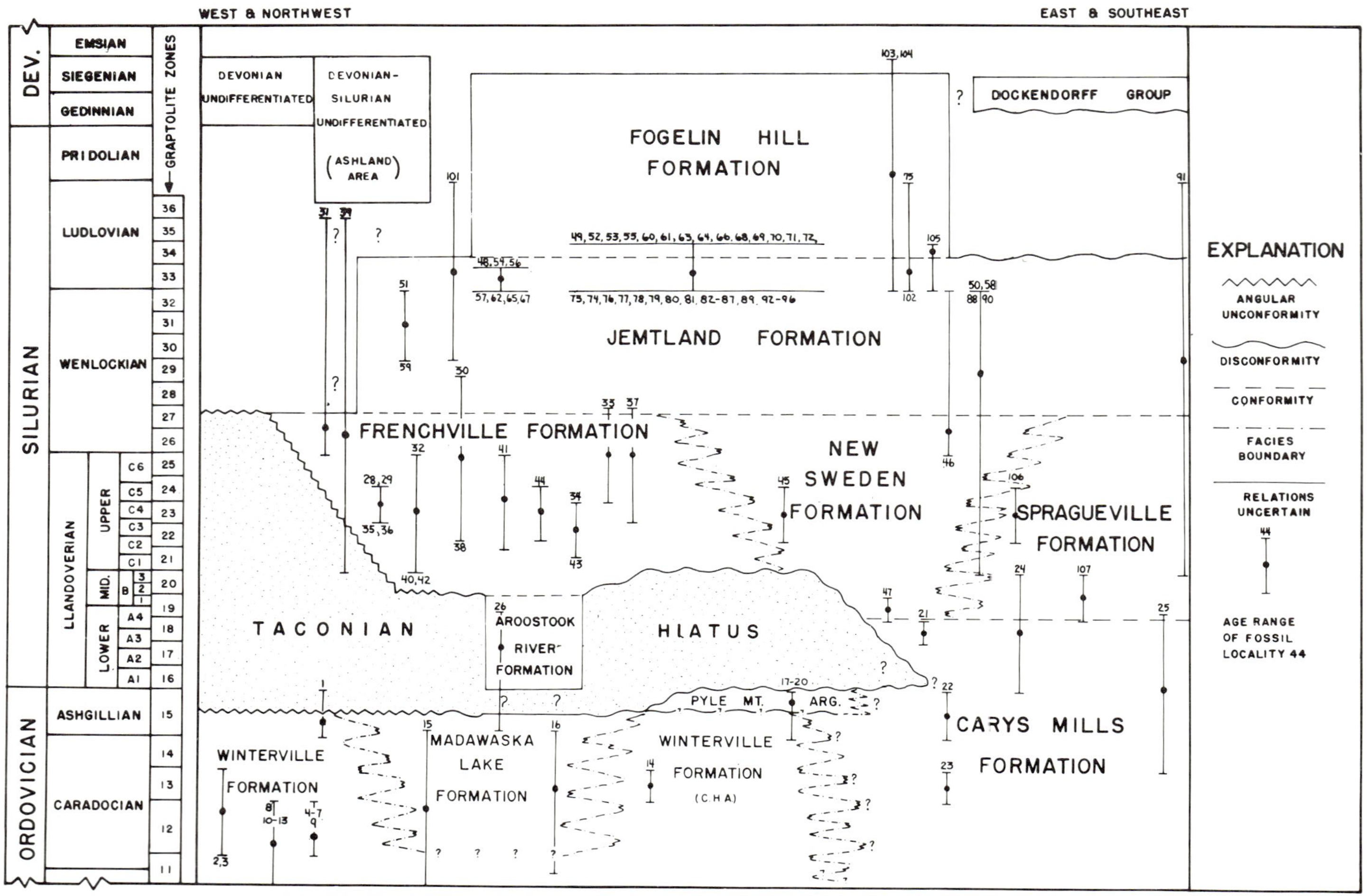

Figure 12. Correlation chart showing the facies relationships of Ordovician and Silurian rock units and the age ranges of fossil localities. Subdivisions of the Llandoverian (Al through A4, Bl through B3, and C1 through C6) are used in chronology based on shelly fauna, especially brachiopods. Fossil localities are discussed in detail by Roy (1970a).

The problem of correlation on the basis of lithology alone is further complicated by the general similarities of some units of differing ages; for example, the Ordovician Medawaska Lake Formation is grossly similar to some slate phases of the Devonian Seboomook Formation and at higher metamorphic grades these slate units might be very difficult to separate.

Great care must also be exercised in the use of "unusual" rock types for correlation purposes. Marker beds have not been found that may be traced for distances greater than a few kilometres along the flanks of folds, and none persists across a major fold. In the region of this study, the laminated manganiferous ironstones do appear to be restricted to the Lower Silurian New Sweden Formation, but ironstones are also present in the Upper Silurian (Pavlides and Berry, 1966) and even the Lower Devonian (Pavlides, 1962) in the vicinity of Houlton, Maine. Banded manganiferous ironstones of Ordovician age have not been documented in northeastern Maine, but there are some ironstone localities of uncertain age (Pavlides, 1962). Micritic limestone, thinly interlayered with slate, or as isolated thin beds and lenses, is found in Ordovician and Silurian units of the Ashland synclinorium and Aroostook-Matapedia anticlinorium but have not been observed in Devonian sequences.

In a recent attempt to synthesize the history of the Appalachian Mobile Belt in terms of the modern concept of plate tectonics, Bird and Dewey (1970) have caused a great deal of attention to be focused on the axial region of the New England Appalachians as the location of an island arc-trench system during the later part of the Ordovician. The apparent belt of Ordovician volcanic rocks along the Bronson Hill-Boundary Mountain anticlinorium and its extensions in northwestern Maine (Naylor, 1968) are interpreted to be the volcanic arc. The trench is inferred to have been located immediately to the east of the belt of volcanic rocks in the largely metasedimentary rocks of the Merrimac synclinorium and extending through the area of this study. Unfortunately, the transition from the volcanic-rich Ordovician sections in the anticlinoria to the more eastern sedimentary sequences is obscured in the Merrimac synclinorium by complex structure, metamorphism, inadequate paleontological control, or a cover of younger rocks. In northern Maine, the pre-Silurian is widely exposed and sufficiently fossiliferous to permit assessment of the presence of an Upper Ordovician trench-melange complex.

The Winterville Formation forms the most northeastern part of the Ordovician volcanic belt. This belt appears to terminate at Square Lake (Fig. 1), but may continue further northward under a cover of Silurian and Devonian rocks. The observed stratigraphic interfingering of the Winterville and Madawaska Lake Formations in the vicinity of Portage and Moose Mountain (Fig. 3), together with the probable, but unexposed, gradation of the Madawaska Lake and Carys Mills Formations in the Ashland synclinorium (Fig. 10), suggests that during the Caradocian-early Llandoverian time span no trench-melange complex developed. The Carys Mills Formation is part of the Aroostook-Matapedia limestone facies (Pavlides and Berry, 1966) interpreted by Bird and Dewey (1970, Fig. 10) as having been deposited on an oceanic plate southeast of the proposed trench. Investigations by Pavlides (1965, 1968, 1971), Neuman (1967), and Ekren and Frischknecht (1967) suggest that a mélange belt is also unlikely to the south of the region of this study.

With regard to the plate tectonic interpretation of the Upper Ordovician system of northeastern Maine, we conclude that no subduction zone was present. It is clear that the Upper Ordovician sedimentary lithofacies developed in a basin between belts of volcanism represented by the Winterville Formation to the west and the volcanic phases of the Tetagouche Group (Skinner, 1974; Davies, 1966) to the east in New Brunswick. It is possible that the basin was ensimatic but this is certainly not presently established since pre-Ordovician rocks have not been recognized within the Aroostook-Matapedia belt. From what we know now, it is possible for the volcanic rocks of both the Winterville and Tetagouche belts to be parts of a complex arc with an intra-arc basin (Aroostook-Matapedia limestone facies) and with a subduction zone located to the southeast of the Tetagouche belt (and dipping west) or to the northwest of the Winterville belt (and dipping east).

In latest Ordovician and earliest Silurian time, regional uplift associated with the Taconic orogeny caused an extensive emergent area to form in the western and southwestern parts of the area of Figure 1, exposing the rocks of the Winterville and Madawaska Lake Formations. Sediment derived from this land area was shed eastward and northeastward into a continuing but possibly shallowed basin in the eastern part of the region. The sediments of the Silurian basin became finer grained and more carbonate-rich eastward from the land area and finer-grained vertically as the Silurian sea advanced generally westward onto the eroding and possibly subsiding land mass. By Late Silurian time (Ludlovian-Pridolian), the western emergent areas appear to have been small and more or less isolated with shallow-water, near-shore deposits forming around them.

ACKNOWLEDGMENTS

The writers are grateful to Lincoln R. Page, Michael Pease, John Peper, and Robert Moench for reviewing the manuscript; their comments greatly improved the presentation, but we are responsible for errors of commission or omission. Our field investigations in northern Maine have been supported by the National Science Foundation (NSF Grant GP-1547) and the Maine Geologic Survey with supplemental funds from the Geoscience Thesis Fund of the Department of Earth and Planetary Sciences, Massachusetts Institute of Technology.

APPENDIX 1
DESCRIPTIONS OF NEW AND REVISED STRATIGRAPHIC UNITS

General

New or revised stratigraphic units in the region are described below and shown in Figure 3. Because of space limitations, these units and their type areas will be described only briefly here and more fully in later papers. New and revised stratigraphic nomenclatures are presented so that stratigraphic relationships and the significance of rock units could be discussed in an understandable and consistent fashion in the body of this paper.

The paleontologic age data presented in Figure 12 is from collections made by us and others (Williams and Gregory, 1900; Berry, 1960, 1964; Pavlides and others, 1961; Neuman, 1963; Boucot and others, 1964; Pavlides and Berry, 1966; Schopf and others, 1966; Pavlides, 1968; Neuman, 1968). We wish to thank the following paleontologists for their prompt and comprehensive analyses of our collections: A. J. Boucot of Oregon State University; W.B.N. Berry of the University of California (Berkeley); G. Klapper of the University of Iowa; C. B. Rexroad of Indiana State Geological Survey; R. B. Neuman, J. W. Huddle, J. M. Berdan, J. M. Schopf, and W. A. Oliver of the U.S. Geological Survey; J. C. Brower of Syracuse University; and H. N. Andrews of the University of Connecticut. Roy (1970a, Appendix) gives detailed locality and faunal descriptions of each locality indicated in Figure 12.

Winterville Formation (*New Name*)

Type Area. The type area of the Winterville Formation is in the vicinity of Winterville, Maine, along Maine Route 11 (Fig. 1). Several outcrops along Route 11 south of Winterville and in the vicinity of Portage provide good exposures of most of the rock types composing the unit. No complete section is available. The Winterville Formation was formerly designated as the Winterville Volcanics by Boone (1970) and Unite O_{vs} by Boucot and others (1964).

Lithology. The Winterville Formation is a complex assemblage of mafic and felsic extrusive rocks (with associated sills and dikes); tuffs and tuffaceous clastic sedimentary rocks; black slate; red, green, and gray cherts; graywacke; and conglomerate. Horodyski (1968) estimated the relative abundances of the principal rock types in the Big Machias Lake Area (Fig. 1), based on their frequency of exposure, as follows: 61 percent andesite, 9 percent basalt, 5 percent quartz keratophyre, 10 percent water-laid tuffs, 2 percent tuffaceous mudstone, 10 percent sandstone and conglomerate, and 3 percent black slate.

Thickness. The formation is at least 1,500 m thick, but probably much thicker.

Age. Fourteen fossil localities have been found within the Winterville Formation (Fig. 12) that indicate the unit spans the Caradocian and Ashgillian stages of the Ordovician (graptolite zones 11 through 15).

Madawaska Lake Formation (*New Formation*)

Type Area. Typical rock types of the Madawaska Lake Formation are found in exposures along the north shore of Madawaska Lake, along Route 161, 3.7 km north of the Little Madawaska River and along lumbering roads just north of the lake (Fig. 3). The vicinity of the north shore of the lake is here designated the type area. The formation also is well exposed along easily accessible lumbering roads west of Blackstone Siding and along similar roads on the east flank of Moose Mountain (Fig. 3).

Lithology. Green-gray to light olive-green, very fine-grained brittle slate with thin beds, generally less than 2.5 cm thick, and laminae of light-gray, orange-weathering siltstone constitute approximately 95 percent of the Madawaska Lake Formation. The thin siltstone beds commonly display cross-lamination and ripple morphology. Thin-bedded argillaceous micritic limestone, conglomerate, and graywacke are present locally in the formation. Maroon and green slate are also found, especially near the contact of the Madawaska Lake Formation with the overlying Frenchville Formation.

Thickness. No reliable basis for assessing the total thickness of the formation is currently available, but it is at least 600 m thick (Hamilton-Smith, 1970) and probably much thicker.

Age. Two fossil localities (Fig. 12) in the Madawaska Lake Formation containing poorly preserved graptolites indicate a possible age span of Caradocian to Ashgillian (graptolite zones 11 through 15). Stratigraphic relationships of the formation in the vicinity of Siegas, New Brunswick, suggest that locally it may be as young as late Llandoverian (Hamilton-Smith, 1970).

Aroostook River Formation (*New Formation*)

Type Area. The type area of the Aroostook River Formation is a 4.0-km stretch of the Aroostook River beginning 8.7 km downstream from the Maine Route 11 crossing in Ashland.

Lithology. Dark-gray to green-gray, fracture-cleared, calcareous slate thinly interlayered with fine-grained calcareous graywacke and coarse siltstone composes most of the Aroostook River Formation. Interbedded with these rock types is medium to coarse-grained lithic graywacke in beds from 2 to 30 cm thick. Lithic graywacke may locally form as much as 50 percent of the unit and appears to be more abundant near the upper conformable contact with the Frenchville Formation.

Thickness. The Aroostook River Formation is approximately 600 m thick.

Age. Shelly fossils recovered from two lithic graywacke beds at one exposure along the Aroostook River suggest a possible age range of Ashgillian to early Llandoverian (Fig. 12; see discussion of U.S. National Museum collection no. 13031 by Neuman, 1968, p. 44). Because the fossils come from the top of the formation near its conformable contact with the well-dated Frenchville Formation (see below), the Aroostook River Formation is here considered to be early Llandoverian in age.

Frenchville Formation (*New Provisional Members*)

Boucot and others (1964) applied the name "Frenchville Formation" to a sequence of sandstone and conglomerate previously called the "Sheridan Sandstone" by Williams and Gregory (1900) and "Sandstone" by White (1943). Roy (1970a) recently subdivided the Frenchville Formation into five complexly interrelated members: Graywacke Member, Conglomerate Member, Feldspathic Sandstone Member, Sandstone-Slate Member, and Quartzose Sandstone Member. The distribution of these members is shown in Figure 3. The Graywacke, Conglomerate, and Feldspathic Sandstone Members compose the formation in its type area (Boucot and others, 1964), around Frenchville. The conglomerate member forms the entire formation along much of the western flank of the Stockholm Mountain Syncline (Fig. 3). The slate-rich Sandstone-Slate and Quartzose Sandstone Members interfinger with the New Sweden Formation to the east.

Type Areas. The type area for the Graywacke, Conglomerate, and Feldspathic Sandstone Members is between Maine Route 227 and the Aroostook River near the village of Frenchville. The type area for the Sandstone-Slate Member is area II of Figure 3 in the north plunge of the Castle Hill anticline. The type area for the Quartzose Sandstone Member is along the Bangor and Aroostook Railroad mainline 3.7 km north of the village of Stockholm.

Lithology of the Graywacke Member. Thin- to thick-bedded, green-gray to green, medium- and coarse-grained lithic graywacke forms most of the Graywacke Member. Much less thin- to medium-bedded pebble conglomerate is present locally. No pelitic beds have been observed.

Lithology of the Conglomerate Member. In the vicinity of Frenchville, the exposed section of the Conglomerate Member consists of approximately 75 to 80 percent pebble conglomerate and 20 to 25 percent lithic graywacke with slate composing no more than two percent. Along the western flank of the Stockholm Mountain syncline, lithic graywacke appears to form 50 percent or more of the member. Slate and pebbly slate locally are minor rock types.

Lithology of the Feldspathic Sandstone Member. Medium- to massive-bedded feldspathic graywacke and arenite with minor gray, green-gray, and red slate constitute the Feldspathic Sandstone Member. Laminated manganiferous ironstone is associated with sandstone and maroon slate at a few exposures.

Lithology of the Sandstone-Slate Member. The Sandstone-Slate Member consists of lithic and feldspathic graywacke interlayered with variable gray and green-gray slate. Graywacke beds generally display sedimentary features of turbidites. Lenticular and persistent thin beds of micritic limestone are locally present.

Lithology of the Quartzose Sandstone Member. The Quartzose Sandstone Member is composed of quartz arenite, quartzofeldspathic arenite and graywacke, and lithic graywacke interbedded with varying amounts of slate and minor conglomerate. The presence of abundant quartz-rich sandstone is the distinctive feature of the unit. Lithic graywacke, compositionally and texturally similar to that in the Conglomerate Member, is approximately equal in abundance to the quartz-rich sandstone.

Thickness. Over most of the area of its occurrence, the Frenchville Formation is 1,050 m to 1,150 m thick, but locally sections are much thinner; for example, east of Ashland (Fig. 3). Rapid variations in total thickness and thicknesses of its members characterize the formation in the Ashland-Frenchville area (Roy, 1970a).

Age. Seventeen fossil localities (Fig. 12) collectively indicate a possible age span, within the Silurian, of late Llandoverian (C-2) to early Ludlovian for the Frenchville Formation. Regional stratigraphic relationships with the overlying Jemtland Formation suggest that the youngest age of the formation is more likely early Wenlockian. Since the undated Graywacke Member is stratigraphically between dated rocks of late Llandoverian (Conglomerate Member) and Ashgillian-early Llandoverian (Aroostook River Formation), it is possible that it represents a middle or lower Llandoverian phase of the Frenchville Formation.

Perham Group (*New Group*)

It is here proposed that the Perham Formation of Boucot and others (1964) be elevated to group rank, and its distinctive and widespread upper and lower members be named respectively the Jemtland Formation and New Sweden Formation. The type section (Woodland Township) of the Perham Formation cited by Boucot and others (1964; Appendix IV) and modified by Roy (1970a) may be taken as the type section of the group; the two formations are well exposed also in Perham Township in the vicinity of Perham (Fig. 3).

New Sweden Formation (*New Name*)

Type Area. The New Sweden Formation underlies most of New Sweden Township and is well exposed along Maine Route 161 in New Sweden (Fig. 3). Red and green slate units containing manganiferous ironstone are found on Gelot hill near New Sweden.

Lithology. Approximately 90 percent of the exposed New Sweden Formation is gray, calcareous, commonly phyllitic, slate containing calcareous siltstone laminae that are typically folded on a small scale. Red, green, and green-gray slates, generally non-calcareous, are abundant locally in association with laminated manganiferous ironstones. The ironstones occur at various stratigraphic levels within the formation and are found to be widely spaced along the New Sweden outcrop belt from Mapleton to Van Buren (Figs. 1 and 3). Sandstone and conglomerate are less than 5 percent of the unit and occur almost exclusively in the western part of the formation.

Thickness. The New Sweden Formation is quite variable in thickness; a range of 180 to 1,200 m is probably a reasonable estimate. The formation is apparently thickest along the eastern margin of the Ashland synclinorium and either thins (Siegas, New Brunswick area, Hamilton-Smith, 1970) or interfingers with the Frenchville (west) and Spragueville (east) formations (Roy, 1970a; Pavlides, 1968).

Age. The paleontologic data currently available (Fig. 12) indicate that the New Sweden spans the time interval of the Llandoverian (zone 19 or 20) to the Wenlockian (probably early Wenlockian).

Jemtland Formation (*New Name*)

Type Area. The Stockholm-Jemtland area (Fig. 3) is here designated the type area for the Jemtland Formation. The formation is named after the community of Jemtland. Excellent exposures are found along Maine Route 161 in Jemtland and in a road metal quarry east of the tracks at Blackstone Siding (Fig. 3).

Lithology. The Jemtland Formation consists predominantly of four thinly interlayered rock types: (1) indistinctly laminated silty shale; (2) nonlaminated silty shale; (3) fine-grained shale; and (4) fine-grained calcareous graywacke. Interlayered with these four rock types, particularly in western and southwestern exposures, are beds of medium- and coarse-grained, calcareous, lithic graywacke; these graywacke beds as well as those of finer grain size display sedimentary structures consistent with the turbidite model of Bouma (1962). Roy (1970b) has discussed the lateral variations of these rock types within the formation. Locally, conglomerate, micritic limestone, limestone breccia, and aquagene tuff are present in the sequence (Roy, 1970a).

Thickness. The Jemtland Formation varies from approximately 850 m thick in the Stockholm Mountain syncline to 610 m thick in the Presque Isle area (Boucot and others, 1964). The thinner section in the vicinity of Presque Isle is possibly the result of post-Jemtland erosion (Fig. 12).

Age. A total of 55 fossil localities in the Jemtland Formation document an age span of late Wenlockian-early Ludlovian (fig. 12).

Fogelin Hill Formation (*New Formation*)

Type Area. The type area for the Fogelin Hill Formation is here designated as the axial region of the Stockholm Mountain syncline between Stockholm and Jemtland, Maine. Excellent exposures of the formation are present in the village of Stockholm and along the north-south road between Stockholm and Jemtland (Fig. 3). The formation is named for Fogelin Hill; the unit is well exposed along the road on the northwest flank of the hill (Fig. 3).

Lithology. The Fogelin Hill Formation consists of light-green and red slate interlayered with variable amounts of calcareous fine-grained sandstone and siltstone.

Thickness. The top of the Fogelin Hill Formation has not been observed, therefore the total thickness of the formation is not known. At least 760 m of the unit is, however, present along the axis of the Stockholm Mountain syncline (Fig. 3).

Age. Three fossil localities in the Fogelin Hill Formation suggest a possible age range of early Ludlovian to the Siegenian(?). The Siegenian upper limit is based on poorly preserved graptolites recovered from high in the formation and is considered tentative (Roy, 1970a).

REFERENCES CITED

Ayrton, W. G., Berry, W.B.N., Boucot, A. J., Lajoie, Jean, Lesperance, P. J., Pavlides, Louis, Skidmore, W. B., 1969, Lower Llandovery of the Northern Appalachians and adjacent regions: Geol. Soc. America Bull., v. 80, p. 459–484.

Berry, W.B.N., 1960, Early Ludlow graptolites from the Ashland area, Maine: Jour. Paleontology, v. 34, p. 1158–1163.

——1964, Early Ludlow graptolites from the Presque Isle quadrangle, Maine: Jour. Paleontology, v. 38, p. 587–599.

——1968, Ordovician paleogeography of New England and adjacent areas based on graptolites, *in* Zen, E-an, White, W. S., Hadley, J. B., and Thompson, J. B., Jr., eds., Studies of Appalachian geology, northern and maritime: New York, Interscience Pubs., Inc., chap. 2, p. 23–34.

Bird, J. M., and Dewey, J. F., 1970, Lithosphere plate: continental margin tectonics and the evolution of the Appalachian orogen: Geol. Soc. America Bull., v. 81, p. 1031–1060.

Boone, G. M., 1958, Geology of the Fish River Lake-Deboullie area, northern Maine (Ph.D. thesis): New Haven, Yale Univ., 186 p.

——1962, Potassic feldspar enrichment in magma: Origin of syenite in Deboullie district, northern Maine: Geol. Soc. America Bull., v. 73, p. 1451-1476.

——1970, The Fish River Lake Formation and its environments of deposition; *in* Shorter contributions to Maine geology: Maine Geol. Survey Bull. 23, p. 27-41.

Boucot, A. J., 1968, Silurian and Devonian of the northern Appalachians, *in* Zen, E-an, White, W. S., Hadley, J. B., and Thompson, J. B., Jr., eds., Studies of Appalachian geology; northern and maritime: New York, Interscience Pubs., Inc., chap. 6, p. 83-94.

Boucot, A. J., and Thompson, J. B., Jr., 1963, Metamorphosed Silurian brachiopods from New Hampshire: Geol. Soc. America Bull., v. 74, p. 1313-1334.

Boucot, A. J., Field, M. T., Fletcher, R., Forbes, W. H., Naylor, R. S., and Pavlides, L., 1964, Reconnaissance bedrock geology of the Presque Isle quadrangle, Maine: Maine Geol. Survey Quad. Mapping Ser., no. 2., 123 p.

Bouma, A. H., 1962, Sedimentology of some flysch deposits: Amsterdam, Elsevier, 168 p.

Davies, J. L., 1966, Geology of Bathhurst-Newcastle area, *in* Poole, W. H., ed., Guidebook: Geology of parts of Atlantic provinces: Geol. Assoc. Canada and Mineralog. Assoc. Canada, 155 p.

Ekren, E. B., and Frischknecht, F. C., 1967, Geological-geophysical investigations of bedrock in the Island Falls quadrangle, Aroostook and Penobscot counties, Maine: U.S. Geol. Survey Prof. Paper 527, 36 p.

Hall, B. A., 1970, Stratigraphy of the southern end of the Munsungun anticlinorium, Maine: Maine Geol. Survey Bull. 22, 63 p.

Hamilton-Smith, Terence, 1970, Stratigraphy and structure of Ordovician and Silurian rocks of the Siegas area, New Brunswick: New Brunswick Dept. Nat. Resources Mineral Resources Br. Rept. Inv. 12, 55 p.

Horodyski, R. J., 1968, Bedrock geology of portions of Fish River Lake, Winterville, Greenlaw, and Mooselouk Lake quadrangles, Aroostook County, Maine (M.S. thesis): Cambridge, Massachusetts Inst. Technology, 192 p.

Miller, R. L., 1947, Manganese deposits of Aroostook County, Maine: Maine Geol. Survey Bull. 4, 77 p.

Naylor, R. S., 1968, Origin and regional relationships of the core-rocks of the Oliverian domes, *in* Zen, E-an, White, W. S., Hadley, J. B., and Thompson, J. B., eds., Studies of Appalachian geology, northern and maritime: New York, Interscience Pubs., Inc. chap. 17, p. 231-240.

Naylor, R. S., and Boucot, A. J., 1965, Origin and distribution of rocks of Ludlow age (Late Silurian) in the northern Appalachians: Am. Jour. Sci., v. 263, p. 153-169.

Neuman, R. B., 1963, Caradocian (Middle Ordovician) fossiliferous rocks near Ashland Maine: U.S. Geol. Survey Prof. Paper 475-B, p. B117-B119.

——1967, Bedrock geology of the Shin Pond and Stacyville quadrangles, Penobscot County, Maine: U.S. Geol. Survey Prof. Paper 524-I, 37 p.

——1968, Paleogeographic implications of Ordovician shelly fossils in the Magog Belt of the northern Appalachian region, *in* Zen, E-an, White, W. S., Hadley, J. B., and Thompson, J. B., Jr., eds., Studies of Appalachian geology, northern and maritime: New York, Interscience Pubs., Inc., chap. 3, p. 35-48.

Pavlides, Louis, 1962, Geology and manganese deposits of the Maple and Hovey Mountains area, Aroostook County, Maine, with a section on lithology and mineralogy of the deposits by Louis Pavlides and Charles Milton: U.S. Geol. Survey Prof. Paper 362, 116 p.

——1965, Meduxnekeag Group and Spragueville Formation of Aroostook County, northeast Maine: U.S. Geol. Survey Bull. 1244-A, p. A52-A57.

——1968, Stratigraphic and facies relationships of the Carys Mills Formation, northeast Maine and adjoining New Brunswick: U.S. Geol. Survey Bull. 1264, 44 p.

——1971, Geologic map of the Houlton quadrangle, Aroostook County, Maine: U.S. Geol. Survey Geol. Quad. Map, GQ-920.

Pavlides, Louis, and Berry, W.B.N., 1966, Graptolite-bearing Silurian rocks of the Houlton-Smyrna Mills area, Aroostook County, Maine: U.S. Geol. Survey Prof. Paper 550-B, p. B51-B61.

Pavlides, Louis, Neuman, R. B., and Berry, W.B.N., 1961, Age of the "ribbon rock" of Aroostook County, Maine: U.S. Geol. Survey Prof. Paper 424-B, p. B65-B67.

Pavlides, Louis, Mencher, Ely, Naylor, R. S., and Boucot, A. J., 1964, Tectonic features of eastern Aroostook County, Maine: U.S. Geol. Survey Prof. Paper 501-C, p. C28-C38.

Rodgers, John, 1971, The Taconic orogeny: Geol. Soc. America Bull., v. 82, p. 1141-1178.

Roy, D. C., 1971a, The Silurian of northeastern Aroostook County, Maine (Ph.D. thesis): Cambridge, Massachusetts Inst. Technology, 483 p.

——1970b, Variations in a thin-bedded turbidite sequence from the Silurian of northeastern Maine: Geol. Soc. America Abs. with Programs (Ann. Mtg.), v. 2, no. 7, p. 669-670.

Schopf, J. M., Mencher, Ely, Boucot, A. J., and Andrews, H. N., 1966, Erect plants in the Early Silurian of Maine: U.S. Geol. Survey Prof. Paper 550-D, p. D69-D75.

Skinner, Ralph, 1974, Geology of the Tetagouche Lakes, Bathhurst, and Nepisiquit Falls map areas, New Brunswick: Canada. Geol. Survey Mem. 371, 131 p.

Twenhofel, W. H., 1941, The Silurian of Aroostook County, northern Maine: Jour. Paleontology, v. 15, p. 166-174.

Van Der Lingen, G. J., 1969, The turbidite problem: New Zealand Jour. Geology and Geophysics, v. 12, p. 7-50.

Walker, R. G., 1967, Turbidite sedimentary structures and their relationship to proximal and distal depositional environments: Jour. Sed. Petrology, v. 37, p. 25-43.

White, W. S., 1943, Occurrence of Manganese in eastern Aroostook County, Maine: U.S. Geol. Survey Bull. 940-E, 40 p.

Williams, H. S., and Gregory, H. E., 1900, Contributions to the geology of Maine: U.S. Geol. Survey Bull. 165, 212 p.

Manuscript Received by the Society April 18, 1974
Revised Manuscript Received January 2, 1975
Manuscript Accepted January 10, 1975

Printed in U.S.A.

Geological Society of America
Memoir 148

Discussion: Ordovician and Silurian Stratigraphy of Northeastern Aroostook County, Maine

LOUIS PAVLIDES
U.S. Geological Survey
Reston, Virginia 22092

Roy and Mencher (this volume) presented an interesting summary of the stratigraphy of northeastern Maine. Counterparts to some of the lithotectonic units they described occur along strike to the south, although in detail the stratigraphy shows some marked differences. For comparative purposes, I will briefly discuss some of the similarities and differences in the stratigraphy on the west side of the Aroostook-Matapedia anticlinorium southward from the Presque Isle area.

The Carys Mills Formation (of Middle Ordovician to Early Silurian age) extends as a continuous unit throughout northeast Maine within the core of the Aroostook-Matapedia anticlinorium. The lithology and the internal stratigraphy of this formation have been described elsewhere (Pavlides, 1968, 1966; Pavlides and others, 1965) and are essentially as described by Roy and Mencher (this volume) except for minor differences; for example, in the Bridgewater area, about 20 mi (32 km) southeast of Presque Isle, the formation contains lenses of slate and of slate and graywacke. Absence of marker beds hinders subdivision of the Carys Mills Formation into subunits, and Roy and Mencher's subdivision of it into a lower and upper part north of Presque Isle is apparently biostratigraphic rather than lithostratigraphic.

The Frenchville Formation, of Llandovery C_3-C_5 age, is absent along the west flank of the Aroostook-Matapedia anticlinorium; in the Mars Hill area (about 13 mi [20.8 km] southeast of Presque Isle), the coeval Spragueville Formation overlies the Carys Mills Formation and underlies the early Ludlow upper part of the Perham or Jemtland Formation of Roy and Mencher (this volume) with conformable contacts (Pavlides, 1968, Pl. 1, p. 18). The Llandovery to Wenlock(?) lower member of the Perham or New Sweden Formation of Roy and Mencher also is absent here, apparently having wedged out to the north of Presque Isle. The upper Perham or Jemtland of Roy and Mencher does not extend beyond the Mars Hill quadrangle as a mappable unit. Apparently it changes in character southward to a micaceous quartzite, siltstone, and slate (Pavlides and others, 1965, p. 27–28), which at one place has yielded a brachiopod fauna of early Ludlow age (Pavlides, 1968, p. 29). At another place within this southern facies, however, an early Ludlow graptolite fauna in orange-weathered siltstone, typical of the upper Perham Formation, is present (Pavlides, 1968, p. 29–30, and Fossil loc. 17, Pl. 1).

The Maple Mountain Formation of Silurian age (Pavlides, 1962; 1964, p. B5-B6) in the Howe Brook quadrangle is chiefly slate, enclosing manganiferous lenses about 5,000 ft (1,520 m) above its base. At the base is a discontinuous conglomerate that is as much as 300 ft (91.2 m) thick, which elsewhere is continuous and thicker along the contact with the underlying volcanic and sedimentary rocks of the Dunn Brook Formation of Silurian and Ordovician age. The lower part of the Dunn Brook may be coeval with the Winterville Formation of Roy and Mencher of the Castle Hill anticline. The Middle Ordovician unit in fault contact east of the Dunn Brook consists primarily of slate and graywacke, with minor volcanic rocks and thin radiolarian chert horizons; at one place these contain Zone 12 graptolites (Pavlides, 1968, p. 24-25). This unit is considered to be a lateral equivalent of the lower part of the Dunn Brook. The stratigraphic sequence at the northeast end of the Weeksboro-Lunksoos Lake anticline (Pavlides, 1968, Pl. 1; 1962, p. 16-17, Pl.1) both differs from, and compares with, that at and near the Castle Hill anticline (Pavlides, 1968, Fig. 3, p. 37). The Maple Mountain Formation and the conglomerate at its base, which rests on volcanic rocks of the Dunn Brook Formation, compare in general with the Lower Perham-Frenchville-Winterville sequence; there are, however, differences in distribution of these rock types and in the absence of some units such as the Ashgillian Pyle Mountain Argillite in the Maple Mountain area. In the Bridgewater area (Pavlides and others, 1965), the Carys Mills Formation thins to the west and overlies the lenticular Chandler Ridge Formation of Ordovician age (Pavlides, 1966, p. A54-A55) that is composed chiefly of slate and graywacke. Volcanic clasts occur in some of the lithic graywacke of the Chandler Ridge (Pavlides, 1962, p. 10, Fig. 4) and may have been derived, in part, from early volcanic rocks of the Dunn Brook Formation to the west or possibly even older volcanic rocks, such as the Shin Brook Formation (Neuman, 1967, p. I7-I10, Pl. 1) that is exposed in nearby areas. Slate of the Burnt Brook Formation of Silurian(?) age locally overlies the Carys Mills in the west-central part of the Bridgewater quadrangle (Pavlides, 1968, p. 11, Pl. 1) and is probably coeval with the Maple Mountain Formation to the west, the lower part of the Perham to the north, and parts of the Smyrna Mills Formation to the south.

Farther south of Maple Mountain (20 mi [32 km]), the stratigraphic section in the Houlton-Smyrna Mills area contrasts with the section west and north of Presque Isle, that is described by Roy and Mencher (this volume). The Smyrna Mills Formation (Pavlides and Berry, 1966, p. B53-B54, Fig. 1; Pavlides, 1971), composed of irregularly distributed quartzite and graywacke (about 60 to 70 percent), green slate (about 35 percent), conglomerate, lithic graywacke and black carbonaceous slate (3 percent), and red and maroon siltstone and slate and manganiferous ironstone (2 percent), is of lower Llandovery to early Ludlow age and is coeval with Frenchville, Spragueville, and Perham rocks to the north. Manganese deposits within the Smyrna Mills Formation occur at various horizons and are of different ages in contrast to their localization within the lower part of the Perham of Llandovery-Wenlock(?) age to the north. The lack of clearly distinguishable map units within the Smyrna Mills Formation contrasts with their presence in the Perham and related rocks but conforms with the generally poorly divisible character of coeval rocks along strike to the southwest.

Faulted blocks of black graptolitic slate of Middle Ordovician age within the Smyrna Mills Formation southeast of Houlton (Pavlides, 1971) appear to be a southeast facies coeval to parts of the Carys Mills Formation as well as the slate and graywacke unit in fault contact with the Weeksboro-Lunksoos Lake anticline at its northeast side (Pavlides, 1968, Pl. 1, Fig. 2).

REFERENCES CITED

Neuman, R. B., 1967, Bedrock geology of the Shin Pond and Stacyville quadrangles, Penobscot County, Maine: U.S. Geol. Survey Prof. Paper 524-I, p I1-I37.

Pavlides, Louis, 1962, Geology and manganese deposits of the Maple and Hovey Mountains area, Aroostook County, Maine: U.S. Geol. Survey Prof. Paper 362, 116 p.

——1964, The Hovey Group of northeastern Maine: U.S. Geol. Survey Bull. 1194-B, p. B1-B6.

——1966, Meduxnekeag Group and Spragueville Formation of Aroostook County, northeastern Maine, *in* Cohee, G. V., and West, W. S., Changes in stratigraphic nomenclature by the U.S. Geological Survey, 1965: U.S. Geol. Survey Bull. 1244-A, p. A52-A57.

——1968, Stratigraphic and facies relationships of the Carys Mills Formation of Ordovician and Silurian age, northeast Maine: U.S. Geol. Survey Bull. 1264, 44 p.

——1971, Geologic map of the Houlton quadrangle, Aroostook County, Maine: U.S. Geol. Survey Geol. Quad. Map GQ-920, scale 1:62,500.

Pavlides, Louis, and Berry, W.B.N., 1966, Graptolite-bearing Silurian rocks of the Houlton-Smyrna Mills area, Aroostook County, Maine: U.S. Geol. Survey Prof. Paper 550-B, p. B51-B61.

Pavlides, Louis, Griscom, Andrew, and Kane, Martin F., 1965, Geology of the Bridgewater quadrangle, Aroostook County, Maine: U.S. Geol. Survey Bull. 1206, 72 p.

Roy, D. C., and Mencher, Ely, 1976, Ordovician and Silurian stratigraphy of northeastern Aroostook County, Maine, *in* Page. L. R., ed., Contributions to the stratigraphy of New England: Geol. Soc. America Mem. 148 (this volume).

MANUSCRIPT RECEIVED BY THE SOCIETY APRIL 18, 1974
REVISED MANUSCRIPT RECEIVED NOVEMBER 8, 1974
MANUSCRIPT ACCEPTED DECEMBER 5, 1974

Printed in U.S.A.

Geological Society of America
Memoir 148

Lower Devonian Seboomook Formation and Matagamon Sandstone, Northern Maine: A Flysch Basin-Margin Delta Complex

Bradford A. Hall
Department of Geology
University of Maine
Orono, Maine 04473

Stephen G. Pollock
18700-80 Yorba Linda Boulevard
Yorba Linda, California 92686

Kerry M. Dolan
Rural District
West Glover, Vermont 05875

ABSTRACT

Detailed sedimentological work in Lower Devonian (primarily Siegenian) rocks of the Matagamon Lake area in northern Maine indicates that mudstone and fine-grained sandstone grades upward from a turbidite-bearing prodelta-marine slope association (Seboomook Formation) to coarser grained associations characteristic of delta-front and delta-plain environments (Matagamon Sandstone). Paleocurrent directions determined from turbidites, channeled fluxoturbidites, and various facies of the shallower water environments indicate sediment transport from east to west. The coarsening-upward sequence coupled with paleocurrent and paleoenvironmental data indicate a westerly prograding delta-slope-basin sequence. It is proposed that the widespread Siegenian deeper water mudstone and fine-grained sandstone facies (Seboomook Formation and equivalents) is a syntectonic shaly flysch derived from an extensive landmass to the east that was uplifted in Early Devonian time.

INTRODUCTION

Interbedded Lower Devonian (primarily Siegenian) mudstone and fine-grained sandstone or siltstone (and their metamorphic equivalents) are present in a belt more than 1,300 km long and as much as 160 km wide from eastern Gaspé Peninsula southwest to Long Island Sound, southern New England. These sparsely fossiliferous rocks form what is probably the most widespread sedimentary unit in the Northern Appalachians. The known distribution, dating, and facies relationships of these rocks have been recently treated by Boucot (1970), Poole and others (1970), and Rodgers (1970) and will only be summarized here. The rocks are variously known as the Gile Mountain and Littleton Formations in Vermont and New Hampshire; the Seboomook Formation in Maine; the Compton and Seboomook Formations, the Temiscouata Group, and parts of the St. Juste and St. Francis Groups in the Eastern Townships of Quebec; and the Fortin Series on the Gaspé Peninsula. Thickness estimates range from 3,000 to 6,000 m.

Facies equivalents, both calcareous and more massively sandy, occur in various places within the belt. Consistently, the calcareous facies lie on the northwest side of the belt, where they are present in such units as the Grande Grève Formation of the Gaspé Limestones that extend from the Gaspé Peninsula southwestward to the Lake Temiscouata area. Fossiliferous massive sandstone units, interpreted as facies of the Seboomook Formation, are present in Maine: (1) in the Moose River synclinorium, the Tarratine Formation; (2) near the eastern edge of the belt at Matagamon Lake, the Matagamon Sandstone; and (3) in several scattered localities between and to the northeast. On the Gaspé Peninsula, the lowermost part of the York River Formation of the Gaspé Sandstone appears in part to grade southwestward into the Fortin Series. The upper part of the Gaspé Sandstone is coarse-grained and largely nonmarine.

The origin and significance of these rocks has been considered by various people in varying degrees of detail. Little information other than that provided by regional stratigraphic mapping is available. Boucot (1969) and Rankin (1968) interpreted the massive Siegenian sandstone bodies of Maine, the Matagamon Sandstone and Tarratine Formation, as having been derived locally from islands. On the Gaspé Penninsula, however, the Gaspé Sandstones are considered by Poole and others (1970) and Rodgers (1970) to be the result of early Acadian uplift to the south of the Gaspé Trough. Rodgers, considering the coarse-grained Gaspé Sandstones, suggested the possibility of a westward migrating delta system.

The origin of the finer grained Siegenian clastic rocks, such as the Seboomook Formation, has not been considered except by Boucot (1962), who tentatively ruled out a westerly source on the basis of regional considerations, relationships within the Moose River synclinorium, and because of the presence of the carbonate facies to the west and the greater angularity of the quartz in the Seboomook relative to that in the carbonates. The Seboomook was interpreted as grading into the Tarratine sandstones, where it is considerably more sandy. Boucot, however, condidered islands to be inadequate sources because of the great regional volume and extent of the Seboomook and equivalent rocks. He thus suggested an easterly or more distant westerly source.

Work in the Matagamon Lake area (Fig. 1) was planned to delineate, describe, and interpret the sedimentary facies present in the Seboomook Formation and Matagamon Sandstone from a sedimentologic rather than strictly stratigraphic point of view and to couple this with paleocurrent data to better understand the regional paleogeographic and tectonic relationships.

SEDIMENTARY ENVIRONMENTS

Approximately 4,000 m of rock in the Matagamon Sandstone and Seboomook Formation within the Matagamon Lake area are assigned to three major sedimentary environments (Fig. 2). From top to bottom, these are (1) delta plain, (2) delta front, (3) prodelta-marine slope. The upper two environments are primarily within the Matagamon Sandstone, and the lower is primarily within the Seboomook. The salient features of each environment are outlined below.

Delta Plain Environment

Delta-plain deposits consist primarily of migrating distributary channels filled with fine- to medium-grained sandstone as much as several meters thick. Such units extend over tens of meters, and their channel morphology is visible locally. The sequence of structures and lithologies commonly encountered upward through a channel is (1) shale-pebble lag conglomerate (uncommon) or massive sandstone or both, (2) trough cross-bedded sandstone, and (3) thin horizontal bedded or current cross-laminated sandstone.

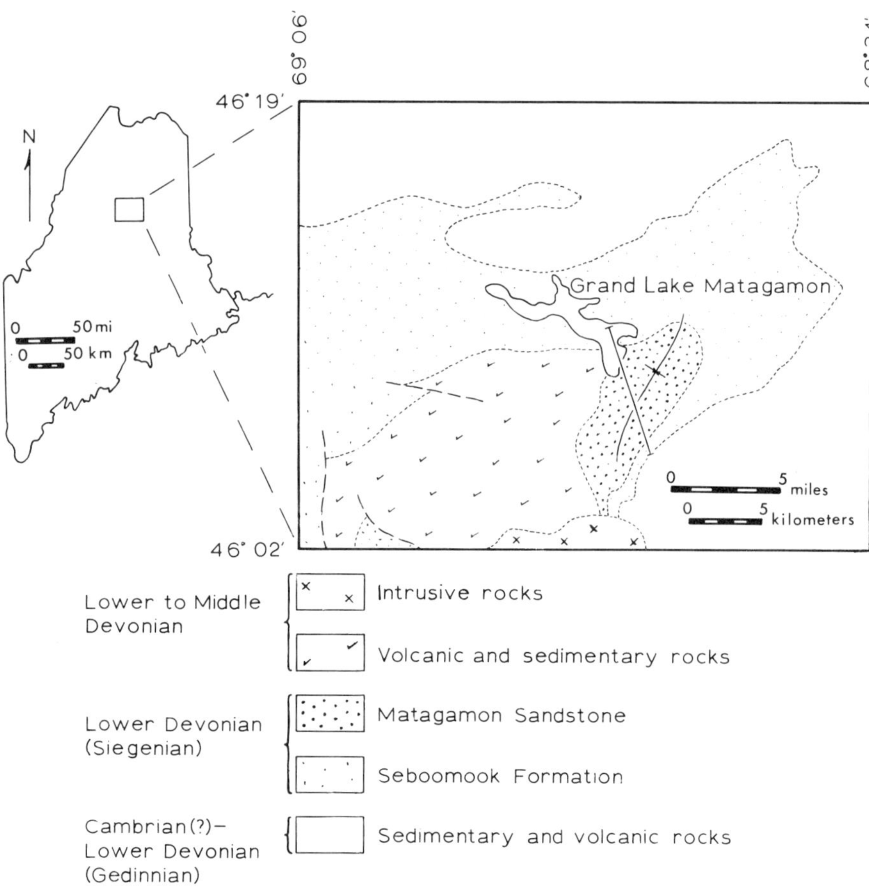

Figure 1. Matagamon Lake area, northern Maine, indicating location of section in Figure 2 (geologic map after Doyle, 1967).

Minor units, as much as 3 m thick of mudstone, plant debris-bearing mudstone, and bioturbated siltstone with uncommon shelly fossils are interbedded and in places are cut by the channel deposits. These are interpreted to be marsh, lake, bay, or lagoon interdistributary deposits. Brachiopods and, rarely, gastropods and pelecypods occur only in the basal part of the distributary channels, attesting to channel erosion of finer grained, possibly brackish-water inter channel deposits.

Several laterally persistent units as much as 2 m thick, possibly beach deposits, are also present. These are very fossiliferous medium-grained sandstone with indistinct horizontal lamination and minor cross-lamination in their upper parts. They are interbedded with the channel deposits and probably represent short periods of beach migration across or into mouths of distributaries on the delta (Allen, 1964; Moore, 1966).

Current directions obtained from rose plots of channel cross-bedding are essentially unidirectional and indicate transport toward 285° (Pollock and Hall, 1973).

Delta Front Environment

Delta-front deposits can be divided into two subenvironments. The most abundant are nearshore, probably distributary-influenced, deposits of trough cross-laminated very fine grained sandstone beds 15 cm to 15 m thick. Accumulations of shelly fossils occur in the bases of troughs. Scattered, isolated tentaculitids are common. Current directions obtained from rose plots of cross-bedding are multi-directional (Pollock and Hall, 1973).

The more distal delta-front facies is present in the basal portions of the Matagamon Sandstone and at the top of the Seboomook Formation adjacent to the western edge of the Matagamon. It consists predominantly of parallel-laminated, rarely cross-laminated, ungraded, very micaceous siltstone and very fine grained sandstone in beds several centimetres to 1 m thick interbedded with mudstone beds (as much as 3 m thick). Bioturbation is extremely common throughout these deposits. The darker laminae are largely accumulations of finely divided plant debris. Shelly fossils are present and locally abundant. Graded, fluted sand beds and ball-and-pillow structure are present. These beds are probably the distal fringes of the distributary-influenced deposits reworked by waves and partly transported seaward by wave surge. Flutes on these beds indicate sediment transport toward 285°.

Prodelta-Marine Slope Environment

Most of the Seboomook Formation in the Matagamon area formed on a prodelta-marine slope and consists of four major types of deposits. The most common is mudstone in beds from 2 to about 20 cm thick. Commonly, each bed is subtly graded with either a slightly more silty basal portion or a discrete basal siltstone layer 0.25 to 1.5 cm thick. The basal siltstone may also grade upward into the basal part of the mudstone layer (Hall and Stanley, 1972). Tracks, trails, horizontal burrows, and cross-lamination may occur in these silt layers. These rocks are probably the result of pulsatory input of fine-grained sediment from the deltaic distributaries, material swept off the delta front by waves, and the reworking of this sediment by weak bottom currents.

A second major type of deposit consists of graded beds of very fine grained sandstone and siltstone 1.5 cm to 1 m thick that displays parallel lamination; type three ripple-drift cross-lamination of Jopling and Walker (1968); and flute casts, tool marks, tracks, and trails. The graded beds are abundant as single beds or packets of beds as much as 20 m thick interspersed within the mudstone and are probably of turbidite origin.

The other types of deposits are fluxoturbidites (Kuenen, 1958; Stanley, 1969b) in erosion chanels (Hall and Stanley, 1973) as much as 29 m thick and pebbly mudstone as much as 3 m thick. The latter contain mixed shelly faunas derived from the delta front and distributary channels.

A distinctive feature associated with all these lithologies is slump structure. This plus the combination of lithologies is indicative of a prodelta-marine slope (Moore, 1966; Allen, 1970; Donaldson and others, 1970), probably the lower steeper portions (Stanley, 1969a, 1969b). Measurements on flutes at the bases of turbidite sandstones indicate a downslope movement toward 280°. The orientation of fluxo-turbidite channels and slump folds is consistent with this direction (Hall, 1973).

DISCUSSION

The mineralogical similarity between Seboomook and Matagamon sedimentary rocks and the intergradational nature and positions of the environmental types within the sequence indicate that Lower Devonian rocks of the Matagamon Lake area formed as a prograding delta-marine slope system. Paleocurrents in both the deltaic and slope environments indicate that progradation was from east to west and that the strike of the paleoslope was approximately north-south. Thickening from east to west (Fig. 2) of the slope and delta-front deposits supports this interpretation.

Derivation of Siegenian sediment from the east was probably not a local phenomenon confined to the Matagamon Lake area. A similar westward thickening of the Seboomook Formation capped by sandstone (Tarratine Formation) similar to the Matagamon Sandstone is present 130 km to the southwest in the Moose River synclinorium (Fig. 2). Reconnaissance work in the course of this study in the Moose River synclinorium and to the northeast in Maine, Quebec, and New Brunswick indicates that the Seboomook and equivalent rocks are similar to the

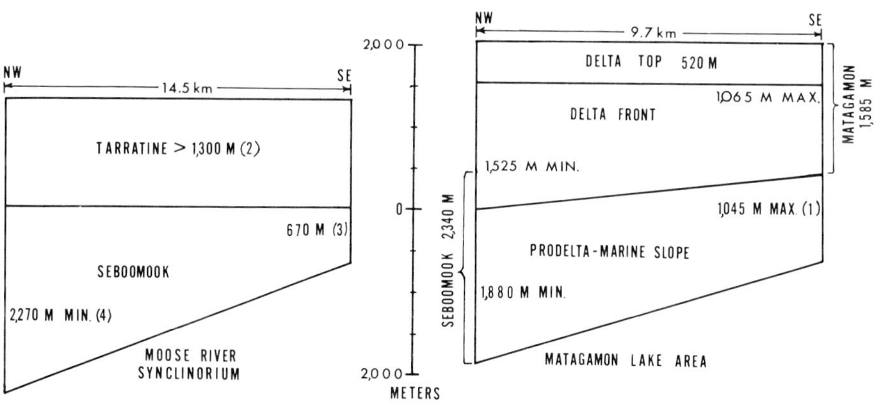

Figure 2. Thicknesses of the Seboomook Formation and associated sandstone (Matagamon and Tarratine Formations) in the Matagamon Lake area and the Moose River synclinorium, Maine. Only those thicknesses are used for which sections have been measured or structure is simple and outcrop control good. Thickness (1) is estimated from Plate 1 of Neuman (1967). Thicknesses (2), (3), and (4) are from Plates 14, 29, and 24, respectively, of Boucot (1969). The latter (4) thickness is a minimum for the greatest thickness of the Seboomook in the area.

marine-slope deposits of the Matagamon area except that pebbly mudstone, fluxo-turbidites, and slump structure are confined or best developed in the more easterly portion of the belt. Such facies delineate the basin margin and location of sediment input (Stanley, 1969a). Preliminary paleocurrent data compiled during this reconnaissance indicate westerly transport on the southeastern margin of the belt and southwesterly and possibly northeasterly transport parallel to the belt away from the margins.

Thus, the Seboomook Formation represents a classical, though shaly, flysch (Cline, 1970; Stanley, 1970) basin in which the sediment was derived marginally. Further out in the basin, turbidite flow was longitudinal as in most turbidite basins (Potter and Pettijohn, 1963). Throughout Silurian time, the area to the southeast of the present belt of Siegenian rocks was a major site of clastic deposition (Boucot, 1968, 1970; Poole and others, 1970) in which sediment was at least locally derived from the west or northwest (Roy and Mencher, this volume). The elevation of this basin of deposition to provide a sediment source for the large Siegenian flysch basin to the west was a major event in the history of sedimentation in the Northern Appalachians (Hall and others, 1972).

ACKNOWLEDGMENTS

This study was supported by National Science Foundation Grant GA-21119 and the Maine Geological Survey.

REFERENCES CITED

Allen, J.R.L., 1964, Sedimentation in the modern delta of the Niger River, West Africa, *in* van Straaten, L.M.J.U., ed., Deltaic and shallow marine deposits: Amsterdam, Elsevier Pub. Co., p. 26-34.

——1970, Sediments of the modern Niger Delta: A summary and review, *in* Morgan, J. P., ed., Deltaic sedimentation: Soc. Econ. Paleontologists and Mineralogists Spec. Pub. 15, p. 138-151.

Boucot, A. J., 1962, Appalachian Siluro-Devonian, *in* Some aspects of the Variscan Fold Belt: Manchester, England, Manchester Univ. Press, p. 155-163.

——1968, Silurian and Devonian of the Northern Appalachians, *in* Zen, E-an, White, W. S., and Hadley, J. B., eds., Studies of Appalachian geology: Northern and maritime: New York, Wiley-Interscience, p. 83-94.

——1969, Geology of the Moose River and Roach River synclinoria, northwestern Maine: Maine Geol. Survey Bull. 21, 115p.

——1970, Devonian slate problems in the Northern Appalachians: Maine Geol. Survey Bull. 23, p. 42-48.

Cline, L. M., 1970, Sedimentary features of late Paleozoic flysch, Ouachita Mountains, Oklahoma: Geol. Assoc. Canada Spec. Pub. 7, p. 85-101.

Donaldson, A. C., Martin, R. H., and Kanes, W. H., 1970, Holocene Guadalupe Delta of Texas Gulf Coast: *in* Morgan, J. P., ed., Deltaic sedimentation: Soc. Econ. Paleontologists and Mineralogists Spec. Pub. 15, p. 107-137.

Doyle, R. G., 1967, Preliminary geologic map of Maine: Maine Geol. Survey, scale 1:500,000.

Hall, B. A., 1973, Slump folds and the determination of paleoslope: Geol. Soc. America Abs. with Programs, v. 5, p. 648.

Hall, B. A., and Stanley, D. J., 1972, Mud turbidites and the downslope transport of fine-grained sediment: Geol. Soc. America Abs. with Programs, v. 4, p. 524.

——1973, Levee-bounded submarine base-of-slope channels in the Lower Devonian Seboomook Formation, northern Maine: Geol. Soc. America Bull., v. 84, p. 2101-2110.

Hall, B. A., Pollock, S. G., and Dolan, K. M., 1972, An early Acadian flysch basin margin-delta complex in the Northern Appalachians; the Lower Devonian Seboomook Formation and Matagamon Sandstone, northern Maine: Internat. Geol. Congress, 24th, Montreal 1972, p. 187.

Jopling, A. V., and Walker, R. G., 1968, Morphology and origin of ripple-drift cross-lamination with examples from the Pleistocene of Massachusetts: Jour. Sed. Petrology, v. 38, p. 971-984.

Kuenen, P. H., 1958, Problems concerning source and transportation of flysch sediments: Geologie en Mijnbouw, v. 20, p. 329-339.

Moore, D., 1966, Deltaic sedimentation. Earth Sci. Rev., v. 1, p. 87-104.

Neuman, R. B., 1967, Bedrock geology of the Shin Pond and Staceyville quadrangles, Penobscot County, Maine: U.S. Geol. Survey Prof. Paper 524-I, p. 11-137.

Pollock, S. G., and Hall, B. A., 1973, Variations in paleocurrent patterns in the Lower Devonian Matagamon Sandstone, Maine, as an indicator of paleoenvironments: Geol. Soc. America Abs. with Programs, v. 5, p. 208.

Poole, W. H., Sanford, B. W., Williams, H., and Kelley, D. G., 1970, Geology of southeastern Canada, *in* Geology and economic minerals of Canada: Canada Dept. Energy, Mines and Resources Econ. Geology Rept. 1, p. 229-304.

Potter, P., and Pettijohn, F. J., 1963, Paleocurrents and basin analysis: Berlin, Academic Press, 296 p.

Rankin, D. W., 1968, Volcanism related to tectonism in the Piscataquis volcanic belt, an island arc of Early Devonian age in north-central Maine, *in* Zen, E-an, White, W. S., and Hadley, J. B., eds., Studies of Appalachian geology: Northern and maritime: New York, Wiley-Interscience, p. 335-369.

Rodgers, John, 1970, The tectonics of the Appalachians: New York, Wiley-Interscience, 271 p.

Roy, D. C., and Mencher, Ely, 1976, Ordovician and Silurian stratigraphy of northeastern Aroostook County, Maine, *in* Page, L. R., ed., Contributions to the stratigraphy of New England: Geol. Soc. America Mem. 148 (this volume).

Stanley, D. J., 1969a, Sedimentation in slope and base-of-slope environments, *in* The new concepts of continental margin sedimentation: Washington, D.C., Am. Geol. Inst., p. DJS-8-1—DJS-8-25.

——1969b, Submarine channel deposits and their fossil analogs ("fluxoturbidities"), *in* The new concepts of continental margin sedimentation: Washington, D.C., Inst., p. DJS-9-1—DJS-9-17.

——1970, Flyschoid sedimentation on the outer Atlantic margin off northeast North America: Geol. Assoc. Canada Spec. Pub. 7, p. 179-210.

MANUSCRIPT RECEIVED BY THE SOCIETY APRIL 18, 1974
REVISED MANUSCRIPT RECEIVED APRIL 14, 1975
MANUSCRIPT ACCEPTED APRIL 30, 1975

Printed in U.S.A.

Geological Society of America
Memoir 148

Fossil-Based Stratigraphy in the Merrimack Synclinorium, Central Maine

ALLAN LUDMAN
Department of Earth and Environmental Sciences
Queens College
Flushing, New York 11367
and
Maine Geological Survey
Augusta, Maine 04330

ABSTRACT

Ages from 17 new graptolite localities have made possible refinements in Silurian and Silurian-Devonian stratigraphy in the Kingsbury and Skowhegan quadrangles and indicate major facies changes in the Silurian metasedimentary rocks of central Maine. Northwest of the Currier Hill syncline, the observed sequence is: Sangerville Formation (upper Llandovery through lower Wenlock)—variably bedded metagraywacke and pelite with minor polymict granule conglomerate, ribbon limestone, and carbonaceous pelite; Eddy Formation (lower Ludlow)—rusty weathering sulfidic carbonaceous pelite and non-carbonaceous quartzose sandstone and granule conglomerate; Brighton Formation [Silurian-Devonian(?)]—massive biotite-quartz granulite with minor calc-silicate granulite beds and pods (lower member) and thinly interbedded biotite-quartz granulite and biotite-muscovite phyllite (upper member); Solon Formation [Lower Devonian(?)]—nearly homogeneous slate with occasional rhythmically interbedded quartzose sandstone and slate.

Southeast of the Currier Hill syncline, the sequence is similar but there is a decrease in grain size in most units accompanied by an increased amount of pelite and a drastic thinning of Sangerville bedding to produce a pinstriped unit (transitional Sangerville). My interpretation of the facies relationships suggests a sequence that becomes finer to the east with a source of sediments to the northwest.

Correlations based on projection of these facies changes and on fossil ages wherever possible yield the following:

West		East
Seboomook Formation	Solon Formation	
Madrid Formation	Brighton Formation	Vassalboro Formation
Smalls Falls Formation	Eddy Formation	
Perry Mountain Formation	Sangerville Formation	Waterville Formation
Rangeley Formation		Mayflower Hill Formation

INTRODUCTION

The Merrimack synclinorium, approximately 100 km wide in central and west-central Maine, has been the object of detailed study for the past 15 yr. The western limb of the structure is best known because much of the early work was near the New Hampshire border (Moench, 1970; Moench and Boudette, 1970; Harwood and Berry, 1967; Guidotti, 1965; Green, 1964). The axial zone of the synclinorium in central Maine lies in what has been referred to as the Central Maine Slate Belt (Rodgers, 1970) and the Monotonous Slate Belt (Neuman, 1967). Espenshade and Boudette (1964, 1967) mapped in the northern part of this belt, but most of the mapping in the past 5 yr has been done by the Maine Geological Survey and much of this data is unpublished (Osberg, 1968; Ludman, 1969a, 1969b, 1970; Pankiwskyj, 1975; Griffin, 1973; Keith, 1970).

The Kingsbury and Skowhegan 15′ quadrangles (Fig. 1) lie across the axial zone of the Merrimack synclinorium between the previously described stratigraphic sequences of the Rangeley (Moench, 1970; Moench and Boudette, 1970) and Waterville (Osberg, 1968) areas and on strike with the more highly metamorphosed rocks in the Buckfield (Warner, 1967), Dixfield (Pankiwskyj, 1964), and Bryant Pond (Guidotti, 1965) quadrangles to the southwest.

Metasedimentary rocks in the Kingsbury-Skowhegan area are, for the most part, in the biotite and chlorite zones and are only slightly recrystallized, except in contact aureoles surrounding granitic plutons correlated with the New Hampshire Magma Series of Billings (1956) where andalusite, sillimanite, and wollastonite occur. Primary bedding features are well preserved in most lithologies, and facing indicators are abundant. Graded bedding, truncated cross-laminations, scour-and-fill features, and flame structures have been used as facing evidence. Fossils have also been preserved, including graptolites from 17 new localities, crinoid columnal segments, and worm trails. Faunal identifications and age assignments for the graptolite-bearing rocks have been made by W.B.N. Berry and will be discussed in detail in a future paper (Ludman and others, in prep.).

The area is structurally complex (Figs. 2, 3), consisting of gently plunging northeast-trending anticlines and synclines with nearly vertical axial surfaces. The minor folds are isoclinal and range from 7.5 cm to 3 km in wavelength. These folds are only slightly modified by two subsequent deformations (Ludman, 1969b) that do not greatly disturb the outcrop patterns. Mapping in the Kingsbury quadrangle indicates the presence of major faults that offset the folds of the first deformation but not the metamorphic isograds. A minimum age of deformation is provided by Middle Devonian radiometric dates for the post-kinematic Hartland pluton in the Skowhegan quadrangle (Faul and others [1963] estimated the age as 356 m.y.; Zartman and others [1970] estimated 376 m.y.). The Hartland pluton truncates folds of all three generations, and contact metamorphic isograds are unaffected by faulting so that the faulting must also be pre-Middle Devonian. A maximum age for the onset of deformation is given by the fact that rocks of probable Early Devonian age are deformed in the first period of folding.

STRATIGRAPHY

The five stratigraphic units recognized in the area (Figs. 3, 4) record a history of continuous sedimentation from late Llandovery through early Devonian(?) time. These stratigraphic units and the lateral variations within them may be traced across several folds. For convenience, the stratigraphy northwest and southeast

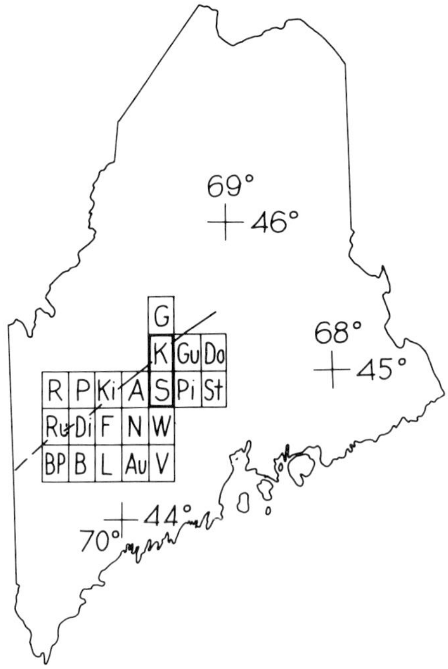

Figure 1. Map showing trace of the axis of the Merrimack synclinorium. Quadrangle index: G-Greenville; K-Kingsbury; Gu-Guilford; Do-Dover-Foxcroft; R-Rangeley; P-Phillips; Ki-Kingfield; A-Anson; S-Skowhegan; Pi-Pittsfield; St-Stetson; Ru-Rumford; Di-Dixfield; F-Farmington; N-Norridgewock; W-Waterville; BP-Bryant Pond; B-Buckfield; L-Livermore; Au-Augusta; V-Vassalboro. Area covered in this work outlined.

of the Currier Hill syncline will be described separately by the informal nomenclature in use by the Maine Geological Survey for the past 4 yr. The prefix *meta* will not be used in the descriptions below because so much of the primary bedding has been preserved and the original textures and clast compositions are readily determined. It should be remembered, however, that all rocks have been metamorphosed to some extent and the least metamorphosed rocks are in the chlorite zone.

WEST OF THE CURRIER HILL SYNCLINE

Sangerville Formation

The Sangerville Formation crops out in a belt 16 km wide in the core of the Chapman Ridge anticline and underlies a large part of the Skowhegan quadrangle (Fig. 3). It is composed for the most part of clastic rocks of graywacke composition interbedded with pelite in turbiditelike graded units. Four distinct lithologies are recognized within the Sangerville, but details of the internal stratigraphy of the formation are not presently known.

Polymict Granule Conglomerate. Several bands of polymict granule conglomerate occur within the Sangerville outcrop belt (Fig. 3) and represent repetition of a single stratigraphic horizon. Beds of granule conglomerate also occur in isolated lenses at several horizons within the Sangerville, but cannot be shown at the scale of the map.

The granule conglomerate occurs in thick, massive beds (as much as 3 m thick) showing little systematic grain size variation, interspersed with well-defined graded sequences (Fig. 5) containing granule conglomerate, sandstone, siltstone, and pelite. The conglomerate units are poorly sorted and contain a wide variety of clasts.

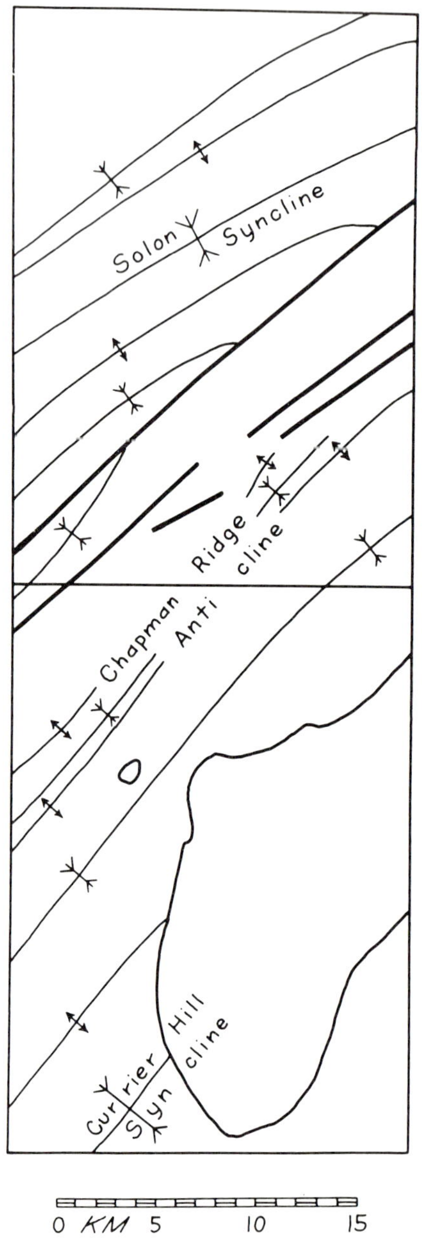

Figure 2. Structural framework of the Kingsbury and Skowhegan quadrangles. Major local fold axes and faults (without dip symbols) indicated. Hartland pluton outlined in Skowhegan quadrangle.

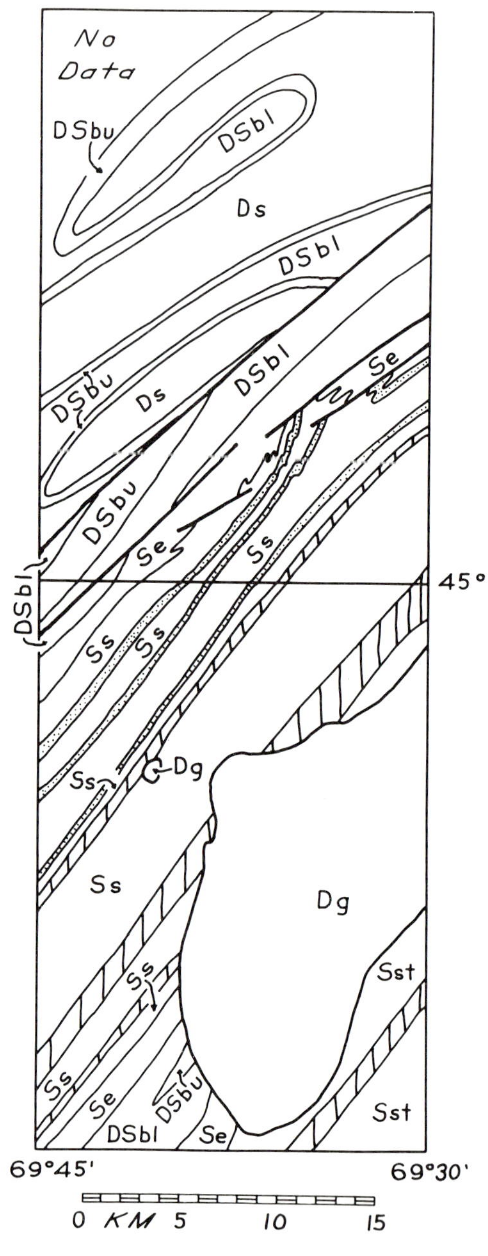

Figure 3. Simplified geologic map of the Kingsbury and Skowhegan quadrangles. Symbols: Ss-Sangerville Formation (stippled pattern-polymict conglomerate; ruled pattern-ribbon limestone); Sst-transitional Sangerville; Se-Eddy Formation; DSb-Brighton Formation (u=upper member, 1=lower member); Ds-Solon Formation; Dg-granitic rocks.

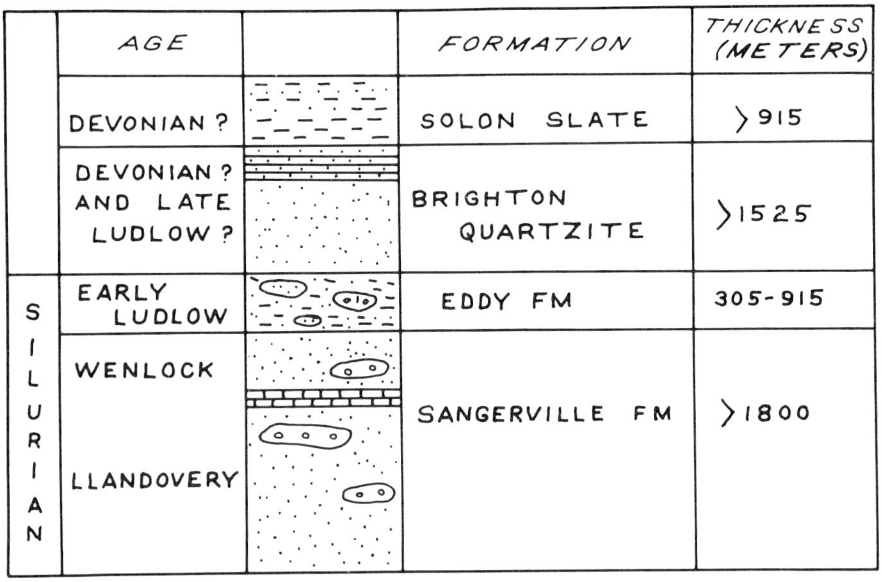

Figure 4. Stratigraphy west of the axis of the Currier Hill syncline.

Included are monomineralic quartz and plagioclase feldspar, felsic and mafic volcanic fragments, slate and schist chips, chertlike grains, quartzite, and granitic hypabyssal igneous fragments (Figs. 6, 7). Blue quartz grains are locally abundant. The clasts are set in a fine-grained detrital matrix with an additional calcareous cement in some instances.

Graywacke and Pelite. These rocks comprise approximately 75 percent of the Sangerville, but their internal stratigraphy is complex and still under examination. Included here are sandstones of graywacke composition, siltstones, and mudstones intercalated in widely variable proportions with pelite. Bedding thicknesses also vary widely, but graded bedding is characteristic of these rocks regardless of thickness (Figs. 8, 9, 10). Evidence for widespread soft-sediment deformation of these rocks includes disharmonic folding, welded contracts, and siltstone and mudstone pull-aparts in graded units (Griffin, 1970). Bedding ranges from 2 mm to 2 m in thickness, but most beds lie in the range of 5 to 40 cm. Calcareous pods (calc-silicate granulite at high metamorphic grade) are abundant particularly near the upper contact of the Sangerville.

Both sandstones and pelites commonly contain an iron-bearing carbonate (ankerite or siderite) and develop a distinctive light red-brown weathering rind. Calcite is also present in most sandstones. All clast types identified in the Sangerville conglomerates have been observed in the sandstones, although the ratio of monomineralic grain to lithic fragment is higher in the sandstones. The matrix contains large amounts of muscovite and either chlorite or biotite depending on metamorphic grade.

Ribbon Limestone. At low metamorphic grades, this distinctive lithology consists of micritic limestone, calcareous and non-calcareous siltstone, sandstone, and some graywacke interbedded on a scale of 2 mm to 30 cm (Fig. 11). At high metamorphic grade, these rocks become thinly interlaminated biotite-quartz granulite and calc-silicate granulite. Crinoid columnal segments have been found in the ribbon limestones

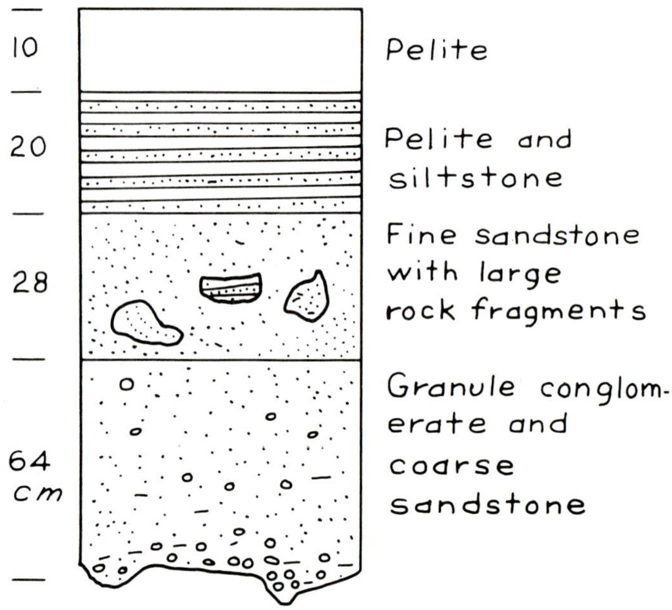

Figure 5. Bedding sequence characteristic of well-graded Sangerville polymict granule conglomerate.

at Athens in the Skowhegan quadrangle and at Dover-Foxcroft in the Dover-Foxcroft quadrangle (Griffin, 1973).

Ribbon limestones occur at several horizons within the Sangerville and occur both as easily mappable outcrop bands and as small isolated lensoid bodies within the Sangerville sandstone and pelite.

Carbonaceous Pelite. Isolated lenses of rusty-weathering sulfidic-carbonaceous pelite occur at several horizons within the Sangerville Formation, but are too small to be shown at the scale of the geologic map. Volumetrically these lenses are of minor importance; their significance lies in the fact that they are the only lithology within the formation containing age-diagnostic fossils.

Age. The Sangerville Formation in the Skowhegan and Kingsbury quadrangles is considered to span the time range from early Llandovery through middle Wenlock. Graptolites of "Silurian age, probably late Llandovery" (W.B.N. Berry, 1968, written commun.) have been found in the Anson quadrangle, and species of middle Wenlock age have been collected in the Skowhegan, Kingsbury, and Guilford quadrangles (Ludman, 1969, 1970; Griffin, 1973). The lower contact of the Sangerville Formation has not been recognized, however, so that the formation may be partly Ordovician in age. The major ribbon limestone horizon lies stratigraphically between rocks containing the Llandovery and Wenlock graptolites, and is thus, in part at least, younger than the ribbon rock of the Carys Mills Formation (Meduxnekeag Group) of Aroostook County, the lower part of which has been shown to be Ordovician (Pavlides, 1965).

Eddy Formation

Conformably overlying the Sangerville on both flanks of the Chapman Ridge anticline are the distinctive rusty-weathering lithologies of the Eddy Formation. The Eddy is composed of a variety of sulfide-bearing lithologies including carbonaceous pelite (shale, slate, phyllite), siltstone and mudstone, non-carbonaceous siltstone and sandstone, and some non-sulfidic lithologies—quartzose sandstone, granule conglomerate, and ribbon limestone. The sulfide phase present is pyrite in the chlorite zone, pyrite and (or) pyrrhotite in higher grades, and occurs as blebs elongated parallel to planes of cleavage.

Within the Eddy, zones of carbonaceous lithologies up to 100 m thick alternate with thinner layers of quartzose sandstone and non-carbonaceous pelite. The ratio of the carbonaceous pelite to the coarser clastic rock is markedly different on the two limbs of the Chapman Ridge anticline: 2:1 on the southeast limb, 2:3 on the northwest. Granule conglomerate and ribbon limestone appear to be confined to the northwest limb.

The sandstones of the Eddy Formation are readily distinguished from those of the Sangerville by their thicker bedding, less well-developed graded bedding, lack of an argillaceous matrix in many instances, and by the dominance of quartz clasts. Even in the granule conglomerates of the Eddy, monomineralic fragments (monocrystalline quartz, chert, coarsely polycrystalline quartz) are dominant, although volcanic fragments and polymineralic clasts have also been identified. Cross-bedding and scour-and-fill features are rare in the Eddy Formation.

Age. The Eddy Formation is considered to be early Ludlow in age. Six graptolite localities have been found in the northwest outcrop belt of the Eddy Formation, but most are in rocks which have been sheared so that only a general age of Silurian can be assigned to them. Berry was able, however, to identify specimens from one locality in the Guilford quadrangle as early Ludlow and this serves as the basis for the age assignment.

Brighton Formation

The Brighton Formation is exposed in and near the axial regions of the two major synclines (Fig. 3). Exposures are far better in the northern region than in the Currier Hill syncline, and cross-sections through the formation are well exposed in the center of the Kingsbury quadrangle. The Brighton Formation is divided into upper and lower members by bedding style and mineralogic differences. The dominant rock type in both members is a well-sorted quartz sandstone with minor amounts of argillaceous matrix expressed as biotite or chlorite flakes depending on metamorphic grade.

Lower Member. The lower member is almost entirely composed of thick-bedded (15 cm to over 4 m) variably calcareous quartz sandstone. Color is dependent on metamorphic grade; the rocks are light purple in the biotite zone, grayish green in the chlorite zone. With the exception of a lens of carbonaceous pelite, pelitic rocks are absent from the lower member, and graded bedding is rare. Bedding in this member is thinner in the Currier Hill syncline than in the Solon syncline. Truncated cross-bedding and biotite-rich laminae are well developed in some of the thicker beds. Pods, beds, and stringers of calc-silicate granulite are abundant in the biotite zone, and the variations in carbonate content are illustrated in the field by a ribbed weathering texture.

The sandstone is characterized by homogeneity of grain size, presence of large poikiloblastic biotite flakes above the chlorite zone, and the apparent absence of plagioclase (in unstained thin sections). Polycrystalline quartz and lithic fragments are unknown. The matrix is often calcareous and may contain anhedral clinozoisite or actinolite, but muscovite is rare.

Upper Member. The upper member consists of thinner bedded (1.5 to 3 cm) sandstones interbedded with pelite (biotite-muscovite-quartz-chlorite phyllite) in a series of graded beds similar to thin-bedded Sangerville lithologies. The ratio

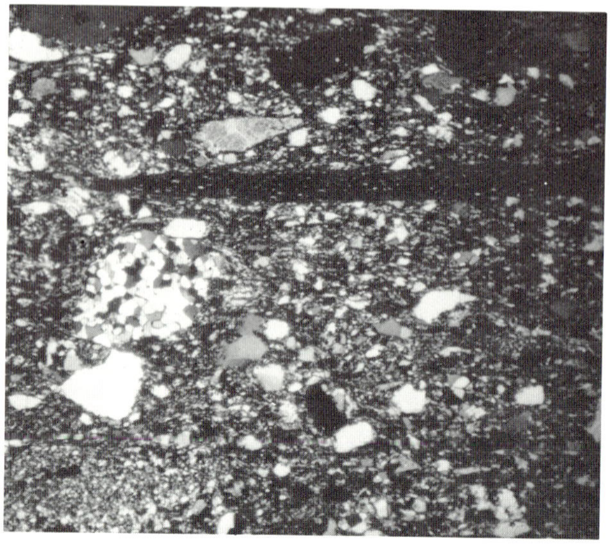

Figure 6. Monocrystalline quartz, polycrystalline quartz, and slate chip in Sangerville polymict granule conglomerate. Slate chip is 3.5 mm long.

Figure 7. Felsic volcanic fragment in Sangerville polymict granule conglomerate. Clast is 2.5 mm long.

Figure 8. Graded bedding in thin-bedded Sangerville sandstone and pelite.

of sandstone to pelite is commonly 1:1, in contrast to the dominance of sandstone in Sangerville rocks of comparable thickness. The sandstone generally contains little feldspar and no rock fragments, but unlike that of the lower member does contain significant amounts of muscovite and chlorite in addition to biotite. As in the lower member, calc-silicate granulite may be present, but the upper member is, in general, less calcareous than the lower.

Age. Diagnostic fossils have not yet been found in the Brighton in either the Kingsbury or Skowhegan quadrangles, and facing evidence is not conclusive at Brighton-Eddy contacts because of shearing in the Eddy Formation. Evidence that the Brighton Formation is older than the Solon Formation is, however, unequivocal. The Brighton-Solon contact zone is well exposed in the Kingsbury quadrangle and is clearly gradational. Graded sets in the upper member of the Brighton Formation thin; grain size in the sandstone portion of the graded units decreases; and the ratio of sandstone to pelite decreases toward the Solon contact. The transition from well-graded sandstone and pelite to massive slate with little evidence of primary bedding takes place in a zone approximately 100 meters thick; the actual contact is drawn at the disappearance of sandstone. Because the Brighton Formation lies above the Eddy Formation, a Late Silurian to Early Devonian(?) age is suggested. Griffin (1973) has collected graptolites of "Silurian" age from a rusty phyllite interbedded with sandstones correlated with the Brighton Formation in the Stetson quadrangle.

Solon Formation

The Solon Formation is not exposed in the Currier Hill syncline, apparently having been removed by erosion, but it does crop out extensively in the centers of small synclines in the Solon syncline. Its upper contact has not been observed, so that only an incomplete picture of the formation can be presented here.

Three rock types have been observed in the Solon Formation, but a detailed study is needed to determine their precise stratigraphic relationships. A bluish-gray weathering, dark-gray homogeneous slate is the dominant rock type, comprising approximately 80 percent of the exposed unit. It is this slate, formerly extensively quarried at Monson in the Greenville quadrangle and Brownville Junction in the Sebec quadrangle, which gave the Central Maine Slate Belt its name. Original bedding is generally obscured by cleavages associated with the first two deformations recorded in the area and when present is indicated by light-gray weathering mudstone layers. These are commonly pulled apart or sheared; deformation is, in some instances, tectonic, in others it is clearly penecontemporaneous with deposition.

Nearly 20 percent of the formation is composed of rhythmically interbedded sandstone and slate in thin graded units (5 to 10 cm). These rocks are similar in their bedding characteristics to the upper member of the Brighton Formation, but are different mineralogically. The Solon sandstones are quartzose, but are poorly sorted, contain little carbonate, and have a far less argillaceous matrix than the sandstones of the Brighton Formation. The large biotite porphyroblasts characteristic of both Brighton sandstones and pelites are absent from Solon lithologies. This is expressed by the difference in color in the two formations: greenish gray or purple for the Brighton, gray to black for the Solon.

A single lens of granule conglomerate has been found at the Solon-Brighton contact in the Kingsbury quadrangle. It is identical in bedding and mineralogy to the massive conglomerates of the Sangerville Formation.

Age. Fossils have not been found in the Solon Formation in the Kingsbury quadrangle. Espenshade and Boudette (1967) reported that Silurian-Devonian bra-

Figure 9. Graded bedding in medium-bedded Sangerville sandstone and pelite.

← Figure 10. Graded bedding in thick-bedded Sangerville sandstone with calc-silicate pod in center of photograph.

Figure 11. Bedding characteristic of Sangerville ribbon limestone; rocks tightly folded with sandstone (light) in core of fold.

chiopods and plant fragments from slate in the Greenville quadrangle correlated with the Solon Formation, but these were not useful in dating the unit. The position of the formation in the stratigraphic section suggests an Early Devonian age. The formation's involvement in regional isoclinal folding, dated as Middle Devonian by the discordant igneous plutons, supports this conclusion.

EAST OF THE CURRIER HILL SYNCLINE

The stratigraphic sequence on the east limb of the Currier Hill syncline is similar in many respects to that just described, particularly in the upper part (Fig. 12), except for the absence of the Solon Formation. The Brighton Formation shows little variation other than a thinning of bedding in the lower member. Emphasis will be placed in description of the older units.

Eddy Formation

Rusty-weathering carbonaceous lithologies that crop out east of the Currier Hill syncline have major differences from their correlatives across this structure. The carbonaceous pelite appears to account for 80 percent of the formation, far more than in either of the outcrop belts to the northwest. This is part of a continuous trend of increasing pelitic content in this formation from northwest to southeast. The interbedded sandstones are finer grained, better sorted, somewhat more calcareous, and more argillaceous than Eddy sandstones in the two belts described above. It is difficult to distinguish between some Eddy and Brighton sandstones in this area. Ribbon limestones and granule conglomerate have not been observed in this belt. Correlation is confirmed by graptolites in the Pittsfield quadrangle identified by Berry (1969, written commun.) as "Silurian, probably early Ludlow." Eddy rocks of this belt interfinger with Sangerville types along strike to the northeast in the Pittsfield quadrangle and are represented only by a few sulfidic interbeds

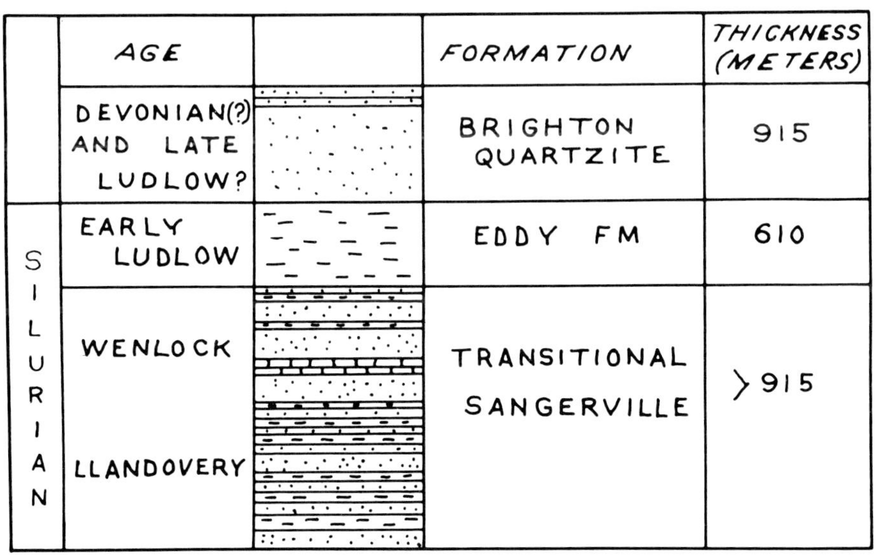

Figure 12. Stratigraphy east of the axis of the Currier Hill syncline.

in the Sangerville Formation in the Dover-Foxcroft quadrangle. Ludlow-aged graptolites are, however, found in these interbeds, so that the Sangerville to the east of the Skowhegan-Kingsbury area ranges in age from Llandovery through early Ludlow.

Transitional Sangerville

The greatest lithologic changes are noted in rocks at the stratigraphic position of the Sangerville Formation. The granule conglomerate of the Chapman Ridge anticline is not present; the carbonaceous pelite is less abundant; and only the ribbon limestones appear essentially unchanged. Most of the rocks in the southeast corner of the Skowhegan quadrangle are correlated with the sandstone and pelite of the Sangerville and are mapped here as the transitional Sangerville.

The sandstones and pelite in the Transitional Sangerville are generally thinner bedded than in the Sangerville to the west; the ratio of sandstone to pelite is lower; and the grain size in the coarsest clastic rock is generally only medium-grained sand. Although graded bedding is as ubiquitous as in the Sangerville, the thickness of each graded set is far less, commonly less than 1.5 cm. Siltstone and pelite interbedded on a 5 mm scale become the dominant rocks in the extreme southeast corner of the Skowhegan quadrangle, alternating only rarely with the 15 to 20 cm graded units characteristic of the Sangerville. Ankerite and siderite have not been found in this area. The thinly interlaminated siltstone and slate have a "pinstriped" appearance, and such outcrops become dominant both across structure to the southeast and along strike to the northeast. Pinstriped rocks interfinger with thicker bedded rocks at the southeastern edge of the Skowhegan quadrangle and make up most of the unit on strike to the northeast in the Pittsfield and Guilford quadrangles.

INTERPRETATION OF FACIES TRANSITIONS

A reconstruction of the pre-deformation facies relationships in the stratigraphic section is presented in Fig. 13. Both vertical and lateral variations in the sequence are noted.

Vertically, the section is characterized by an increase in maturity of clastics from late Landovery (Sangerville) through Early Devonian (Brighton) times. This is indicated by the decrease in content of lithic fragments and feldspar, and the

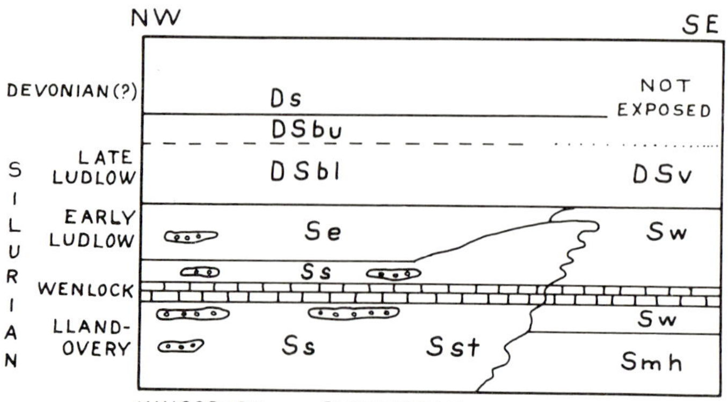

Figure 13. Reconstructed facies relationships for the area Kingsbury-Skowhegan-Waterville. Symbols for Waterville area; Smh-Mayflower Hill Formation; Sw-Waterville Formation; DSv-Vassalboro Formation.

TABLE 1. REGIONAL CORRELATION OF KINGSBURY-SKOWHEGAN STRATIGRAPHY

	Age	Rangeley/ Phillips	Kingsbury/ Skowhegan	Waterville/ Vassalboro
Devonian(?)		Seboomook Fm.*	Solon Fm.*	Not exposed
Devonian and Late Ludlow(?)		Madrid Fm.	Brighton Fm.	Vassalboro Fm.*
Silurian	Early Ludlow	Smalls Falls Fm.	Eddy Fm.*†	
	Wenlock	Perry Mountain Fm.	Sangerville Fm.*†	Waterville Fm.*
	Llandovery	Rangeley Fm.*		Mayflower Hill Fm.*

*Fossiliferous outside the Skowhegan/Kingsbury area.
†Fossiliferous within the Skowhegan/Kingsbury area.

increase in degree of sorting upward in the section. This trend is masked somewhat, particularly toward the northwest, by the dominant lateral transition.

REGIONAL CORRELATION

Extension of the lithofacies transitions to west and south results in the correlations shown in Table 1. These correlations essentially confirm those proposed by Osberg and others (1968) and provide fossil-based dates for the section. The Rangeley (proximal)-Sangerville (intermediate)-Waterville (distal) transition involves the most drastic change, but seems well documented. The mineralogy and bedding style of the Sangerville conglomerates is nearly identical to those of Rangeley B of Moench and Boudette (1970), and the Perry Mountain-Smalls Falls contact shown by Moench and Boudette (1970) is mirrored by the Sangerville-Eddy contact at the type locality for the Eddy Formation in the Skowhegan quadrangle.

REFERENCES CITED

Billings, M. P., 1956, The geology of New Hampshire, Part II—Bedrock geology: New Hampshire State Plan. Devel. Comm., p. 1-200.

Espenshade, G. M., and Boudette, E. L., 1964, Geology of the Greenville quadrangle, Maine: U.S. Geol. Survey Geol. Quad. Map GQ-330.

——1967, Geology and petrology of the Greenville quadrangle, Piscataquis and Somerset Counties, Maine; U.S. Geol. Survey Bull. 1241-f.

Faul, H., Stern, T. W., Thomas, H. H., and Elmore, P.L.D., 1963, Ages of intrusion and metamorphism in the northern Appalachians: Am. Jour. Sci., v. 261, p. 1-19.

Green, J. C., 1964, Stratigraphy and structure in the Boundary Mountain anticlinorium in the Errol quadrangle, New Hampshire-Maine: Geol. Soc. America Spec. Paper 77, 78 p.

Griffin, John, 1970, Prelithification slump structures in Middle Silurian rocks of central Maine: Geol. Soc. America Abs. with Programs, v. 7, p. 562.

——1973, Geology of the Dover-Foxcroft quadrangle, Maine [PhD. dissert.]: Univ. California at Riverside.

Guidotti, C. V., 1965, Geology of the Bryant Pond quadrangle, Maine: Maine Geol. Survey Quad. Mapping Ser. 3, 116 p.

Harwood, D. S., and Berry, W.B.N., 1967, Fossiliferous Lower Paleozoic rocks in the Cupsuptic quadrangle, west-central Maine: U.S. Geol. Survey Prof. Paper 575-D, p. D16-D23.

Keith, Brian, 1970, Geology of the southern third of the Bingham quadrangle, Maine [Master's thesis]: Syracuse, Syracuse Univ.

Ludman, Allan, 1969a, Geology of the Skowhegan quadrangle, Maine [Ph.D. dissert.]: Philadelphia, Univ. Pennsylvania, 230 p.

——1969b, Structure and stratigraphy of Silurian rocks in the Skowhegan area, south-central Maine: Geol. Soc. America Abs. with Programs, pt. 1, p. 37-38.

——1970, Revised geologic map of the Kingsbury quadrangle, Maine: Maine Geol. Survey Open-File Rep.

Moench, R. H., 1970, Geologic map of the Rangeley and Phillips quadrangles, Franklin and Oxford Counties, Maine: U.S. Geol. Survey Misc. Gelo. Inv. Map I-605.

Moench, R. H., and Boudette, E. L., 1970, Stratigraphy of the northwest limb of the Merrimack synclinorium in the Kennebago Lake, Rangeley, and Phillips quadrangles, western Maine, *in* Boone, G., ed., New England Intercoll. Geol. Conf., Guidebook for field trips in the Rangeley lakes-Dead River basin region, western Maine, 1970: p. D1-D12.

Neuman, R. B., 1967, Bedrock geology of the Shin Pond and Stacyville quadrangles, Penobscot County, Maine: U.S. Geol. Survey Prof. Paper 524-I, 34 p.

Osberg, P. H., 1968, Stratigraphy, structural geology, and metamorphism of the Waterville-Vassalboro area, Maine: Maine Geol. Survey Bull. 20, 64 p.

Osberg, P. H., Moench, R. H., and Warner, J., 1968, Stratigraphy of the Merrimack synclinorium in west-central Maine, *in* Zen, E-an, White, W., Hadley, J., Thompson, J. B., eds., Studies of Appalachian geology, northern and maritime: New York, Interscience Pubs., Inc., p. 241-253.

Pankiwskyj, Kost, 1964, Geology of the Dixfield quadrangle, Maine [Ph.D. dissert.]: Cambridge, Harvard Univ., 224 p.

——1975, Geology of the Anson, Kingfield, Farmington, Norridgewock, and Dixfield quadrangles, Maine: Maine Geol. Survey Bull.

Pavlides, Louis, 1965, Meduxnekeag Group and Spragueville Formation of Aroostook County, northeast maine, *in* Cohee, G. V., and West, W. S., eds., Changes in stratigraphic nomenclature by the U.S. Geological Survey, U.S. Geol. Survey Bull. 1244-A, p. A52-A57.

Rodgers, John, 1970, The tectonics of the Appalachians: New York, Interscience Pub., Inc., 271 p.

Warner, Jeffrey, 1967, Geology of the Buckfield quadrangle, Maine [Ph.D. dissert.]: Cambridge, Mass., Harvard Univ. 232 p.

Zartman, R. E., Hurley, P. M., Krueger, H. W., Giletti, B. J., (1970), A Permian disturbance of K-Ar radiometric ages in New England: Its occurrence and cause: Geol. Soc. America Bull., v. 81, p. 3359-3374.

Manuscript Received by the Society April 18, 1974
Revised Manuscript Received December 23, 1974
Manuscript Accepted January 10, 1975

Printed in U.S.A.

Geological Society of America
Memoir 148

Pre-Silurian Stratigraphic Succession in Central Western Maine

EUGENE L. BOUDETTE
U.S. Geological Survey
Reston, Virginia 22092

AND

GARY M. BOONE
Department of Geology, Syracuse University
Syracuse, New York 13210

ABSTRACT

Three major pre-Silurian stratigraphic divisions are recognized in the rocks of west-central Maine: Division I, Precambrian crystalline rocks of the Chain Lakes massif; Division II, Cambrian(?) to Lower Ordovician(?) eugeosynclinal rocks; and Division III, Middle Ordovician to Upper Ordovician(?) eugeosynclinal rocks. The rocks are sparsely fossiliferous and are cut by four plutonic series, the oldest of which predates the Highlandcroft Plutonic Series. Division I consists of massive-bedded granofels and thin- to massive-bedded gneiss, metavolcanic rocks, schist, metasandstone, and amphibolite (~3,100 m). The granofels and gneiss are characterized by an unusual fragmental texture whose origin is obscure. Division II has four lithologic units that represent a transition from the sedimentary and volcanic components of an ophiolite sequence in the lower part to a flysch sequence in the upper part. The four units, in ascending stratigraphic order, are (1) greenstone displaying relict pillow structure, amphibolite, keratophyric metavolcanic rocks, and iron formation (~1,500 m); (2) quartz-rich metaquartzwacke and metagrawacke, magnesian metapelite, and greenstone displaying relict pillow structure (~1,100 m); (3) dark sulfidic interbedded metapelite, metagraywacke, feldspathic metasandstone, and quartzite (~600 m); and (4) green and red magnesian metapelite, metagraywacke, and quartzite, all with characteristic pinstripe lamination (>800 m).

Rocks in New Hampshire equivalent to those of Division II are intruded by the Highlandcroft Plutonic Series, which is similar to the Attean Quartz Monzonite, provisionally dated at about 470 m.y. The constancy of facing directions at contacts between units has made possible the definition of major fold axes and faults, and the definition of these structures has enabled us to make reasonable estimates of regional stratigraphic thicknesses. The stratigraphic succession established in this paper is believed to be reasonably comparable to ophiolite-flysch successions indicated from geophysical data and drilling in modern oceanic environments. If the comparison is valid, then the process of generating oceanic crust can be projected back to early Paleozoic time in this region.

Units 3 and 4 of Division II have been traced directly to the Maine-New Hampshire border, where they are continuous with the Dixville and Albee Formations. This correlation involves either inversion of part of the section or overthrusting from either the southeast or northwest of the section where units 3, 2, and part of 1 would be sturcturally superposed on 4. Division III, which has been described elsewhere, probably conformably overlies unit 4 in the southwest part of the region.

INTRODUCTION

A regionally extensive Precambrian through Lower Devonian stratigraphic sequence is exposed in west-central Maine. The region is tectonically complex. The sedimentary and volcanic rocks making up the sequence have been variably metamorphosed and folded along northeast-trending axes, then faulted. Local secondary folds within units are common. Major fold structures and numerous faults have been mapped. The region has been divided into two paleosedimentologic domains by a northeast-trending tectonic hinge zone (Fig. 1): (1) essentially uninterrupted deposition of units ranging in age from (at least) Middle Ordovician to Early Devonian southeast of the tectonic hinge zone and (2) transgressive onlap of Lower Silurian to Lower Devonian rocks from southeast to northwest over older rocks, northwest of the tectonic hinge zone. The lower age limit of the latter units is not known, but it could be Cambrian, the age of the oldest Paleozoic rocks exposed northwest of the zone.

Northwest of the hinge zone, pre-Silurian, Silurian, and Devonian metasedimentary rocks are separated by an angular unconformity; southeast of the zone, an unconformity between the two sequences has not been found, and the subdivision is arbitrary. The pre-Silurian rocks are not as extensively exposed southeast of the zone. Northwest of the hinge zone, we have recognized two major pre-Silurian stratigraphic divisions (Fig. 2; see also Fig. 1 of Boone and others, 1970): Division I, Precambrian crystalline rocks of the Chain Lakes massif, and Division II, Cambrian(?) to Lower Ordovician(?) eugeosynclinal rocks. A third major stratigraphic division is exposed only in the southwestern part of the region, along the hinge zone but mainly southeast of it. Division III consists of Middle Ordovician to Upper Ordovician(?) eugeosynclinal rocks. Rocks assigned by us to the base of Division III, but less than 10 km outside the area shown on Figure 2, are the only pre-Silurian rocks accurately dated by fossils (see Harwood and Berry, 1968). Harwood and Berry (1968) lithologically correlated these dated rocks across regional strike with rocks we assign to the base of Division II. We disagree with their correlation and propose an alternative, which is amplified in the sections on Middle Ordovician and Upper(?) Ordovician rocks and correlation of pre-Silurian rocks. Because the rocks we assign to Division III are described in detail elsewhere (Moench, 1969; Moench and Boudette, 1970; Harwood, 1973) and because it cannot be clearly demonstrated that the dated Middle Ordovician rocks have local regional correlatives, we focus herein on the stratigraphy of the rocks we assign to Divisions I and II. Division II is north and northwest of the regional strike of Division III rocks and northwest of the tectonic hinge zone. The succession of Divisions I and II is schematically represented and summarized in Figure 3. Igneous rocks of varied chemistry and emplacement mechanics, tentatively assigned to four plutonic episodes (Table 1), intruded the pre-Silurian rocks prior to the injection of an extensive mafic dike system.

The major plutonic series are (1) Cambrian(?) to Lower Ordovician(?) ultramafic, mafic, and granitic rocks that predate the Highlandcroft Plutonic Series; (2) Lower Ordovician granitic rocks, including the 470-m.y.-old Attean Quartz Monzonite (Table 1), which is correlative with the Highlandcroft; (3) Late Ordovician(?) quartz porphyry; and (4) Lower Devonian to Middle Devonian(?) gabbroic and granitic rocks correlative with the New Hampshire Plutonic Series (see Boone and others, 1970).

DIVISION I

Division I, consisting of the Precambrian[1] Chain Lakes massif, forms the core of the Boundary Mountains anticlinorium and is arranged in an elongate dome coincident with, but not necessarily conformable to, the outer geometry of the anticlinorium (Fig. 1). The rocks of the massif were originally metamorphosed to at least sillimanite grade prior to Cambrian (?) time and subsequently variously altered by hydrothermal and metamorphic events. They aggregate 3,100 m or more in thickness and are divisible into a lower unit of predominantly massive-bedded granofels and gneiss at least 600 m thick and an upper unit about 2,500 m thick of rocks of more varied internal lithology (including gneissic granofels and schist of both volcanic and sedimentary origins).

The entire lower granofels unit and the gneissic part of the upper unit are characterized by unusual fragmental textures and structures, a feature unique to these rocks. These textures and structures vary from coarse angular breccia, which occurs only locally, to angular and rounded clasts enclosed in a quartzofeldspathic matrix. The clasts make up a varied lithologic suite that includes quartz nodules and plutonic, volcanic, and metamorphic rocks. The lower unit is dominantly massive, light- to medium-gray, medium- to coarse-grained chloritic and sericitic quartz-feldspar-sillimanite-muscovite-biotite (±cordierite ± garnet) granofels or gneiss that contains thin layers and laminations. Lithic fragments as much as a metre across but more commonly a few centimetres, some individually foliated, and quartz nodules as much as several centimetres across occur within the granofels and gneiss, although nowhere can the protolith of this rock be interpreted as a conglomerate, breccia, or typical bouldery mudstone. Lenses within the gneiss are contorted complexly with swirls and are dislocated into segments, possibly as a result of the mechanics of original deposition. Where this contorted rock occurs with massive or layered gneiss or granofels, clear gradation between the two is observed. Leucosomes of quartz and feldspar are common. In the northeast, massive dark mafic breccia makes up part of the lower unit.

The granofels and gneiss of the lower unit grade upward into the more systematically bedded, lithologically heterogeneous upper unit. Gneiss that is compositionally comparable to that in the lower unit is abundant in the upper unit. This gneiss is evenly layered; layers range from 1 to 10 cm in thickness and are internally laminated. It is nowhere contorted, but lithic and quartz fragments similar to those in the lower unit are common, although characteristically smaller in size and sparsely distributed. The upper unit contains, in addition to the bedded quartzofeldspathic gneiss, chloritic and sericitic quartzofeldspathic schist, thick-bedded feldspathic quartzite, felsic metavolcanic rocks, thick-bedded and laminated quartzite, black

[1]Isotopic dating by R. S. Naylor (Naylor and others, 1973) indicates an apparent age greater than 950 m.y. for the Chain Lakes massif.

schist, and finely laminated amphibolite. In some large exposures, two or more rock types are interlayered and separated by sharp boundaries. The gneiss, schist, and quartzite commonly are intergradational.

Where rocks of the massif are gneissic or schistose, the foliation is parallel to compositional layering. Relatively abundant small lithic clasts of metamorphic rocks within these layers have a foliation that, from one clast to another, varies in angular divergence from the regional foliation, which probably postdates that in the clasts. The bedding foliation has, in some cases, modified foliation in the perimeters of the clasts.

The lithologic sequence within the massif in the area represented by the tectonic and geologic maps (Figs. 1, 2) is interpreted to be upright, as shown by graded bedding and cross-lamination. The sequence dips away from the axis of the dome forming the massif at angles generally less than 40°. We conclude that the rocks of the massif are the products of aquagene volcanism mixed with detritus produced by weathering of a crystalline-metamorphic terrane. They were deformed and

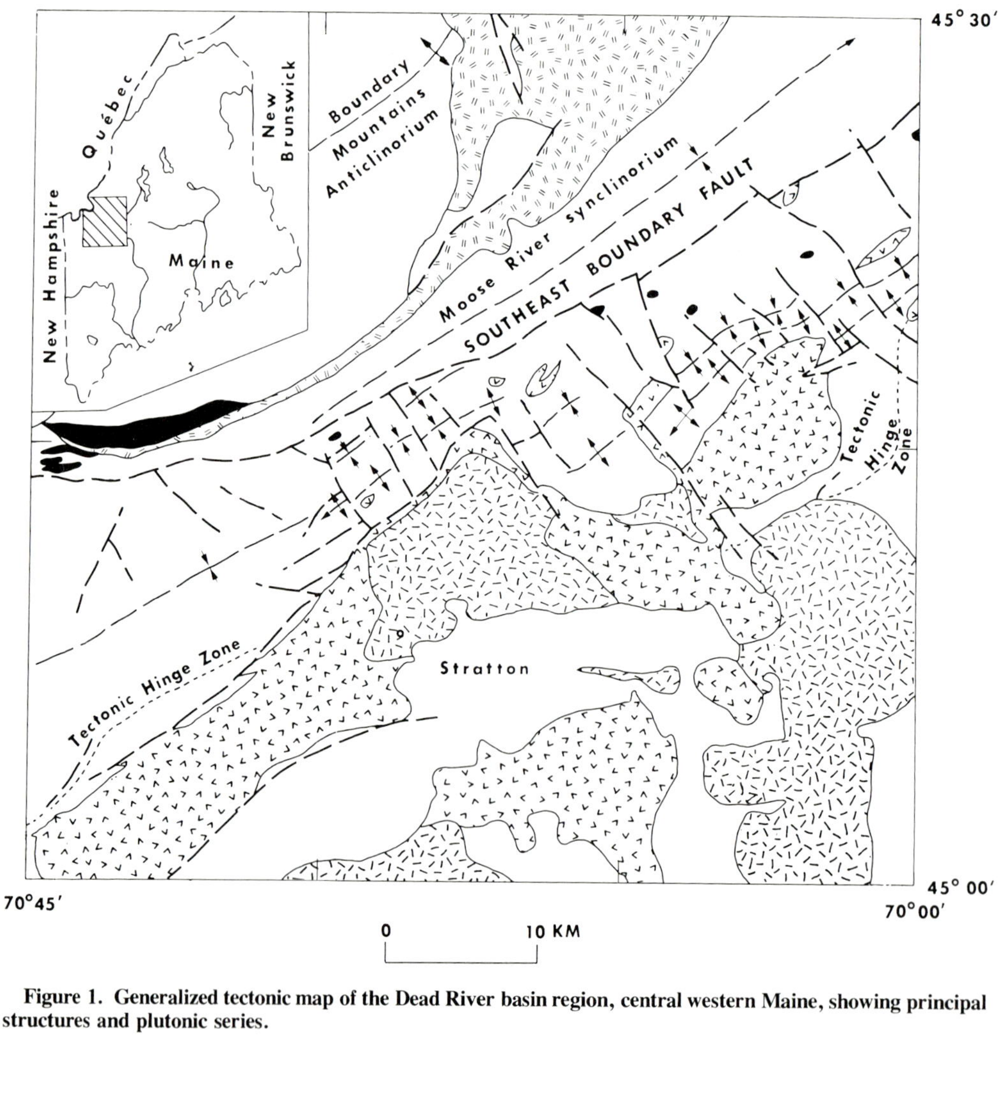

Figure 1. Generalized tectonic map of the Dead River basin region, central western Maine, showing principal structures and plutonic series.

metamorphosed to sillimanite rank, partly retrograded, and eroded prior to deposition of Division II. Moreover, rocks of the massif subsequently behaved as a rigid buttress that resisted penetrative deformation during the Acadian orogeny.

The rocks of the massif are unfossiliferous. We have assigned them to the Precambrian Y on the basis of stratigraphic position, metamorphic history, and radiogenic age (Boudette, 1970; Boone and others, 1970; Naylor and others, 1973).

Division I is separated from Division II by pre-Highlandcroft ultramafic and related rocks and by the Attean Quartz Monzonite (Fig. 2; Table 1). Neither strike discordance nor brittle fault deformation can be demonstrated between Divisions I and II. Ductile faulting in a low-rank metamorphic environment is a likely explanation for the juxtaposition. The dip of internal layering in the massif along the southern boundary is about 20° less than that in Division II and the adjacent ultramafic units. A structural, as well as stratigraphic, hiatus is therefore suggested on the basis of divergence of dips and discordance of metamorphic history.

DIVISION II

Division II comprises a series of metamorphic rocks that flank the south side of the dome forming the Chain Lakes massif and that have been traced northeastward (Fig. 2). In the west-central part of the area, Division II makes up an upright south- and southeast-facing, essentially homoclinal sequence repeated by the Southeast Boundary fault. In the northeast, Division II is deformed into a series of complementary folds (Fig. 1).

The total thickness of Division II and its principal units is varied; it ranges from about 1,500 m northwest of the Southeast Boundary fault to about 5,700 m in the repeated section to the southeast along the North Branch of the Dead River and about 760 m to the northeast, along the main course of the Dead River. The sections are incomplete in the southwest part of the region, where faults

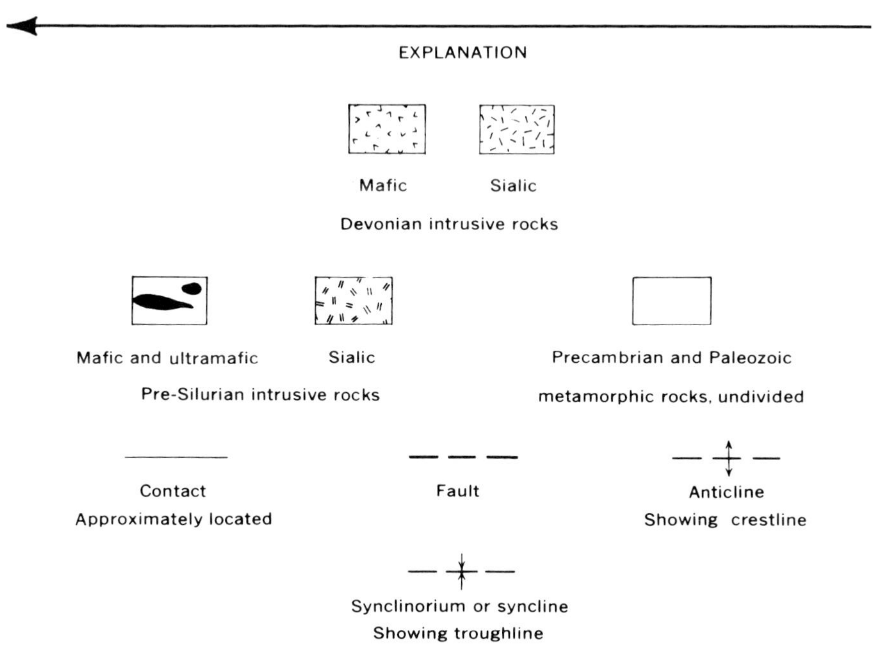

cut out part of the top and bottom. In the northeast, the lowest part of the section is truncated by the Southeast Boundary fault, the top by an unconformity. We estimate an average regional thickness of about 4,000 m, which probably increases considerably to the southeast beneath Divisions III and IV. The upper part of Division II is relatively uniform in internal lithologic ratios in the region despite its range in thickness. The lower part is characterized by distinct facies changes both along strike and in repeated sections (Fig.2).

The principal units of Division II and their average regional thicknesses are, in ascending stratigraphic order: unit 1, pillowed greenstone and amphibolite, keratophyric metavolcanic rocks, and iron formation with subordinate metaquartz-wacke and metagraywacke, about 1,500 m thick; unit 2, metaquartzwacke, quartz-rich metagraywacke, volcaniclastic metasandstone, Mg-rich (magnesian) metapelite, black slate, and pillowed greenstone, about 1,100 m thick; unit 3, dark sulfidic interbedded metapelite, metasiltstone, metaquartzwacke, feldspathic metasandstone, and quartzite, about 600 m thick; and unit 4, green and red magnesian metapelite,

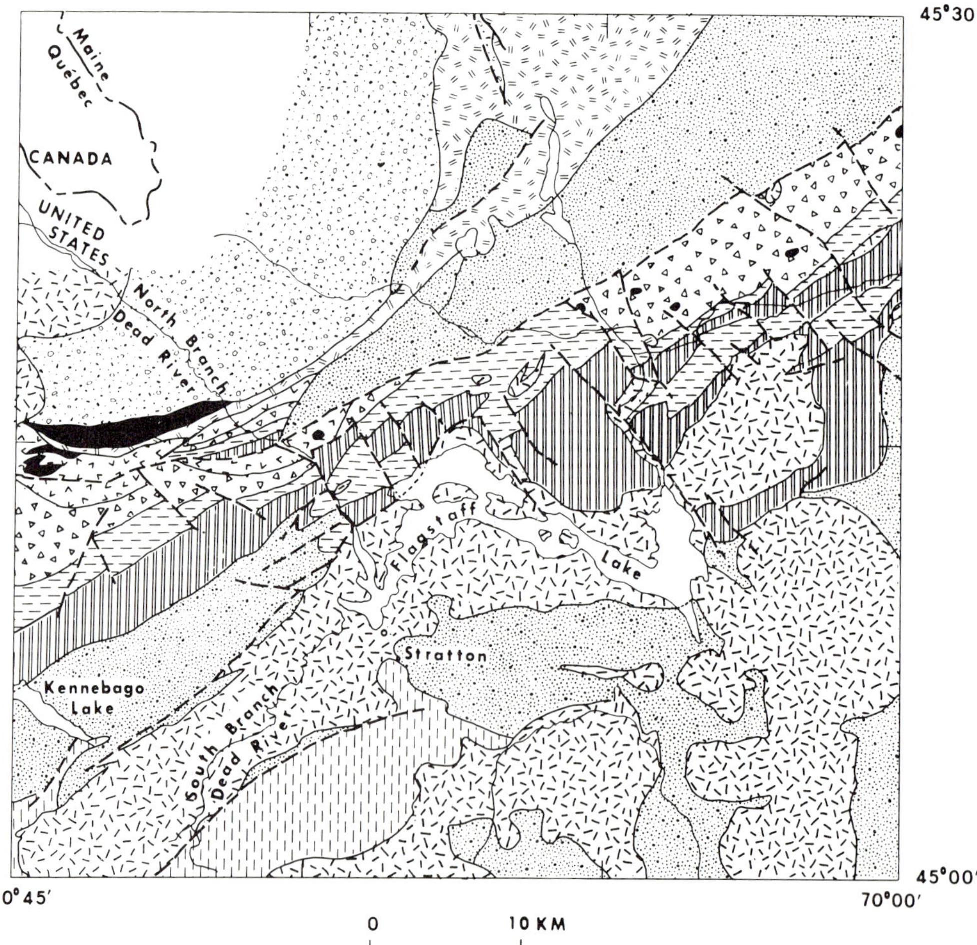

Figure 2. Generalized geologic map of the Dead River basin region, central western Maine, showing distribution of pre-Silurian units.

EXPLANATION — FIGURE 2

Metamorphic rocks

Division IV

Metamorphosed stratified rocks, undivided — Lower Silurian to Lower Devonian — SILURIAN AND DEVONIAN

Division III

Metagraywacke, felsic metavolcanic rocks, metapelite, and metasandstone of the Quimby and Greenvale Cove Formations; greenstone, dark sulfidic slate, metagraywacke, and lenses of metaconglomerate of unnamed formation — Middle Ordovician to Upper Ordovician(?) — ORDOVICIAN AND ORDOVICIAN(?)

Division II — Cambrian(?) to Lower Ordovician(?) — CAMBRIAN(?) AND ORDOVICIAN(?)

Unit 4

Green, locally variegated red and green metapelite, gray-green metagraywacke, and metaquartzwacke, characteristically pinstriped; laminated to thickly bedded

Unit 3

Dark sulfidic slate, phyllite, and metaquartzwacke; subordinate metagraywacke, feldspathic metasandstone, and quartzite

Unit 2

Dominantly metaquartzwacke, metagraywacke, and volcaniclastic metasandstone with interlayered chloritic metapelite and black slate; subordinate greenstone, red slate, and polymictic metaconglomerate

Unit 1

Dark and light greenstone, keratophyric metavolcanic rock, and metagraywacke; minor cherty iron-formation

Tectonic break

Division I

Gneissic granofels, gneiss, schist, gneissic quartzite, and thinly layered amphibolite of the Chain Lakes massif — PRECAMBRIAN

Igneous rocks

Mafic and sialic intrusive rocks, undivided — Lower and Middle Devonian — DEVONIAN

Attean Quartz Monzonite — Lower Ordovician — ORDOVICIAN

Serpentinized pyroxenite

Epidiorite, gabbro, pyroxenite, serpentinite, and quartz diorite of the ultramafic complex — CAMBRIAN OR ORDOVICIAN

Contact

Dotted where concealed

Fault

Unclassified

metagraywacke, and quartzite, commonly showing a characteristic pinstripe lamination, more than 760 m thick.

Unit 1

Unit 1 is best exposed in the two belts separated by the Southeast Boundary fault along the North Branch, Dead River. Keratophyric metavolcanic rocks and iron formation appear only in the northwestern belt, where the iron formation is at the top in conformable contact with a much-thinned section of unit 2 (see below). The northwestern belt contains a pillowed greenstone, the oldest volcanic unit. This greenstone is separated from the ultramafic complex by the Attean Quartz Monzonite, as along the North Branch, Dead River, and locally is in gradational contact with epidiorite of the complex to the southwest. Keratophyre constitutes about 30 percent of unit 1 along the North Branch and southwest on strike in the northwestern belt. Its extent to the northeast is unknown because of younger cover.

In the northwestern belt, unit 1 is composed of the following rock types, given with average thicknesses in ascending stratigraphic order: (1) greenstone (260 m; includes metamorphosed pillowed spilitic lava and mafic volcaniclastic rocks and local patches of altered amphibolite where the unit is in contact with the Attean Quartz Monzonite and epidiorite of the ultramafic complex) with minor amounts of metagraywacke, (2) keratophyric metavolcanic rock with minor amounts of metaquartzwacke and metagraywacke (330 m), and (3) lenses of cherty iron formation (to 50 m).

In the southeastern belt, unit 1 is composed almost entirely of pillowed greenstone with lenses of mafic metavolcaniclastic rocks (as much as 2,450 m thick); maroon

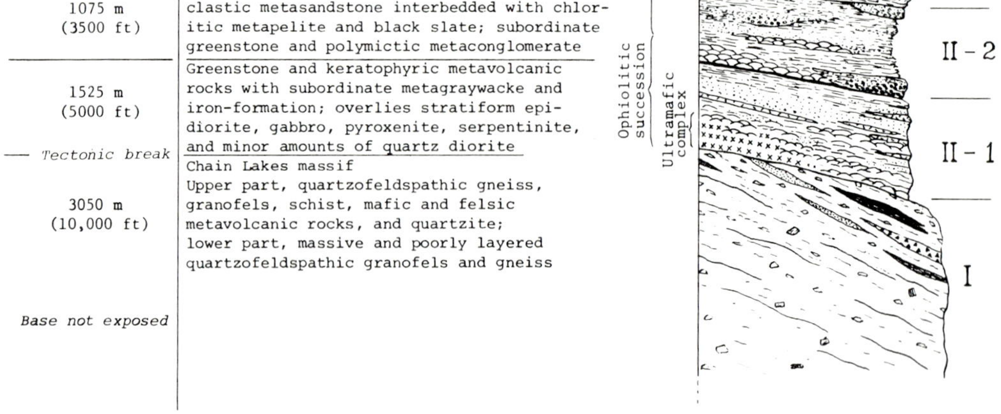

Figure 3. Stratigraphic column of Divisions I and II northwest of the tectonic hinge zone, central western Maine.

metapelite lenses as much as 60 m thick occur locally within the greenstone and at its top. The metapelite could in part be equivalent to the iron formation of the northwestern belt. There is more than a twofold stratigraphic thickening of the unit in the southeastern belt, but this occurs only along the North Branch of the Dead River and 15 km to the west (Fig. 2). Where unit 1 is missing in the southeastern belt, its spatial position appears to be taken up by unit 2. We interpret a large part of units 1 and 2 to be facies equivalents.

Greenstone and mafic clastic rocks of unit 1 on either side of the North Branch in the southeastern belt appear to make up a shield volcano that has been beveled after tilting (clockwise viewed to the northeast along strike) following burial by clastic rocks of unit 2. As shown on the geologic map, the North Branch nearly bisects this feature.

The greenstone (chlorite-albite-epidote assemblage) and, locally, the altered amphibolite (chlorite-albite-epidote assemblage with most chlorite after relict am-

TABLE 1. MAJOR PLUTONIC SERIES IN THE DEAD RIVER BASIN REGION, CENTRAL WESTERN MAINE

	Plutonic series and lithology	Correlation	Radiogenic age
Lower Devonian to Lower Middle Devonian(?)	(4) Gabbro, granodiorite, and quartz monzonite	New Hampshire Plutonic Series (Billings, 1956)	360 to 393 m.y. (Faul and others, 1963; Lyons and Faul, 1968)
Upper Ordovician(?)	(3) Quartz porphyry and aplitic quartz monzonite (small epizonal bodies cutting Attean Quartz Monzonite and older rocks)	Felsic volcanic rocks in the Quimby Formation (Moench, 1969)	433 to 457 m.y. (F. C. Canney, unpub. data)
Lower Ordovician	(2) Attean Quartz Monzonite	Highlandcroft Plutonic Series (Billings, 1956)	~470 m.y. on Pb/U (Leon T. Silver, 1970, oral commun.)
Cambrian(?) to Lower Ordovician(?)	(1) Altered peridotite, pyroxenite, gabbro, quartz diorite, and epidiorite of the ultramafic complex	Not established	Not established. Series is arbitrarily assigned a Cambrian(?) to Early Ordovician(?) age on the basis of stratigraphic position

phibole) of unit 1 commonly show relict amygdaloidal structure and less well developed pillow structure. Metagraywacke beds constitute less than 10 percent of the section. These beds range in thickness from about 1 to 15 m, show grading and cross-lamination, and are composed principally of lapilli, feldspar, and quartz. They are interlayered with the greenstone. Some of the quartz clasts in the metagraywacke are blue, similar in color to quartz in keratophyric intrusive rocks associated with the ultramafic complex and suggesting a coupled genesis (see Boudette, 1970, p. C-9). The upper greenstone contact is marked by the appearance of interlayered light-gray keratophyric metavolcanic rocks composed of sodic quartz latite, ash flows, and breccia. These felsic layers average about 15 m thick and occur in a zone 100 to 200 m thick. Boundaries between greenstone and keratophyre are sharp. Southwest of the North Branch, a repetitive sequence of the two volcanic rock types occurs where individual flows average about 30 m in thickness. The top of the keratophyric metavolcanic rocks is marked by the appearance of interlayered quartz-rich metagraywacke and metaquartzwacke in layers about 5 to 10 m thick (described below) associated with a thin-banded cherty iron formation. The iron formation is laminated, dark red or dark metallic gray, very fine grained, and occurs in layers 10 to more than 20 m thick.

The top of unit 1 is arbitrarily taken as the top of the highest iron formation. The upper part of the unit is either in sharply defined contact with greenstone of unit 2 identical to that at the base of unit 1 or gradational into metaquartzwacke interlayered with greenstone. A combination of pillow tops in the metavolcanic rocks and graded beds and cross-lamination in the clastic rocks establish the stratigraphic facing directions in unit 1 and the sequence at the contact between unit 1 and unit 2. The only basis for distinguishing the greenstone and some layers of the metaquartzwacke of unit 2 from those of unit 1 is stratigraphic position. The metaquartzwacke of unit 2 tends to be somewhat coarser, more poorly sorted, thicker bedded, and locally conglomeratic. Alternatively, units 1 and 2 could be facies equivalents in the northwestern belt separated by a fault of similar geometry to that of the Southeastern Boundary fault, which partly repeats the section. This alternative is unattractive because neither keratophyre nor iron formation appears at the top of unit 2 in the northwest belt, where it would be expected in such a relatively narrow depositional zone, if in fact it was unit 1 repeated by faulting.

Unit 2

Unit 2 occurs southeast of the Southeast Boundary fault in the northeastern part of the region, where it is known only from reconnaissance. Along the North Branch of the Dead River, it lies in two parallel belts on either side of the fault. The thickness of the unit in the northeast is probably 1,100 m. Closer limits can be set in the southwest, where detailed mapping has been done (Boudette, 1970). Unit 2 in the North Branch sequence differs in thickness and the proportion of lithic constituents on either side of the Southeast Boundary fault. Northwest of the fault, unit 2 averages about 430 m thick. Southwest of the fault, the thickness is uncertain because of folding but is estimated to be a minimum of 760 m. Northwest of the fault, the ratio of metaquartzwacke to greenstone is about 1:4; southeast of the fault, the proportion reverses markedly through probable facies changes to more than 10:1, and the greenstone is restricted to small lenses.

Greenstone of unit 2 northwest of the Southeast Boundary fault is texturally and mineralogically identical to greenstone at the base of unit 1. The greenstone to the southeast is similar, but it tends to show better development of pillow structure with fewer interlayers of lapilli-bearing metagraywacke.

The metaquartzwacke is a light-greenish-brown, coarse- to fine-grained mixture of quartz, plagioclase, alkali feldspar, amphibole, sparse pyroxene, rock fragments, and accessories set in a very fine grained cloudy matrix predominantly of chlorite and epidote. This lithology is characterized by great variation in the thickness of beds, from a few centimetres to several metres. Polymictic conglomerate occurs locally, especially where the beds are relatively massive. Where beds are relatively thin (less than 3 m thick), it is usually possible to determine facing directions from a combination of grading and cross-lamination. Locally, the metaquartzwacke grades into lenses of magnesian metapelite and volcaniclastic feldspathic metasandstone that resemble the finer grained fractions of the thick graded beds within the metaquartzwacke. The contacts between the metasedimentary rocks and pillowed greenstone within the unit are sharp, suggesting relatively rapid submarine deposition because their assemblages of clasts are the same. Maroon to brick-red metapelite similar to that in the southeastern belt of unit 1 is commonly associated with the lenses of greenstone. The contact between unit 2 and the overlying unit 3 is sharply defined by a topographic break over the region that coincides with the first appearance of sulfidic metapelite, which is relatively erodable. Facing directions are shown by pillow tops in greenstone and graded beds in unit 1 and by graded beds and cross-lamination in unit 3.

Unit 3

The base of unit 3, in the northeast part of the area, is a fairly abrupt transition from interlayered felsic metavolcanic rocks, metasiltstone, and chert to the lithic assemblage: sulfidic metapelite and metasiltstone (commonly a phyllite) with cherty lenses and light- to dark-gray, variably sulfidic and sulfide-free metaquartzwacke. This assemblage is typical of the unit along and across strike throughout the entire area. Typical also is the generally fine to medium flaser bedding (Pettijohn and Potter, 1964, Pl. 17), wavy bedding, and sparse graded bedding. Conspicuously absent from the unit are any significant sections of large-scale parallel bedding as much as a metre or more in thickness. Parallel laminations, however, commonly form thin zones between, and grade into, flaser-bedded and wavy-bedded zones.

These characteristic features are interrupted by bouldery mudstone in two zones near the top of the unit (argille á bloc facies). These zones appear to be 100 to 150 m thick separated by 50 to 80 m of phyllite; the enclosed boulders are generally less than 2 m thick and 4 m long; some exotic slabs may exceed 15 m in length. They are generally composed of dark-blue, coarse- to fine-grained orthoquartzite, but small fragments of equigranular granite and well-foliated amphibolite have been found. Near the top of the unit, a locally thick sequence (150 m) of medium- to thick-bedded light-gray quartzite and metaquartzwacke forms much of the rock along the crestline of the anticline in the eastern part of the area (Fig. 2) and is found along strike farther southwest. Small lenses, laminae, and lapilli of light-gray tuff are scattered throughout the phyllite of the uppermost part of the unit.

Unit 3 is probably less than 1,500 m thick in the eastern part of the area. In the western part of the area, in the North Branch sequence, it is about 310 m thick in the belt northwest of the Southeast Boundary fault and is estimated to be at least 690 m thick to the southeast, where folding repeats the section.

The transition from unit 3 to 4 is marked by a zone of interbedding in which medium to thick beds of dark-gray, pyritic, rusty-weathering metasiltstone typical of unit 3 alternate with beds of similar thickness of light-greenish-gray, nonsulfidic metasiltstone and phyllite. This contact zone is 30 m thick where exposed in folds north and south of the Dead River in the eastern part of the region, but locally it may be thicker in the north-central part of the Dead River basin. The top of unit 3 is arbitrarily placed at the uppermost occurrence of laterally extensive pyritic rusty-weathering phyllite. Relative stratigraphic order of the two units has been determined by graded bedding in metasiltstone-phyllite and metaquartzwacke-phyllite sets across the transitional contact. This has been documented in three localities in the northeast, where the contact zone is repeated across the major fold structures, and in the southwest in the homoclinal sequence. E. V. Post (unpub. data) reported the same stratigraphic order for these units along strike northeast of the area shown in the tectonic and geologic maps (Figs. 1, 2).

Unit 4

The lower half of unit 4 is composed of magnesian metapelite containing about 15 percent metagraywacke in beds about 5 to 10 cm thick. The amount of metagraywacke and thickness of beds gradually increase upward so that the upper half contains 50 percent or more metagraywacke and metaquartzwacke beds 10 to 50 cm thick. Sparse, small calcareous lenses occur near the base of the unit. These and rocks intermediate between metagraywacke and calcareous rocks are associated in a variety of depositional structures ranging from parallel lamination to wavy and flaser bedding to thickly bedded, graded sets. In general, parallel bedding and uninterrupted graded bedding are more common in the lower half of the unit and become less common in the upper half as the zones of small-scale chaotic structure, convolute structure, and other well-preserved soft-sediment deformational features increase (Boone, 1970). Despite the wide variety of bedding, a characteristic feature of metapelite throughout the unit is a pinstriped appearance produced by quartz-rich and quartz-feldspar-rich laminae parallel to cleavage and relict bedding. Metagraywacke and metaquartzwacke beds commonly contain quartz-rich laminae developed along fracture-cleavage and slip-cleavage surfaces. These pinstripe structures persist through a wide range of metamorphic conditions.

Neither mafic volcanic nor iron-magnesium-poor aluminous metasedimentary rocks have been found in unit 4 as end members of the characteristic iron-magnesium-aluminum-rich composition. A comparison of chemical compositions of metapelitic rocks in unit 4 with those of average shale and basaltic rocks (Boone, 1973) suggests that the magnesian metapelite of unit 4 originates as a mixture of terrigenous mud and weathered mafic volcanic debris.

The top of unit 4 is not seen in the Dead River region, where rocks of Division IV (Fig. 2) rest unconformably on those of Division II. To the west of the region, about 10 km southwest of Kennebago Lake (Fig. 2), Harwood and Berry (1968) reported rocks equavalent to the upper part of unit 4 in conformable contact with a sulfidic metapelite-greenstone-metagraywacke unit of Middle Ordovician age at the base of Division III. None of these rocks have been identified to the northeast on regional strike in the Dead River region.

DIVISION III

The Middle Ordovician rocks of Harwood and Berry (1968), which we assigned to the base of Division III in the southwest, along the tectonic hinge zone, are succeeded conformably by the Quimby and Greenvale Cove Formations (Moench, 1969, 1971), which we assigned to the top of Division III. The Middle Ordovician rocks of Harwood and Berry (1968), assigned by them to part of the Dixville

Formation (Hatch, 1963), are assigned by us to the lower part of Division III. We do not concur with the extension of the name to the Middle Ordovician rocks, for reasons given in the section on pre-Silurian correlation. The Middle Ordovician rocks of Harwood and Berry (1968) consist predominantly of dark sulfidic slate and phyllite interbedded with feldspathic quartzite, greenstone, and dark lithic metagraywacke (see Harwood, 1973). These rocks can be observed to grade into the base of the Quimby Formation. The Quimby Formation consists predominantly of metagraywacke and metaconglomerate, sulfidic metapelite with sparse layers of felsic metavolcanic rock, and calc-silicate rock. The overlying Greenvale Cove Formation consists of interbedded metasandstone and metapelite, and laminated metasandstone and metapelite with layers of calc-silicate rock. The Quimby and Greenvale Cove Formations have been assigned a Late Ordovician(?) age by Moench (1969). The top of the Greenvale Cove is conformable with and observed to be gradational into the lowermost units of Division IV in the southwest, which are dated as Early Silurian (Moench and Boudette, 1970). Division III is of only indirect concern here, but its transitional status between Divisions II and IV is significant, and this status is the principal basis for recognition of the tectonic hinge zone. Unit 3, rocks of Middle Ordovician age described by Harwood and Berry (1968), and the Quimby Formation are all characterized by appreciable amounts of sulfidic metapelite. These units superficially resemble one another, but differences in lithology, such as grain size and mode, and relict bedding features, such as lamination habit and thickness, allow each to be distinguished in the field.

ENVIRONMENTS OF DEPOSITION AND SEDIMENTOLOGY OF THE PRE-SILURIAN ROCKS

Division I

The quartzo-feldspathic, relatively alumina-rich matrix and polymictic fragments of the lower granofels unit of the Chain Lakes massif, as well as the varied lithologic suite within the upper unit, indicate a complex depositional environment for Division I. The rocks of Division I preserve their original sedimentary structures. All these structures are consistent with our interpretation that Division I accumulated in a shallow marine trough. The varied lithologic suite represented by the fragments, which are especially schistose and granitic, implies a mixed provenance that must have been in part a crystalline terrane. The upper part of Division I contains rocks whose protoliths are interpreted to include felsic submarine ash flows and laminated mafic water-laid tuffs. These rocks are interlayered with rocks whose protoliths ranged from graywacke to arkose to arkosic quartzite. Volcanic protoliths are not as obvious in the lower unit as in the upper, but many of the fragments are derived from felsic volcanic rocks, and much of the material in the matrix could be similarly derived. Division I is relatively aluminous. It is not known whether this indicates an induction of a saprolitic regolith developed by weathering from the crystalline provenance or is an effect of allochemical metamorphism. Division I appears to owe its origin to submarine eruptive events in an environment in which relatively immature clastic rocks were being contributed. The accumulation of these rocks was probably interrupted by large-scale slumps that locally produced the chaotic structures observed. Indicators of paleocurrent directions have not been observed, and source areas cannot be designated. We speculate that the crystalline clastic material originated to the northwest and that the volcanic materials were contributed essentially in situ. Laminated volcanic, arkosic, and graywacke protoliths probably accumulated slowly under relatively quiet marine conditions that were periodically interrupted by emplacement of felsic ejecta flows and turbidity flows of arkose and graywacke. Thick clastic beds greater than 10 cm are mostly poorly sorted but here and there show grading. Reworking of the sediment during deposition was apparently minimal. Large-scale bedding interruptions are attributed to occasional slumping.

Divisions II and III

Pillow structure in units 1 and 2 of Division II and bedding characteristics of clastic and calcareous rock types throughout Divisions II and III imply accumulation under conditions that ranged, respectively, from volcanic arc, to abyssal, and to continental or island-arc rise environments.

Units 1 and 2 of Division II accumulated in situ from materials contributed from volcanism and intrusive rocks associated with the ultramafic complex (Table 1). The source of blue quartz in clastic layers within unit 1 probably is a quartz diorite facies of the complex (Boudette, 1970, p. C-9; Moench and Boudette, 1970, p. A-1-15). Most of the clastic material was probably contributed from subaerial weathering of emergent parts of the irruptive, as would occur in volcanic islands. One volcanic center (see above) is interpreted to be preserved within unit 1.

Units 3 and 4 of Division II accumulated in deep marine environments that varied from closed, poorly aerated basins that produced the sulfidic rocks of unit 3 to well-aerated continental or island rises and abyssal floors that produced unit 4. The rocks of units 3 and 4 are generally well sorted and contain an abundance of sedimentary structures such as sole casts and markings and climbing-ripple lamination. Turbidity currents that produced graded-bed sets were an imporant means of transport. Much of the material in units 3 and 4, however, appears to have been repeatedly reworked by bottom currents. Locally, quiescent conditions from time to time during the formation of units 3 and 4 allowed small lenses of limestone and chert to accumulate. The provenance of units 3 and 4 probably involved simultaneous contributions from the weathering of both volcanic and crystalline terranes (Boone, 1973).

The depositional environment of Division III, described by Moench (1969, p. L16-L17) and Harwood (1970, 1973), also is marine but transitional into a shallower environment. The lower part of Division III (Middle Ordovician rocks of Harwood and Berry [1968] and the Quimby Formation) maintains the essential character of metasedimentary rocks in Division II, and the lithology of the upper part (Greenvale Cove Formation) indicates a rapid change from quiet marine surroundings to a great influx of coarse clastic sediment marking the emergence of terrane now underlain by Divisions I and II (see also Moench, 1969; Harwood, 1970, 1973).

CORRELATION OF PRE-SILURIAN ROCKS

Division I has no close regional counterparts. Lithologically, it does resemble some of the rocks and structures of the Appalachians within the Fleur de Lys Supergroup of western Newfoundland (Kennedy, 1971), the Fish Brook Gneiss of eastern Massachusetts (D. Alvord, 1970, oral commun.), and bouldery gneiss near Quantico, Virginia (L. Pavlides, 1973, oral commun.). Unit I and the Fleur de Lys Supergroup apparently occupy the same relative stratigraphic position beneath Lower Ordovician rocks. We do not, however, imply that an age correlation exists, because Division I could be considerably older than the Fleur de Lys.

Rocks we assign to Division II are not dated paleontologically. Trace fossils tentatively ascribed to *Oldhamia*, which suggests a Cambrian(?) to Early Ordovician(?) age, occur in unit 4 (S. Serra and W. Forbes, 1972, oral commun.). Poorly preserved internal molds of fragmentary brachiopod shells have been found in a highly metamorphosed calc-silicate rock lens near the uppermost part of unit 4. Unit 1 is intruded by the Attean Quartz Monzonite (Table 1) provisionally dated at about 470 m.y. B.P. Rocks on strike with, and equivalent to units 3 and 4 near the Maine-New Hampshire boundary, are intruded by the Highlandcroft Plutonic Series, which is equavalent to the Attean Quartz Monzonite. This strongly suggests that the minimum age of Division II is Early Ordovician. We prefer this assignment of minimum age but recognize that a problem exists west of the region where the upper part of unit 4 and the Middle Ordovician rocks of Harwood and Berry (1968) are interpreted to be gradational (see also Harwood, 1973). Either unit 4 ranges in age from at least Early Ordovician to Middle Ordovician, or the contact southwest of Kennabago Lake between Divisions II and III is a fault (D. S. Harwood, 1973, written commun.) or an unconformity.

Even though northwest-striking faults offset the upper parts of the section, unit 4 can be traced from the east margin of the area shown in Figures 1 and 2 to the Maine-New Hampshire boundary, where Harwood (1973) and Green (1964) have correlated this unit with the Albee Formation of New Hampshire (Billings, 1956; see also Harwood and others, 1970). Unit 3 can likewise be traced to the southwestward-plunging axial portion of the Boundary Mountains anticlinorium, where Green (1964, 1968) and Harwood (1973) have named it the Dixville Formation (see also Harwood and others, 1970).

Harwood and Berry (1968; see also Harwood, 1973) extended the Dixville Formation southeastward across the strike belt composed of rocks that we assign to unit 4 (Fig. 2) to include the dated Middle Ordovician rocks of Division III. The basis of this correlation is entirely lithologic and is the method by which sulfidic metapelites are equated. There is a northwestern and southeastern belt of the Dixville Formation in the literature. We prefer, as indicated, to restrict correlations of Division II only within the northwestern belt. We do not concur with the correlation of Harwood and Berry (1968) for two reasons: (1) the euxinic metapelite of unit 3 and the sulfidic rocks in the lower part of Division III are not rigorously comparable on the basis of lithology and depositional structures; and (2) we believe that we have documented, by sedimentary structural criteria, that the stratigraphic sequence of unit 3 (northwestern part of the Dixville Formation) is older than unit 4 (Albee Formation). Therefore, if the correlation of Harwood and Berry (1968) is correct, then either (1) no part of unit 4 can be older than Middle Ordovician, and this is contrary to the dates obtained from rocks that intrude it that indicate an older age; or (2) the application of the criteria of sedimentary structures is invalid, and we have the sequence of units in Division II inverted. Harwood (1973, p. 11) reasoned independently that the maximum age of unit 4 is at least Early(?) Ordovician.

It is not possible at this time to resolve the disparity between the section along the Maine-New Hampshire boundary (Harwood and others, 1970) and the section in the Dead River basin region. A reconstruction involving overthrusting or sliding, northwest over southeast, possibly with subsequent folding and transcurrent faulting (Fig. 4, hypothesis B), offers a tectonic approach to the problem of correlation. We favor hypothesis B rather than A because we recognize that much of the internal stratigraphic order from the upper part of unit 1 to the top of 3 is similar to Green's lithic sequence (1964) in the Dixville Formation, and the Aziscohos Formation (Green, 1964) lithologically resembles the lower half of unit 4. Observed facings are available between all units in Division II in the Dead River basin; although contact relations are not so well exposed to the southwest, the few contacts mapped by Harwood (1973) are more compatible with solution B.

Resolution of this problem must involve recognition of subtle, formationwide differences between commonly recurrent eugeosynclinal nonfossiliferous lithic assemblages, particularly where affected by varied degrees of regional metamorphism.

INTERPRETATION

The stratigraphy for the pre-Silurian rocks of central western Maine is essentially similar to the sequence of rock types expected in a coupled ophiolite-flysch succession within a eugeosynclinal environment. The character of the ophiolitic rock embodies in many respects the criteria illustrated by Upadhyay and others (1971) and reviewed by Church (1972), and that of the flysch corresponds with the criteria established by Dzulynski and Walton (1965) and summarized with regard to unit 4 by Boone (1973). The rocks of the ultramafic complex (Table 1) combined with the metavolcanic rocks and iron-formation rock of unit 1 and the greenstone of unit 2 make up the ophiolitic suite. The other metasedimentary rocks of units 2, 3, and 4 constitute the flysch sequence. The four principal units of Division II define an orderly stratigraphic transition from the sedimentary and volcanic components of an ophiolitic sequence in the lower part, to a flysch sequence in the upper part.

Critical exposures located at azimuth 322° about 19 km from Stratton show the transition between ultramafic plutonic rocks at the base of the ophiolitic sequence and rocks of the Chain Lakes massif (Division I). Slight coarsening of muscovite in Division I near the contact, and protoclastic fabrics in these, as well as in

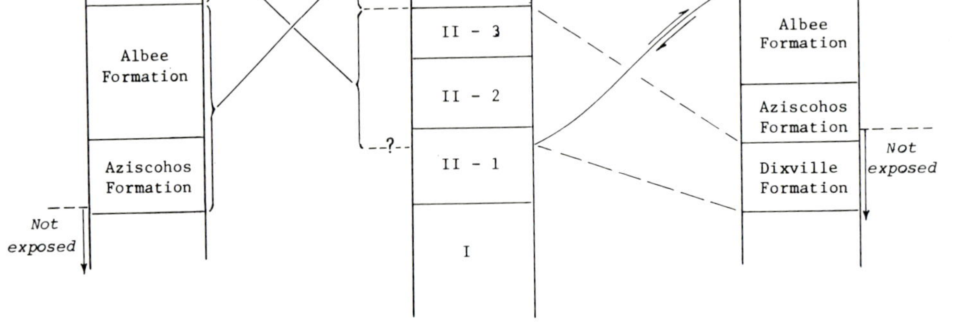

Figure 4. Diagram of two possible relations of rock units in the Dead River basin region to contiguous rock units in the vicinity of the Maine-New Hampshire boundary. Hypothesis A involves inverting the stratigraphic order of part of the section. Stratigraphic orders between the Aziscohos and Albee Formations (left column) and between units 2 and 3 (center column) are not changed in hypothesis A. Hypothesis B involves overthrusting.

the ultramafic rocks at the base of the ophiolitic sequence are observed. These criteria may indicate thrust dislocation, in present-day northwestward sense, of oceanic crust against Division I. The ophiolitic sequence is thin (~3 km), and a sheeted dike horizon has not been found. Apparently this ophiolite represents an accumulation that was distal to a spreading zone. This accumulation probably occurred in close proximity to a continental rise or island-arc trench floored by rocks of Division I.

We interpret the units of Division II to have accumulated conformably upon the lower (intrusive) part of the ophiolitic sequence during and after this sequence was obducted onto an older sialic crustal block composing Division I. The contact between the ophiolitic rocks and Division I (but not necessarily between those of the ophiolite and Division II) may therefore be regarded as a tectonic surface that nonetheless expresses the essential aspects of intrusion and unconformity. In tectonic and metamorphic history, and probably age, the discordance across this contact between Divisions I and II is profound.

ACKNOWLEDGMENTS

We thank J. B. Lyons, L. R. Page, J. H. Peck, S. Serra, and D. S. Harwood for comments and suggestions that have materially improved the presentation. We acknowledge map information on the central part of the pre-Silurian outcrop belt north of Flagstaff Lake, generalized from the work of S. Serra (1973, written commun.); on the Moose River synclinorium from A. J. Boucot (1961); and on the Stratton area (A. Griscom, unpub. map). G. M. Boone's field program is supported by the Maine Geological Survey.

REFERENCES CITED

Billings, M. P., 1956, The geology of New Hampshire; Part II, bedrock geology: Concord, N.H., New Hampshire State Planning and Devel. Comm., 203 p.

Boone, G. M., 1970, Pre-Silurian flysch relict structures in cordierite-K-feldspar granofelses, Long Falls of Dead River, Somerset County, western Maine, *in* Boone, G. M., ed., New England Intercollegiate Geol. Conf. guidebook for field trips in the Rangeley Lakes-Dead River basin region, western Maine: Augusta, M., Kennebec Jour. Press, p. B-1, 1-16.

——1973, Metamorphic stratigraphy, petrology, and structural geology in the Little Bigelow Mountain map area, western Maine: Maine Geol. Survey Bull. 24, 136 P.

Boone, G. M., Boudette, E. L., and Moench, R. H., 1970, Bedrock geology of the Rangeley Lakes-Dead River basin region, western Maine, *in* Boone, G. M., ed., New England Intercollegiate Geol. Conf. guidebook for field trips in the Rangeley Lakes-Dead River basin region, western Maine: Augusta, M., Kennebec Jour. Press, p. 1-24.

Boucot, A. J., 1961, Stratigraphy of the Moose River synclinorium, Maine: U.S. Geol. Survey Bull. 1111-E, p. 153-188.

Boudette, E. L., 1970, Pre-Silurian rocks in the Boundary Mountains anticlinorium, northwestern Maine, *in* Boone, G. M., ed., New England Intercollegiate Geol. Conf. guidebook for field trips in the Rangeley Lakes-Dead River basin region, western Maine: Augusta, M., Kennebec Jour. Press, p. C-1-21.

Church, W. R., 1972, Ophiolite—Its definition, origin as oceanic crust, and mode of emplacement in orogenic belts, with special reference to the Appalachians, *in* Irving, E., ed., The ancient oceanic lithosphere: Canada Dept. Energy, Mines and Resources Earth Phys. Br. Pub., v. 42, p. 71-85.

Dzulynski, Stanislaw, and Walton, E. K., 1965, Sedimentary features of flysch and graywackes:

Amsterdam, Elsevier Pub. Co., 274 p.

Faul, Henry, Stern, T. W., Thomas, H. H., and Elmore, P.L.D., 1963, Ages of intrusion and metamorphism in the northern Appalachians: Am. Jour. Sci., v. 261, p. 1-19.

Green, J. C., 1964, Stratigraphy and structure of the Boundary Mountain anticlinorium in the Errol quadrangle, New Hampshire-Maine: Geol. Soc. America Spec. Paper 77, 78 p.

——1968, Geology of the Connecticut Lakes-Parmachenee area, New Hampshire and Maine: Geol. Soc. America Bull., v. 79, p. 1601-1638.

Harwood, D. S., 1970, Nature of the Taconic orogeny in the Cupsuptic quadrangle, west-central Maine, *in* Boone, G. M., ed., New England Intercollegiate Geol. Conf. guidebook for field trips in the Rangeley Lakes-Dead River basin region, western Maine: Augusta, M., Kennebec Jour. Press, p. H-1-17.

——1973, Bedrock geology of the Cupsuptic and Arnold Pond quadrangles, west-central Maine: U.S. Geol. Survey Bull. 1346, 90 p.

Harwood, D. S., and Berry, W.B.N., 1967, Fossiliferous lower Paleozoic rocks in the Cupsuptic quadrangle, west-central Maine: U.S. Geol. Survey Prof. Paper 575-D, p. D16-D23.

Harwood, D. S., Green, J. C., and Guidotti, C. V., 1970, Geology of the lower Paleozoic rocks in the Boundary Mountain anticlinorium, *in* Boone, G. M., ed., New England Intercollegiate Geol. Conf. guidebook for field trips in the Rangeley Lakes-Dead River basin region, western Maine: Augusta, M., Kennebec Jour. Press, p. A-3, 1-19.

Hatch, N. L., Jr., 1963, The geology of the Dixville quadrangle, New Hampshire: New Hampshire Dept. Resources and Econ. Devel. Bull. 1, 81 p.

Kennedy, M. J., 1971, Structure and stratigraphy of the Fleur de Lys Supergroup in the Fleur de Lys area, Burlington peninsula, Newfoundland: Geol. Assoc. Canada Proc., v. 24, p. 59-71.

Lyons, J. B., and Faul, Henry, 1968, Isotope geochronology of the northern Appalachians, *in* Zen, E-an, White, W. S., Hadley, J. B., and Thompson, J. B., Jr., eds., Studies of Appalachian geology—Northern and maritime: New York, Interscience Pubs., p. 305-318.

Moench, R. H., 1969, The Quimby and Greenvale Cove Formations in western Maine: U.S. Geol. Survey Bull. 1274-L, 17 p.

——1971, Geologic map of the Rangeley and Phillips quadrangles, Franklin and Oxford Counties, Maine: U.S. Geol. Survey Misc. Geol. Inv. Map 1-605.

Moench, R. H., and Boudette, E. L., 1970, Stratigraphy of the northwest limb of the Merrimack synclinorium in the Kennebago Lake, Rangeley, and Phillips quadrangles, western Maine, *in* Boone, G. M., ed., New England Intercollegiate Geol. Conf. guidebook for field trips in the Rangeley Lakes-Dead River basin region, western Maine: Augusta, Me., Kennebec Jour. Press, p. A-1, 1-25.

Naylor, R. S., Boone, G. M., Boudette, E. L., Ashenden, D. D., and Robinson, Peter, 1973, Pre-Ordovician rocks in the Bronson Hill and Boundary Mountains anticlinorium, New England, U.S.A. [abs.]: EOS (Am. Geophys. Union Trans.), v. 54, p. 495.

Pettijohn, F. J., and Potter, P. E., 1964, Atlas and glossary of primary sedimentary structures: New York, Springer-Verlag, Inc., 370 p.

Upadhyay, H. D., Dewey, J. F., and Neale, E.R.W., 1971, The Betts Cove ophiolite complex, Newfoundland—Appalachian oceanic crust and mantle: Geol. Assoc. Canada Proc., v. 24, p. 27-34.

MANUSCRIPT RECEIVED BY THE SOCIETY APRIL 18, 1974
REVISED MANUSCRIPT RECEIVED FEBRUARY 21, 1975
MANUSCRIPT ACCEPTED MARCH 17, 1975

Printed in U.S.A.

Geological Society of America
Memoir 148

Stratigraphy of the Belfast Quadrangle, Maine

Charles E. Bickel
Department of Geology
San Francisco State University
San Francisco, California 94132

ABSTRACT

The Belfast quadrangle is located west of Penobscot Bay in east-central Maine. The stratified rocks of the quadrangle have been regionally metamorphosed, multiply folded, and intruded by plutonic rocks. They are exposed in two major structural blocks separated by the St. George longitudinal fault.

The oldest rocks in the Passagassawakeag block, the Passagassawakeag Formation, are Precambrian(?) in age and are exposed in the Liberty-Orrington anticline. These rocks consist primarily of quartz-plagioclase-biotite gneiss; minor amphibolite and gneiss with sillimanite and muscovite are present. A calc-silicate member and a fine-grained gneiss unit composed almost entirely of quartz, oligoclase, and microcline (the metavolcanic Mixer Pond Member) are mapped separately. Unconformably overlying the Passagassawakeag Formation northwest of the Liberty-Orrington anticline is the Hogback Schist of probable Cambrian or Ordovician age. Aluminous schist, micaceous quartz-feldspar gneiss, biotite gneiss, and calc-silicate gneiss are included in the Hogback Schist. Southeast of the Liberty-Orrington anticline, the Bucksport Formation of probable Late Silurian or Early Devonian age is interpreted to overlie unconformably the Passagassawakeag Formation. The Bucksport Formation is composed of calc-silicate gneiss, biotite gneiss, and minor biotite schist. It is overlain conformably by the Appleton Ridge Formation of probable Early Devonian age. The Appleton Ridge Formation consists of phyllite or staurolite schist cyclically interbedded with quartzite; at the top of the exposed sequence there is schist rich in garnet and cummingtonite (Muzzy Ridge Member). The stratigraphic section of the Passagassawakeag block is bounded on the southeast by the St. George fault.

The oldest rocks exposed in the Penobscot block that lies southeast of the St. George fault are marble and other poorly exposed rocks found in the core of the Coleman antiform. They are correlated with the Rockport sequence of the Camden-Rockland area. Also exposed in the Coleman antiform are quartzite, quartz-mica schist, retrograded sillimanite gneiss, feldspar-quartz-biotite gneiss, and calc-silicate quartzite correlated with the upper part of the Battie Formation in the Camden area. The age of these rocks is believed to be Cambrian or Ordovician. The Penobscot Formation of Ordovician age overlies the Battie Formation and constitutes most of the Penobscot block. The Penobscot Formation consists dominantly of phyllite, andalusite schist, sillimanite gneiss and granofels, quartz-mica schist, and quartzite; most of these rocks are sulfidic. There are rare marble, calc-silicate gneiss, amphibolite, garnet-cummingtonite schist, and plagioclase-quartz granofels of volcanic origin. Metamorphosed volcanic rocks of mafic and felsic composition (Gushee Member) are largely restricted to the upper part of the formation.

The Jam Brook Formation is exposed between the Penobscot Formation and the St. George fault but thins and terminates along strike to the northeast. The Jam Brook contains quartz-pebble metaconglomerate, quartzite, quartz-mica schist, andalusite schist, calc-silicate gneiss, and marble, in generalized order of appearance from east to west. There are changes of facies within the formation. The age of the Jam Brook is believed to be Cambrian or Ordovician. The Jam Brook probably correlates with the Rockport sequence, and its contact with the Penobscot Formation is inferred to be a thrust fault.

INTRODUCTION

C. H. Hitchcock (1863, p. 227) once wrote of the region west of Penobscot Bay, "I may say truly that I never before found a region where it was more difficult to ascertain the true position of the strata than here." The Belfast 15-minute quadrangle (Fig. 1) is located in this belt of metamorphic rocks that until recently had been studied only by reconnaissance. I spent 40 weeks mapping the quadrangle during the summers of 1967, 1968, 1969, and 1970 (Bickel, 1971). Although there has been much recent geologic study of the region around Penobscot Bay (Boucot and others, 1972; Brookins and others, 1973; Osberg and Guidotti, 1974; Stewart, 1974; Stewart and Wones, 1974), those most familiar with the geology of east-central Maine do not yet agree as to the correlation, age, or even the direction of the stratigraphic tops (topping sense) of rocks on strike with the rocks described herein. For this reason and because no identifiable fossils were found in the Belfast quadrangle, descriptions of the rocks are separated from and emphasized at the expense of speculations as to their correlation and ages.

STRATIGRAPHY

The stratified rocks of the Belfast quadrangle belong to two major structural blocks: the Passagassawakeag block that lies northwest of the St. George longitudinal fault and the Penobscot block that lies southeast of the fault (Fig. 2).

The stratified rocks have been regionally metamorphosed at low pressure and over a wide range of temperature. A northeastward-striking tract of rocks in the garnet zone crosses the central part of the quadrangle. The metamorphic grade rises to the northwest and southeast. Staurolite is present in the andalusite zone only northwest of the garnet zone, in the Appleton Ridge Formation. Andalusite is abundant, and the sillimanite is rare and fibrolitic in the lower sillimanite zone; andalusite is rare, and (in general) coarse sillimanite first appears in the upper sillimanite zone. The highest grade rocks of the area belong to the sillimanite-K-feldspar zone. The metamorphic grade drops from the sillimanite-K-feldspar zone to the andalusite zone in the northwestern and southeastern corners of the quadrangle.

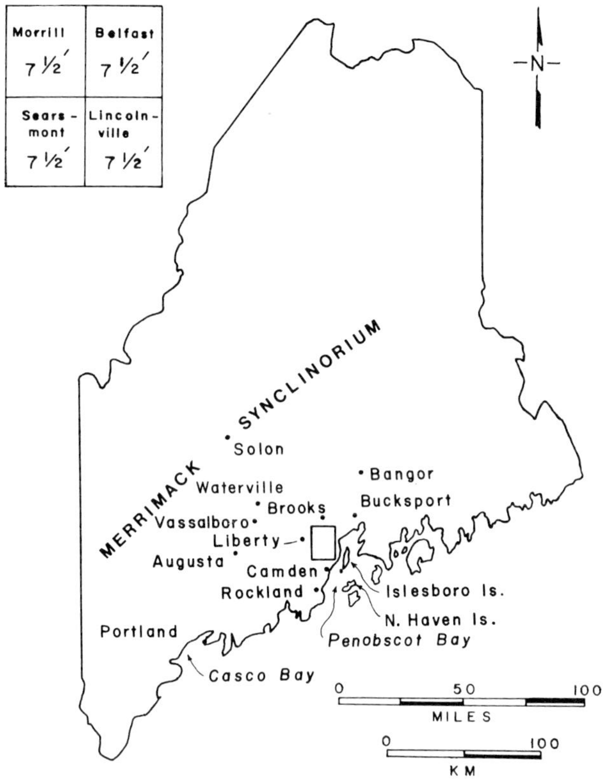

Figure 1. Location of the Belfast 15-minute quadrangle, Maine.

In most of the area southeast of the garnet zone, much of the biotite was replaced by chlorite, and much of the andalusite or sillimanite was replaced by sericite during retrograde metamorphism. The mineralogy and texture of the retrograded rocks indicate that partial re-equilibration occurred under conditions of garnet- or biotite-grade metamorphism after the termination of folding and faulting (Bickel, 1974).

Passagassawakeag Block

The Passagassawakeag Formation of probable Precambrian age occupies the core of the northeast-trending Liberty-Orrington anticline (Osberg, 1974). The Passagassawakeag Formation is overlain unconformably on the northwest by the Hogback Schist of probable Cambrian or Ordovician age and on the southeast by the Bucksport Formation of probable Late Silurian or Early Devonian age. The Appleton Ridge Formation of probable Early Devonian age conformably overlies the Bucksport Formation and is bounded on the southeast by the St. George fault.

Passagassawakeag Formation. The name "Passagassawakeag Formation" is here proposed for gneissic rocks of probable Precambrian age (Stewart and Wones, 1974) that crop out at the headwaters of the Passagassawakeag River in the north-central part of the Belfast quadrangle. This formation was named the Knox Gneiss by Perkins and Smith (1925), but "Knox" had already been assigned as a stratigraphic name (Keroher and others, 1966). Nearly continuous exposures in Wescot Stream are designated as the type area of the formation. The topping sense of these rocks is not known. The average thickness of the Passagassawakeag

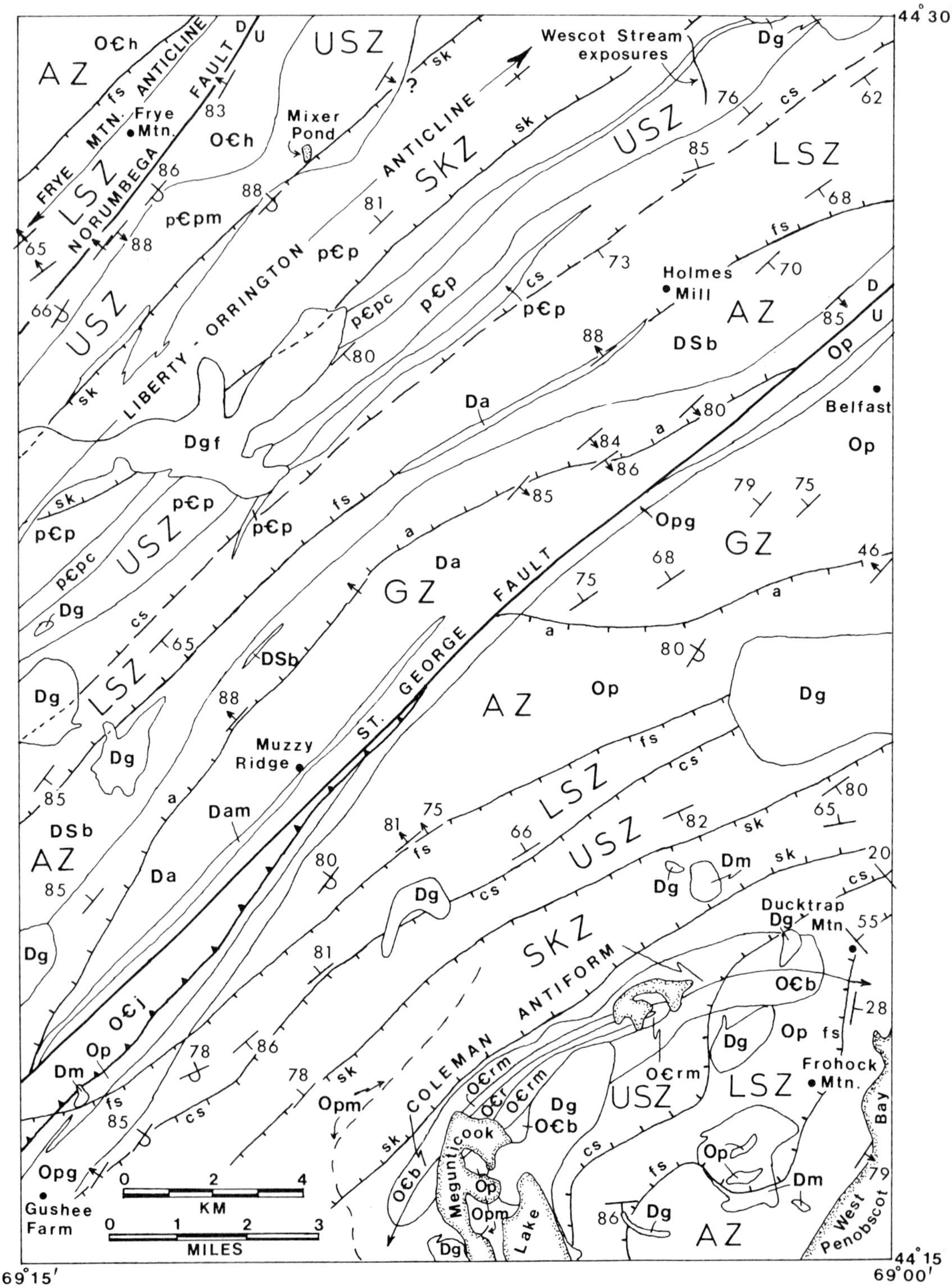

Figure 2. Geologic map of the Belfast quadrangle, Maine. Explanation on facing page.

INTRUSIVE ROCKS

Dg

Dg — Unfoliated or weakly foliated binary granitic rocks

Dgf

Dgf — Foliated binary granitic rocks

Dm

Dm — Metamorphosed mafic and ultramafic rocks

STRATIFIED ROCKS

PASSAGASSAWAKEAG BLOCK

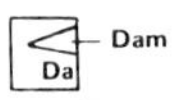

Appleton Ridge Formation

Dam — *Muzzy Ridge Member.* Cummingtonite-garnet schist with or without hornblende

Da — *Garnet zone:* quartz-muscovite-chlorite phyllite with minor garnet
Andalusite zone: staurolite schist and staurolite-andalusite schist interbedded with quartzite

DSb

Bucksport Formation

DSb — Calc-silicate gneiss interbedded with biotite gneiss and minor biotite schist

O€h

Hogback Schist

O€h — Aluminous schist; micaceous quartz-feldspar gneiss with partings and beds of mica schist; biotite gneiss; calc-silicate gneiss

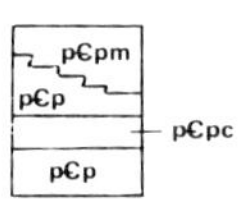

Passagassawakeag Formation

p€p — Quartz-plagioclase-biotite gneiss with granulose and schistose layers; augen gneiss; minor gneiss with sillimanite and muscovite; calc-silicate gneiss, and amphibolite

p€pm — *Mixer Pond Member.* Fine-grained quartz-oligoclase-microcline gneiss with minor magnetite, biotite, hornblende, or ferrosalite

p€pc — *Calc-Silicate Member.* Calc-silicate gneiss and biotite gneiss

PENOBSCOT BLOCK

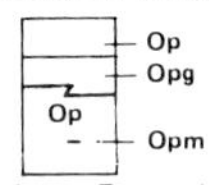

Penobscot Formation

Opg — *Gushee Member:* sulfidic anthophyllite-cordierite rock; amphibolite with relic scoria and pillows; albite-quartz granofels with relic phenocrysts of quartz and albite; minor mica schist

Op — *Garnet zone:* Quartz-muscovite-chlorite phyllite interbedded with quartzite; in part sulfidic
Andalusite zone and lower sillimanite zone: Andalusite schist interbedded with quartz-mica schist and quartzite; massive andalusite schist; in part sulfidic
Upper sillimanite zone and sillimanite-K-feldspar zone: sillimanite gneiss and granofels interbedded with quartzite; massive sillimanite gneiss and granofels; in part sulfidic
Andalusite through sillimanite-K-feldspar zones: Minor amounts of calc-silicate gneiss in beds and pods, garnet-cummingtonite schist, amphibolite, and plagioclase-quartz granofels with or without phenocrysts of quartz

Opm — *Marble lentil:* Calcite and dolomite marble

O€b

Battie Formation (?)

O€b — Quartzite, quartz-mica schist, retrograded sillimanite gneiss, feldspar-quartz-biotite gneiss, and calc-silicate quartzite

O€rm
O€r

Rockport Sequence

O€rm — Calcite and dolomite marble

O€r — Biotite schist, quartzite, and quartz-feldspar-biotite gneiss

O€j

Jam Brook Formation

O€j — Quartz-mica schist, calc-silicate gneiss, quartz-pebble metaconglomerate, quartzite, marble, andalusite schist, and fine-grained plagioclase-quartz granofels with relic phenocrysts of quartz and albite

SYMBOLS THAT INDICATE METAMORPHIC GRADE

The isograds of regional metamorphism are hachured on the high-intensity side; long dashes indicate that there is very poor control on the location of the isograd; short dashes indicate that it is not certain whether the isograd cuts a granitic body

The isograds are labeled to indicate first appearance of a mineral or mineral pair:

- a — andalusite
- fs — fibrolitic sillimanite
- cs — coarse sillimanite
- sk — sillimanite-K-feldspar

The zones of regional metamorphism are labeled:

- GZ — Garnet Zone
- AZ — Andalusite Zone
- LSZ — Lower Sillimanite Zone
- USZ — Upper Sillimanite Zone
- SKZ — Sillimanite-K-Feldspar Zone

STRUCTURAL SYMBOLS

Contact

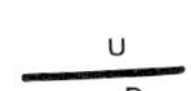

High-angle fault

Inferred thrust fault

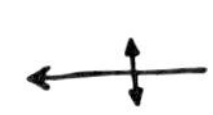

Plunging anticline

Inclined and vertical bedding; topping not determined

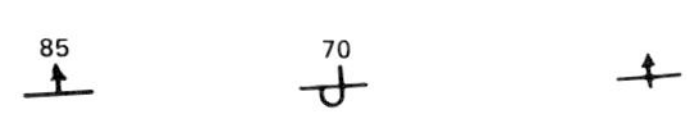

Inclined, vertical, and overturned bedding; topping based on primary features

Formation exposed in the mapped area is estimated to be about 2,300 m (7,600 ft).

Most of the Passagassawakeag Formation is gneiss composed of plagioclase, quartz, and biotite in alternating granulose and schistose layers. Strongly schistose layers are less abundant than layers in which granular minerals predominate. The thickness of individual layers ranges from about 1 cm to 3 m. Some of the layering is interpreted to be bedding, but much of it displays a pinch-and-swell structure, is discordant, and appears to have been produced during recrystallization, granulation, and shearing of the rock.

The biotite gneiss is purplish gray to gray and is weakly schistose (Table 1,

TABLE 1. ESTIMATED MODES OF THE PASSAGASSAWAKEAG FORMATION

	Sample number						
	1	2	3	4	5	6	7
Quartz	43	30	33	23	25	5	6
Plagioclase	40	45	46	56	62	45	77
Microcline	x	x	4	4	5	24	x
Muscovite	. .	1	3	. .	. .	. .	. .
Biotite	14	23	13	10	5	2	. .
Chlorite*	x	. .	x	x	. .	x	. .
Garnet	3	x	. .	. .	. .	. .	. .
Sillimanite	. .	. .	1	. .	. .	. .	. .
Cummingtonite†	. .	. .	. .	. .	. .	4	. .
Hornblende	. .	. .	. .	. .	. .	14	4
Calcic amphibole	. .	. .	. .	6	. .	. .	. .
Ferrosalite	. .	. .	. .	. .	. .	. .	10
Epidote	. .	. .	. .	. .	1	3	x
Sphene	x	. .	. .	x	1	2	3
Opaque minerals:							
Oxide(s)	x	1	x	. .	1	1	. .
Sulfide(s)	x	x	. .	x	. .	. .	. .
Graphite	. .	. .	. .	. .	. .	. .	x
Other accessory minerals	x	x	x	1	x	x	x
Approximate An content of plagioclase	28	40	28	40	23	22	25
Grain size	0.2-1.5	0.1-3.0	0.1-1.0	0.2-3.0	0.1-0.8	0.1-0.4	0.1-1.0

Note: Modes are given in percent; *x* indicates mode less than 1%. Grain sizes are given in millimetres. Description and location of samples (see Fig. 1 for locations of the four 7-1/2-minute quadrangles that compose the Belfast 15-minute quadrangle):

1. Gray biotite gneiss. Belfast 7-1/2-minute quadrangle; Route 7, 7,500 ft southeast of Evans Corner.
2. Dark-purplish-gray biotite schist. Belfast 7-1/2-minute quadrangle; Route 131, 600 ft southwest of Evans Corner.
3. Gray micaceous quartz-feldspar gneiss. Morrill 7-1/2-minute quadrangle; summit of hill 602, southwest of Morey Hill.
4. Calc-Silicate Member; gray calc-silicate gneiss. Morrill 7-1/2-minute quadrangle; summit of hill 284, southwest of Whitcomb Road.
5. Mixer Pond Member; light-gray to yellowish-gray magnetitic biotite gneiss. Morrill 7-1/2-minute quadrangle; summit of hill 534, southwest of Halfmoon Pond.
6. Mixer Pond Member; greenish-gray hornblende gneiss. Morrill 7-1/2-minute quadrangle; 440-ft elevation on road east of Thompson Brook and north of Center Montville Road.
7. Mixer Pond Member; gray to white ferrosalite-hornblende gneiss. Morrill 7-1/2-minute quadrangle; jeep trail junction 380, north of Mixer Pond.

*Retrograde.
†Not present in layers with microcline.

col. 1). Some of the gneiss contains garnet, and commonly graphite is present in trace amounts. The biotite schist contains the same minerals but has a higher proportion of biotite (Table 1, col. 2). Megacrysts of plagioclase as much as 2 cm in diameter or augen of granular plagioclase and quartz occur in the granulose layers but are most characteristic of the strongly schistose layers. Coarse- and fine-grained varieties of gneiss occur together in the same outcrops.

Gneiss with minor amounts of fresh or replaced coarse-grained or fibrolitic sillimanite makes up about 10 percent of the Passagassawakeag Formation (Table 1, col. 3). In the upper sillimanite zone, the biotite is reddish brown, and microcline is not found with sillimanite and muscovite. In the sillimanite-K-feldspar zone, the biotite is greenish brown; microcline occurs with sillimanite and muscovite but is most abundant in the granulose laminae rich in quartz and plagioclase.

Discontinuous, thin layers of calc-silicate gneiss, amphibolite, and fine-grained, leucocratic magnetitic biotite gneiss similar to rocks of the Mixer Pond Member are present throughout the formation. Numerous small bodies of granitic rock intrude the gneisses of the Passagassawakeag Formation. Thin pegmatitic veins of coarse-grained quartz and plagioclase have a pinch-and-swell structure and are locally concordant or discordant with the schistosity.

Calc-Silicate Member. The Calc-silicate member (Table 1, col. 4) of the Passagassawakeag Formation ranges from about 30 to 150 m (100 to 500 ft) in thickness. In part, the member consists of purplish-gray biotite gneiss interbedded irregularly with greenish-gray, diopside- and amphibole-bearing calc-silicate gneiss that makes up about 30 percent of the rock. More typically, the rock is gray gneiss with biotite and calcic amphibole (amphibole with composition intermediate between members of the tremolite-actinolite series and the hornblende series). This rock contains layers 1 to 5 mm thick and enriched in biotite or calcic amphibole that alternate with layers of quartz and feldspar. Part of the gneiss contains very little amphibole, and within the Calc-silicate member there are a few outcrops without any calc-silicate minerals. Feldspathic rocks with as much as three percent graphite or containing seams of sillimanite, massive rock composed almost wholly of calcic amphibole, and marble are rare constituents of the member.

Mixer Pond Member. The northwesternmost portion of the Passagassawakeag Formation is fine-grained gneiss mapped as the Mixer Pond Member. The member is here named for Mixer Pond, near which most of its constituent rocks are exposed. The thickness of this unit is estimated to average about 300 m (1,000 ft).

The gneiss of the Mixer Pond Member is particularly feldspathic (Table 1, cols. 5, 6, and 7). These light-gray to white rocks are massive or indistinctly layered on a scale of a few metres. The compositional layering, variable mafic mineralogy, and uniform, fine grain size of these rocks distinguish them from the granitic rocks of the area.

Hornblende, green biotite, or, less commonly, ferrosalite is concentrated in laminae about 1 mm thick in about 40 percent of the gneiss. Magnetite, with or without green biotite, is the mafic mineral in the remainder of the gneiss. In some of the gneiss, the felsic minerals are segregated into layers from 0.2 to 1 mm in thickness that are rich in quartz, oligoclase, or microcline.

Contact Relationships with the Hogback Schist and the Bucksport Formation. The contact between the Mixer Pond Member of the Passagassawakeag Formation and the Hogback Schist can be located to within a few metres, but the contact itself is not exposed. The strike of the lithologic layering in the Mixer Pond Member is extremely variable and is not parallel to the contact. Bedding in the Hogback Schist is concordant with the contact. My mapping in the Brooks quadrangle (north of the Belfast quadrangle) indicates that the Mixer Pond Member of the Passagassawakeag Formation terminates along strike to the northeast. Farther northeast, the contact between the Passagassawakeag Formation and the Hogback Schist is intersected by the Norumbega fault. In the Bucksport quadrangle (Stewart and Wones, 1974), gneiss of the Passagassawakeag Formation at least in the upper sillimanite zone is separated by the Norumbega fault from rocks in the chlorite zone that may correlate with the Hogback Schist. These relationships suggest that the contact in the Belfast quadrangle is an unconformity and that in this part of Maine there was a high-grade metamorphic event before deposition of the Hogback Schist and a second metamorphism of variable intensity after deposition of the Hogback Schist. Possibly the contact in the Belfast quadrangle is a fault along which movement occurred between these two metamorphic events.

In the Belfast quadrangle, biotite gneiss accompanied by calc-silicate gneiss is found in both the Passagassawakeag Formation and the Bucksport Formation near their contact. Therefore, this contact is less easy to locate than that between the Mixer Pond Member and the Hogback Schist. Location of the contact is based on the fact that more calc-silicate gneiss is present in the Bucksport Formation than in the Passagassawakeag Formation, and calcium-free minerals rich in aluminum are restricted to the Passagassawakeag Formation. In Figure 2 the contact is drawn so that all outcrops included in the Bucksport Formation contain a minimum of 5 to 10 percent calc-silicate gneiss and none of them contain garnet, muscovite, or sillimanite. In the Bucksport quadrangle, this contact marks a discontinuity between the sillimanite-grade rocks of the Passagassawakeag and biotite-grade rocks of the Bucksport Formation. Here, the contact is interpreted to be an unconformity. Stewart and Wones (1974) interpreted the contact to be a fault.

Hogback Schist. The Hogback Schist of probable Cambrian or Ordovician age underlies the Frye Mountain Range in the northwestern corner of the Belfast quadrangle. The formation was named by Perkins and Smith (1925) for Hogback Mountain, about a mile to the west; the Hogback Schist is better exposed in the Frye Mountain Range, which is here redefined as the type area of the formation. The thickness of the formation exposed in the Belfast quadrangle is estimated to be about 1,700 m (5,600 ft). Its upper contact is not exposed, and an unknown thickness is not exposed because of faulting.

The Hogback Schist includes four distinct types of rocks: aluminous schist, micaceous quartz-feldspar gneiss, biotite gneiss, and calc-silicate gneiss. These rocks are interbedded, but their proportions differ greatly across the strike of the formation. Figure 3 shows map units of (1) micaceous quartz-feldspar gneiss with interbedded mica schist (O€hg), (2) mica schist and andalusite schist (O€hs), (3) biotite gneiss with minor calc-silicate gneiss and mica schist (O€hb), and (4) calc-silicate gneiss (O€hc). Separate map units of micaceous quartz-feldspar gneiss, of schist, and of calc-silicate gneiss are distinguished in Figure 3 with numerical subscripts. The northwesternmost map units are poorly exposed.

About 35 percent of the exposed Hogback Schist is micaceous quartz-feldspar gneiss with interbedded mica schist (O€$hg_{1,2}$). Its bedding style ranges from massive quartz-feldspar gneiss with irregularly spaced schistose partings and sparce beds of mica schist to sets of graded gneiss-schist beds in which schist predominates. These graded beds commonly range from 0.2 to 0.6 m in thickness, but thinner beds are present. The micaceous quartz-feldspar gneiss is typically light gray on both fresh and weathered surfaces. The mica appears as lustrous patches on surfaces parallel to the schistosity and as thin "pin stripes" on surfaces perpendicular to the schistosity. The gneiss consists mostly of quartz and oligoclase (Table 2, col. 1). Most of the mica schist interbedded with the micaceous quartz-feldspar gneiss contains garnet but not andalusite.

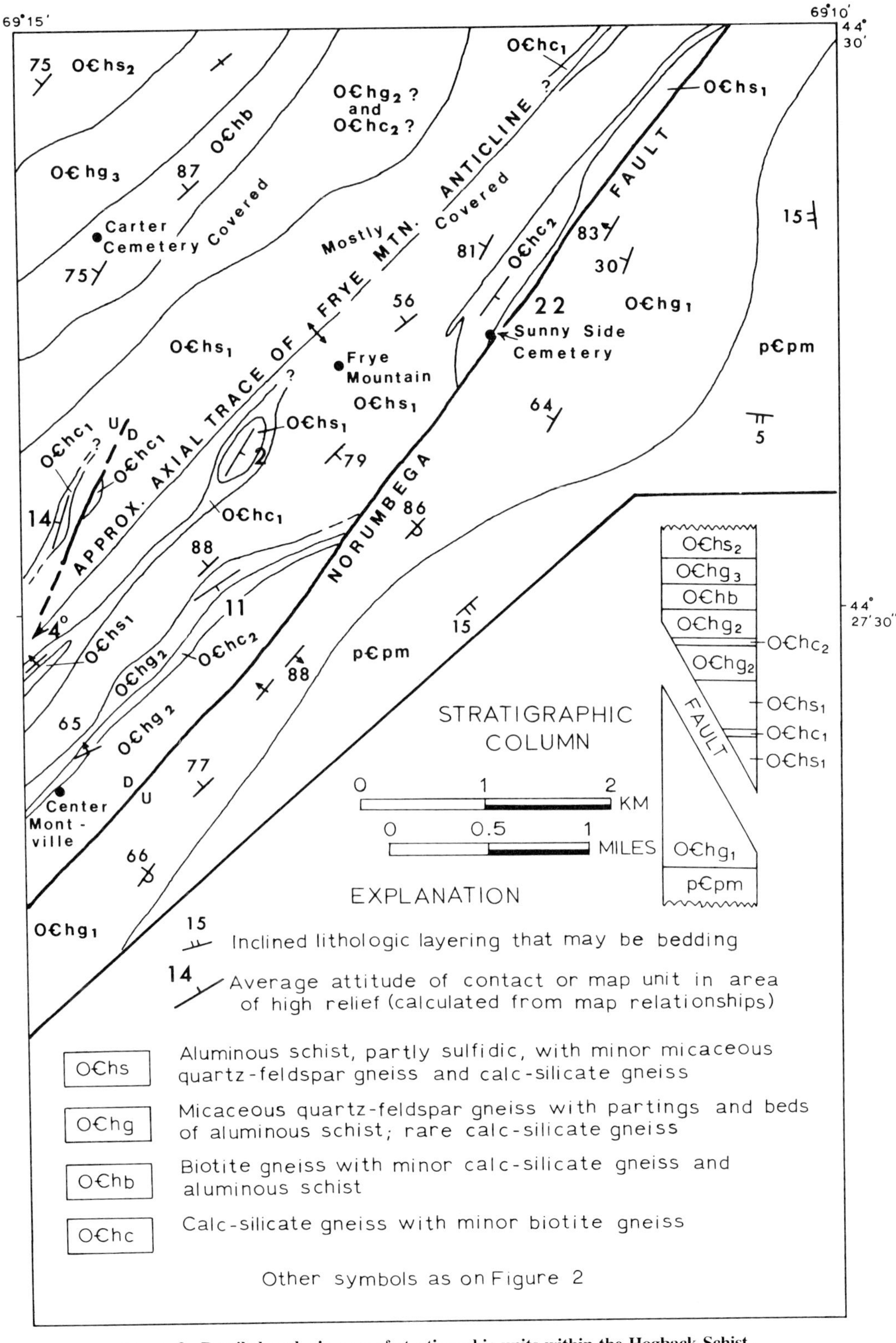

Figure 3. Detailed geologic map of stratigraphic units within the Hogback Schist.

Gray mica schist and bluish-gray andalusite schist (O€$hs_{1,2}$) make up about 50 percent of the Hogback Schist exposed in the Belfast quadrangle. Most of these schists weather rusty; the others weather gray. Commonly, mica schist and andalusite schist are interlayered in beds from 3 to 10 cm in thickness. The mode of an andalusite schist sample is given in Table 2 (col. 2). Beds of micaceous quartz-feldspar gneiss and calc-silicate gneiss are rare in these map units.

Biotite gneiss (O€hb) containing minor calc-silicate gneiss and mica schist makes up about 10 percent of the Hogback Schist. The biotite gneiss is very fine grained, weakly schistose, bluish-gray, and weathers chalky (Table 2, col. 3). Most outcrops exhibit distinct beds 2 to 4 cm thick. Near its northwestern contact, the map unit contains beds of mica schist 0.2 to 1.2 m thick; most are from 0.3 to 0.6 m thick. The mica schist makes up about 5 percent of the map unit.

Mappable units of calc-silicate gneiss (O€$hc_{1,2}$) make up about 5 percent of

TABLE 2. ESTIMATED MODES OF THE HOGBACK SCHIST

	Sample number			
	1	2	3	4
Quartz	65	25	35	30
Plagioclase	23	15	30	24
Microcline	3	. .	. .	15
Muscovite	2	10	. .	. .
Biotite	5	30	35	x
Chlorite*	x	4	x	. .
Garnet	1	7	. .	. .
Andalusite	. .	6†	. .	. .
Sillimanite	. .	1	. .	. .
Calcic amphibole	. .	. .	. .	20†
Diopside	. .	. .	. .	4†
Clinozoisite	. .	. .	. .	4
Sphene	. .	. .	. .	3
Tourmaline	x	1	x	. .
Opaque minerals:				
Oxide(s)	x	1	. .	x
Sulfide(s)	. .	. .§	. .	. .
Other accessory minerals:	1	x	x	x
Approximate An content of plagioclase	24	20	20	40
Grain size:				
Groundmass	0.2–0.4	0.05–0.6	0.03–0.05	0.03–0.2
Porphyroblasts	. .	1.0–3.0	. .	0.1–0.4

Note: Modes are given in percent; x indicates mode less than 1%. Grain sizes are given in millimetres. Description and location of samples:

1. Light-gray micaceous quartz-feldspar gneiss. Morrill 7-1/2-minute quadrangle; 1,290 ft S20°W from southwesternmost corner of Frye Mountain State Game Management Area.
2. Bluish-gray andalusite schist. Morrill 7-1/2-minute quadrangle; 120 ft north-northeast of Game Management Boundary Corner 1.2 mi northeast of Frye Mountain.
3. Dark-purplish-gray, schistose biotite gneiss. Morrill 7-1/2-minute quadrangle; 150 ft east of summit 895, southeast of Walker Ridge Road.
4. Laminated, purplish-gray to olive-green calc-silicate gneiss. Morrill 7-1/2-minute quadrangle; 691-ft elevation on Game Management Boundary, at extreme western edge of the quadrangle, north of Bartlett Stream.

*Retrograde.
†Porphyroblastic.
§Sulfide and graphite are abundant in andalusite schist and mica schist not represented on this table.

the Hogback Schist; they are separated on Figure 3 within unit O€hs$_1$ of schist and unit O€hg$_2$ of micaceous quartz-feldspar gneiss. The true thickness of the map units of calc-silicate gneiss is nowhere more than a few tens of metres, although the width of exposure of unit O€hc$_2$ is about 300 m along the dip slopes north of Sunny Side Cemetery.

The calc-silicate gneisses are greenish gray, bluish gray, or light gray, and are dominantly gray weathering; those with abundant biotite are light purplish gray. Generally, the beds range from 1 to 10 cm in thickness; some are as much as 20 cm thick.

Table 2 (col. 4) gives the mode of a typical calc-silicate gneiss sample. Some of these rocks contain considerable diopside, biotite, or chlorite. Many display discontinuous laminae that range from 0.5 to 3 mm across. Dark-greenish-gray laminae of calcic amphibole alternate with lighter colored laminae composed mostly of quartz and feldspar.

Bucksport Formation. The Bucksport Formation, of probable Late Silurian or Early Devonian age, was named by Trefethen (1950, p. 1) and was defined as "limey siltstone, phyllite, calcareous, laminated quartzites and their metamorphic equivalents, limesilicate gneisses and hornfels." Trefethen mapped the formation from the Bucksport quadrangle into the northeastern corner of the Belfast quadrangle (Fig. 2). Primary sedimentary features such as cross-lamination, scour-and-fill structures, and graded beds were not observed in this formation in the Belfast quadrangle. However, because the Bucksport Formation is interpreted to underlie conformably the Appleton Ridge Formation, in which there is evidence that the top lies to the southeast, the Bucksport Formation is also taken to have its top to the southeast. Its estimated thickness in the Belfast quadrangle is about 1,200 m (4,000 ft).

The Bucksport Formation is dominantly well-bedded, purplish-gray and greenish-gray biotite and calc-silicate gneiss rich in quartz and plagioclase that ranges from An_{34} to An_{88}. Rare beds of purple biotite schist in part weather rusty. The essential minerals and estimated abundance of the different rock types are given in Table 3.

Both biotite gneiss and calc-silicate gneiss are present in all exposures. Calc-silicate gneiss dominates in the southeastern part of the outcrop area; biotite gneiss dominates in the northwestern part, where the rocks are coarser grained and the bedding is less distinct. The biotite gneiss has weathered more completely than the calc-silicate gneiss. Consequently, most outcrops are strongly ribbed. The beds are mostly from 3 to 10 cm in thickness (Fig. 4). In the southeastern part of the outcrop area, lithologic layers are about 0.6 to 2 cm thick. These layers may have been produced during metamorphism and deformation of the rock and may not be bedding features. Generally, a gross layering is distinguished by the different shades of color resulting from a variation in the proportions of biotite gneiss and calc-silicate gneiss. The thickness of this layering generally ranges from 0.6 to 1 m.

The modes of three gneisses from the Bucksport Formation are given in Table 4. Porphyroblasts of diopside and calcic amphibole in the calc-silicate gneisses are as much as 2 mm in diameter. In some of the gneisses, these minerals are segregated into laminae 1 or 2 mm in thickness. The biotite schists contain the same assemblage of minerals as the biotite gneisses, but biotite is the dominant mineral in the schists.

Contact Relationships with the Appleton Ridge Formation. The contact between the Bucksport and Appleton Ridge Formations is exposed in a syncline southwest of Homes Mill and appears to be conformable. Within a zone a few metres thick, calc-silicate gneiss of the Bucksport Formation is interlayered with staurolite-bearing

TABLE 3. ROCK TYPES OF THE BUCKSPORT FORMATION

Rock name	Color	Essential minerals	Approximate abundance (%)
Biotite schist	Purple	Quartz-plagioclase-biotite ± microcline ± sphene	2
Biotite gneiss	Purplish-gray	Quartz-plagioclase-biotite ± microcline ± sphene	28
Calc-silicate gneiss	Purplish-gray to greenish gray	Quartz-plagioclase-biotite-calcic amphibole-sphene ± microcline	45
	Greenish-gray to purplish-gray	Quartz-plagioclase-calcic amphibole-biotite-sphene ± microcline	
	Greenish-gray	Quartz-plagioclase-diopside-calcic amphibole-sphene ± biotite ± calcite	23
	Greenish-gray	Quartz-plagioclase-diopside-sphene ± calcite	2

schist and quartzite of the Appleton Ridge Formation.

Appleton Ridge Formation. The name "Appleton Ridge Formation" is here proposed for a stratigraphic unit of probable Early Devonian age that strikes northeasterly across the central part of the quadrangle. Cheney (1967) referred to this unit as the "Appleton Formation," but this name has been assigned previously (Keroher and others, 1966). The type area of the formation is on Appleton Ridge. Consistent graded bedding in the region southwest of Belfast indicates that the top of formation is to the southeast (Fig. 2). Its estimated thickness is about 1,200 m (4,000 ft).

The lower part of the formation is in the andalusite zone and consists of staurolite schist that underlies Appleton Ridge. Generally, beds of schist, from 5 to 60 cm thick, are interlayered with schistose quartzite beds about 3 cm thick. Where quartzite is not present, there is both bedded and massive schist. Graded beds are not common. Quartzite makes up about 10 percent of the formation and is most abundant in the northeast and near the contact with the Bucksport Formation.

The staurolite schist is lustrous gray and weathers gray (Table 5, col. 1). Some of the staurolite schist contains poikiloblastic andalusite that is inconspicuous in hand specimen (Table 5, col. 2). The schistose quartzite interbedded with the staurolite schist is purplish-gray to bluish-gray (Table 5, col. 3).

In the garnet zone the Appleton Ridge Formation consists of muscovite phyllite with minor or abundant chlorite (Table 5, col. 4). Most bedding features apparently have been destroyed by shearing, but where they are preserved, phyllite beds rich in quartz and about 3 cm thick alternate with mica-rich beds about 1 cm

Figure 4. Distinctly bedded rocks of the Bucksport Formation. The light beds are calc-silicate gneiss, and the dark beds are biotite gneiss.

TABLE 4. ESTIMATED MODES OF THE BUCKSPORT FORMATION

	Sample number		
	1	2	3
Quartz	26	34	23
Plagioclase	36	41	50
Microcline	3	. .	. .
Biotite	4	16	27
Chlorite*	x	. .	x
Calcic amphibole	8†	9	. .
Diopside	22†	. .	. .
Clinozoisite	x	. .	x
Calcite	x	. .	. .
Sphene	1	x	. .
Tourmaline	x	. .	. .
Opaque minerals:			
Oxide(s)	. .	x	. .
Sulfide(s)	x	. .	x
Other accessory minerals	x	x	x
Approximate An content of plagioclase	44–88§	55	35
Grain size:			
Groundmass	0.08–0.15	0.1–1.2	0.15–0.4
Porphyroblasts	0.2–0.6	. .	. .

Note: Modes are given in percent; x indicates mode less than 1%. Grain sizes are given in millimetres. Description and location of samples:

1. Finely layered greenish-gray calc-silicate gneiss and purplish-gray biotite gneiss. Belfast 7-1/2-minute quadrangle; cliff on northeastern slope of hill 271, west-northwest of Shermans Corner.
2. Greenish-gray to purplish-gray calc-silicate gneiss. Morrill 7-1/2-minute quadrangle; 290-ft elevation on north ridge of hill 305, west of Quantabacook Lake.
3. Purplish-gray biotite gneiss. Searsmont 7-1/2-minute quadrangle; 288-ft elevation in roadbed of Route 173, near its intersection with Stearns Brook.

*Retrograde.
†Porphyroblastic.
§Ranges from An_{44} in layers of biotite gneiss to An_{88} in layers of calc-silicate gneiss.

thick. These rocks range from greenish gray to bluish gray, steel gray, or silver gray depending on their relative proportions of chlorite, muscovite, and biotite. Immediately northwest of the Muzzy Ridge Member, the phyllite weathers rusty and is rich in garnet and biotite (Table 5, col. 5).

Muzzy Ridge Member. The Muzzy Ridge Member consists of fine-grained, weakly schistose, noticeably dense, gray to black, and predominantly rusty-weathering rocks rich in mafic minerals. On Muzzy Ridge, the type locality, the rocks display layering that ranges from about 0.7 to 3 cm in thickness. Elsewhere the rocks generally are not layered. Most of the rock in the Muzzy Ridge Member is composed almost entirely of cummingtonite and garnet with minor biotite, plagioclase, and quartz, and with or without abundant hornblende (Table 5, col. 6). Rare beds contain only quartz, plagioclase, microcline, hornblende, and biotite.

Contact Relationships with the Jam Brook and Penobscot Formations. The Appleton Ridge Formation is bounded on the southeast by the St. George fault. To the southwest, the formation contacts the Jam Brook Formation, but to the northeast, it contacts the Penobscot Formation. The Muzzy Ridge Member is interpreted to terminate against the fault near the western border of the map area. The Appleton Ridge Formation becomes distinctly thinner to the northeast, and

TABLE 5. ESTIMATED MODES OF THE APPLETON RIDGE FORMATION

	Sample number					
	1	2	3	4	5	6
Quartz	58	55	85	58	60	10
Plagioclase	5	1	2	. .	. .	. .
Muscovite	10	4	3	35	15	. .
Biotite	7*	22*	5*	x	18	x
Chlorite	8†	3†	1†	7	2	x
Garnet	3	x	x	x	5*	34
Staurolite*	8	9	4	. .	. .	. .
Andalusite*	. .	6	. .	. .	. .	. .
Cummingtonite	. .	. .	. .	. .	. .	34
Hornblende	. .	. .	. .	. .	. .	20
Clinozoisite	. .	x	. .	. .	. .	. .
Tourmaline	x	x	x	x	x	. .
Opaque minerals:						
Oxide(s)	1	x	x	x	x	. .
Sulfide(s)	. .	. .	. .	. .	. .	2
Graphite	. .	. .	. .	. .	x(?)	x
Approximate An content of plagioclase	20	. .	20	. .	. .	. .
Grain size:						
Groundmass	0.04–0.1	0.15–0.5	0.1–0.3	0.06–0.1	0.02–0.06	0.07–0.4
Porphyroblasts	0.5–5.0	1.0–14.0	0.5–2.0	. .	0.08–0.6	. .

Note: Modes are given in percent; x indicates mode less than 1%. Grain sizes are given in millimetres. Description and location of samples (all locations are within the Searsmont 7-1/2-minute quadrangle):

1. Gray staurolite schist. Field 500 ft north of junction 238 of Route 131 and the road to Pitmans Corner.
2. Gray staurolite-andalusite schist. 1,700 ft S21°E from summit 554 of Appleton Ridge.
3. Gray, schistose quartzite. Appleton Ridge Road, 470 ft northeast of its junction with the road to Ghent.
4. Bluish-gray to greenish-gray phyllite. St. George River at Ghent.
5. Dark-gray phyllite. Benchmark 233 at Bickfords Corner.
6. Muzzy Ridge Member; dark-gray, banded cummingtonite-garnet-hornblende schist. 1,230 ft S57°E from the intersection of Routes 105 and 131.

*Porphyroblastic or poikiloblastic.
†Retrograde.

near Bucksport, the Bucksport Formation is mapped in fault contact with the Penobscot Formation (Stewart and Wones, 1974).

Penobscot Block

The Penobscot block is bounded on the northwest by the St. George fault and on the southeast by the Turtle Head fault zone in West Penobscot Bay (Stewart and Wones, 1974). Rocks correlated with the Rockport sequence of the Camden-Rockland area are exposed in the core of the Coleman antiform. Their age is believed to be Cambrian or Ordovician. Rocks correlated with the Battie Formation of the Camden area enclose the Rockport sequence in the Coleman antiform. The probable age of the Battie Formation also is Cambrian or Ordovician. The Penobscot Formation of probable Ordovician age conformably overlies the Battie Formation and forms most of the Penobscot block. The Jam Brook Formation is exposed between the Penobscot Formation and the St. George fault in the southwestern part of the Belfast quadrangle, but the Jam Brook terminates along strike to the northeast. Possibly the Jam Brook Formation unconformably overlies the Penobscot Formation, but more likely their contact is a fault. The Jam Brook Formation may correlate with the Rockport sequence, and its probable age is Cambrian or Ordovician.

Rockport Sequence. Marble and underlying rocks in the Coleman antiform (Fig. 2) are correlated with part of the Rockport sequence of the Camden-Rockland area (Osberg and Guidotti, 1974). The Rockport sequence was originally named the Rockland Formation by Bastin (1908). A Cambrian or Ordovician age is assigned to the rocks after the work of Osberg and Guidotti. About 200 m (700 ft) of this sequence is exposed in the Coleman antiform.

The lowest exposed rocks in this sequence are very poorly represented by a few small outcrops of biotite schist, quartzite, and quartz-feldspar-biotite gneiss in the core of the antiform. Marble about 150 m thick makes up the uppermost part of the sequence exposed in the antiform. The fine- to medium-grained rock displays layers of calcite and dolomite marble. Minor forsterite, phlogopite, spinel, chlorite, serpentine, and talc are present in some layers. Rare calc-silicate gneiss occurs with the marble.

Contact Relationships with the Battie Formation. In the Camden-Rockland region, where large areas are underlain by the Rockport sequence, detailed mapping by Osberg and Guidotti (1974) indicates that the Rockport sequence is in contact with the Battie Formation only along high-angle faults. Although the contact is not exposed in the Coleman antiform, it is folded by the antiform, and there it cannot be a high-angle fault. It is represented as a normal stratigraphic contact in Figure 2, but because the lower, conglomeratic part of the Battie Formation is missing in the Coleman antiform, possibly the contact is a thrust fault.

Battie Formation. Quartzite and associated rocks in the Coleman antiform (Fig. 2) are here correlated with the Battie Formation of Bastin (1908). The type area of this formation is on Mount Battie, west of Camden. Conglomeratic quartzite such as is found in the lower part of the Battie Formation in the type area is not present in the Coleman antiform. The Battie is the lowermost unit of the Megunticook sequence of Osberg and Guidotti (1974), who have assigned it a Cambrian or Ordovician age. The Battie Formation is about 150 m (500 ft) thick in the Coleman antiform.

About 40 percent of the Battie Formation exposed in the Coleman antiform is quartzite and quartz-mica schist. Beds of fine-grained, white to light-gray massive quartzite (Table 6, col. 1) as much as 30 m thick form whole outcrops but do not have continuity along strike. Some of these beds weather chalky and are probably feldspathic. More commonly, beds of quartzite about 3 to 30 cm thick alternate with somewhat thicker beds of quartz-mica schist (Table 6, col. 2) in which the minerals are segregated into layers about 2 mm thick enriched in micas or in quartz and plagioclase. About 30 percent of the formation consists of gray sericite gneiss (Table 6, col. 3) in which the texture is granoblastic and the sericite is pseudomorphous after sillimanite. Feldspar-quartz-biotite gneiss makes up about 25 percent of the formation. The gneiss contains granoblastic layers about 0.1 to 0.6 m thick that alternate with schistose layers rich in biotite. The remaining 5 percent of the formation is calc-silicate quartzite (Table 6, col. 4) and gneiss that is greenish gray, massive, and very fine grained. Plagioclase and microcline are present in the calc-silicate gneiss.

Contact Relationships with the Penobscot Formation. The Battie and Penobscot Formations are distinguished on the basis of differences between their dominant lithologies. Near their contact, beds of massive quartzite are restricted to the Battie Formation, which also contains a greater proportion of quartz-mica schist and

TABLE 6. ESTIMATED MODES OF THE BATTIE FORMATION

	Sample number			
	1	2	3	4
Quartz	79	53	27	59
Plagioclase	7	9	30	. .
Muscovite	2	14	2	. .
Sericite*	6	6	25	. .
Biotite†	3	12	5	. .
Chlorite*	3	4	5	. .
Garnet	x	. .	5	27§
Sillimanite	. .#	1	. .**	. .
Calcic amphibole	. .	. .	. .	1
Diopside	. .	. .	. .	4
Clinozoisite	. .	. .	. .	8
Calcite	. .	. .	. .	1
Sphene	. .	. .	. .	x
Tourmaline	x	x	x	. .
Opaque minerals:				
Oxide(s)	x	1	1	x
Sulfide(s)	. .	. .	. .	x
Approximate An content of plagioclase	6	20	27	. .
Grain size:				
Groundmass	0.05–4.0	0.03–0.5	0.01–1.0	0.05–0.3
Porphyroblasts	. .	. .	2.0–8.0	. .

Note: Modes are given in percent; x indicates mode less than 1%. Grain sizes are given in millimetres. Description and location of samples (all locations are within the Searsmont 7-1/2-minute quadrangle):

1. Bluish-gray quartzite with patches of sericite. Peninsula that separates the northwest shore of Norton Pond from the outlet channel.
2. Gray quartz-mica schist with sillimanite partly replaced by sericite. Slab City Road, at height of land northeast of Coleman Pond.
3. Gray gneiss with coarse patches of sericite pseudomorphous after sillimanite. Slab City Road, 5,000 ft southeast of Clarks Corner.
4. Greenish-gray, massive calc-silicate quartzite. 70-ft elevation in Ducktrap River.

*Secondary.
†Additional biotite has been replaced by chlorite.
§Grossularite, 91%; andradite, 5%; spessartite, 2%; other end members, 2% (electron microprobe analysis recalculated to estimate ferric iron).
#Replaced by sericite.
**Porphyroblastic; replaced by sericite.

calc-silicate rocks than does the Penobscot Formation. The Penobscot Formation near the contact is more sulfidic and has a greater proportion of aluminous schist and gneiss than does the Battie Formation. Both in the Coleman antiform and in the Camden area, the Battie Formation appears to be conformable with the overlying rocks (Osberg and Guidotti, 1974).

Penobscot Formation. The Penobscot Formation, at least part of which is probably Ordovician, was named by Smith and others (1907, p. 3) and was imprecisely defined as "metamorphosed shaly sediments typically developed along nearly the whole length of the west shore of Penobscot Bay." Rocks included in the Penobscot Formation underlie most of the southeastern half of the mapped area (Fig. 2). They strike into the upper two units of the Megunticook sequence of the Camden area (Osberg and Guidotti, 1974). The estimated thickness of the formation in the Belfast area is about 3,150 m (10,000 ft).

The Penobscot Formation includes a range of lithologies in the Belfast quadrangle, partly because of differences in metamorphic grade. In the garnet zone, it is approximately 45 percent muscovite-chlorite phyllite, 25 percent quartz-muscovite-chlorite phyllite (Table 7, col. 1), and 30 percent fine-grained quartzite (Table 7, col. 2) and feldspathic quartzite. The phyllites are dark gray, bluish gray, or greenish gray. Most contain fine-grained carbonaceous material and pyrrhotite or pyrite; these rocks weather rusty. Commonly, the phyllites contain rounded, greenish spots, 1 to 4 mm in diameter, composed of decussate chlorite, decussate muscovite, or aggregates of chlorite and quartz with minor muscovite and limonite. Much of the chlorite probably replaces biotite. The compositions of these rocks are not suitable for the development of garnet.

Interbedded with the phyllites are beds of light-greenish-gray to dark-gray quartzite from 15 cm to 6 m in thickness. The quartzites appear cherty in hand specimens, but most contain a few megascopic relict sedimentary grains of quartz as much as 2 mm in diameter.

In the andalusite and lower sillimanite zones, most of the rocks of the Penobscot Formation contain or are interbedded with rocks that contain andalusite or sericite that replaces andalusite. Approximately 35 percent of these rocks are distinctly bedded quartzite, quartz-mica schist, and andalusite schist; massive andalusite schist forms about 63 percent of these rocks. Massive quartzite is about 2 percent of these rocks. Most beds range from about 3 cm to 30 cm in thickness. Generally, andalusite schist dominates over quartz-mica schist and quartzite, but in some exposures the opposite relationship was observed. Graded beds, in which the schist layers are thicker than the quartzite layers, are not common. Locally, there are quartzite beds as thick as 7 m. Some outcrops contain pods and boudinaged beds of calc-silicate gneiss that originated as calcareous concretions and beds (Fig. 5).

Between Ducktrap and Frohock Mountains, but nowhere northwest of the Coleman antiform, andalusite schist without obvious bedding is more abundant than bedded rock. Most of it is black and contains abundant, randomly oriented porphyroblasts of andalusite, partly replaced by sericite (Table 7, col. 3). Minor graphite and abundant pyrrhotite account for the black color before and rusty color after weathering of this rock.

Most of the andalusite schist is gray on fresh surfaces, but portions of the rock particularly rich in andalusite or sericite pseudomorphous after andalusite are bluish gray or light gray. Immediately east of Megunticook Lake and northwest of the Gushee Member in the southwestern part of the quadrangle, there is massive, gray andalusite schist that weathers gray. Elsewhere, most of the andalusite schist weathers rusty; interbedded quartzite and quartz-mica schist weathers less rusty than the andalusite schist.

Prisms of andalusite average about 1 cm and range up to 5 cm in length. Completely fresh andalusite is rare. Generally, the andalusite is partly or completely replaced by patches of sericite with or without coarse, nearly equant, randomly oriented crystals of muscovite. In some of the most strongly schistose rocks, the sericite patches are flattened in the plane of the schistosity. Abundant, coarse, randomly oriented muscovite cuts across a weak schistosity in many of the rocks and probably represents andalusite porphyroblasts for which no other textural evidence remains.

In the lower sillimanite zone, many andalusite schists contain traces of fibrolitic sillimanite preserved, where included within quartz or other grains, from replacement by sericite (Table 7, col. 4).

The quartz-mica schist in the andalusite and lower sillimanite zones is light- to dark-gray and generally weathers rusty. Some of the schist contains minor, indistinct spots of sericite pseudomorphous after andalusite, flattened in the plane

TABLE 7. ESTIMATED MODES OF THE PENOBSCOT FORMATION

	Sample number								
	1	2	3	4	5	6	7	8	9
Quartz	65	85	29	34	38	10	5	3	50
Plagioclase	10	3	3	18	10	48	48	29	40
Microcline	. .	. .	. .	. .	. .	18*	. .	. .	. .
Muscovite	7	9	42	9	2	2	. .	. .	1
Sericite†	. .	. .	5	22	. .	. .	. .	. .	. .
Biotite	. .	. .	4	3	22	5	2	x	2
Chlorite	12	2	x†	8†	x†	. .	12†	15	7
Talc	. .	. .	. .	. .	. .	. .	x†	. .	. .
Garnet	. .	. .	. .	5	. .	. .	. .	1	. .
Cordierite	. .	. .	. .	. .	12*	. .	15*	. .	. .
Pinite†	. .	. .	. .	. .	1	. .	1	. .	. .
Andalusite	. .	. .	12§	. .§	3	. .	. .	. .	. .
Sillimanite	. .	. .	. .	. .	10*	15	. .	. .	. .
Anthophyllite	. .	. .	. .	. .	. .	. .	15*	. .	. .
Cummingtonite	. .	. .	. .	. .	. .	. .	. .	10*	. .
Hornblende	. .	. .	. .	. .	. .	. .	. .	40*	. .
Clinozoisite	. .	x	. .	. .	. .	. .	. .	. .	x
Tourmaline	1	x	. .	x	. .	. .	. .	. .	. .
Opaque minerals:									
Oxide(s)	. .	. .	. .	1	x	1	1	2	x
Sulfide(s)	x	1	5	. .	2	1	1	. .	. .
Graphite	5	x	x	x(?)	. .	. .	. .	. .	. .
Approximate An content of plagioclase	15	20	. .	25	25	33	32	. .	5
Grain size:									
Groundmass	0.03–0.3	0.02–0.2	0.02–0.5	0.02–1.0	0.1–0.8	0.2–1.0	0.1–0.6	0.01–0.2	0.03–0.1
Porphyroblasts, phenocrysts, and relict sedimentary grains	. .	1.0–2.0	1.0–20.0	3.0–10.0	1.0–4.0	1.0–3.0	1.0–5.0	0.3–4.5	0.5–3.5

Note: Modes are given in percent; x indicates mode less than 1%. Grain sizes are given in millimetres. Description and location of samples:

1. Rusty-weathering, dark-gray phyllite. Belfast 7-1/2-minute quadrangle; Dog Island Corner.
2. Gray, phyllitic quartzite with relict sedimentary grains. Belfast 7-1/2-minute quadrangle; bend in Little River, 750 ft southeast of 140-ft contour.
3. Rusty-weathering, black andalusite schist. Lincolnville 7-1/2-minute quadrangle; Ducktrap River, 300 ft north of U.S. Route 1.
4. Gray mica schist with patches of sericite pseudomorphous after andalusite. Lincolnville 7-1/2-minute quadrangle; summit 922, northwest of Millerite Ledges.
5. Rusty-weathering, gray cordierite-sillimanite granofels. Lincolnville 7-1/2-minute quadrangle; Route 52 at Heals Corner.
6. Rusty-weathering, gray sillimanite granofels with porphyroblasts of microcline. Lincolnville 7-1/2-minute quadrangle; Back Belfast Road, at 520-ft contour on southwest slope of Beech Hill.
7. Gushee Member; rusty-weathering, bluish-gray cordierite-anthophyllite schist. Searsmont 7-1/2-minute quadrangle; 335-ft elevation in stream south of junction 233, extreme southwestern corner of quadrangle.
8. Gushee Member; greenish-gray, fascicular hornblende-cummingtonite rock; matrix of metamorphosed agglomerate with relict, amygdaloidal scoria. Searsmont 7-1/2-minute quadrangle; southeast shore of Lawry Pond.
9. Gushee Member; gray, flinty granofels with relict phenocrysts of quartz and albite. Lincolnville 7-1/2-minute quadrangle; height of land northwest of Tilden Pond.

*Porphyroblastic or poikiloblastic.

†Retrograde.

§Porphyroblastic; replaced partly or wholly by sericite.

Figure 5. Distinctly bedded rocks of the Penobscot Formation. The dark beds are rich in sericite pseudomorphous after andalusite. The light beds are quartzite, and the boudinaged, discontinuous bed is calc-silicate gneiss.

of the schistosity. In some of these rocks, granular or decussate textures predominate over schistose textures.

In the upper sillimanite and sillimanite-K-feldspar zones, the Penobscot Formation is typically a rusty-weathering sillimanite granofels or sillimanite gneiss. However, on the peninsula that divides Megunticook Lake at the southern boundary of the quadrangle, there is sillimanite granofels that weathers gray. The texture of the sillimanite granofels is dominated by granular quartz and plagioclase. Some of the sillimanite gneiss is strongly layered, but much of it is a weakly foliated, massive rock that is transitional to sillimanite granofels. Both of these rocks are less strongly foliated than the lower grade rocks of the Penobscot Formation.

Bedding is poorly preserved in the sillimanite granofels and sillimanite gneiss. Locally, beds from 2 to 20 cm in thickness can be distinguished by their proportions of quartz and feldspar, mica, and sillimanite (or sericite pseudomorphous after sillimanite). Many outcrops contain boudinaged and generally complexly folded beds of quartzite from 5 to 10 cm in thickness. Thicker beds of quartzite are as much as 6 m across and are generally continuous across outcrops. Pods and boudinaged beds of calc-silicate gneiss are rare.

Perhaps 80 percent of the rocks in the upper sillimanite and sillimanite-K-feldspar zones contain sillimanite, sharply-bounded aggregates of sericite pseudomorphous after sillimanite, or indistinct patches of muscovite and sericite that are inferred to have replaced sillimanite (Table 7, cols. 5 and 6). Muscovite tends to be less abundant in these rocks than in the schists and phyllites of the lower metamorphic grades. Cordierite first appears in aluminous rocks of the Penobscot Formation in the sillimanite-K-feldspar zone.

A lentil of calcitic and dolomitic marble (Opm), represented by only seven exposures, is mapped within the Penobscot Formation on Figure 2. The thickness of exposed marble ranges from about 2 to 100 m. Unmapped zones from 2 to 30 m thick of calc-silicate gneiss, garnet-cummingtonite schist, amphibolite, and plagioclase-quartz granofels with or without relict phenocrysts of quartz make up a small proportion of the Penobscot Formation. These rocks are found only northwest of the Coleman antiform; except for the calc-silicate gneiss, they are largely restricted to the area between the lentil of marble and the Gushee Member.

Gushee Member. A belt of amphibole-bearing rocks and fine-grained plagioclase-quartz granofels at or near the northwestern contact of the Penobscot Formation is here named the Gushee Member of the Penobscot Formation. The type area of the member is on the Roland Gushee farm in Appleton Township. Its estimated thickness is about 500 m (1,600 ft).

About 65 percent of the rocks in the Gushee Member are mafic. Near the type area, the rocks weather rusty and are rich in anthophyllite and cordierite (Table 7, col. 7). They range from fine- to coarse-grained; there are conspicuous "sunbursts" of anthophyllite in the coarser varieties. Elsewhere, mafic rocks of the Gushee Member contain rounded structures about 0.3 to 0.7 m across that are interpreted to be relict pillows. There are also agglomeratic beds that contain fragments of flattened and metamorphosed amygdaloidal scoria (Table 7, col. 8).

About 30 percent of the member consists of very fine grained, chalky-weathering, gray quartz-albite granofels with relict phenocrysts of quartz and albite (Table 7, col. 9). Rusty-weathering, dark-gray mica schist makes up about 5 percent; biotite schist, quartz-mica schist, and calc-silicate gneiss each make up less than one percent of the member.

Contact Relationships with the Jam Brook and Appleton Ridge Formations. The Penobscot Formation is in contact with the Appleton Ridge Formation along the St. George fault. Possibly the Jam Brook Formation unconformably overlies the Penobscot Formation. But because map units of both the Penobscot and Jam Brook Formations terminate against the contact, the contact is interpreted to be a thrust fault that is truncated by the St. George fault northeast of Lawry Pond (Fig. 6).

Jam Brook Formation. The name "Jam Brook Formation" is here proposed for a stratigraphic unit of probable Cambrian or Ordovician age that is confined to the southwestern corner of the Belfast quadrangle (Figs. 2 and 6) and may correlate with the Rockport Sequence of Osberg and Guidotti (1974). The unit was named the St. George Formation by Cheney (1967), but this name has been assigned previously (Keroher and others, 1966). All the rocks of the formation are exposed in the bed of Jam Brook, its type area. Features indicative of topping sense were not observed in these rocks, but the lithologic sequence in the formation suggests that the top may be to the northwest. Quartz-pebble metaconglomerate is restricted to the southeastern part of the formation. A map unit of quartzite lies about 100 m northwest of the metaconglomerate. Andalusite schist is restricted to the middle, and calc-silicate gneiss and marble occupy the middle and northwestern parts of the formation. Based on the premise that complete regressive sedimentary sequences are more likely to be removed by erosion than are complete transgressive sequences, it is proposed that the Jam Brook Formation, as exposed in the Belfast quadrangle, represents a transgressive sedimentary cycle that began with deposition of quartz-pebble conglomerate. The formation terminates along strike to the northeast and thickens to about 1,000 m (3,000 ft) to the southwest.

In decreasing order of abundance, the Jam Brook Formation includes quartz-mica schist, calc-silicate gneiss, quartzite, quartz-pebble metaconglomerate, marble, andalusite schist, and plagioclase-quartz granofels with relict phenocrysts. One rock type dominates in each map unit separated in Figure 6. The rocks are described below in their generalized order of appearance, from southeast to northwest.

Quartz-pebble metaconglomerate (Table 8, col. 1) is restricted to unit O€jc. This gray rock consists of ellipsoidal pebbles of polycrystalline quartz, flattened in the plane of the schistosity of the quartz-rich matrix. The axial dimensions of the largest pebbles are about 8 by 5 by 2 cm. The matrix of some of the metaconglomerate is rich in calcic amphibole.

Massive, granular, virtually pure, gray to white quartzite occurs in a unit (O€ jq) separate from and northwest of the quartz-pebble metaconglomerate. Locally, the quartzite unit includes beds of quartz-mica schist and less abundant andalusite schist.

Gray quartz-mica schist and phyllite (O€ j) encloses all the other map units

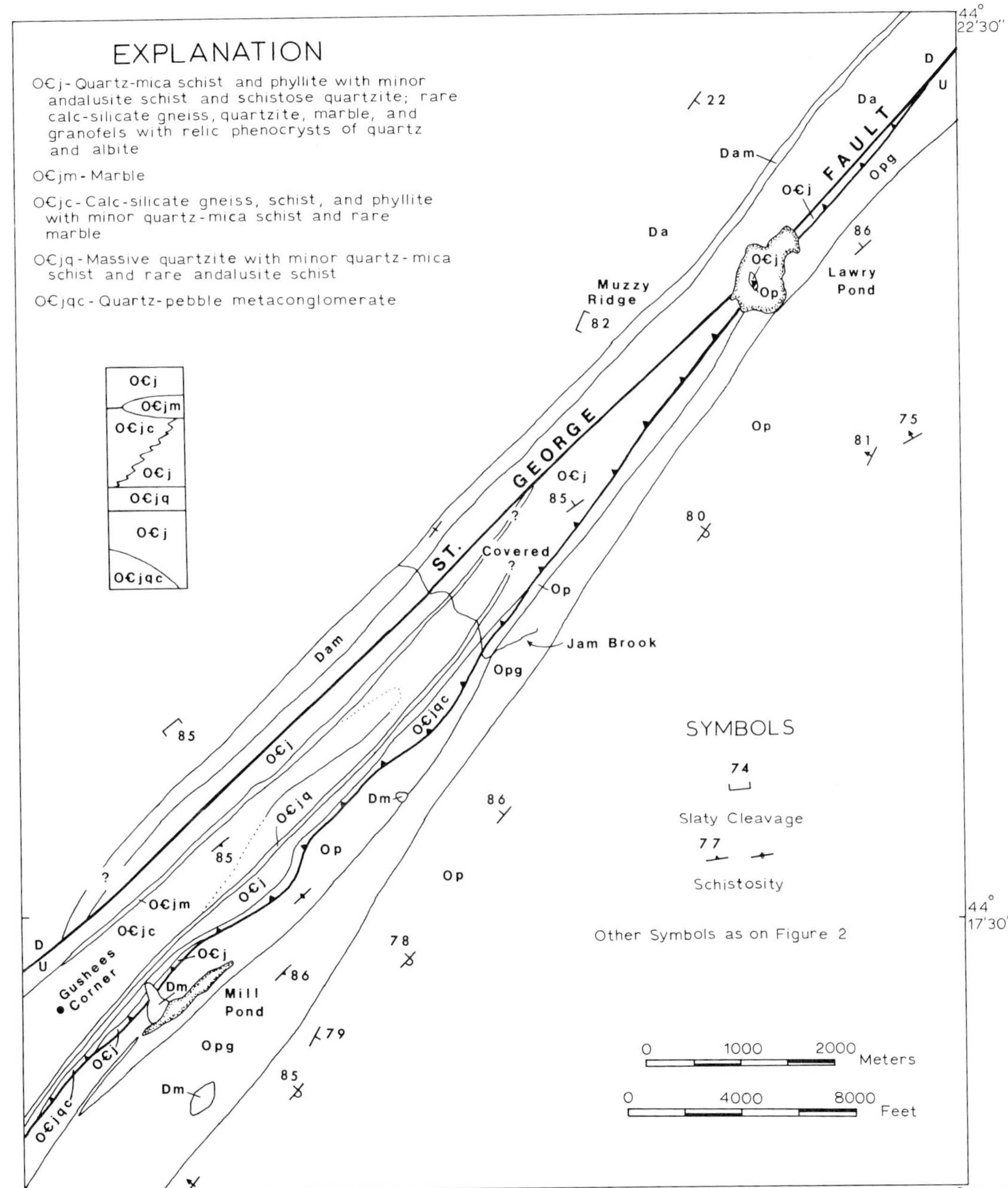

Figure 6. Detailed geologic map of stratigraphic units within the Jam Brook Formation.

TABLE 8. ESTIMATED MODES OF THE JAM BROOK FORMATION

	Sample number			
	1	2	3	4
Quartz	83	64	39	15
Plagioclase	3	5	. .	30
Muscovite	x*	15	8	. .
Sericite†	. .	. .	12	. .
Biotite	4	15	13	2
Chlorite	4	x†	6†	. .
Garnet	x	. .	. .	. .
Andalusite	. .	. .	20§	. .
Calcic amphibole	4	. .	. .	37
Clinozoisite	1	1	. .	8
Calcite	x	. .	. .	5
Sphene	x	. .	. .	2
Tourmaline	x	x	x	. .
Opaque minerals:				
Oxide(s)	x	x	2	. .
Sulfide(s)	1	. .	. .	1
Approximate An content of plagioclase	45	60	. .	40
Grain size:				
Groundmass	0.1-3.0	0.02-0.5	0.01-0.2	0.05-0.6
Porphyroblasts and relict sedimentary grains	8.0-80.0	. .	1.0-5.0	. .

Note: Modes given in percent; x indicates mode less than 1%. Grain sizes are in millimetres. Description and location of samples (all locations are within the Searsmont 7-1/2-minute quadrangle).

1. Quartz-pebble metaconglomerate with gray, schistose matrix. 3,600 ft S19°W from crossroads at North Appleton.
2. Gray quartz-mica phyllite. Road from North Appleton to Gushees Corner, 4,100 ft southwest of North Appleton.
3. Bluish-gray quartz-mica phyllite with andalusite partly replaced by sericite. Knox-Waldo county line, 750 ft southeast of the St. George River.
4. Greenish-gray to purplish-gray calc-silicate phyllite. 4,790 ft S51°W from crossroads at North Appleton.

*Not present in the matrix.
†Retrograde.
§Porphyroblastic.

except for the quartz-pebble metaconglomerate. Phyllite or schist containing abundant quartz, muscovite, and biotite (Table 8, col. 2) is interbedded with minor amounts of phyllitic or schistose quartzite and much less commonly with massive quartzite. Locally, andalusite is present in the schist or phyllite (Table 8, col. 3). A few beds and pods of calc-silicate gneiss occur within the map unit. Immediately northwest of the quartzite unit (O€jq), there are at least seven distinct layers from 1 to 8 m in thickness of fine-grained plagioclase-quartz granofels with relict phenocrysts of quartz and albite. Isolated exposures of similar rock are present in the area of quartz-mica schist southeast of the quartzite unit.

A map unit composed dominantly of calc-silicate gneiss, phyllite, and schist (O€ jc) extends from the western boundary of the quadrangle northeastward, where it grades along strike into quartz-mica schist. This unit contains no andalusite schist and only a small proportion of quartz-mica schist. The rocks range from purplish-gray biotite gneiss and schist to green decussate or phyllitic rocks rich in calcic amphibole (Table 8, col. 4).

The marble unit (O€ jm) ranges from fine-grained, white marble rich in dolomite to coarser, bluish-gray marble rich in calcite. Tremolite is a minor constituent of the marble. A lentil of ribbon marble northwest of the marble map unit and two other lentils of marble within the calc-silicate gneiss map unit to the southeast are not separated in Figure 6.

Protoliths of the Stratified Rocks

The precursors of the metasedimentary and metavolcanic rocks of the Belfast quadrangle are given in Table 9, together with the estimated cumulative thickness and inferred environment of deposition of each major stratigraphic unit. The precursors were determined by comparing the preserved primary features and compositions (expressed in modes) of the metamorphic rocks with primary features and compositions of common sedimentary and volcanic rocks.

AGES OF METAMORPHISM AND OF THE STRATIFIED ROCKS

No identifiable fossils have been found in the Belfast quadrangle. Therefore, the proposed ages of the rocks described in this paper are based on correlation with other rocks in east-central Maine (Fig. 7), most of which also are unfossiliferous. Because assignments of ages to the stratified rocks have been based partly on the recognition of multiple metamorphic events in east-central Maine, the age of metamorphism is discussed first below. The ages of the rocks are discussed according to the major structural block in which they lie.

Age of Metamorphism

The distribution of the metamorphic zones in the Belfast quadrangle may reflect the summation of between two and four episodes of prograde metamorphism. In the Bucksport area there is evidence, discussed below, that the Passagassawakeag Formation was metamorphosed to high grade during Precambrian time. On North Haven Island, in Penobscot Bay, and to the east of the bay, rocks of Cambrian or Ordovician age were metamorphosed to low grade before deposition of rocks ranging in age from Middle Silurian to Early Devonian (Dow, 1965; Wingard, 1961; Stewart and Wones, 1974). But evidence for an early Paleozoic metamorphism in the Belfast quadrangle or elsewhere west of Penobscot Bay is indirect and inconclusive.

Structures and textures of most of the rocks in the Belfast quadrangle indicate that prograde metamorphism was essentially contemporaneous with folding of Acadian(?) age. For this reason, I believe that the isograds primarily indicate gradients of Devonian metamorphism. However, D. B. Stewart (1975, written commun.; Stewart and Wones, 1974) has observed that the postmetamorphic St. George fault is intruded by the Mount Waldo pluton, for which a 390 ± 10 m.y. Rb-Sr whole-rock age has been determined (Brookins, this volume). Furthermore, along the Turtle Head fault zone in West Penobscot Bay, andalusite-grade metamorphic rocks of the Penobscot block are in contact with rock at chlorite grade. The Turtle Head fault zone is intruded by the Lucerne pluton, for which a 402 ± 13 m.y. Rb-Sr whole-rock age has been determined (D. B. Brookins, cited in Wones, 1974). Stewart has questioned the likelihood that an episode of Devonian regional metamorphism, followed by faulting, could have preceded the Early Devonian intrusion of these plutons. The alternative—that there was an episode of high-grade metamorphism

TABLE 9. APPROXIMATE CUMULATIVE THICKNESS, PROTOLITH,

Approximate cumulative thickness in metres (feet)	Approximate thickness in metres (feet)*	Map unit	Present lithology
PASSAGASSAWAKEAG BLOCK			
4,700 (15,600)	1,200 (4,000)	Appleton Ridge Formation	Cummingtonite-garnet rocks at top of sequence; phyllite; interbedded staurolite schist and quartzite at bottom of sequence
3,500 (11,600)	1,200 (4,000)	Bucksport Formation	Calc-silicate gneiss interbedded with quartz-plagioclase-biotite gneiss
Unconformity: The Bucksport and Appleton Ridge Formations are not present northwest of the Liberty-Orrington anticline			
4,000 (13,200)	1,700 (5,600)	Hogback schist	Micaceous quartz-feldspar gneiss; aluminous schist, in part sulfidic; biotite gneiss; calc-silicate gneiss
Unconformity: The Hogback Schist is not present southeast of the Liberty-Orrington anticline			
2,300 (7,600)	300 (1,000)	Passagassawakeag Formation, Mixer Pond Member	Quartz-oligoclase-microcline gneiss with minor magnetite, biotite, amphibole, or ferrosalite
2,000 (6,600)	2,000 (6,600)	Passagassawakeag Formation	Primarily quartz-plagioclase-biotite gneiss
PENOBSCOT BLOCK			
4,300 (13,900)	150 (500)	Penobscot Formation, upper pelitic unit	Primarily sulfidic phyllite and schist; quartz-mica schist and quartzite
4,150 (13,400)	500 (1,600)	Penobscot Formation, Gushee Member	Sulfidic anthophyllite-cordierite rock; amphibolite; plagioclase-quartz granofels with relict phenocrysts of quartz and albite
3,650 (11,800)	2,500 (8,000)	Penobscot Formation	Primarily sulfidic phyllite, schist, and gneiss; quartz-mica schist and quartzite
1,150 (3,800)	150 (500)	Battie Formation	Quartzite, quartz-mica schist, aluminous gneiss, quartz-feldspar-biotite gneiss; minor calc-silicate gneiss
Possible break in stratigraphic sequence due to faulting			
	200 (700)	Rockport Sequence	Marble, underlain by poorly exposed biotite schist, quartzite, and quartz-feldspar-biotite gneiss
Rockport Sequence is probably equivalent to the Jam Brook Formation, which is interpreted to be thrust over the other rocks of the Penobscot block			
1,000 (3,300)	1,000 (3,300)	Jam Brook Formation	Quartz-mica schist, calc-silicate gneiss, quartzite, quartz-pebble metaconglomerate, marble, andalusite schist; minor quartz-plagioclase granofels with relict phenocrysts

*An attempt has been made to correct thicknesses for repetition by folding of a smaller scale than can be represented on Figure 2. Inferred thickness of some map units are much less than their width of exposure, despite the fact that most beds dip steeply.

Original lithology	Critical primary structures or features	Environment of deposition; comments
Tuffaceous mafic volcanic rocks at top of sequence; siliceous shale cyclically bedded with quartz silt and very fine quartz sand	In quartzite, rounded tourmaline grains with overgrowths indicate that coarsest grained sediment was very fine sand	Marine
Clastic, feldspathic, pelitic sediment with variable, but substantial amounts of carbonate	Bedding is 0.6 to 2.0 cm in thickness near top of sequence, generally 3 to 10 cm in thickness; variable gross lithologic layering averages 0.6 to 1 m	Marine; aluminous material was consumed in reaction with carbonate to make calcic plagioclase
Graywacke and quartz-feldspar silt with minor shale beds; thick units of shale, partly black shale; clastic sediments with carbonate	Graded beds, 0.2 to 0.6 m thick, of quartz-feldspar gneiss grading into mica schist; schistose partings in quartz-feldspar gneiss; massive mica schist	Marine, in part anaerobic
Feldspathic volcanic rocks	Rocks undersaturated in alumina; thick layered units; low abundance of quartz	Volcanic; marine(?)
Primarily sediment rich in feldspar and poor in clay; perhaps primarily felsic volcanic rocks	Aluminous minerals rare; few amphibolite and calc-silicate beds	Marine; volcanic(?)
Primarily black, siliceous shale; quartz silt and fine quartz sand		Marine, primarily anaerobic
Basaltic and altered basaltic rocks, mostly tuffaceous; tuffaceous felsic volcanic rocks	Rare pillows and fragments of scoria in mafic rocks; phenocrysts and flattened fragments in felsic rocks	Primarily pyroclastic rocks with some submarine flows
Primarily black shale interbedded with quartz silt and fine quartz sand	Bedding is 3 to 20 cm thick, with aluminous rock predominating over quartzite; unbedded zones of aluminous rock; zones of quartzite as much as a few metres thick	Marine, primarily anaerobic
Quartz sand and silt; shale; feldspathic sediment; minor calcareous silt	Few primary features preserved; beds of quartzite 3 to 30 cm thick in quartz-mica schist; possible abrupt changes of facies	Marine (nearshore?)
Limestone and dolomite, underlain by clastic rocks		Marine
Quartz silt, calcareous silt, quartz sand, quartz-pebble conglomerate, limestone, dolomite, siliceous shale; felsic tuff	Quartz-pebble metaconglomerate at base of exposed sequence; massive quartzite restricted to near base; calc-silicate gneiss and marble generally near top; quartz-mica schist throughout sequence	Marine, near shore; transgressive sequence with abrupt lateral and vertical changes of facies

AREA / AGE	CASCO BAY AREA (HUSSEY, 1968)	WATERVILLE-VASSALBORO AREA (OSBERG, 1968)	BUCKSPORT AREA (PASSAGASSAWAKEAG AND BUCKSPORT BLOCKS) (STEWART AND WONES, 1974)	BELFAST QUADRANGLE: PASSAGASSAWAKEAG BLOCK	BELFAST QUADRANGLE: PENOBSCOT BLOCK	CAMDEN-ROCKLAND AREA (OSBERG AND GUIDOTTI, 1974)	PENOBSCOT BAY, EAST OF TURTLE HEAD FAULT ZONE (STEWART, 1974)
EARLY DEVONIAN			APPLETON RIDGE FM.	APPLETON RIDGE FM.			CASTINE VOLCANICS* VINALHAVEN RHYOLITE THOROFARE ANDESITE*
LATE SILURIAN OR EARLY DEVONIAN	BERWICK FM.	VASSALBORO FM.	BUCKSPORT FM.	BUCKSPORT FM.			AMES KNOB FM. *
SILURIAN	ELIOT FM. KITTERY FM.	WATERVILLE FM. * MAYFLOWER HILL FM. *					
ORDOVICIAN					PENOBSCOT FM.	UPPER MEGUNTICOOK SEQUENCE; BENNER HILL SEQUENCE (*) IN ANOTHER BLOCK	
CAMBRIAN OR ORDOVICIAN	CASCO BAY GROUP		SCHIST OF RYDER BLUFF COPELAND FM.	HOGBACK SCHIST	BATTIE FM. ? ROCKPORT SEQUENCE (= JAM BROOK FM.)	LOWER MEGUNTICOOK SEQUENCE ? ROCKPORT SEQUENCE	ISLESBORO FM.
CAMBRIAN(?)							ELLSWORTH SCHIST (NORTH HAVEN GREENSTONE)
PRECAMBRIAN			PASSAGASSAWAKEAG FM.	PASSAGASSAWAKEAG FM.			UNNAMED ROCKS ON SPRUCE HEAD ISLAND

Figure 7. Correlation chart for east-central Maine. Fossils are present in formations with an asterisk. Stewart and Wones interpreted the Appleton Ridge and Bucksport Formations to be of Ordovician age.

west of Penobscot Bay earlier in the Paleozoic—is not attractive, for a pre-Devonian, Paleozoic regional metamorphism of sillimanite-K-feldspar grade is not recognized elsewhere in New England.

In contrast, Osberg (1968, 1974) theorized that regional metamorphism in east-central Maine is Permian (or younger) because the K-Ar ages of Devonian plutonic rocks are preferentially reset to Permian ages in areas affected by high-grade regional metamorphism (Zartman and others, 1970). If the latest regional prograde metamorphism in the Belfast quadrangle was in fact Devonian, retrograde metamorphism may have been the Permian event that Osberg proposed.

Passagassawakeag Block

A Precambrian age is assigned to the Passagassawakeag Formation based on the findings of Wones (1974) that in the Bucksport area the Passagassawakeag was multiply deformed and recrystallized to high metamorphic grade before the intrusion of granitic rocks, of which the youngest, nonfoliate representatives are Ordovician. Wones cited a preliminary Pb^{206}-Pb^{207} age determination by Zartman of 430 ± 10 m.y. for zircons from the Stricklen Ridge pluton, which cuts the Passagassawakeag Formation. A preliminary Rb-Sr age of 600 ± 80 m.y. for the Passagassawakeag Formation further establishes its Precambrian age but "is weighted by samples with favorable Rb/Sr ratios which may or may not be closed systems" (D. G. Brookins, 1975, written commun.).

The age of the Hogback Schist is believed to be Cambrian or Ordovician. The Hogback has been traced into the Copeland Formation of the Bucksport area (Trefethen, 1950), to which Stewart and Wones (1974) assigned a tentative Cambrian age. The dominant lithologies of the Hogback resemble those of the Aziscohos and Albee Formations, northwest of the Merrimack Synclinorium (Green and Guidotti, 1968), the age of which ranges from Late Cambrian(?) to Middle Ordovician. Kost Pankiwskyj (1974, written commun.) traced the Hogback Schist into the Cape Elizabeth Formation, part of the Casco Bay Group of Hussey (1968). Hussey (1975, written commun.) indicated that the Casco Bay Group "is Ordovician in age, possibly in part Cambrian," and preliminary Rb-Sr studies by D. G. Brookins (1975, written commun.) also imply a Cambrian or Ordovician age for the Cape Elizabeth Formation.

There is presently less agreement about the age of the Bucksport Formation and the overlying Appleton Ridge Formation. Osberg (1974) thought the Bucksport Formation to be Silurian and the Appleton Ridge Formation to be Devonian. But Stewart and Wones (1974) proposed that both of these formations are Ordovician. Stewart rejected a Devonian age for the Appleton Ridge Formation, based on his evidence, already presented, against a Devonian age of regional prograde metamorphism west of Penobscot Bay. But the Bucksport Formation and the dominant, pelitic rocks of the Appleton Ridge Formation do not resemble any other rocks in central Maine for which an Ordovician age has been proposed. The interpretation accepted here is that the Bucksport Formation is of Late Silurian or Early Devonian age and the Appleton Ridge Formation is of Early Devonian age. The lithology of the Bucksport Formation is similar to the dominant lithology, at high metamorphic grade, of the Silurian-Devonian Vassalboro Formation in the Waterville-Vassalboro area (Osberg, 1968). In both formations, biotite and calc-silicate gneiss are dominant, but it should be noted that, in comparison with the Vassalboro Formation in its type area, the Bucksport Formation is much thinner bedded, richer in plagioclase, and has a narrower range of rock composition. The

Bucksport Formation does not resemble any of the Silurian rocks that lie below the Vassalboro Formation in the Merrimack synclinorium.

The lithology and bedding features of the Appleton Ridge Formation closely resemble the Lower Devonian Seboomook Formation of northwestern Maine (Boucot, 1961). The closest exposures of similar rocks belong to the Solon Formation, approximately 85 km (50 mi) to the northwest, across strike (Fig. 1; Ludman and Griffin, 1974).

Penobscot Block

Marble and underlying rocks in the core of the Coleman antiform are correlated with the Rockport sequence of the Camden-Rockland area (Osberg and Guidotti, 1974). This correlation is uncertain but is accepted in preference to proposing a new name for these poorly exposed rocks. The Islesboro Formation on Islesboro Island in western Penobscot Bay (Stewart, 1974) probably correlates with the Rockport sequence. The age of the Rockport sequence and Islesboro Formation is believed to be Cambrian or Ordovician.

There is excellent correspondence between the rocks of the Jam Brook Formation and units 2 through 5 of the Rockport sequence in the Camden-Rockland area (Osberg and Guidotti, 1974), and therefore, it is likely that the Jam Brook correlates with the Rockport sequence. The formation is separately named here because it is well exposed in the Belfast quadrangle, and the nearest exposures of the Rockport sequence occur 13 km (8 mi) across strike. They are believed to be separated from the Jam Brook by two major faults. The proposed Cambrian or Ordovician age of the Jam Brook Formation is based on its correlation with the Rockport sequence. If the contact between the Jam Brook and Penobscot Formations is not a fault but an unconformity, the age of the Jam Brook is Ordovician or younger.

The dark, sulfidic, rusty-weathering rocks of the Penobscot Formation are lithologically similar to Middle Ordovician rocks of the Danforth area of Maine (Larrabee and Spencer, 1963) and to Lower Ordovician rocks exposed at St. John and St. Stephen, New Brunswick (Stewart, 1974). The age of the Penobscot Formation is presumed to be Ordovician, based on these possible correlations. Rb-Sr whole-rock studies by Brookins (this volume) also indicate an Ordovician age of the Penobscot Formation. The age of the underlying Battie Formation may be Ordovician or Cambrian.

INTRUSIVE ROCKS

Many bodies of intrusive igneous rocks cut the stratified rocks. The larger bodies are shown in Figure 2. Some consist of pretectonic or early syntectonic metadiabases (Dm) that have the mineralogy of amphibolite, but locally have a poorly preserved relict subophitic texture, and rarely have unaltered pyroxenes and zoned intermediate plagioclase. Some of the largest bodies also contain metamorphosed ultramafic rock. The mafic and ultramafic intrusive rocks are restricted to the Penobscot block.

Other large intrusive bodies consist of binary granitic rocks that range from foliated syntectonic to unfoliated posttectonic. Sodic plagioclase is more abundant than potassium feldspar in the granitic rocks. Bodies of granitic rocks too small to show on Figure 2 are common near some of the mapped stocks.

STRUCTURE

Faults

The St. George longitudinal, high-angle fault divides the stratified rocks of the Belfast quadrangle into the Passagassawakeag and Penobscot blocks (Fig. 2). The fault is nowhere exposed in the mapped area, but map units of the Appleton Ridge, Jam Brook, and Penobscot Formations terminate against it (Figs. 2 and 6). Because isograds of regional metamorphism in the Belfast quadrangle are cut by the fault, movement on it must have occurred after the cessation of prograde (possibly Acadian) metamorphism. The Passagassawakeag block is inferred to have moved downward relative to the Penobscot block.

The Norumbega fault (Stewart and Wones, 1974) lies entirely within the Hogback Schist in the Belfast quadrangle. For this reason, rocks that lie across the fault are considered to belong to the same structural block for the purposes of this paper. However, different map units of the Hogback Schist are exposed across the fault, and some map units terminate against it (Fig. 3). Kost Pankiwskyj (1974, written commun.) mapped the fault entirely across the Liberty quadrangle (west of the Belfast quadrangle), where Hogback Schist also is found on both sides of the fault. Further southwest, the fault is mapped in the Wiscasset quadrangle (Hatheway, 1971) and in the Portland area (the Cape Elizabeth fault zone of Hussey, 1971). My mapping in the Brooks quadrangle, where isograds of regional metamorphism are cut by the fault, extends it northeastward to the Bucksport quadrangle. Farther to the northeast, Middle Devonian postorogenic conglomerates are found along the trace of the fault; this establishes the age of its latest movement as post-Middle Devonian (Stewart and Wones, 1974; Stoesser, 1966). Microscopic textures in rocks of the Belfast quadrangle indicate that retrograde metamorphism of possible Permian age occurred after shearing associated with the fault zone. In the Belfast quadrangle, rocks northwest of the fault have been displaced downward a minimum of 460 m relative to the rocks southeast of the fault.

Because map units of both the Penobscot and Jam Brook Formations terminate at their contact, the contact is interpreted to be a thrust fault. Where exposed in Jam Brook, the contact between quartz-pebble metaconglomerate of the Jam Brook Formation and retrograded andalusite schist of the Penobscot Formation dips steeply to the southeast. Textures of the rocks adjacent to the inferred fault are not cataclastic. The trace of the inferred fault is cut by a pretectonic or early syntectonic body of mafic and ultramafic rock, and by the sillimanite isograd. The fault is truncated by the St. George fault. These observations indicate that displacement on the fault must have ceased before the termination of prograde (possibly Acadian) metamorphism.

Folds

The Liberty-Orrington anticline is the major structural feature of the Passagassawakeag block, and of east-central Maine (Osberg, 1974). To the northeast, Precambrian high-grade metamorphic rocks of the Passagassawakeag Formation plunge gently to the northeast under weakly metamorphosed Lower Paleozoic(?) rocks that can be traced partly around the nose of the fold northeast of Bucksport (Stewart and Wones, 1974). In the Belfast quadrangle, the axial trace of the fold coincides with a region of high-grade Acadian(?) metamorphism. Consequently, evidence for the fold is not obvious in the Belfast quadrangle. Furthermore, dissimilar rocks are exposed on the opposite limbs of the fold. On the southeast limb, the Bucksport Formation of probable Silurian or Devonian age is interpreted to lie unconformably on the Passagassawakeag Formation. But to the northwest, the Passagassawakeag Formation is overlain unconformably by the Cambrian or Ordovician Hogback Schist and an unnamed sequence of Silurian(?) rocks in the Brooks quadrangle. These relationships suggest that the Liberty-Orrington anticline had developed as an elevated area during earliest Paleozoic time.

Also in the Passagassawakeag block, the Passagassawakeag, Bucksport, and Appleton Ridge Formations are repeated in what are interpreted to be tight, doubly-plunging folds that formed concurrently with Acadian(?) regional metamorphism. The axial traces of these folds are omitted from Figure 2 to avoid crowding of symbols.

Northwest of the Norumbega fault is an open anticline, here named the Frye Mountain anticline. Within the Frye Mountain anticline, measured bedding dips steeply, but the beds are crumpled by minor folds. Deflections of contacts and changes in the width of map units in the area of steep topography underlain by the Frye Mountain anticline indicate that the limbs of the fold dip gently and that its axis plunges gently to the southwest. The fold is cut by the Norumbega fault but is younger than the regional metamorphism.

The only major fold recognized in the Penobscot block is the Coleman antiform. Because both bedding and schistosity wrap around the nose of the fold between Ducktrap and Frohock Mountains, the fold is a late tectonic feature.

Minor folds are common in the stratified rocks. The axial planes of most of the minor folds strike northeasterly and dip steeply to the northwest or to the southeast. The axes of most minor folds plunge gently or moderately to the northeast or to the southwest. Steeply plunging folds generally are restricted to the Passagassawakeag Formation, which was deformed and metamorphosed before the deposition of the other rocks of the area. Two generations of tight minor folds are interpreted to have formed during Acadian(?) deformation and metamorphism. Open crenulations of schistosity formed after regional metamorphism but before Permian(?) retrograde metamorphism.

DISCUSSION

Rocks from Precambrian to Early Devonian(?) age southeast of the St. George fault bear no similarity to rocks with the same span of age northwest of the fault (Figs. 2 and 7). Precambrian rocks exposed on Spruce Island, immediately south of Islesboro Island in western Penobscot Bay, do not resemble lithologies in the Precambrian Passagassawakeag Formation in the Liberty-Orrington anticline (Stewart and Wones, 1974). The lower Paleozoic Rockport sequence and its probable correlatives the Islesboro and Jam Brook Formations are found only southeast of the St. George fault. Neither these rocks nor the Ordovician Penobscot Formation and Benner Hill sequence (of the Camden-Rockland area) resemble the lower Paleozoic Hogback Schist or its correlatives northwest of the St. George fault. These rocks may have been deposited in different basins or on different lithospheric plates. The latter explanation is supported by differences in the Precambrian rocks across the St. George fault.

A thick sequence of Silurian rocks lies in the Merrimack synclinorium northwest of the Belfast quadrangle. Although rocks of probable Silurian age are present between the Hogback Schist and the Silurian-Devonian Vassalboro Formation in the Brooks quadrangle, the Silurian-Devonian(?) Bucksport Formation is interpreted to lie unconformably on Precambrian rocks east of the Liberty-Orrington anticline. Nondeposition of Silurian rocks east of the Liberty-Orrington anticline is consistent with the interpretation of Ludman and Griffin (1974) that there was an eastern source for the Silurian Kenduskeag Formation, which underlies and intertongues with the Vassalboro Formation southeast of the axis of the Merrimack synclinorium. According to my interpretations, the Bucksport and Appleton Ridge Formations were deposited east of the Liberty-Orrington anticline during latest Silurian and Early Devonian time.

On North Haven Island, east of the Penobscot block, the Silurian and Lower Devonian are represented by a distinctly different sequence of rocks. The Ames Knob Formation, a thin clastic sequence that contains fossils ranging from Middle Silurian to Early Devonian age, lies above an unconformity. Lower Devonian volcanic rocks interfinger with and overlie the clastic rocks and are also exposed elsewhere in and east of Penobscot Bay (Brookins and others, 1973). These Silurian and Devonian rocks are separated from the inferred Silurian and Devonian rocks of the Belfast quadrangle by at least two major faults, along which there may have been considerable lateral displacement (Stewart, 1974). For this reason, the paleogeographic relationship between the inferred Silurian and Devonian rocks of the Belfast quadrangle and those exposed in Penobscot Bay is not yet clear.

ACKNOWLEDGMENTS

I gratefully acknowledge the financial support of the Maine Geological Survey and the Department of Geological Sciences of Harvard University. Marland P. Billings and James B. Thompson, Jr., supervised the field and laboratory work on which this paper is based. Robert G. Doyle, Arthur M. Hussey II, Philip H. Osberg, Jeffrey Warner, and David B. Stewart introduced me to the geology of east-central Maine. I have benefited from discussions with Douglas Brookins, Charles V. Guidotti, David Wones, and numerous other geologists who are unraveling the stratigraphy of eastern New England. This paper has benefited from the review and criticism of Lincoln Page, Robert Moench, Stephen Norton, David B. Stewart, and Don Alvord. Not all these reviewers agree with all the contentions of this paper, however.

REFERENCES CITED

Bastin, E. S., 1908, Description of the Rockland quadrangle: U.S. Geol. Survey Geologic Atlas of the United States, Folio 158, 15 p.

Bickel, C. E., 1971, Bedrock geology of the Belfast quadrangle, Maine [Ph.D. dissert.]: Cambridge, Mass., Harvard Univ., 342 p.

——1974, Metamorphism in the Belfast area, Maine, *in* Osberg, P. H., ed., Geology of east-central and north-central Maine: New England Intercoll. Geol. Conf., guidebook, p. 1-22.

Boucot, A. J., 1961, Stratigraphy of the Moose River synclinorium, Maine: U.S. Geol. Survey Bull. 1111-E, p. 153-188.

Boucot, A. J., Brookins, D. G., Forbes, W., and Guidotti, C. V., 1972, Staurolite zone Caradoc (Middle-Late Ordovician) age, Old World Province brachiopods from Penobscot Bay, Maine: Geol. Soc. America Bull., v. 83, p. 1953-1960.

Brookins, D. G., 1975, Geochronologic contributions to stratigraphic interpretation and correlation in the Penobscot Bay area, eastern Maine, *in* Page, L. R., ed., Contributions to the stratigraphy of New England: Geol. Soc. America Mem. 148 (this volume).

Brookins, D. G., Berdan, J. M., and Stewart, D. B., 1973, Isotopic and paleontologic evidence for correlating three volcanic sequences in the Maine coastal volcanic belt: Geol. Soc. America Bull., v. 84, p. 1619-1628.

Cheney, E. S., 1967, Reconnaissance and economic geology of the northwestern Knox County marble belt: Maine Geol. Survey Spec. Econ. Studies Ser., no. 7, Bull. no. 19, 32 p.

Dow, G. M., 1965, Petrology and structure of North Haven Island and vicinity, Maine [Ph.D. dissert.]: Urbana, Illinois Univ., 146 p.
Green, J. C., and Guidotti, C. V., 1968, The Boundary Mountains anticlinorium in northern New Hampshire and northwestern Maine, *in* Zen, E., and others, eds., Studies of Appalachian geology: Northern and Maritime: New York, Wiley-Interscience, p. 255-266.
Hatheway, R. B., 1971, Evidence for major faulting in south-central Maine: Geol. Soc. America Bull., v. 82, p. 253-258.
Hitchcock, C. H., 1863, Geology of Maine: 2nd Annual Report upon the Natural History and Geology of the State of Maine, p. 223-430.
Hussey, A. M., II, 1968, Stratigraphy and structure of southwestern Maine, *in* Zen, E., and others, eds., Studies of Appalachian geology: Northern and Maritime: New York, Wiley-Interscience, p. 291-301.
——1971, Geologic map of the Portland quadrangle, Maine: Maine Geol. Survey, GM-1, scale 1:62,500.
Keroher, G. C., and others, 1966, Lexicon of geologic names of the United States for 1936-1960: U.S. Geol. Survey Bull. 1200, 3 parts, 4341 p.
Larrabee, D. M., and Spencer, C. W., 1963, Bedrock geology of the Danforth quadrangle, Maine: U.S. Geol. Survey GQ-221, scale 1:62,500.
Ludman, A., and Griffin, J. R., 1974, Stratigraphy and structure of central Maine, *in* Osberg, P. H., ed., Geology of east-central and north-central Maine: New England Intercoll. Geol. Conf., guidebook, p. 154-179.
Osberg, P. H., 1968, Stratigraphy, structural geology, and metamorphism of the Waterville-Vassalboro area, Maine: Maine Geol. Survey Bull. 20, 64 p.
——1974, Foreword, *in* Osberg, P. H., ed., Geology of east-central and north-central Maine: New England Intercoll. Geol. Conf., guidebook, p. iii-xiii.
Osberg, P. H., and Guidotti, C. V., 1974, The geology of the Camden-Rockland area, *in* Osberg, P. H., ed., Geology of east-central and north-central Maine: New England Intercoll. Geol. Conf. guidebook, p. 48-60.
Perkins, E. H., and Smith, E.S.C., 1925, Contributions to the geology of Maine, no. 1; A geological section from the Kennebec River to Penobscot Bay: Am. Jour. Sci., v. 9, p. 204-228.
Smith, G. O., Bastin, E. S., and Brown, C. W., 1907, Description of the Penobscot Bay quadrangle: U.S. Geol. Survey Geologic Atlas of the United States, Folio 149, 14 p.
Stewart, D. B., 1974, Precambrian rocks of Seven Hundred Acre Island and development of cleavage in the Islesboro Formation, *in* Osberg, P. H., ed., Geology of east-central and north-central Maine: New England Intercoll. Geol. Conf., guidebook, p. 86-98.
Stewart, D. B., and Wones, D. R., 1974, Bedrock geology of northern Penobscot Bay area, *in* Osberg, P. H., ed., Geology of east-central and north-central Maine: New England Intercoll. Geol. Conf., guidebook, p. 223-239.
Stoesser, D. B., 1966, Geology of a portion of the Great Pond quadrangle, Maine [M.A. dissert.]: Orono, Maine Univ., 87 p.
Trefethen, J. M., 1950, Bucksport-Orland map area: New England Intercoll. Geol. Conf., guidebook, 5 p.
Wingard, P. S., 1961, Geology of the Castine-Blue Hill area, Maine [Ph.D. dissert.]: Urbana, Illinois Univ., 138 p.
Wones, D. R., 1974, Igneous petrology of some plutons in the northern part of the Penobscot Bay area, *in* Osberg, P. H., ed., Geology of east-central and north-central Maine: New England Intercoll. Geol. Conf., guidebook, p. 99-125.
Zartman, R. E., Hurley, P. M., Krueger, H. W., and Giletti, B. J., 1970, A Permian disturbance of K-Ar radiometric ages in New England: Its occurrence and cause: Geol. Soc. America Bull., v. 81, p. 3319-3374.

MANUSCRIPT RECEIVED BY THE SOCIETY APRIL 18, 1974
REVISED MANUSCRIPT RECEIVED NOVEMBER 10, 1975
MANUSCRIPT ACCEPTED DECEMBER 16, 1975

Printed in U.S.A.

Geological Society of America
Memoir 148

Geochronologic Contributions to Stratigraphic Interpretation and Correlation in the Penobscot Bay Area, Eastern Maine

Douglas G. Brookins
Department of Geology
University of New Mexico
Albuquerque, New Mexico 87131

ABSTRACT

The Rb-Sr whole-rock method has been used to test stratigraphic correlations in the Penobscot Bay area, eastern Maine. The Castine Volcanics, Thorofare Andesite-Vinalhaven Rhyolite, and Cranberry Islands Volcanics are consanguineous and were probably formed 390 ± 5 m.y. ago; this Devonian age is supported by new paleontologic evidence. Calcite-bearing rocks from these three formations usually yield spurious results and are considered unsuitable for Rb-Sr work; careful petrographic and x-ray studies are necessary for selection of suitable samples.

Rb-Sr ages of Cambrian-Ordovician (or possibly older) rocks were determined for the Ellsworth Schist that unconformably underlies the Devonian Castine Volcanics and Middle Silurian Ames Knob Formation on the east side of Penobscot Bay. The part of the Penobscot Formation on the west side of Penobscot Bay, informally called Benner Hill by C. V. Guidotti, yielded an age of 460 ± 10 m.y., in good agreement with new fossil evidence for a Middle to Late Ordovician age. Silurian or older ages were obtained for the Penobscot, Appleton Ridge, and Battie Formations; however, the overlap in error limits leaves their relative ages in doubt.

Retrograded high-grade rocks in a horst in Islesboro Township have been dated as Precambrian; a muscovite pegmatite cutting these rocks yields a late Precambrian mineral date. These are the first Precambrian rocks reported from Maine. The major faults mapped by D. B. Stewart on either side of Islesboro Township occurred before Middle Devonian time, according to dates determined for the stratigraphic units and for the plutons that cut and seal the faults.

INTRODUCTION

Radiometric data determined by the U, Th-Pb, K-Ar, and Rb-Sr methods, coupled with geologic mapping, have aided greatly in the understanding of Appalachian geology. Many of these data and their utilization are summarized in Faul and others (1963), Lyons and Faul (1968), Tilton (1965), Zartman and others (1965, 1970), Brookins and Hurley (1965), Bottino and Fullagar (1966), Fullagar and Bottino (1968), and Spooner and Fairbairn (1970). Dating of minerals has defined limits of periods of metamorphism, and Rb-Sr whole-rock studies have proved useful

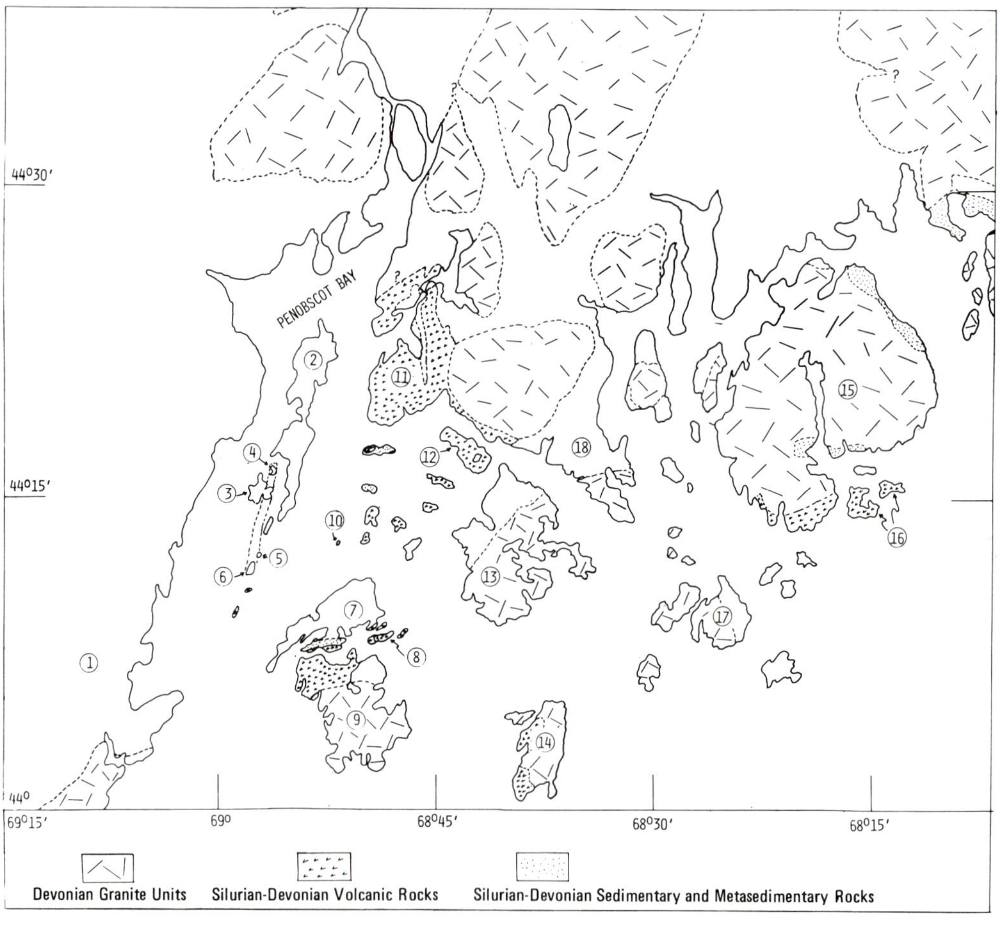

Figure 1. Location map of the Penobscot Bay region, Maine, showing selected outcrops (modified from Bastin, 1908; Stewart, 1956, and unpub. material; Wingard, 1961; McGregor, 1964; Dow, 1965; Doyle, 1967; Chapman, 1968; Karner, 1968). Numerous islands have been omitted for clarity. 1, Middle to Late Ordovician fossil site; 2, Islesboro Island; 3, Seven Hundred Acre Island; 4, Spruce Island; 5, Lime Island; 6, Lasell Island; 7, North Haven Island; 8, Stimpsons Island; 9, Vinalhaven Island; 10, Compass Island; 11, Harborside Mine (site of most samples of Castine Volcanics); 12, Little Deer Isle; 13, Deer Isle; 14, Isle au Haut; 15, Mount Desert Island; 16, Cranberry Islands; 17, Swans Island; 18, Blue Hill Peninsula. The dashed line around Spruce, Lime, Lasell, and the eastern part of Seven Hundred Acre Islands outlines the Precambrian rocks found in the area. No attempt is made to show map units in those areas left blank because of uncertainties in the available data; the reader is referred to Doyle (1967) for the available reconnaissance geologic mapping.

in pinpointing geologic boundaries on the absolute time scale. In many places, radiometric dating has confirmed stratigraphic interpretations that were based on field work, and the absolute ages have been applied to various units; in other places, such dating has caused revisions of the geologic interpretations.

In only a few areas (for example, Faul and others, 1963) has a systematic geochronologic study been an integral part of the detailed geologic mapping investigations as has been the case in the Penobscot Bay region of Maine (Fig. 1). Reconnaissance geologic mapping by Doyle (1967) and areal mapping by D. B. Stewart, C. E. Bickel, C. V. Guidotti, P. H. Osberg, D. R. Wones (1971, oral communs.), as well as the work of earlier investigators, allow detailed comparison of the geochronologic results reported here with the results of field work. The locations and brief description of the samples that were studied geochronologically are given in Table 1.

METHODS

The background information for K-Ar dating is given by Faul and others (1963) and for Rb-Sr whole-rock dating by Brookins (1963). Rb-Sr whole-rock dating has the advantage over K-Ar dating in that an age of the rocks can be determined by the isochron method, providing that the whole-rock system has remained closed during any postformational disturbance. A group of unaltered samples with a spread in Rb/Sr greater than three is required in order that the initial Sr^{87}/Sr^{86} ratio can be determined by extrapolation. Some samples with very high Rb/Sr (and therefore high Sr^{87}/Sr^{86} ratio for Paleozoic rocks) can be used by assuming an initial Sr^{87}/Sr^{86} ratio; such rocks are commonly granitic varieties although such favorable ratios are found in some metasedimentary rocks. In most metasedimentary rocks, the problems of excess inherited radiogenic Sr^{87} and of anomalous resetting of the Rb-Sr system made the use of an assumed initial Sr^{87}/Sr^{86} ratio extremely risky. Rb and Sr can be measured to ± 1 percent by x-ray spectrography for samples with more than 10 ppm of each; isotope dilution must be used to determine low Sr concentrations in the presence of high Rb concentrations. Sr^{87}/Sr^{86} ratios are routinely measured to better than ± 0.0002 in the laboratory at the University of New Mexico. The decay constant for Rb^{87} of 1.39×10^{-11}/yr has been used (for justification, see Spooner and Fairbairn, 1970; Brookins and Spooner, 1970). All isochron ages, initial Sr^{87}/Sr^{86} ratios, and errors for both, except where noted, have been calculated using the least-squares method described by York (1966).

The wide scatter of data about some whole-rock isochrons may be caused by erroneous sampling that results from faulty or insufficient geologic control, incipient alteration, comparison of mineral as opposed to whole-rock systems, or real differences in the samples. It is because of such uncertainties that relatively large errors have been assigned to the ages of some of the rocks. Considerable overlap in possible ages are noted in some units, but the ages of many formations can be clearly resolved.

There is no absolutely positive way to state that a K-Ar mineral age is either older or younger than an age obtained by the Rb-Sr whole-rock method. However, a single K-Ar mineral age is less often representative of the real age than is an isochron age obtained from several samples with the required spread in Rb^{87}/Sr^{86}, unless it can be unequivocally demonstrated that some factor or factors have caused perturbance of the Rb-Sr whole-rock system(s) without affecting the K-Ar system of individual minerals (an unlikely case). Ar is usually quite mobile relative to radiogenic Sr^{87}, and K-Ar ages are commonly younger than Rb-Sr whole-rock

TABLE 1. LOCATIONS AND BRIEF DESCRIPTIONS OF SAMPLES

Sample	Rock type	Quadrangle	Approximate location	North latitude	West longitude
			Castine Volcanics		
*Me 2, 5	Very fine grained volcanic-sedimentary rock	Castine	Callahan mine, pit	42°21.2'	68°48.5'
Me 102	Devitrified rhyolite glass	Castine	Callahan mine, 1.22-meter diameter drill core recovered from shaft-sinking operation: 3.05-meter depth	44°19.9'	68°48'
Me 104	Devitrified rhyolite glass	Castine	10.6-meter depth	44°19.9'	68°48'
Me 106	Lithic fragmental rhyolitic tuff	Castine	19.8-meter depth	44°19.9'	68°48'
Me 112	Strongly banded rhyolite flow	Castine	34.8-meter depth	44°19.9'	68°48'
Me 115	Strongly banded rhyolite flow	Castine	47.9-meter depth	44°19.9'	68°48'
Me 116	Devitrified (spherulitic in part) rhyolite	Castine	48.5-meter depth	44°19.9'	68°48'
Me 117	Devitrified rhyolite flow	Castine	78.1-meter depth	44°19.9'	68°48'
Me 118	Devitrified rhyolite with pyrite	Castine	79.3-meter depth	44°19.9'	68°48'
*Me 6	Volcanic sedimentary rock	Castine	Callahan Mine drainage ditch	44°19.8'	68°47.6'
Me 7	Silty, volcanic fragment-rich, sedimentary rock	Castine	Callahan Mine drainage ditch	44°19.8'	68°47.6'
Me 8	Metarhyolite	Castine	Callahan Mine drainage ditch	44°19.8'	68°47.6'
Me 9	Fine-grained metarhyolite	Castine	W. side Blake Point	44°18.5'	68°48.3'
*Me 10	Metadiorite or mafic meta-trachyte	Castine	Orr Cove	44°19.5'	68°49.3'
*Me 13	Garnet-grade metasedimentary rock	Castine	E. side Smith Cove	44°22.5'	68°46'
*Me 14	Metadiorite (greenschist)	Blue Hill	N. of Buck Harbor	44°20.6'	68°44.8'
Me 15	Volcanic-sedimentary rock	Blue Hill	Buck Harbor	44°20.5'	68°44.5'
*Me 16	Schistose metavolcanic rock	Blue Hill	Grays Hill Cemetery, Little Deer Isle	44°17.9'	68°43.7'
*Me 17	Metamorphic rock of uncertain origin	Blue Hill	Little Deer Isle	44°18.5'	68°44.2'
			Thorofare Andesite and Vinalhaven Rhyolite		
*Me 22	Volcanic-sedimentary rock in Thorofare Andesite	Vinalhaven	SE side of Southern Harbor, North Haven Island	44°7.8'	68°53.4'
*Me 23	Quartz-rich sedimentary rock	Vinalhaven	Fox Rocks, Vinalhaven Island	44°6.3'	68°53'
Me 24	Devitrified rhyolite with abundant opaque minerals	Vinalhaven	Fox Rocks, Vinalhaven Island	44°6.3'	68°53'

Me 25	Devitrified rhyolite	Vinalhaven	Along road from North Haven-Vinalhaven Ferry Landing to Fox Rocks, Vinalhaven Island	44°6.4'	68°53.6'
Me 27	Rhyolite tuff	Vinalhaven	Along road from North Haven-Vinalhaven Ferry Landing to Fox Rocks, Vinalhaven Island	44°7.1'	68°53.7'
Me 28	Spherulitic rhyolite tuff with collapsed pumice shards	Vinalhaven	Along road from North Haven-Vinalhaven Ferry Landing to Fox Rocks, Vinalhaven Island	44°7.2'	68°53.6'
Me 29	Rhyolite tuff	Vinalhaven	Along road from North Haven-Vinalhaven Ferry Landing to Fox Rocks, Vinalhaven Island	44°7.3'	68°53.5'
Me 30	Rhyolite tuff	Vinalhaven	Along road from North Haven-Vinalhaven Ferry Landing to Fox Rocks, Vinalhaven Island	44°7.3'	68°53.5'
Me 31	Coarse, fragmental latite tuff	Vinalhaven	Along road from North Haven-Vinalhaven Ferry Landing to Fox Rocks, Vinalhaven Island	44°7.3'	68°53.4'
Me 32	Vitric tuff	Vinalhaven	Along road from North Haven-Vinalhaven Ferry Landing to Fox Rocks, Vinalhaven Island	44°7.2'	68°53'
*Me 33	Vitric tuff with some bedded parts	Vinalhaven	Along road from North Haven-Vinalhaven Ferry Landing to Fox Rocks, Vinalhaven Island	44°7.3'	68°53.3'
			Cranberry Island Series		
Me 134a	Devitrified glass containing feldspar and trachyte fragments	Bar Harbor	Little Cranberry Island	44°15.2' to 44°15.8'	68°14.5'
Me 134b	Vitric crystal tuff	Bar Harbor	Little Cranberry Island	44°15.2' to 44°15.8'	68°14.5'
Me 134c	Coarse-grained latite	Bar Harbor	Little Cranberry Island	44°15.2' to 44°15.8'	68°14.5'
*Me 134d	Volcanic-sedimentary rock	Bar Harbor	Little Cranberry Island	44°15.2' to 44°15.8'	68°14.5'
*Me 134e	Quartz-feldspar sedimentary rock or recrystallized glass	Bar Harbor	Little Cranberry Island	44°15.2' to 44°15.8'	68°14.5'
*Me 134f	Metamorphic rock of uncertain origin	Bar Harbor	Little Cranberry Island	44°15.2' to 44°15.8'	68°14.5'
Me 134g	Coarse-grained volcanic sedimentary rock	Bar Harbor	Little Cranberry Island	44°15.2' to 44°15.8'	68°14.5'
Me 134h	Rhyolite flow	Bar Harbor	Little Cranberry Island	44°15.2' to 44°15.8'	68°14.5'
Me 134i	Volcanic-rich, lithic sedimentary rock	Bar Harbor	Little Cranberry Island	44°15.2' to 44°15.8'	68°14.5'
*Me 134j	Rhyolitic volcanic-sedimentary rock	Bar Harbor	Little Cranberry Island	44°15.2' to 44°15.8'	68°14.5'

Continued on next page.

TABLE 1. *(Continued)*

Sample	Rock type	Quadrangle	Approximate location	North latitude	West longitude
			Cranberry Island Series Continued		
Me 135a	Originally spherulitic rhyolite glass	Mount Desert	Preble Cove, Great Cranberry Island	44°15.1' to 44°15.8'	68°16.3'
Me 135b	Devitrified rhyolite containing some feldspar crystals and spherulites	Mount Desert	Preble Cove, Great Cranberry Island	44°15.1' to 44°15.8'	68°16.3'
Me 135c	Devitrified rhyolite	Mount Desert	Preble Cove, Great Cranberry Island	44°15.1' to 44°15.8'	68°16.3'
Me 135d	Devitrified spherulitic rhyolite glass	Mount Desert	Preble Cove, Great Cranberry Island	44°15.1' to 44°15.8'	68°16.3'
Me 135e	Fine-grained devitrified rhyolite	Mount Desert	Preble Cove, Great Cranberry Island	44°15.1' to 44°15.8'	68°16.3'
Me 135f	Volcanic-sedimentary rock	Mount Desert	Preble Cove, Great Cranberry	44°15.1' to 44°15.8'	68°16.3'
			Ellsworth Schist		
Me 19	Dacite tuff	Deer Isle		44°17'	68°40.5'
Me 177	Greenschist	Blue Hill		44°18'	68°37.9'
Me 178	Greenschist	Blue Hill		44°18'	68°37.9'
Me 179	Felsic tuff	Blue Hill		44°18'	68°37.9'
Me 180	Felsic tuff	Blue Hill		44°18'	68°37.9'
Me 181	Latite tuff	Blue Hill		44°17'	68°37.2'
Me 183	Fissil greenstone	Blue Hill		44°18'	68°31.8'
Me 185	Felsic tuff	Blue Hill		44°18.3'	68°37.4'
Me 186	Contact metamorphosed felsic tuff (cordierite-garnet)	Blue Hill		44°23'	68°34.1'
Me 188	Contact metamorphosed felsic tuff (cordierite-garnet)	Blue Hill		44°29.8'	68°32.5'
Me 189	Felsic tuff	Blue Hill		44°29.7'	68°31'
Me 191	Dacitic greenschist	Ellsworth		44°31.5'	68°25.5'
			Waldo Granite		
Me 39a	Biotite granite	Bucksport		44°30.4'	68°50.13'
Me 39b	Biotite granite	Bucksport		44°30.4'	68°50.13'
Me 39c	Biotite granite	Bucksport		44°30.4'	68°50.13'
			Wallamatogus Granite		
R 1064a-c	2-mica granite	Castine		44°29.8'	68°45.4'
			Cak Point Granite		
R 1102	Rapakivi granite	Blue Hill		44°15.9'	68°31.2'
Me 151a, b	Rapakivi granite	Deer Isle		44°12.9'	68°39.1'
Me 152a, b	Aplite	Deer Isle		44°12.9'	68°39.1'

		Stonington Granite		
Me 153a-d	Rapakivi granite	Deer Isle	44°10.8'	68°39.1'
Me 154a-b	Rapakivi granite	Deer Isle	44°9'	68°40.4'
		Vinalhaven Granite		
Me VHa-e		Vinalhaven	44°2.8'	68°50.8'
		South Penobscot Granite		
R 1063a	Biotite granite (2)	Blue Hill	44°26'	68°42.1'
R 1063b	Biotite granite (2)	Blue Hill	44°27.3'	68°43'
		Sedgwick Granite		
R 1057	Biotite granite	Blue Hill	44°21.2'	68°44.2'
R 1058a, b	Biotite granite	Blue Hill	44°20.8'	68°42.5'
R 1059a, b	Biotite granite	Blue Hill	44°21.0'	68°35.2'
		East Blue Hill Granite		
R 1061a, b	Biotite granite	Blue Hill	44°24.9'	68°31.0'
R 1062a, b	Biotite granite	Blue Hill	44°24.8'	68°33.5'
		Appleton Ridge Formation		
Me 158	Phyllite	Belfast	44°20.4'	69°11.6'
Me 160	Amphibolite	Belfast	44°19.1'	69°12.3'
		High-grade rocks on Spruce Island		
Me 168	Pegmatite	Castine	44°16.5'	68°56.2'
Me 169	Pegmatite	Castine	44°16.5'	68°56.2'
Me 170	Marble	Castine	44°16.5'	68°56.2'
Me 171	Garnet amphibolite	Castine	44°16.5'	68°56.2'
Me 172	Meta-andesite	Castine	44°16.5'	68°56.2'
Me 173	Mica quartzite	Castine	44°16.5'	68°56.4'
		Penobscot Formation		
Me 36	Sulfidic shale	Castine	44°27'	68°51.1'
Me 37	Sulfidic, sandy shale	Castine	44°28.2'	68°49.1'
Me 44	Andesite schist, dikes	Castine	44°21.9'	68°58.6'
Me 47	Silicic schist	Castine	44°18.5'	68°59.7'
Me 48	Andesitic schist	Castine	44°18.5'	68°58.4'
		Battie Formation		
Me 50(a-f)	6 samples of mica quartzite	Camden	44°13.1'	69°4.2'
Me 150(a-g)	7 samples of mica quartzite	Camden	44°13.2'	69°4.0'

* Volcanic rocks containing calcite or minerals indicative of a Ca-rich protolith.

ages owing to diffusion of radiogenic Ar^{40}, although some anomalously old K-Ar ages have been reported. The number of samples and types of minerals used for the K-Ar determinations is critical; just as one must select several samples for Rb-Sr work, so should one determine K-Ar ages for several samples of the same mineral and, if possible, on several different minerals in the host rock. This is mentioned here only because it has not been the common practice in the area of study; the reader will note in Table 1 that a single K-Ar age on a single mineral (without statement of error) has been used to date enormous volumes of rock. Similarly, Rb-Sr ages may not be representative of a unit if all of the whole-rock samples have been taken from a single outcrop. For this reason, I have increased the error in the ages of some units reported in Tables 1 to 4 and 6 if the number of samples is less than five or if the spread in Rb^{87}/Sr^{86} is less than three. For these ages, the error (indicated by an asterisk) has been calculated by constructing error limits about each point and then constructing maximum and minimum isochrons. A set of three samples, closely collinear and with a range of Rb^{87}/Sr^{86} of less than three, may yield a lower error by the York (1966) method than by this rather arbitrary "error box" method; however, the latter method is more realistic from a geologist's point of view in that by so increasing the error limits, one is not guilty of attempting to pinpoint some event by the use of statistics alone. Fortunately, this had to be done for only a few of the rock units studied in the Penobscot Bay area, and a more comprehensive study of these is planned.

DEVONIAN GRANITE PLUTONS

More geochronologic data have been determined from the Maine coastal Devonian granite plutons than from any other group of rocks in Maine (see Faul and others, 1963; Brookins, 1968; Lyons and Faul, 1968; Spooner and Fairbairn, 1970; Brookins and Spooner, 1970; and this paper). These data are summarized in Table 2. In addition, the petrologic, mineralogic, and chemical contents of many of these plutons have been discussed in detail by Stewart (1956), Wingard (1961), Chapman (1968), Page (1968), and Karner (1968).

A bimodal distribution of granitic emplacement in Maine at about 400 m.y. and 360 m.y. ago was argued for by Faul and others (1963) and Spooner and Fairbairn (1970). However, I (Brookins, 1968, and this paper) have found several plutons of intermediate age. Page (1968) argued for a Late Devonian age for these plutons—a view that is in conflict with the presence of 400-m.y.-old plutons (see Table 2).

The ages of a pluton determined by several methods may disagree with each other (Table 2). The ages of plutons so dated must also be viewed with caution because the dates by single minerals do not always agree within a single pluton or with Rb-Sr whole-rock isochrons. In at least two plutons (Waldo and Wallamatogus), the K-Ar biotite ages are younger than the Rb-Sr whole-rock ages, and in at least two others (Sedgwick and Vinalhaven plutons), the K-Ar biotite ages are older than the Rb-Sr whole-rock ages. The plutons clearly intrude their country rocks and, thus, provide a cross-check between radiometric determinative methods.

The rapakivi granite plutons (Oak Point, Stonington, Vinalhaven, part of Tunk Lake) are younger than most of the nonrapakivi granite plutons (see Table 2). Only the Wallamatogus granite contains muscovite, and the K-Ar (and Rb-Sr) muscovite age is older than the ages obtained from biotite and the whole rock. The reasons for this are not clear.

Initial Sr^{87}/Sr^{86} ratios in the plutons range from about 0.704 to 0.712 (Brookins, 1968; Spooner and Fairbairn, 1970), and this emphasizes the need for establishing

TABLE 2. K-Ar AND Rb-Sr AGES OF DEVONIAN GRANITES FROM SOUTHEASTERN COASTAL MAINE

Pluton	Geochronologic age (m.y.)	Method	Material	Source
Stonington (rapakivi)	350, 360	K-Ar	2 biotite	1
Stonington (rapakivi)	341 ± 21	Rb-Sr	6 whole rock	2
Oak Point (rapakivi)	343 ± 15	K-Ar	1 biotite	3
Oak Point (rapakivi)	354	K-Ar	1 biotite	1
Oak Point (rapakivi)	357 ± 1	Rb-Sr	6 whole rock	2
Vinalhaven (rapakivi)	399	K-Ar	1 biotite	1
Vinalhaven (rapakivi)	361 ± 7	Rb-Sr	5 whole rock	5
Sedgwick	413 ± 15	K-Ar	1 biotite	3
Sedgwick	395 ± 15	Rb-Sr	5 whole rock	4
East Blue Hill	387	K-Ar	1 biotite	1
East Blue Hill	391	Rb-Sr	1 biotite	1
East Blue Hill	382 ± 10	Rb-Sr	4 whole rock	4
South Penobscot	360 ± 15	K-Ar	1 biotite	3
South Penobscot	361 ± 12	Rb-Sr	4 whole rock	4
Wallamatogus	379	K-Ar	1 muscovite	1
Wallamatogus	413	Rb-Sr	1 muscovite	1
Wallamatogus	330	K-Ar	1 biotite	1
Wallamatogus	349	Rb-Sr	1 biotite	1
Wallamatogus	371 ± 15	Rb-Sr	3 whole rock	4
Waldo	390 ± 10	Rb-Sr	3 whole rock	5
Waldo	325	K-Ar	1 biotite	6
Tunk Lake	357 ± 10	K-Ar	1 biotite	7

Note: Sources are 1, Faul and others (1963); 2, Brookins and Spooner (1970); 3, Wingard and Brookins (1964); 4, Brookins (1968); 5, D. G. Brookins (Unpub. data); 6, Zartman and others (1970); 7, F. R. Karner (1968, written commun.).

an isochron rather than assuming an initial value. Use of initial Sr^{87}/Sr^{86} ratio data alone for correlation has yet to be convincingly demonstrated (Lyons and Faul, 1968).

VOLCANIC ROCKS

Many of the volcanic rocks in the Penobscot Bay region and elsewhere along the Atlantic seaboard from Newburyport, Massachusetts, to Nova Scotia were thought to be of Silurian age on the basis of early biostratigraphic work (see summary in Gates, 1969), but more recent work has demonstrated that at least some of these rocks are Devonian. New and previously reported geochronologic data are presented in Table 3.

A correlation of the Castine Volcanics with the volcanic rocks on North Haven and Vinalhaven (Vinalhaven Rhyolite plus Thorofare Andesite) and Cranberry Islands Volcanics was proposed by D. B. Stewart (1968, 1969, written communs.; see also Brookins and others, 1969). Presumably many of the volcanic rocks of similar appearance on the islands of Penobscot Bay are also coeval with these rocks. A Silurian age previously was assigned to these volcanic rocks on the basis of paleontologic and geochronologic control. Dow (1965) mentioned that the Thorofare Andesite underlies the Vinalhaven Rhyolite and overlies the Ames Knob Formation on North Haven Island; abundant Silurian fossils have been recovered from the

TABLE 3. GEOCHRONOLOGIC DATA FROM SOUTHEASTERN COASTAL MAINE VOLCANIC ROCKS

Rock unit	Number of samples	Rb-Sr whole-rock isochron age (m.y.)	Possible geologic age	Spread in Rb^{87}/Sr^{86}	Initial Sr^{87}/Sr^{86}
Vinalhaven Rhyolite Thorofare Andesite	8	393 ± 6	E. Devonian	4-16	0.7070 ± 0.0012
Castine Volcanics	11	390 ± 5	E. Devonian	0.7-34	0.7079 ± 0.0015
Cranberry Islands Volcanics	12	387 ± 9	E. Devonian	1.8-12	0.7061 ± 0.0015
Quoddy Formation		420 ± 15	L. Silurian		
Denny's Formation		415 ± 15	L. Silurian		
Pembrook Formation		410 ± 15	E. Devonian	from Fullagar and Bottino (1968)	
Arisaig Formation		430 ± 15	M. Silurian(?)		
Eastport Formation		412 ± 5	E. Devonian		
Hedgehog Formation*		413 ± 10	E. Devonian		

* Not coastal Maine but is exposed near Presque Isle.

Ames Knob Formation (see Gates, 1969). Further, although the error limits for the K-Ar ages are not given in Table 1, Faul and others (1963) reported Early Devonian or pre-Devonian K-Ar ages from the Sedgwick Granite that intrudes the Castine Volcanics and from the Vinalhaven Granite that intrudes the Vinalhaven Rhyolite.

More recent paleontologic and Rb-Sr whole-rock ages suggest a Devonian age for the above rocks. The whole-rock isochrons from the Sedgwick and Vinalhaven Granite yield ages of 395 ± 15 m.y. and 361 ± 7 m.y., respectively. Thus, at least the Vinalhaven Granite is certainly of Devonian age; as pointed out by Stewart (1956), the Vinalhaven Granite has close affinities to the Stonington Granite. They are used interchangably for building purposes, and the Stonington Granite also has been dated (Faul and others, 1963) by the K-Ar method at 350 and 360 m.y. and by the Rb-Sr whole-rock method at 341 ± 21 m.y. (Brookins and Spooner, 1970). Hence, the Rb-Sr age does not necessitate that the intruded Vinalhaven Rhyolite be of Silurian age. The Rb-Sr whole-rock age for the Sedgwick Granite is also consistent with a Devonian age. The data on samples from the Castine, Vinalhaven, and Cranberry Islands volcanic rocks yield Rb-Sr whole-rock ages (Table 3) of 390 ± 5 m.y., 393 ± 6 m.y., and 387 ± 9 m.y., respectively. These are Early Devonian rather than Silurian ages (presuming the Silurian-Devonian boundary is close to 410 to 415 m.y., which is older than the value of 400 m.y. in use by the U.S. Geological Survey). In addition, J. Berdan (1970, written commun.; 1971) reported an occurrence of Early Devonian fauna from the Ames Knob Formation on Stimpsons and Vinalhaven Islands and possibly Devonian ostracodes from a thin limestone interbedded between rhyolite and andesite on Compass Island. She also recognized an Early Devonian fauna in the Eastport quadrangle in rocks intercalated with volcanic rocks that yielded many of the samples dated by Bottino and Fullagar (1966). The fossil and isotopic data indicate that at least these parts of the volcanic rocks in Penobscot Bay area are probably Devonian rather than Silurian.

SUITABILITY OF VOLCANIC ROCKS FOR Rb-Sr GEOCHRONOLOGIC STUDIES

There has been considerable controversy over the use of volcanic rocks for Rb-Sr whole-rock dating. Cormier (1969) showed that samples from the late Precambrian Coldbrook Formation (of southern New Brunswick, Canada; Gates, 1969) decrease in age with increasing Rb/Sr; only those samples with low Rb/Sr yield the most probable age, whereas the remainder of the data yields an erroneous Ordovician age. Incipient weathering along with cation exchange during devitrification of glassy portions of volcanic rocks has been suspected of causing such anomalies (Brookins, unpub. data).

The effects of alkali exchange during devitrification are not well understood, but, as proposed by Lipman and others (1968), if the process is principally one involving volatile loss or if ionic exchange occurs soon after initial formation (that is, a short time interval relative to the age of the rock) then the whole-rock system(s) will not be greatly affected. Brooks and others (1969) indicated that the presence of 3 percent or less finely distributed calcite (presumably deposited at the time of formation) produces a ±40-m.y. error but does not drastically affect the age of some Archean metavolcanic rocks. Such an error for Archean rocks is obviously less serious than a similar one for Paleozoic rocks.

Certain volcanic formations in the Penobscot area either contain calcite (both finely disseminated calcite and vein or replacement calcite) or were calcite-bearing volcanic-sedimentary rocks before metamorphism or other forms of alteration. Samples from these volcanic formations were examined with the close cooperation and assistance of D. B. Stewart by x-ray and (or) petrographic methods to determine the distribution of calcium-bearing minerals and their compositions. Plagioclase in samples suitable for age determination by the Rb-Sr whole-rock method (Table 4) is commonly oligoclase, some of which may be retrograded from andesine. Such retrogradation can yield calcite and epidote, which may have a great effect in disturbing the Rb-Sr system. Examination of the data of Tables 3 and 4 for calcite-free and calcite-bearing volcanic rocks allows one to compare the effect of at least calcite on the Rb-Sr system. Most of calcite-bearing samples (see Table 4) were selected from the same horizon as calcite-free samples. The presence of calcite in some samples may offer evidence for open-system conditions and thus perturbance of the Rb-Sr system. Certainly, reliable fossil control is required to state whether the age for even the calcite-free volcanic rocks is correct; this has been tested for the Vinalhaven Rhyolite and Castine Volcanics, and because the same care in sampling and other control was used for the Cranberry Island units, there is no good reason to suspect that the three volcanic units are not coeval.

TABLE 4. GEOCHRONOLOGIC DATA FOR CALCITE-BEARING VOLCANIC ROCKS FROM SOUTHEASTERN COASTAL MAINE

Rock unit	Number of samples	Rb-Sr whole-rock isochron age (m.y.)	Possible geologic age	Spread in Rb^{87}/Sr^{86}	Initial Sr^{87}/Sr^{86}
Vinalhaven Rhyolite	3	350 ± 21	E. Devonian	1-10	0.711 ± 0.0018
Castine Volcanics	8	423 ± 34	E. Devonian	1-7	0.7082 ± 0.0029
Cranberry Islands Volcanics	4	368 ± 33	E. Devonian	0.8-4	0.7049 ± 0.0017

Table 4 summarizes the results for calcite-bearing volcanic rocks. The ages for the calcite-bearing samples differ by as much as ±40 m.y. from the ages for calcite-free samples from the same formations, and the errors reported are three to five times larger. The initial ratios derived from calcite-containing samples vary from about 0.705 to 0.711, whereas the initial ratios derived from the calcite-free samples vary much less, from 0.706 to 0.708. Although the number of calcite-bearing samples is small (most such samples were purposely avoided in the collecting), the results definitely demonstrate that extreme care is necessary in selecting samples for Rb-Sr whole-rock work. It should be emphasized here that, owing to the nature of exposure in this area (that is, possible exchange with sea water), calcite-bearing samples may be incorporated into geochronologic studies unless one carries out a thorough petrographic and x-ray investigation.

ROCKS OF PRE-EARLY DEVONIAN AGE

The pre-Early Devonian stratigraphy of the Penobscot Bay area is not well defined. There are virtually as many stratigraphic interpretations proposed as there have been investigators (see summary in Gates, 1969). On the east side of Penobscot Bay, the formation that exhibits the most intense regional deformation is the Ellsworth Schist. This formation is separated from the Castine Volcanics by a pronounced unconformity (Wingard, 1961). Fragments of Ellsworth Schist, as well as undeformed Middle Silurian fossils, are found on North Haven Island in the Ames Knob Formation, which also contains Early Devonian fossils (Berdan, 1971); this indicates that the Ellsworth Schist is at least of pre-Middle Silurian age (Stewart, 1956) and that one regional metamorphism occurred before Middle Silurian time. The various rock types found in the Ellsworth Schist were described by Stewart (1956), Wingard (1961), and McGregor (1964) and were summarized by Gates (1969). Reconnaissance collecting was carried out in the summer of 1970 in order to date the formation radiometrically. The rocks collected were initially classified as green schist or tan felsite. These two varieties, plus fissile schist, are intimately intermingled; thus separation of the green schist and the felsite into separate units is not warranted. This conflicts with an earlier statement of mine that a possible division between the two varieties could be made (which, in turn, might imply the existence of separate members of a formation, and so forth). Previously unpublished data for 12 samples (five of felsite and seven of green schist; Table 5) indicate that

TABLE 5. GEOCHRONOLOGIC DATA FOR PRE-EARLY DEVONIAN ROCKS FROM SOUTHEASTERN COASTAL MAINE

Rock unit	Number of samples	Rb-Sr whole-rock isochron age (m.y.)	Possible geologic age	Spread in Rb^{87}/Sr^{86}	Initial Sr^{87}/Sr^{86}
Battie Quartzite	13	425 ± 35	Silurian(?)	1-10	0.720 ± 0.006
Penobscot Formation (Castine quadrangle)	6	440 ± 20	Ordovician(?)	1-5	0.710 ± 0.003
Penobscot Formation (Benner Hill map unit)	7	460 ± 10	Ordovician(?)	2-21	0.706 ± 0.002
Appleton Ridge Formation	4	455 ± 25	Ordovician(?)	1-23	0.712 ± 0.003
Ellsworth Schist	12	510 ± 15	Ordovician(?)	0.2-5.0	0.707 ± 0.003

both varieties are part of one coeval rock unit. The twelve samples define a 510 ± 15-m.y. isochron. Because the initial ratio is less than 0.71 and the spread in Rb/Sr is 0.2 to 5.0, these preliminary ages may be considered a minimum. Thus, the Ellsworth Schist is significantly older than the Ames Knob Formation and the Castine Volcanics, a fact well recognized by field geologists.

Rocks of pre-Early Devonian age on the mainland west of Penobscot Bay are mostly quite different from the Ellsworth Schist, and the rocks on Islesboro and nearby islands (Fig. 1) are apparently different from either. Bastin (1908) mapped an extensive area west of Penobscot Bay as the Penobscot Formation (see Doyle, 1967). Some of the rock types within the Penobscot Formation appear similar to some found on Islesboro, but evidence for correlation is lacking. D. B. Stewart (1971, written commun.) pointed out that the main distinction between the Islesboro rocks and the rocks to the west is the metamorphic grade and the sequence of rock types. Quartz conglomerate, orthoquartzite, and impure dolomitic limestone of uncertain sequence are common on Islesboro but rare on the mainland except in the area mapped by C. E. Bickel (Jam Brook Formation; this volume). Otherwise, where these rocks are found on the mainland, the sequence is apparently different. Sulfidic slate is rare on Islesboro but common on the mainland, whereas intercalated sandy and shaly metasedimentary rocks are common to both areas. Furthermore, the style of deformation is different in the two terranes, and whereas chlorite-grade rocks occur on Islesboro, few rocks below biotite grade are found on the mainland. This is not surprising because Islesboro lies between two major faults (D. B. Stewart, unpub. map).

Along the northeastern side of Penobscot Bay, the Penobscot Formation is in fault contact (note "?" for contact north of loc. 11 in Fig. 1) with the Ellsworth Schist and the Castine Volcanics near the towns of Penobscot, Brooksville, and Castine. C. E. Bickel (this volume and 1975, written commun.) indicated that some of the units he has mapped in the Belfast quadrangle can be traced south-southwestward into the areas near Mount Battie and Rockland that are being mapped by P. H. Osberg and C. V. Guidotti. Some samples originally collected as Appleton Ridge Formation are now known to represent part of the Jam Brook Formation (C. E. Bickel, 1975, written commun.); yet the 455 ± 25-m.y. age reported at the 1971 symposium on the stratigraphy of New England for the Appleton Ridge Formation was based primarily on four samples from this formation. It must be pointed out that at the 1971 symposium, the original stratigraphic interpretation proposed by Bickel was questioned by P. H. Osberg (1971, oral commun.) and that Bickel's revised interpretation (this volume) is different from his 1971 version. There is disagreement on the correlation of some of the rocks mapped by Bickel with those mapped by Osberg to the south-southwest and with those mapped by D. R. Wones to the northeast (C. E. Bickel, 1975, written commun.).

Table 5 lists preliminary geochronologic data of the rocks of pre-Early Devonian age west of and at the north end of Penobscot Bay. Because of a narrow spread in Rb/Sr, samples of the Penobscot Formation from the Castine quadrangle define a poor 440 ± 40-m.y. isochron with an initial Sr^{87}/Sr^{86} ratio of 0.710 ± 0.003. Samples of the informally named Benner Hill map unit of the Penobscot Formation (C. V. Guidotti, 1971, oral commun.) yield a 460 ± 10-m.y. isochron with an initial Sr^{87}/Sr^{86} ratio of 0.7060 ± 0.002. Middle to Late Ordovician brachiopods of European affinity have been found in beds of mixed rock types that underly the Benner Hill in the northeast 1/9 of the Thomaston 7.5-minute quadrangle (loc. 1 in Fig. 1). These fossils were found by C. V. Guidotti, collected by William Forbes, and identified by A. J. Boucot. A report that discusses the paleontologic, geologic, and geochronologic aspects of the locality has been published (Boucot and others, 1972). The European affinities of these fossils and the fact that they occur in garnet-andalusite-grade rocks make them of extreme interest. The fossils occur in a 50- to 75-mm-thick weathered shelly bed within thick-bedded quartzite. The Benner Hill map unit (see Boucot and others, 1972) is characterized by quartzose pelite, biotite- and garnet-rich amphibolite, graphitic phyllite, and schist.

The paleontologic age of the fossils is in close agreement with the geochronologic age. These whole-rock systems were apparently unchanged by the Permian thermal event described by Zartman and others (1970), by regional metamorphism during the Acadian orogeny, or by Devonian plutonism. From the data in Table 4, it is not possible to distinguish the age differences between the Appleton Ridge Formation and either the Battie Formation or Penobscot Formation because the error limits on the ages overlap.

PRECAMBRIAN ROCKS IN THE TOWN OF ISLESBORO

The sequence of rocks on Islesboro and Seven Hundred Acre (and smaller) Islands is not correlated now with rocks on either side of Penobscot Bay, although some of the units in the town of Islesboro appear to be similar to some encountered to the south and west (D. B. Stewart and P. H. Osberg, 1970, oral communs.). P. C. Bateman, D. B. Stewart, and H. E. Belkin mapped a horst of retrograded garnet-andalusite gneiss and schist, garnet amphibolite gneiss, dolomitic marble, and mica quartzite on Spruce, Lime, Lasell, and the eastern side of Seven Hundred Acre Islands in Islesboro Township. This high-grade block yields a possible 750 ± 100(?)-m.y. isochron (Table 6) with an initial Sr^{87}/Sr^{86} ratio near 0.71 ± 0.01, *if* the rocks are coeval.

The sequence of rock types on Spruce Island (loc. 4 in Fig. 1) from east to west is dolomitic marble, garnet amphibolite, meta-andesite, garnet-andalusite gneiss, and mica quartzite. The samples (see Table 6) were taken from several rock types; hence, a large error was assigned to the possible 750-m.y. age. Much more work is needed on these rocks before an accurate age can be established for them. At the northern end of Spruce Island, the dolomitic marble is cut by muscovite pegmatite. The pegmatite yields (Table 6) a 620 ± 20-m.y. mineral isochron age (initial Sr^{87}/Sr^{86} ratio of 0.730 ± 0.005) and yields a K-Ar muscovite age of 594 ± 18 m.y. (R. L. Armstrong, 1971, written commun.).

STRUCTURAL AND REGIONAL IMPLICATIONS[1]

D. B. Stewart (1971, written commun.) mapped major faults on either side of Islesboro and on nearby islands. These faults explain (1) the difference in metamorphic grade between low-grade Paleozoic rocks in a horst containing Precambrian rocks on Islesboro Island and similar high-grade rocks on the west side of Penobscot Bay and (2) the position of the Castine Volcanics adjacent to the much older Ellsworth Schist. The preliminary geochronologic data support Stewart's structural interpretation.

The faults are cut and sealed by granitic plutons of Devonian age (D. B. Stewart, 1971, oral commun. and unpub. map). The approximate position of part of one of these faults is indicated by the "?" north of location 11 (Fig. 1), whereas

[1]Since preparation of this paper, some of the material has been published elsewhere (see articles by Boucot and others, 1972; Brookins and others, 1973).

TABLE 6. GEOCHRONOLOGIC DATA FOR PRECAMBRIAN ROCKS FROM SOUTHEASTERN COASTAL MAINE

Rock unit	Number of samples	Rb-Sr whole-rock isochron age (m.y.)	Possible geologic age	Spread in Rb^{87}/Sr^{86}	Initial Sr^{87}/Sr^{86}
Pegmatite on Spruce Island	3	620 ± 20 (594 ± 18 on muscovite by K-Ar)	Precambrian	0.4-4.0	0.730 ± 0.005
Garnet-grade rocks containing pegmatite	5	750 ± 100(?)	Precambrian	1-12	0.72 ± 0.01

the trace of the other is uncertain. Examination of the *Preliminary Geologic Map of Maine* (Doyle, 1967) shows a nearly knife-edge contact between the Lucerne Granite (the granite north-northwest of loc. 11 in Fig. 1; the outcrops extend beyond the northern limit of Fig. 1) and the undifferentiated metamorphic rocks at its west contact. This may tempt some to postulate a fault along the west side of the Lucerne Granite, yet D. R. Wones (1972, oral commun.) showed me outcrops where the granite exhibits a chilled margin against the metamorphic rocks; hence, the contact is most probably intrusive and not the result of faulting. Major post-Acadian faults are evident in the Gulf of Maine (M. F. Kane, 1972, written commun.) as well but have not been positively identified in the Penobscot Bay area (D. B. Stewart, 1971, written commun.). A strong angular unconformity below the Ames Knob Formation indicates a period of folding or faulting followed by uplift and erosion that affected the Ellsworth Schist and its correlatives (North Haven Greenstone) in pre-Early Devonian time.

The Precambrian mineral dates from the pegmatite on Spruce Island may indicate that the heat associated with the Devonian plutonism was quite local and insufficient to reset any mineral systems except those in the immediate plutonic aureoles. Relatively shallow emplacement of some of the Devonian plutons is indicated by the fact that they intrude Lower Devonian volcanic rocks. The pegmatite could have been foliated in any one of a number of pre-Devonian faulting episodes.

The presence of fossils of European affinity in this area suggests that a clearer picture of the relationships between the coastal, geosynclinal, and plate-hinge aspects of Paleozoic rocks will be forthcoming when more field and geochronologic data are available.

ACKNOWLEDGMENTS

Partial financial support for this study was provided by the National Science Foundation (Grants GA-10839 and GA-31139), the U.S. Geological Survey, and the Maine Geological Survey. R. L. Armstrong determined the K-Ar dates on muscovite from Spruce Island. Field guidance and communication with D. B. Stewart, J. Berdan, C. V. Guidotti, P. H. Osberg, C. E. Bickel, and A. M. Hussey have made this study possible. F. Beck kindly allowed me to sample a 122-m core from the Callahan Mining Company, Harborside Mine, to obtain fresh samples of the Castine Volcanics.

REFERENCES CITED

Bastin, E. S., 1908, Rockland folio, Maine: U.S. Geol. Survey Folio 158.

Berdan, J., 1971, Silurian to Early Devonian ostrocodes of European aspect from the Eastport quadrangle, Maine: Geol. Soc. America Abs. with Programs, v. 3, p. 18.

Bottino, M. L., and Fullagar, P. D., 1966, Whole-rock rubidium-strontium age of the Silurian-Devonian boundary in northeastern North America: Geol. Soc. America Bull., v. 77, p. 1167-1176.

Boucot, A. J., Brookins, D. G., Forbes, W. F., and Guidotti, C. V., 1972, Staurolite zone Caradoc (Middle-Late Ordovician) age, Old World Province brachiopods from Penobscot Bay, Maine: Geol. Soc. America Bull., v. 83, p. 1953-1960.

Brookins, D. G., 1963, Rubidium-strontium age investigations in the Middle Haddam and Glastonbury quadrangles, Connecticut [Ph.D thesis]: Cambridge, Massachusetts Inst. Technology, 213 p.

——1968, Whole-rock and mineral ages of granites and related rocks from the Castine-Blue Hill area, Maine: Am. Geophys. Union Trans., v. 49, p. 346.

Brookins, D. G., and Hurley, P. M., 1965, Rb-Sr whole rock studies in the Middle Haddam and Glastonbury quadrangles, Connecticut: Am. Jour. Sci., v. 263, p. 1-16.

Brookins, D. G., and Spooner, C. M., 1970, The isotopic ages of the Oak Point and Stonington granites, eastern Penobscot Bay, Maine: Jour. Geology, v. 78, p. 570-576.

Brookins, D. G., Stewart, D. B., and Chaudhuri, S., 1969, Comparison of the Rb-Sr ages of the Thorofare andesite and Vinalhaven rhyolite with the Castine volcanics, Penobscot Bay, Maine: EOS (Am. Geophys. Union Trans.), v. 50, p. 331.

Brookins, D. G., Berdan, J. M., Stewart, D. B., 1973, Isotopic and paleontologic evidence for correlating three volcanic sequences in the Maine coastal volcanic belt: Geol. Soc. America Bull., v. 84, p. 1619-1628.

Brooks, C., Hart, S. R., Krogh, T. E., and Davis, G. L., 1969, Carbonate contents and Sr^{87}/Sr^{86} ratios of calcites from Archean meta-volcanics: Earth and Planetary Sci. Letters, v. 6, p. 35-38.

Chapman, C. A., 1968, A comparison of the Maine coastal plutons and magmatic central complexes of New Hampshire, *in* Zen, E-an, and others, eds., Studies of Appalachian geology, northern and maritime: New York, Interscience Pubs., Inc., p. 385-398.

Cormier, R. F., 1969, Radiometric dating of the Coldbrook Group of southern New Brunswick, Canada: Canadian Jour. Earth Sci., v. 6, p. 393-398.

Dow, G. M., 1965, Petrology and structure of North Haven Island and vicinity, Maine [Ph.D. thesis]: Urbana, Illinois Univ., 146 p.

Doyle, R. G., 1967, Preliminary geologic map of Maine: Maine Geol. Survey, scale 1:500,000.

Faul, H., Stern, T. W., Thomas, H. H., and Elmore, P.L.D., 1963, Ages in intrusion and metamorphism in the northern Appalachians: Am. Jour. Sci., v. 261, p. 1-19.

Fullagar, P. D., and Bottino, M. L., 1968, Rb-Sr whole-rock ages of Silurian-Devonian volcanics from eastern Maine: EOS Am. Geophys. Union Trans., v. 49, p. 346.

Gates, O., 1969, Lower Silurian-Lower Devonian volcanic rocks of New England coast and southern New Brunswick, *in* Kay, Marshall, ed., North Atlantic—Geology and continental drift: Am. Assoc. Petroleum Geologists Mem. 12, p. 484-503.

Karner, F. R., 1968, Compositional variation in the Tunk Lake granite pluton, southeastern Maine: Geol. Soc. America Bull., v. 79, p. 193-222.

Lipman, P. W., Christiansen, R. L., and Van Alstine, R. E., 1968, Retention of alkalis by calc-alkalic rhyolites during crystallization and hydration: Am. Mineralogist, v. 54, p. 286-291.

Lyons, J. B., and Faul, H., 1968, Isotope geochronology of the northern Appalachians, *in* Zen, E-an, and others, eds., Studies of Appalachian geology, northern and maritime: New York, Interscience Pubs., Inc., p. 305-318.

McGregor, J. D., 1964, Geology of the Ellsworth quadrangle and vicinity, Maine [Ph.D. thesis]: Urbana, Illinois Univ., 155 p.

Page, L. R., 1968, Devonian plutonic rocks in New England, *in* Zen, E-an, and others, eds., Studies of Appalachian geology, northern and maritime: New York, Interscience Pubs., Inc., p. 371-384.

Spooner, C. M., and Fairbairn, H. W., 1970, Relation of radiometric age of granitic rocks near Calais, Maine, to the time of Acadian orogeny: Geol. Soc. America Bull., v. 81, p. 3663-3670.

Stewart, D. B., 1956, Rapakivi granite of the Deer Isle region, Maine [Ph.D. thesis]: Cambridge, Harvard Univ., 320 p.

Tilton, G. R., 1965, Compilation of Phanerozoic geochronological data for eastern North America, *in* Wetherill, G. W., ed., Geochronology of North America: Natl. Acad. Sci.—Natl. Research Council Nuclear Sci. Ser. Rept. 41 (NAS-NRC Pub. 1276), p. 181-220.

Wingard, P. S., 1961, The geology of the Castine-Blue Hill area, Maine [Ph.D. thesis]: Urbana, Illinois Univ., 266 p.

Wingard, P. S., and Brookins, D. G., 1964, Age of basic dikes on the Castine-Blue Hill Peninsula, Maine: Geol. Soc. America Spec. Paper 82, p. 226.

York, D., 1966, Least-squares fitting of a straight line: Canadian Jour. Physics, v. 44, p. 1079-1086.

Zartman, R. E., Snyder, G., Stern, T. W., Marvin, R. F., and Buckman, R. C., 1965, Implications of new radiometric ages in eastern Connecticut and Massachusetts: U.S. Geol. Survey Prof. Paper 525-D, p. D1-D10.

Zartman, R. E., Hurley, P. M., Krueger, H. W., and Giletti, B. J., 1970, A Permian disturbance of K-Ar radiometric ages in New England: Its occurrence and cause: Geol. Soc. America Bull., v. 81, p. 3359-3374.

MANUSCRIPT RECEIVED BY THE SOCIETY APRIL 18, 1974
REVISED MANUSCRIPT RECEIVED MARCH 10, 1975
MANUSCRIPT ACCEPTED MARCH 18, 1975

Printed in U.S.A.

Geological Society of America
Memoir 148

Stratigraphy and Correlation of the Newbury Volcanic Complex, Northeastern Massachusetts

ANDREW F. SHRIDE
U.S. Geological Survey
Denver, Colorado 80225

ABSTRACT

The Newbury Volcanic Complex of northeasternmost Massachusetts is encompassed, except for one small occurrence, in a northeast-trending wedge-shaped belt bounded by regional faults. Ten stratified members, aggregating more than 4,400 m in thickness and intruded by subconcordant podlike masses of micrographic rhyolite, are disposed in a homoclinal sequence that is overturned and broken into blocks by minor faults. In contrast to previous descriptions, this report proposes that the sequence has been inverted and that it neither unconformably overlies the granitic pluton that borders it on the southeast nor is intruded by that pluton, as previously has been suggested. The lower 2,900 m of the complex are entirely volcanic materials. In this interval, the great volume of strata is contained in three of the members: a fine-grained basalt flow member; a thick flow-banded rhyolite vitrophyre member that locally includes at its base as much as 120 m of ash-flow tuff, and a porphyritic andesite member that contains numerous flows but is dominated by flow breccias and tuffs. Most of the andesites accumulated subaerially, but andesitic detritus was also laid down in a shallow marine environment. Two lesser members of rhyolitic tuff, a minor lens of dense basalt(?), and a fine-grained andesite(?) member partly of epiclastic materials complete the lower part of the Newbury. The upper 1,500 m of the complex is composed of three very poorly delineated sedimentary members: a siliceous siltstone, a red sandy mudstone, and a calcareous mudstone. The red mudstone member contains some volcanic detritus.

The calcareous mudstone member is ostracodal, and the porphyritic andesite member contains at several sites a shelly fauna that provides specific correlation with the Leighton Gray Shale Member of the Pembroke Formation of the Eastport area of Maine. In the latter locality, ostracodes indicate affinities with fauna of the Baltic-British region rather than those of the Appalachian region and provide the basis for suggesting that the porphyritic andesite member of the Newbury may be wholly of Pridoli (latest Silurian) age. Ostracodes of the calcareous mudstone member may be longer ranging; therefore, the possibility that higher strata of the Newbury are as young as Gedinnian (earliest Devonian) cannot be ruled out. Recent paleontologic and radiometric dating indicates that the uppermost part of the Ames Knob Formation and the Thorofare Andesite in the southern Penobscot Bay area and the Castine Volcanics in the northern Penobscot Bay area (all more than 200 km up the coast in Maine) are also temporal equivalents.

The Newbury was not affected by mild regional metamorphism, nor was it invaded by granitic plutons during the Acadian orogeny, as were some of the Silurian-Devonian terranes of coastal Maine. The lithologically comparable Lynn and Mattapan Volcanic Complexes, of the nearby Boston area, apparently were similarly unaffected by metamorphism; correlation of the Newbury with these formations is not yet proved.

INTRODUCTION

The Newbury Volcanic Complex—also referred to as the Newbury Formation (Toulmin, 1964; Cuppels, 1967) or Newbury Volcanics (Boucot, 1968)—is exposed mainly in a northeast-trending, wedge-shaped belt in northeastern Essex County. The belt is 16 km long and is about 3 km wide in its easternmost exposures 1.5 to 3 km inland from the coast between Rowley and Newburyport; to the southwest, the belt arcs southward and pinches out in Topsfield (Fig. 1). Six kilometres farther southwest, in the northwest corner of the Salem quadrangle, a few small outcrops and rubbly exposures of comparable rock types have been confirmed on the basis of fossils to be part of the Newbury (Toulmin, 1964, Pl. 1, p. A14-A17). These apparently are exposures of a fragmented sliver of the Newbury isolated within or adjacent to a major northeast-trending fault zone. Northeast of the principal area of exposure shown on Figure 1, no outcrops project above the tidal marsh deposits that extend to the west behind the Plum Island barrier bar.

As the Newbury is one of the few formations in eastern Massachusetts to have been paleontologically dated and the only one thus established as Silurian and Devonian in age, its relationships to adjacent formations and the postdepositional events that can be recognized in the Newbury rocks are factors to be considered in interpreting the effects of Acadian orogeny in this part of New England. For want of definitive descriptions, earlier interpretations based on the Newbury have been speculative. Therefore, the petrography of individual members, their positions in the stratigraphic sequence, and their origins are considered in greater detail than would ordinarily be accorded such an areally restricted formation. Some of the members so delineated previously were erroneously lumped with other parts of the formation that are quite different in petrography and stratigraphic position.

STRUCTURAL SETTING

Structure within the Newbury Volcanic Complex

Early workers (for example, Clapp, 1921, p. 31) regarded strata of the principal Newbury belt as disposed in a syncline or otherwise broadly folded. Apparently, the tendency to correlate isolated outcrops as strata duplicated in folds delayed realization that Newbury strata are homoclinally disposed. Dos Santos (1960) seems to have been the first to note the lack of duplication by folding. He also, for the first time, estimated a reasonable thickness of 3,530 m and enumerated the stratigraphic succession cited in later regional correlations (Boucot, 1968; Gates,

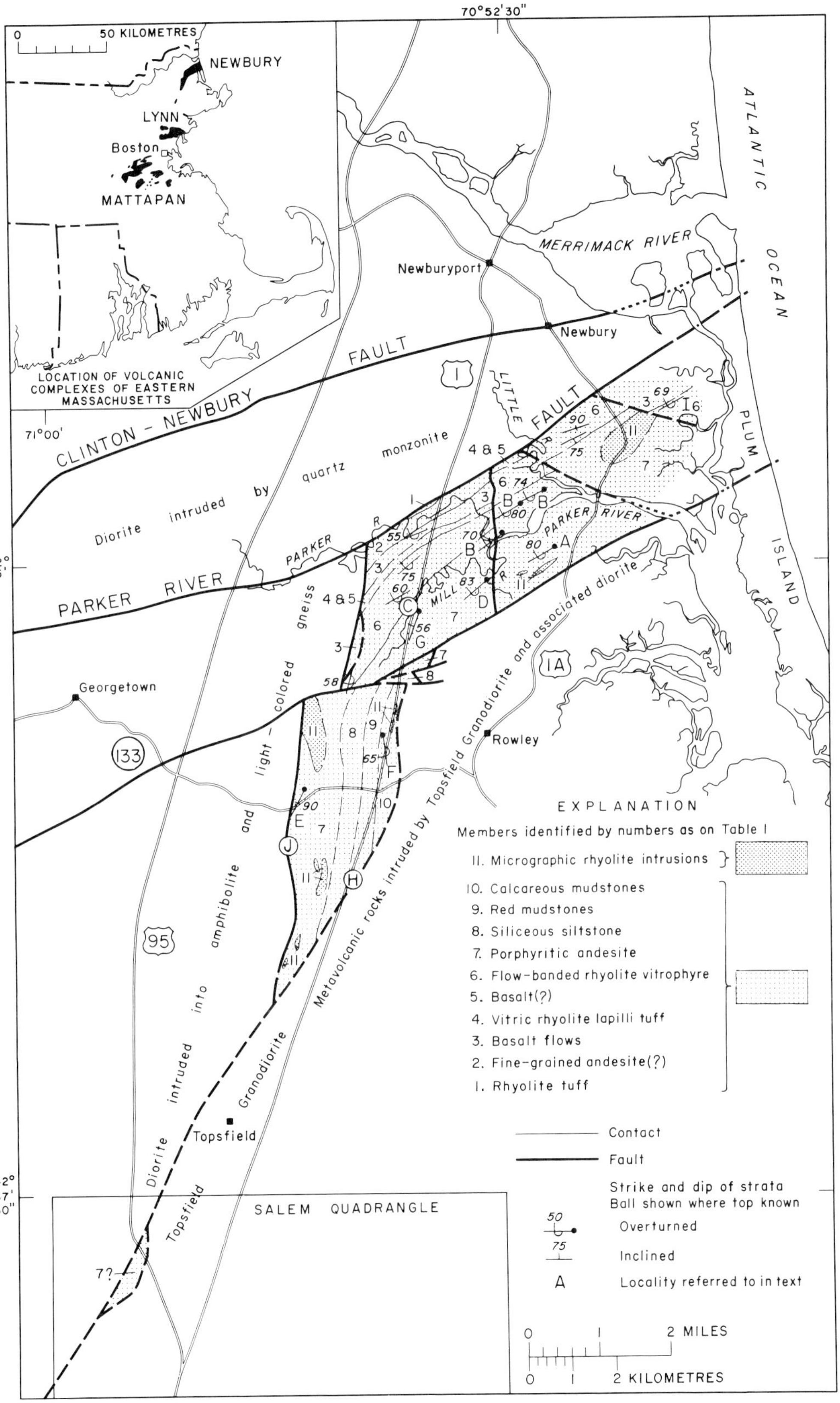

Figure 1. Geologic map of Newbury Volcanic Complex.

1969). Here, contrary to the interpretation of dos Santos and earlier observers, the succession is considered overturned, which leads to still other changes in interpretation.

The main belt is made up of two segments separated by a northeast-trending fault (Fig. 1). In the northern segment, strata strike northeast and generally dip moderately to steeply northwest. In this segment, which is further dislocated by lesser cross faults, exposures are adequate for straightforward reconstruction of the stratigraphic sequence. In the southern segment, however, exposures are few; therefore, the composition and interrelationships of the members in that sequence can be outlined only crudely. Generally, strata of the southern segment strike northward and dip steeply. Strata along the western border of the southern segment fortunately are comparable to those of the southern part of the northern segment, thus permitting correlation between the northern and southern sequences.

Features that demonstrate facing direction are rarely seen in outcrop, and some are not entirely satisfactory. Two separate occurrences of cross-stratification and graded bedding in sandstones within the porphyritic andesite member (loc. A, Fig. 1) positively define bed tops that face southeast. Well-rounded pebbles of flow-banded and spherulitic rhyolite in a conglomerate zone within the same andesitic member (loc. B, Fig. 1) seemingly could have been derived only from the adjacent flow-banded rhyolite member. Indeed, in one area (loc. C, Fig. 1), angular to rounded detrital clasts of rhyolite locally mantled the rhyolite member before the andesites were laid down. Flow tops at two localities (locs. D and E, Fig. 1) are best interpreted as overturned to the southeast. Graded bedding in the calcareous mudstone member (loc. F, Fig. 1) further confirms east-facing tops in the southern segment. On the other hand, at the first-discovered fossil site (loc. G, Fig. 1), a shell hash of brachiopods and crinoids is caught up in a flow top in such a way as to permit the interpretation that beds face northwest; speculatively, certain poorly preserved flow tops in other areas might be interpreted to face similarly. The only clear-cut features (at locs. A to F), however, indicate that the section is everywhere overturned to the southeast or east.

Relationships to Adjacent Formations

Nowhere can the Newbury be seen in contact with adjacent formations; but in places, particularly where contacts are obviously defined by faults, the gaps between formations are only a few metres wide. Dikes of Newbury-type rocks are unknown in adjacent terranes. Apophyses from adjacent plutonic masses have been reported in the Newbury; careful search has failed to confirm them. Even the so-called "trap dikes" of presumed Triassic age—which are ubiquitous and locally form swarms in formations northwest of the Newbury belt—are not known within the complex. Detritus from adjacent formations has not been recognized in the Newbury.

On the northwest side of the belt, a regional fault of northeast strike defines the boundary between the Newbury and fine-grained diorite or pink quartz monzonite that intrudes the diorite. The fault zone is exposed in several localities.[1] The west side of the belt also is bounded by a fault of northerly trend, as evidenced by the abrupt truncation of volcanic members. The fine-grained diorite, intruded

[1]To forestall confusion, this is not the publicized Clinton-Newbury fault, which is 1 to 2 km farther north (Fig. 1) and apparently is the master fault of the region.

into a terrane of quartz-plagioclase gneiss and amphibolite, is also the dominant rock on the west side of the belt.

The relationships between the Newbury and an entirely different group of formations along the southeast boundary of the wedge cannot be defined as succinctly. Nonresistant units of the Newbury are only sparsely exposed near equally sparse outcrops of a distinctive porphyritic granodiorite (long erroneously reported as Newburyport Quartz Diorite) and medium- to coarse-grained diorite, which in turn engulf metavolcanic strata composed of dacitic(?) to basaltic(?) flows and bedded volcaniclastic rock. Textures and bedding structures in the metavolcanic rocks are locally well preserved, but in most exposures, these rocks are thoroughly recrystallized—especially in halos adjacent to the granodiorite, where coarse hornblende may be obvious. Several kilometres to the southwest, in the Salem quadrangle, the porphyritic granodiorite was designated Topsfield Granodiorite and, on an admittedly tenuous basis, was assumed to be intrusive into the Newbury Complex (Toulmin, 1964, p. A17, A24). This interpretation permitted the speculation that the metavolcanic rocks represent Newbury strata that are thermally metamorphosed adjacent to the Topsfield Granodiorite and the associated diorite. This possibility seems refuted by the following observations. Andesites that are unequivocally part of the Newbury are not similarly hornfelsed in the vicinity of the intrusions. Furthermore, the Newbury rocks lack the cataclasis that characterizes some broad expanses of the Topsfield Granodiorite and enclosed metavolcanic rocks and that suggests a more complex and longer structural history for the plutonic rocks. The alternative, here preferred, is that the metavolcanic and intrusive rocks are parts of an older terrane whose affinities are not yet well understood. A K-Ar age of 640 m.y. obtained on hornblende from the coarse-grained diorite (Zartman and Naylor, 1972; R. E. Zartman, 1972, oral commun.) tends to confirm this interpretation.

An interpretation held longer than either of the above proposed that the Newbury Complex unconformably overlies the plutonic rocks along this southeast boundary (Clapp, 1921, p. 31, 51; LaForge, 1932, p. 28-30, 69-70; Toulmin, 1964, p. A13, quoting C. J. Schneer). Indeed, for some parts of the boundary, the idea of an unconformity might be entertained. For example, in a temporary exposure in the northwest corner of the Salem quadrangle, Topsfield Granodiorite cropped out within a few metres of Newbury rocks as granular, much reddened, partly decomposed rock suggestive of a fossil regolith zone. In opposition to the interpretation of unconformity, much of the boundary is linear, and its trace southwestward beyond the wedge of Newbury (see Fig. 1) defines a boundary between two quite different groups of plutonic rocks. This suggests a fault. Furthermore, the probability that Newbury strata are generally overturned and that bedding tops face southeast virtually forces the interpretation that this boundary is everywhere a fault. Thus, all boundaries of the Newbury wedge are faults.

Significantly, the Newbury does not serve to directly define ages of contiguous formations, as previously supposed. The relationships do, however, elucidate regional understanding, in that formations with quite different histories of deformation and metamorphism are juxtaposed against the Newbury, which was overturned but not dynamically metamorphosed during the Acadian orogeny, as has been postulated generally for rocks of eastern Massachusetts. The undated plutonic rocks just north and west of the Newbury wedge are not regionally metamorphosed. With the constraints of supposed metamorphism during the Acadian orogeny removed, these intrusions need not postdate that event and could be at least as old as the Newbury.

STRATIGRAPHY AND PETROGRAPHY

With the disposition of contiguous units established by mapping the separate fault blocks, a composite succession of 10 stratified members, which are intruded by sill-like rhyolite pods that constitute the eleventh and youngest member, can be pieced together as shown on Table 1. Inasmuch as the highest and lowest strata are truncated by faults and the matching offset parts are nowhere recognized, the makeup of other parts of the original volcanic pile is unknown. The strata within the fault wedge aggregate at least 3,770 m and almost certainly somewhat more than 4,400 m. The porphyritic andesite member, sliced by the fault that separates the north and south segments of the wedge, is the only member within the pile that is likely to be incompletely represented because of faulting. For lack of sufficient outcrops, the top of this member and all boundaries of succeeding members (8, 9, and 10 on Table 1) are very poorly defined, but the overall stratigraphic interval estimated for this part of the section is believed to be reasonable. Outcrops are completely lacking for two intervals near the bottom of the sequence—between members 1 and 2, and between members 2 and 3, as identified on Table 1; because these gaps are linear parallel to the strike of strata in adjacent outcrops and apparently represent nonresistant strata, they are included in the total thickness. (On Fig. 1 and Table 1, these gaps are arbitrarily assigned to the fine-grained andesite member.) Wider but otherwise similar gaps within the outcrop belt of the porphyritic andesite member, as exposed across the marshlands of the southeast part of the north segment, are even less likely sources of error in the overall estimate.

Outcrops are adequate to demonstrate that the boundaries between the members identified as 3, 4, 5, 6, and 7 on Table 1 are sharp rather than transitional. By analogy of rock types, the same is likely for contacts between members lower in the sequence, but contacts between members above the porphyritic andesite may be gradational.

The members are taken up in chronologic sequence, as numbered on Table 1. As viewed on Figure 1, the members are younger to the southeast in the northern fault segment and to the east in the southern segment.

Member 1. Rhyolite Tuff (Sheared Sliver at Base of Section)

Stratigraphically, the lowest exposure of the Newbury is a 6-m-wide outcrop of much sheared, poorly welded, vitroclastic tuff plastered against the Parker River fault (Fig. 1). Angular lithic fragments, as much as 8 cm across but generally less than 0.5 cm, sparsely mottle the yellowish-brown to brownish-gray,[2] dense, flinty, devitrified rock that constitutes most of the outcrop. Minute felted plagioclase laths distinguish these darker colored, undeformed dacitic(?) or andesitic(?) fragments. Rather similar fragments occur very sparsely in higher rhyolite tuffs, but nowhere in the pile is this sort of rock seen except as such clasts. Despite pervasive shearing, the crescentic outlines of shards that once made up the matrix are apparent in thin section. The shards are slightly flattened and tend to be aligned; some exhibit marked molding around the angular chips. This member is surmised to be rhyolitic by analogy with rhyolite tuffs higher in the section.

[2]Nomenclature of the *Rock-Color Chart* (Goddard and others, 1948) is followed in this paper.

Member 2. Fine-grained Andesite(?)

Overlying the rhyolite tuff member and first exposed in outcrops 45 m distant stratigraphically are very fine grained olive to olive-brown andesitic(?) rocks, so thoroughly propylitized as to obscure most primary textures and structures. Outcrops occur sparsely and only in the marshes immediately west of U.S. Highway 1 and north of the Parker River, where they span a stratigraphic interval of 80 m. Weathered surfaces of many outcrops are vaguely to conspicuously ribbed in a fashion that suggests bedding. Orientation of minute plagioclase laths parallel with the ribs suggests that some of these layers are flow-banded lavas; other layers that are rather similar in texture may be bedded crystal tuffs. In other outcrops, delicate laminae are intricately convoluted as though deformed before consolidation; these outcrops probably represent mud derived from andesitic detritus. Some massive outcrops with vestiges of probable amygdules apparently are thin flows.

Rock is not exposed in the next 150-m interval. By analogy with other parts of the formation with few outcrops, this stratigraphic gap and that in the 45-m-interval above the rhyolite tuff member may represent very friable, easily eroded sedimentary rock made up of epiclastic andesitic or more mafic detritus. If so, this member is largely composed of volcanic mudstones, which crop out only where the sparse but more resistant thin lava flows will support exposures.

Member 3. Basalt Flows

The next higher member is made up of fine-grained basalt flows, each 30 m or more in thickness, that tend to crop out as prominent topographic ridges separated by narrow depressions, which reflect scoria borders and thin fossil soil zones between flows. Original mafic minerals are completely obliterated in the thoroughly propylitized greenish-gray flow rock, but the relict outlines of plagioclase laths preserve intersertal or intergranular textures that suggest basaltic composition. Fluxion structure is characteristic. The marked elongation and felted habit of the plagioclase and its uniform fine grain size causes this basalt member everywhere to contrast texturally with the coarser porphyritic andesite member higher in the formation. Flow borders were largely obliterated, apparently by weathering before the next flow was laid down, but remnants of minutely vesiculated selvages can be recognized locally. The skeletal frames of voids in the scoria are supported by fillings of calcite or chlorite. Most of the rock visible between flows is greenish black and without notable texture or structure except a vague layering and a tendency toward friable partings; some is very dense and tough; some is perhaps sparsely tuffaceous. These fossil soils(?) are almost free of detrital fragments. Indeed, the lack of any coarsely clastic layers in the member also sets this basalt quite apart from the higher porphyritic andesite member. The basalt member is seen especially well along the south bank of the Parker River adjacent to U.S. Highway 1, where outcrops exist through at least 80 percent of the 250- to 300-m stratigraphic interval.

Member 4. Vitric Rhyolite Lapilli Tuff

In the vicinity of U.S. Highway 1, a grayish-green, hackly fracturing, friable, siliceous, vitric lapilli tuff makes sparse outcrops through the 52-m stratigraphic interval above the basalt member. Joint surfaces tend to weather rusty. Though coarsely devitrified, the outlines of thick-walled, cusp- and lune-shaped shards are strikingly preserved throughout the matrix as seen in thin sections (Fig. 2A).

TABLE 1. STRATIGRAPHIC SEQUENCE IN NEWBURY VOLCANIC COMPLEX

Member	Description	Thickness (m)	Mode of origin
	Top defined by faults		
11. Micrographic rhyolite intrusions	Podlike bodies of brownish-gray to orange-pink, aphanitic to sugary-textured massive felsite. Characterized by micrographic and spherulitic intergrowths; spherulites visible in some outcrops	100–600 m in thickness and as much as 1,600 m in length	Intrusions shallowly emplaced subparallel to strata throughout sequence
10. Calcareous mudstone	Laminated gray limestone and mudstone, very thinly interbedded. Large ostracodes characteristic	90+ (possibly 300+)	Tidal-flat deposits
9. Red mudstone	Grayish-red, friable, micaceous sandy mudstone	Boundaries indeterminate. With (8) and (10) totals 1,500	Mudflat or similar deposit; probably largely of epiclastic volcanic detritus
8. Siliceous siltstone	Dusky yellow green to olive-black, dense, flinty rock; parallel stratification inconspicuous	Indeterminate, as (9)	Shallow marine(?) deposit
7. Porphyritic andesite	Propylitized grayish-green to dark gray andesite, typified by plagioclase phenocrysts. Nonstratified volcaniclastic layers, ranging from fine-grained tuffs to boulder breccias, are much more voluminous than intercalated flows; stratified graywacke is subordinate. Sparsely fossiliferous	1,650+	Mainly deposited as subaerial ash flows and lavas; repeatedly submerged so as to alternate with air-fall and epiclastic andesitic detritus accumulated in quiet, shallow marine waters

TABLE 1. (*Continued*)

Member	Description	Thickness (m)	Mode of origin
	Local(?) erosion		
6. Flow-banded rhyolite vitrophyre	Dense, lithoidal vitrophyre, mostly grayish red and conspicuously laminated; large parts not porphyritic; includes sparse lenses of vitric tuff; locally, basal 120 m is pumiceous tuff	580–670	Laid down as flow-banded obsidian, commonly porphyritic, with minor intercalations of ash-fall tuffs. Basal tuffs are ash-flow(?) deposits
5. Basalt(?)	Dense, tough, dark gray, structureless rock	0–90	Local flow or intrusion
4. Vitric rhyolite lapilli tuff	Grayish-green, friable, hackly fracturing tuff, in which flattened pumice fragments are abundant in shard-rich matrix	0–52	Ash flow
3. Basalt flows	Uniformly fine-grained propylitized flows, devoid of fragmented materials; each 30 m or more thick, and separated by thin lithified soil(?) zones	250–300	Terrestrial flows, individually eroded, before next flow laid down
2. Fine-grained andesite(?)	(Unexposed interval)	150	(Inferred to be mudstones as below)
	Thoroughly propylitized, very fine grained, olive to olive-brown rock; part conspicuously to vaguely laminated, part massive and amygdaloidal(?)	80	Epiclastic mudstones deposited with occasional thin lavas
	(Unexposed interval)	45	(Inferred to be mudstones as above)
1. Rhyolite tuff	Flinty yellow-brown to brownish-gray vitroclastic rock, studded with darker fragments that are felted in texture	6	Ash flow(?). Much sheared adjacent to bounding fault
	Bottom defined by fault		

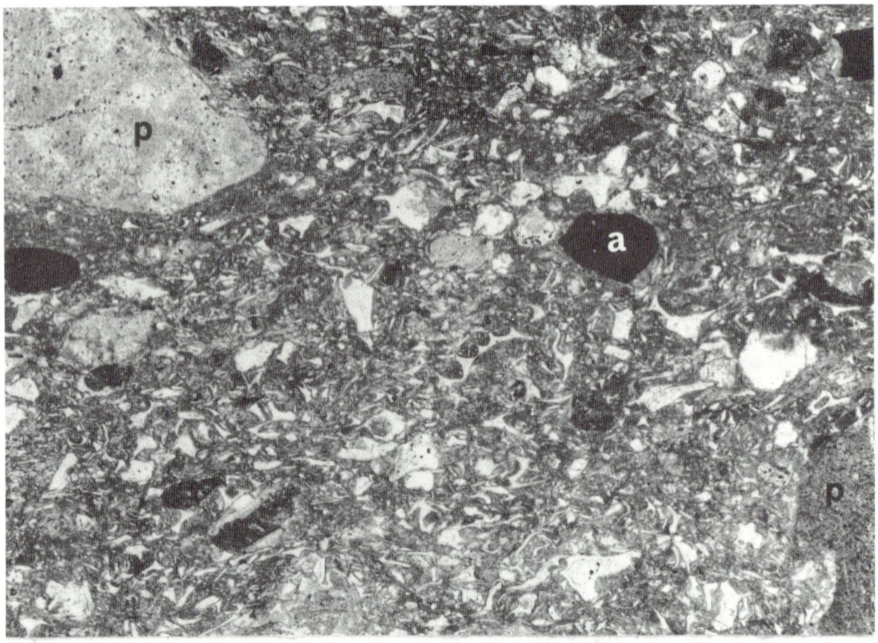

A

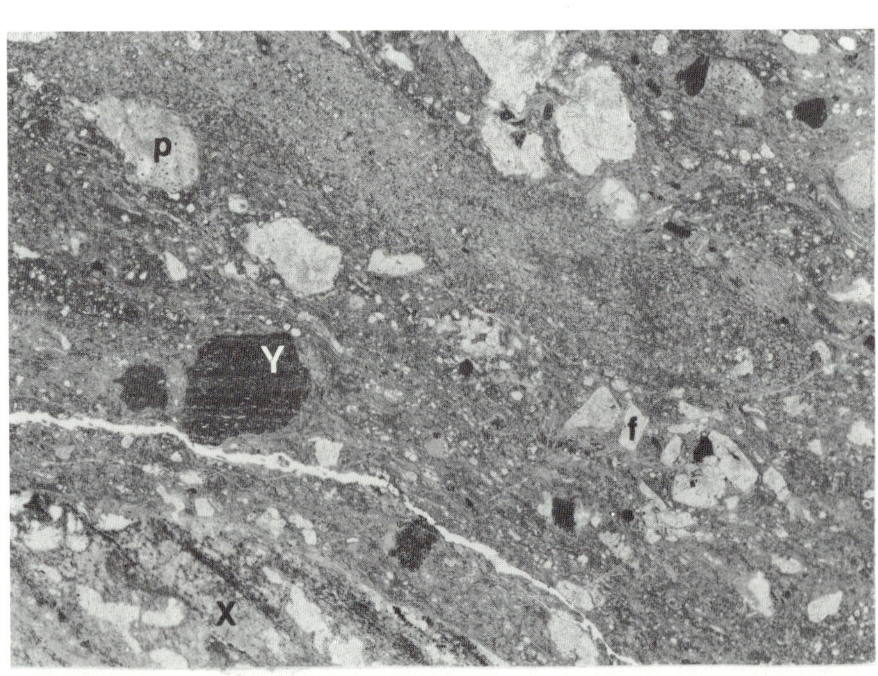

B

Angular to lenticular lapilli, as much as 8 cm across but mostly less than 1 cm, of recrystallized pumice and obsidian are abundantly peppered through this matrix and, depending on differences in texture or the amount of included hematitic coloring, may be conspicuous in outcrops. Bedding structures cannot be distinguished, but the platy fragments are rudely aligned parallel to partings that apparently reflect bedding. In some outcrops, hints of textures suggesting welding were noted but have not been confirmed in thin sections. Thoroughly compressed pumice fragments seem to have collapsed before coming to rest in the matrix. The shards may exhibit obvious planar alignment in the bedding direction and are nicely accommodated to one another and to larger clasts; this indicates some compaction. The shards do, however, retain original shapes almost perfectly and show almost no molding against the pumice fragments. Indeed, in some specimens, rounded, later filled, intershard pores were sustained in the shardy matrix.

Member 5. Basalt(?)

The rhyolite lapilli tuffs can be traced by discontinuous outcrops only 900 m along strike and along that length are separated from rather similar tuffs of the next higher rhyolite member by a 90-m thickness of basaltic(?) rock. The latter is a dense, tough, dark-gray rock seen only in a few massive outcrops. In thin section, only vague outlines of plagioclase laths, as much as 1.5 mm in length, can be recognized. These appear to be very sparsely set, and perhaps aligned, in a cryptocrystalline matrix; the textural features are largely obscured by aggregations of very fine grained alteration products, among which a carbonate is particularly abundant. Additionally, the rock is dusted through with minute opaque grains. Continuity of the topographic swale in which the outcrops occur and sparse glacial erratics apparently plucked from rock now covered in the swale and dropped on ridge-forming strata nearby suggest that the rhyolite tuff member and the overlying basalt(?) member are both continuous southwest beyond the belt of outcrop to the western boundary fault—as depicted on Figure 1. To the northeast, the rhyolite lapilli tuff thins and may pinch out about at the Parker River. In the fault block between the Parker and Little Rivers (Fig. 1), the stratigraphic interval that should include these members does not crop out. East of U.S. Highway 1A, both are definitely missing from the sequence.

Member 6. Flow-banded Rhyolite Vitrophyre

The next higher member, 580 to 670 m thick, is mostly dense, flow-banded rhyolite vitrophyre, which forms the most continuous outcrops of any of the members. Details of textures and primary structures are especially well exposed

Figure 2. Rhyolite tuffs of Newbury Volcanic Complex. Note lack of modification attributable to regional metamorphism. (A) Vitric lapilli tuff member. Typical except that coarser clasts, of variously vesiculated pumice (p), cannot be included in field of view. Generally, the vitroclastic textures are not as well preserved in the basal tuffs of the banded vitrophyre member; the dark andesite(?) clasts (a), however, are more characteristic of that member. Plane-polarized light. (B) Lithic tuff from midway in banded vitrophyre member. Large clast (X) at lower left is typical of laminated flow rock that comprises most of member. Clearing of hematite from coarsely devitrified (white) spots is apparent in this clast but not in the similar dark clasts (Y), which therefore appear to be more thinly laminated. Angular white fragments at lower right are feldspar (f); white fragments at upper left are pumice (p). Plane-polarized light.

on glacially scoured pavements, which are the highest surfaces of flat-topped ridges that stand 15 to 30 m above adjacent lowlands and are therefore consistently somewhat higher than exposures of other members. In the vicinity of U.S. Highway 1, outcrops of the basal 120 m of this member are vitric tuffs somewhat similar to those lower in the complex. Owing to lack of outcrops, these relatively friable tuffs cannot be said to exist everywhere along this interval. Large boulders of the tuff occur as far west as the westernmost outcrop of flow-banded vitrophyre. These might also be erratics from the lower rhyolite lapilli tuff member. The tuffs are almost certainly absent in certain localities where the base of the vitrophyre crops out within a few metres of basalt, as it does 800 m east of U.S. Highway 1 and at Pine Island at the far east end of the Newbury outcrop belt. They were either not deposited or were eroded before the vitrophyre was laid down.

The basal tuff is grayish yellow green to pale red purple; in some of the deepest artificial cuts and at a few other exposures, the flow-banded vitrophyre is similarly light colored. But the bulk of the member is dark-colored—grayish red to very dusky red purple.

The basal tuff of this member differs from the rhyolite tuffs in lower members of the Newbury in the greater variety of clasts it contains and perhaps in showing a slightly greater degree of welding. The clasts are either platy chips with angular edges or are lensoid and are as much as 9 cm in maximum dimension; most are less than 1 cm across. Marked parallel disposition of the clasts is typical, but a particular attitude may not persist more than a few metres. Pumice clasts, in considerable variety owing to differences in vesiculation, are probably the most abundant. Clasts of vitrophyre like that forming the bulk of the member also are abundant. Devitrified obsidian, which retains the outlines of perlitic fractures, makes distinctive clasts, as do even sparser finely crystalline chips with felty or trachoid textures. Furthermore, some vitric clasts were markedly devitrified before incorporation in the tuff, as indicated by sharp boundaries around clasts that show a different form or greater degree of recrystallization than the matrix. Postdepositional devitrification widely obliterated all vitroclastic textures, but where shards can be resolved, slight flattening and locally obvious molding around lithic fragments indicate incipient to moderate welding. Overall textural relationships suggest that welding was effected more by flowage than by compaction.

The flow-banding that marks the megascopically dense rock that makes up the bulk of this member sets it quite apart from the other members. The flow-banding is defined by laminae that mostly range from 0.25 to 2 mm in thickness but are as much as 1 cm thick. The laminae are easily overlooked in the lighter colored outcrops but are very conspicuous in most of the dark-colored vitrophyres, where they are defined by differing contents of hematite dust (Fig. 2B). Apparently, the flows were originally obsidian and contained only sparsely scattered crystals. Perlitic cracks are outlined by hematite in some layers least modified by recrystallization. Laminae occurring in distinctive combinations commonly can be traced for many metres and, exceptionally, for many tens of metres. In most exposures of some extent, however, this continuity is locally interrupted: the laminae are intricately contorted, smeared out, or brecciated in a fashion that indicates slumping of part of a locally chilled but still partly plastic flow. This slumping apparently occurred before the next increment of glassy flow was laid down. In some sites, slumping almost obliterated the lamination. Even where not thus erased, the laminae may be obscured by a profusion of quartz veinlets; most of these are oriented nearly normal to the lamination. The lamination is more delicate, persistent, and sharply defined toward the base of the vitrophyre pile. Conversely, where laminae are thicker and more vague toward the top, the rock seems more pervasively impregnated with iron oxides, and at least in a few areas, quartz veins occur in special abundance.

Still other textural variations occur within the vitrophyre. Of these, only spherulites seem to mark a particular stratigraphic interval. About midsection in the member, one or two spherulitic zones, from 1 to 10 m thick, occur discontinuously along strike. In some localities, the spherulites are microscopic, less than 0.5 mm in diameter; in others they are as much as 2 cm in diameter and in a few instances are thoroughly brecciated. Through large parts of the vitrophyre, phenocrysts are essentially absent or sparse and at best minuscule—no more than 0.1 to 0.2 mm across. In general, albite or oligoclase—and perhaps anorthoclase(?) in some instances—occurs as a scattering of blocky phenocrysts 1 to 2 mm in length that deflect flow laminae. In some localities, they are as much as 4 mm in diameter, and in a few localities (as in outcrops along the north edge of Pine Island; loc. I, Fig. 1), they are abundant enough to qualify the rock as a porphyry. In fewer localities, bipyramidal phenocrysts of quartz of like size, partly resorbed, are abundant and may occur alone or with plagioclase counterparts. Tuffaceous rock—ranging in texture from tuff entirely of shards to tuff formed by various mixtures of shards, pumice fragments, and angular chips of flow-banded rock (Fig. 2B)—makes up a few lenses as much as 6 m thick. If these tuffs were more resistant, such lenses might prove to be more abundant than is now apparent through the vitrophyre.

The boundary between this rhyolite member and the overlying andesite member is sharp where it can be seen, and it is probably an erosion surface. A conglomerate that is at least 4.5 m and locally possibly as much as 15 m thick and that contains angular to subrounded clasts ranging from granule to boulder size marks the upper margin of the rhyolite member in one locality immediately west of U.S. Highway 1. Clasts and matrix apparently are entirely detritus from the vitrophyre. Exposures are nowhere adequate to prove conclusively the plausible interpretation that this is residuum left in a pocket on an erosion surface. No other unit of flow-banded rhyolite is known within the Newbury; hence, pebbles and cobbles of similar rhyolite in conglomerates well up in the andesite member are quite probably relicts from this surface.

Member 7. Porphyritic Andesite

The porphyritic andesite member is nowhere complete, owing to truncation by faults (Fig. 1). The principal exposures are in the southeast half of the northern fault segment, between U.S. Highways 1 and 1A. There, parallel ribs of outcrop that stand above the tidal marshes permit a fairly complete piecing together of the resistant stratigraphic elements—especially those dominated by flows—but only a partial view of more friable volcaniclastic and sedimentary elements. In this segment, the fullest succession of andesitic strata probably uninterrupted by faults is 975 m thick, and plausible reconstructions of at least 1,650 m can be made. Andesitic volcaniclastic units and flows that are comparable—even as to fossils—are poorly exposed in widely scattered outcrops along the western border of the southern fault segment. Although specific details are lacking that would permit estimation of the amount of the andesite member omitted or duplicated between the two fault blocks, the southern exposures do provide the key to show that other members of the southern segment are to be assigned to still higher intervals in the stratigraphic section.

The porphyritic andesite member, in addition to being the thickest member of the Newbury, is petrographically the most diverse. About 1,200 m northeast from the U.S. Highway 1 crossing of Mill River and farther on to the northeast, thin flows dominate the basal 180 m of the member. The bottom of another 60- to 150-m interval consisting mostly of flows is 370 to 460 m above the base of the member. Elsewhere, flows are common but much subordinate to and interspersed with a variety of andesitic volcaniclastic layers. Most of the volcaniclastic bodies lack internal stratification: these massive subunits are recognized as layers only because they are sandwiched between other tabular volcaniclastic bodies that are texturally distinctive or because similar layers are separated by flows or by relatively thin units of well-stratified andesitic detritus. These last-noted graywacke strata are seemingly much subordinate in volume to the other andesitic units. They are the least resistant units, however, and their volume may be much underestimated if they make up considerable parts of the section that is hidden beneath the covered linear gaps, some 50 to 300 m wide, that separate riblike outcrops of the more resistant rock types.

Fine- to medium-grained holocrystalline andesite forms the south bank of the Parker River channel 240 to 760 m east of U.S. Highway 1A and makes lesser outcrops within 80 m to the south. This body is somewhat of an enigma. It lacks scoriaceous borders and other indicators of flows and is not seen in contact with other variants of the member. Although aphanitic selvages, very sparsely exposed, do suggest the chilled borders of sills, they are not compelling evidence that these occurrences are intrusions rather than atypically massive flows. Elsewhere, thin bodies that have similar interior textures certainly are flows.

The effects of propylitization are pervasive throughout this andesite member and obscure the primary minerals. Some of the breccias and coarser tuffs are least affected. In many localities, andesite flows with scoria zones or other primary vugginess or abundant fractures are thoroughly epidotized and may be much veined by calcite. Fresh exposures of the propylite, typically drab and grayish green to dark gray, rapidly weather grayish red or grayish purple. In some occurrences, the reddish oxide stains predate the present weathering cycle and mark soils or rock surfaces weathered before succeeding andesites were laid down. Where the plagioclase can be determined, it is andesine or albite; the latter is apparently secondary. The mafic minerals and their outlines generally are so thoroughly obliterated as to prevent speculation as to the primary mineral species. In the relatively few andesites with relicts of primary mafic minerals in the groundmass, only augite has been identified with certainty; in still fewer flows or clastic layers, augite occurs abundantly as phenocrysts. Although few are recognizable, andesites with pyroxene in varietal abundance exist in sufficient numbers throughout the member to permit the speculation that the member may be typified as pyroxene andesite.

Plagioclase phenocrysts set this member apart from all others in the Newbury; the textural variety noted in the flows (Fig. 3) is equally represented in the clastic layers. Flows and clastic layers of the basal 100 m of the member are those least characterized by phenocrysts. Higher in the member, most contain at least scattered phenocrysts, and many are porphyries. In some of the thicker flows, scoriaceous borders 1 to 2 m thick are well preserved, but in most flows, the vesicular selvages were largely weathered to soils or eroded away before the next andesite unit was laid down. Nonetheless, as indicated on Figure 3, adjacent flows generally are readily distinguished. But commonly only one boundary is visible, so it is difficult to typify flow thicknesses; most flows seem to range from 10 to 30 m in thickness. At an extreme, thin amygdaloidal borders define flows in one locality that are 1.5 to 3.5 m thick through a stratigraphic interval of 50 m.

The volcaniclastic elements of the member range from fine-grained tuffs to boulder

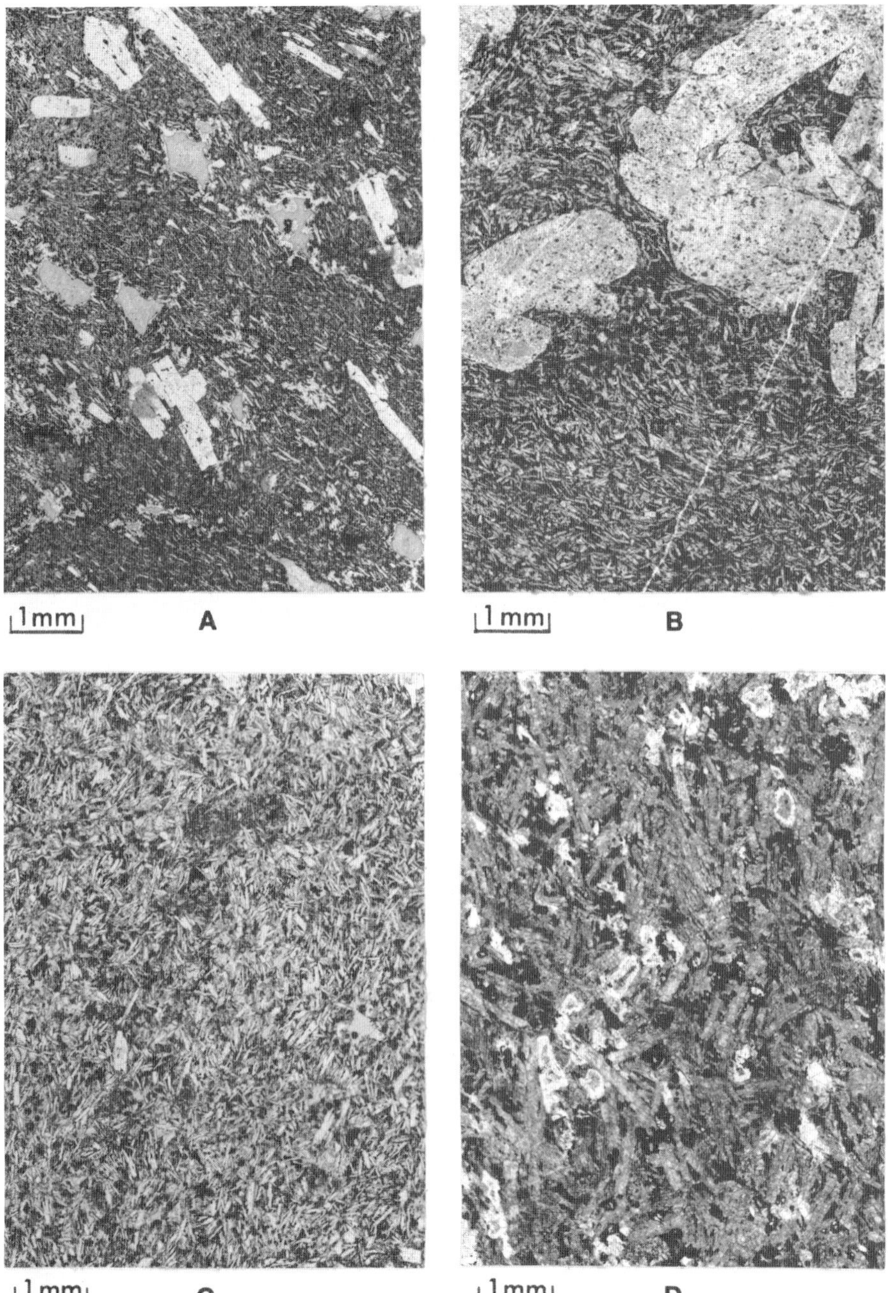

Figure 3. Some textures distinctive of individual flows in porphyritic andesite member. Phenocrysts occur in different combinations of shape, size, abundance, and distribution to individualize separate flows. Trachytic texture is common to most flows. (A) Well-formed plagioclase phenocrysts of moderate size uniformly distributed. (B) Corroded, coarser phenocrysts, as much as 8 mm long, widely separated in cumulophyric clusters. In other flows, single large laths or equant tablets, either euhedral or anhedral, are similarly distributed. (C) Small plagioclase phenocrysts, approaching equant form, sparsely set in matrix of thin laths comparable in length. (D) Coarser, nonporphyritic andesite. Augite—completely obliterated in this thoroughly propylitized specimen (light spots are carbonate)—commonly is a varietal mineral in flows of this texture; in still fewer flows, with or without plagioclase phenocrysts, augite occurs as phenocrysts.

breccias and occur in massive layers 2 m to perhaps more than 30 m thick. A given layer commonly exhibits a different range of clast sizes than adjacent layers, but lacks even crude internal sorting. Some breccias have a large content of angular to subrounded blocks 15 cm to 1 m in diameter and occasionally even larger. The breccia layers of greatest volume, however, are composed mainly of clasts less than 15 cm in diameter. Tuffaceous detritus (clasts less than 2 mm) constitutes large parts of the finer grained breccias; where lapilli (clasts 2 to 64 mm) dominate the coarse fraction, the tuffaceous matrix may make up 30 to 50 percent of the rock (hereafter termed "tuff-breccia"). Ordinarily, the still coarser block breccias contain relatively little tuffaceous material. In most of the tuffaceous rock—whether it occurs as separate layers or as matrix in the breccias—shards, pumice fragments, or other features indicative of original vitric material are absent or only vaguely decipherable, possibly in part as a result of devitrification and propylitization. Ordinarily, relicts of vitric materials are much more easily recognized in the lapilli or coarser clasts. In many layers of tuff or tuff-breccia, crystal fragments make up a large part of the fine fraction; but even more commonly lithic fragments make up the bulk of the tuff and dominate even the finest fractions.

Flow-top breccias of blocks 0.3 to 1 m and exceptionally more than 3 m across locally make up the greater part of a few flows. Overall, such breccias are minor elements of the member. In poor exposures, such breccias can be difficult to differentiate from the breccias that have a matrix of granular andesitic detritus.

In at least two localities, a much altered rock that is distinctively mottled by relicts of large amygdules makes a monolithic rubble that defines a tabular body about 10 m in thickness. This rock apparently represents a coarsely vesiculated glass or aphanite, but it is not certain whether the scoria was laid down as a flow or as breccia. Similar rock occurs as rare fragments in some of the tuff-breccias.

The proportions of the several textural varieties of andesite (Fig. 3) differ from one volcaniclastic layer to another. In some layers, finely vesiculated or cinderlike clasts are abundant. In other breccias with clasts of comparable size and form, vitric materials are almost lacking; this suggests either that the rock was mostly crystalline when it was erupted and first deposited or that there was much mechanical attrition of glass during transport after initial deposition. Some tuff and breccia layers are composed of andesite fragments that are almost wholly of one type, as though they were produced by a single eruptive episode. Other layers are mixtures representing the gamut of andesites seen in the flows; even blocks of earlier consolidated breccias are apparent in some layers.

The internally stratified layers—with one exception—are entirely of andesitic detritus, and most show well-defined bedding structures indicative of subaqueous deposition. All are very poorly exposed. The mudstones and finer grained sandstones are seen mainly in artifical exposures; indeed, strata of the kinds that include structures diagnostic of bedding tops or fossils have been seen only in man-made exposures or in friable erratics dug out of till. Individual stratified units range from 0.2 m to 25 m in thickness; most are apparently less than 10 m thick. Laminae, better marked by concentrations of dark grains than by textural differences, define the bedding style, which is prevalently parallel. Cross-lamination and graded bedding are rare. The graywackes that make up the most obviously bedded and probably the thickest units range from coarse sandstones to granule conglomerates. These graywackes and some of the mudstones locally enclose lenses of pebbly detritus. Mostly angular but in some occurrences well rounded andesite clasts of pebble to boulder size are sparsely dispersed through a few of the sandstone units. In some instances, these conglomeratic sandstones may be laterally transitional into massive conglomerates in which the stratification is best marked by thin seams or lenses of sandstone or mudstone.

The noteworthy compositional exception among the stratified layers are pebble to cobble conglomerates that occur about 210 m above the base of the member and that can be traced from a point near the confluence of the Parker and Mill Rivers northeast 1.5 km to the Little River. In an exceptional exposure near the river junction, two 2-m conglomerate beds, separated by a 15-m andesite flow, are composed largely of well-rounded discoid pebbles and cobbles of dense rhyolite derived (as already discussed) from the banded rhyolite vitrophyre. In other exposures along strike, the conglomerate is in one unit 5 to 15 m thick. Some of these conglomerate exposures exhibit comparable clasts, but most are of poorly rounded to angular small pebbles and granules of rhyolite in a matrix that ranges from granules of andesite to red sandy mudstone. This is the only zone in the member in which clasts clearly not of andesitic composition have been recognized.

For lack of exposure, the nature of the top boundary of the andesite member is unknown. Large glacial erratics, not far removed from their source and almost certainly from this part of the section, contain some scoriaceous volcaniclastic rock and much andesitic sandstone and mudstone. Some of the latter is abundantly fossiliferous. These hint that the top of the andesite member is mainly of bedded epiclastic materials that may be transitional to the overlying siliceous siltstone member.

The scattered exposures of Newbury rocks in the isolated block 6 km to the south of the main belt (Fig. 1) in the northwestern corner of the Salem quadrangle (Toulmin, 1964, p. A14–A17) contain andesite flows but are dominated by sandy mudstones. These exposures conceivably represent the upper part of the andesite member and some of the strata immediately above the member. Another correlation with gritty sandstones low in the member, such as those exposed in the bed of Mill River at and south of U.S. Highway 1, is less appealing, owing to the small proportion of flow rocks or coarse volcaniclastic layers in the southern occurrence. K. G. Bell (1974, oral commun.) reported extensive temporary exposures, open during road construction, almost exclusively of limy mudstone that was somewhat comparable to that of the calcareous mudstone member (see member 10, Table 1). Alternatively, and quite as reasonably, the southernmost block may be made up of juxtaposed fault slivers of various parts of the Newbury.

Member 8. Siliceous Siltstone

Details of the 1,500-m-thick upper part of the Newbury sequence, which is seen only to the east of the andesite outcrops in the southern fault segment, are very poorly understood. Lithologies of the upper three stratified members are determined partly from outcrops, but their boundaries and thicknesses must be speculatively interpreted almost entirely from the distribution of rubble in till. The thicknesses in Table 1 for these members are only crude estimates.

The siliceous siltstone member, immediately above the andesite member, is seen mainly in angular glacial erratics. These are scattered for 700 m along a south-trending ridge, which is 700 m west of U.S. Highway 1 and just south of the fault that separates the north and south segments. In places these erratics are 3 to 10 m in longest dimension, and they are in boulder fields largely lacking in other lithologies; this indicates proximity to the bedrock from which they were plucked. The siltstone is dense, in part flinty, and has parallel stratification that is apparent only where delicate laminae defined by coarser silt fractions are suitably etched by weathering. Internally laminated beds so defined are mostly 0.5 to 10 cm thick; some are perhaps as much as 60 cm thick. The rock ranges from dusky yellow green or dark greenish gray to very dusky purple or olive black; surfaces weather to pale tints of purple and of greenish or yellowish gray. The siltstone of the principal, and perhaps only, bona fide outcrop is dark in color. In thin section, silt- to fine-sand-sized angular grains of quartz and plagioclase constitute 10 to 40 percent of the rock and lie in a matrix in which individual mineral grains are mostly indistinguishable. In some sections, opaque dust outlines finely mosaicked aggregates of quartz and feldspar(?) in forms that hint of shard outlines; none are well enough defined to confirm that volcanic glass may once have been a significant part of the rock. Quartz grains dominate the coarser fractions; plagioclase content ranges from occasional grains to perhaps 20 percent of the silt fraction. If the quartz found in rhyolite detritus is excepted, this is the lowest sedimentary rock in the Newbury that contains detrital quartz.

Member 9. Red Mudstone

The overlying grayish-red mudstone was first seen only in friable, slabby-parting boulders—some 1 to 2.5 m on a side—and in concentrations of shaly to flaggy debris; their distribution suggests closely underlying bedrock. The red mudstone differs from the siliceous siltstone in that angular quartz and plagioclase grains of fine-sand size make up at least 50 percent of the rock, cementation is weaker, and mica is obvious in hand specimens. The mica seen in thin sections ranges from shreds of sericite that are just resolvable under the microscope to aggregates assembled as ill-formed crinkly books or lenses of muscovite that are visible to the unaided eye. Hematite pervasively opaques the matrix.

In 1974, a quarrying operation at the south end (loc. H, Fig. 1) of the inferred belt of mudstone uncovered the only known outcrop. This outcrop has been examined only briefly and in bad light. A poorly exposed andesite(?) flow and sandstone with pebbles and sparse cobbles of similar andesite(?) probably are intercalated with the mudstone of this locality. Lenses, as much as 2 m thick and 6 m long, of medium- to coarse-grained pale yellow-brown graywacke are definitely encompassed within the better exposures of typical red mudstone. On re-examination of the larger concentrations of mudstone debris at the north end of the inferred belt, chips of like graywacke were recognized in the soils. But it is still not evident that flow rock or derived coarse clasts, which may have been the source of the mudstone, occur north of the quarry locality.

Member 10. Calcareous Mudstone

The uppermost member of the stratified sequence is made up of thinly interlayered silty limestone and calcareous mudstone (silty claystone). Beds are parallel, as much as 5 cm thick but generally less than 15 mm thick, and delicately laminated internally. The strata are medium to dark gray on freshly broken surfaces, which assume an olive-gray color within a few months of exposure. Ostracodes large enough (2 to 18 mm long) to be easily seen, especially in limestone that has parted after exposure, are locally abundant. Silt-sized angular grains of the coarsest detrital fraction include sparse plagioclase but are mostly quartz. The dark (organic?) matter is mainly concentrated along microstyolites.

The boundaries of the calcareous mudstone member are especially vague. Even the firmest limestones tend to develop shaly partings when exposed and break down into platy chips that do not survive in the soils. The member has been temporarily exposed in two small areas (loc. F, Fig. 1) along the east side of U.S. Highway 1 about 1 km north of its intersection with State Highway 133 (Kent Corner), where building sites were scraped free of soil. Narrow riblike exposures of limestone that were briefly uncovered 450 m east of Kent Corner may have represented bedrock. From the distribution of limestone rubble in soils, the member need not be more than 90 m thick, but could exceed 300 m.

Member 11. Micrographic Rhyolite Intrusions

The separate bodies of the Newbury Complex remaining for discussion are here recognized as all of one rhyolite type and as petrographically distinct from all other members. Hence, the several occurrences are here interpreted as individual intrusions rather than, as in previous descriptions, stratified extrusive elements variously correlated with the rhyolites described above.

This rhyolite is a brownish-gray to orange-pink aphanitic to sugary-textured rock that occurs in massive bodies scattered without pattern throughout the complex. These masses, tightly jointed and more resistant to glacial plucking than most units in the Newbury, tended to form buttresses against which glacial till was plastered. Consequently, the seemingly largest (northeasternmost) mass—the rock core of a drumlin complex—is little exposed, and the margins of most masses are broadly obscured by till. Where contacts can be most narrowly delimited, the masses are elongate pods, as much as 1,600 m in length and 100 to 600 m in maximum thickness, disposed subparallel to stratification of adjacent members. No hint of planar internal features, such as flow banding or stratification, is seen on any outcrops of these rhyolites. At two localities, andesite 3 to 5 m distant from the rhyolite lacks indications of metamorphism related to intrusion.

Quartz and potassium feldspar, arranged in a characteristic fabric defined by both spherulitic and micrographic intergrowths (Fig. 4), constitute all but a small fraction of this rhyolite. Muscovite, which makes up 1 to 3 percent of some specimens, and sparsely disseminated minute grains of pyrite, which is readily leached out and commonly represented only in rusty stains and grains of opaque oxides near joint and fracture surfaces, are the only other primary minerals ordinarily seen. Albite may be introduced; it is seen very sparsely in some occurrences and there possibly only adjacent to quartz veinlets. In some specimens, the spherulites are the usual radial aggregates of acicular grains, but in most, the spherulites are radially arranged feathery aggregates of micrographic intergrowths (Fig. 4). The spherulites, individually 0.1 to 2 mm in diameter and in clusters as much as 5 mm in diameter, are ordinarily apparent to the unaided eye only where their color contrasts with the matrix. Locally, overlays of sericitic or calcitic alteration obliterate the micrographic textures, but the spherulites at exposed surfaces of such rock are megascopically better defined than elsewhere, because weathering enhances the color contrast.

The spherulites and micrographic intergrowths exist in various proportions and degrees of perfection. In some specimens, spherulites are few, mostly nongraphic, and in a matrix in which the cuneiform outlines of quartz against feldspar are poorly defined or mostly absent. In other specimens, the cuneiform outlines are clearly defined through much of the matrix and in spherulites that occur in abundance. In still other specimens, polygonal-form spherulites occur in clusters that occupy 50 to 95 percent of a microscope field; consequently, continuity of cuneiform outlines in the interstitial matrix may not be striking. In a few places along the margins of the rhyolite pods, a slight diminishing of grain size possibly does occur. Otherwise, the full range of variable textures noted above seems to occur haphazardly from core to margin through each of the separate bodies.

The complete absence of any structures or textures that would suggest an extrusive origin, the erratic distribution of elongate pods through an otherwise straightforward stratigraphic sequence, and the lack of persistence or repetition along strike suggests that these rhyolite masses are sill-form intrusions. They were probably emplaced near the surface at a late stage in the accumulation of what was once a much thicker volcanic pile. So interpreted, they postdate any of the elements now remnant.

FOSSILS AND AGE

The fossils most useful in defining the age of the Newbury have been found in thin layers of friable sedimentary rock intercalated in the porphyritic andesite member; these occurrences have come to view only in artificial excavations. Such fossils, found in 1915 by Arthur Keith, were first reported by LaForge (in Emerson, 1917, p. 163-164) from a site near Glen Mills, apparently in a now-obliterated exposure at the intersection of Center Street and the Newburyport Turnpike (U.S. Highway 1) in Rowley, 2.5 km south of the Parker River bridge (loc. G, Fig. 1). Pockets of shelly fragments, including abundant brachiopods, are still exposed about 25 m east of this junction, in a mudstone zone 15 to 35 cm thick that locally separates a porphyritic andesite flow from an andesite tuff unit. In 1964 N. P. Cuppels discovered a second very fossiliferous exposure 5.5 km to the southwest near Wilson Pond (loc. J, Fig. 1) in a bed of rusty-weathering clayey mudstone about 60 cm thick that also is disposed between andesitic volcanic elements. About 2 km northeast of the Wilson Pond outcrop, he found the same faunal assemblage better preserved in mudstone seams in large drift boulders; these boulders were temporarily exposed at several places in gravel pits that lie north of the

Figure 4. Micrographic rhyolite typical of podlike intrusions in Newbury Volcanic Complex. Graphic intergrowths, radially arranged as spherulites (s), typically are finer grained than the mosaic (m) of quartz and feldspar—partly also graphic textured—that surrounds the spherulites. Crossed nicols.

Hunsley Hills and border the north side of State Highway 133 along an interval 300 to 1,200 m west of U.S. Highway 1.

The paleontologic data presented below for the more recently found localities are based on collections made in 1964 and 1967 by N. P. Cuppels. I am much indebted to him for guidance to the sites from which the fossils were collected and for use of a manuscript he prepared to describe the settings of the collections; that paper includes a discussion by J. M. Berdan of the paleontologic materials available from the Newbury. R. B. Neuman, John Pojeta, Jr., and J. M. Berdan of the U.S. Geological Survey have identified, respectively, the brachiopods, pelecypods, and ostracodes, and have collaborated in comments on their significance (1965 to 1968, written communs. to Cuppels). Independently, A. J. Boucot, then of the California Institute of Technology, identified brachiopods. The information now available on stratigraphic sequence is more complete and permits reinterpretation of the data available to Cuppels. Therefore, I have paraphrased the earlier discussion by Berdan to collate the paleontologic data properly within the framework of members enumerated in the present report. This restatement permits, with Berdan's advice, using conclusions that result from her later studies of similar assemblages in correlative formations along the Maine coast.

The fossiliferous strata near Glen Mills cannot be rigorously related stratigraphically to the Wilson Pond and Hunsley Hills (gravel pit) localities to the southwest. The Glen Mills fossils are at least 305 m—and possibly somewhat more, if minor faults affect the estimate—above the base of the porphyritic andesite member. The Wilson Pond exposure is south of the major fault that separates the two segments of the Newbury outcrop belt; therefore, similar estimates of stratigraphic interval are not feasible. Judging from their size and friability, the fossiliferous boulders in the gravel pits were not moved far from outcrop sources also in the southern segment. Near the southwestern fossil sites, the various volcanic elements are very poorly exposed, but they do exhibit petrographic details in common with flows and breccias of the Glen Mills locality, so all fossil localities in the andesite member might conceivably fall within a narrow stratigraphic interval. However, the southwestern fossil sites are included in a sequence that seems to contain a greater volume of sandstone and mudstone beds than is typical in the lower half of the member. More plausibly, such abundance of water-deposited materials indicates stratigraphic levels toward the top of the member and notably higher than the Glen Mills zone.

The fossils first collected (U.S. Geol. Survey Colln. 467K-OS) from the Glen Mills locality were identified by E. O. Ulrich (in Emerson, 1917, p. 163), without generic identification, as "marine types" comprising "one or more species of brachiopods, a species of gastropod, fragments of crinoids, and probably a pelecypod." Ulrich surmised this fauna to be most nearly contemporaneous with those of the Pembroke and Eastport Formations of easternmost coastal Maine, and with uncertainty considered the Newbury to be either Late Silurian or Early Devonian in age.

More recent paleontologic studies of the Newbury and of various similar terranes scattered along the coast of Maine sharpens correlation with the Pembroke and all but negates correlation with the Eastport.

Reviews of the original and later collections from the Glen Mills locality add little (J. M. Berdan, 1967, written commun. to N. P. Cuppels). The brachiopods are rhynchonellids of indeterminate genus, and the gastropods are mostly high-spired types—a faunal association recognized in the Leighton Gray Shale Member, the lower member of the Pembroke Formation. The most abundant rhynchonellid is similar to *Camarotoechia leightoni* Williams from the Leighton (R. B. Neuman,

1968, written commun. to N. P. Cuppels). Berdan additionally noted poorly preserved beyrichiacean ostracodes; one specimen might be *Macrypsilon*, a genus common to the Leighton Gray Shale Member and to the lower part of the Stonehouse Formation of Nova Scotia.

Preservation of diagnostic details is better in fossils from the Wilson Pond locality (U.S. Geol. Survey Colln. 7348-SD) and is especially favorable from certain of the drift boulder collections (7347-SD, 7346-SD, 7345-SD, 7344-SD). All collections from the andesite of the southern fault segment represent very similar if not identical faunal assemblages. Sampling is adequate from the several localities to permit the following composite listing: brachiopods—*Salopina submedia* (McLearn), *Protochonetes* cf. *P. bastini* (Williams), rhychonellids, genus indet., "*Leptaena rhomboidalis* Wilckens," and a small ?howellellid; pelecypods—*Nuculites* sp., ?*Pteronitella* sp., *Actinopteria* sp. indet., ?*Goniophora* sp., ?*Eurymyella* sp., *Grammysia* sp. indet., ?*Sphenotus* sp., a large ?ambonychiid, and palaeotaxodonts; ostracodes—*Hoburgiella*? sp., *Hemisiella* sp., and *Crytolopholobus* sp. The last two ostracodes, beryichiacean genera recognized in assemblages of the Baltic region, were identified by Anders Martinsson of the University of Uppsala, Sweden (1965, written commun. to J. M. Berdan). Additionally, A. J. Boucot (1965, written commun. to N. P. Cuppels) has reported a calymenid trilobite in collections from the Newbury.

The *Salopina* and *Protochonites*, where best preserved, compare closely with species of the Pembroke Formation. The ostracode *Hoburgiella*? and calymenid trilobites also occur in the Leighton Gray Shale Member of the Pembroke Formation. A. J. Boucot (1966, written commun. to Cuppels) regarded the brachiopods as characteristic of a "*Salopina* Community," an association now reasonably established (Berry and Boucot, 1970, p. 105-106) as occurring in relict bodies of volcanic strata scattered along a belt that extends northeastward from the Newbury area along the coast of Maine and into New Brunswick. Boucot, partly as a result of studies since 1966, interpreted the *Salopina* Community as indicative of a range in age from Pridoli to early Gedinnian (latest Silurian to earliest Devonian). Although the particular *Salopina* fauna of the Newbury could range into the Gedinnian, none of the forms are diagnostic of that age (Boucot, 1975, oral commun. to Shride). R. B. Neuman, J. Pojeta, and J. M. Berdan concurred in 1968 (written commun. to N. P. Cuppels) that this assemblage from the andesites of the southern fault segment should be interpreted as Late Silurian in age.

In still other parts of the Newbury, assemblages entirely of ostracodes have been recognized. In 1894, Foerste (1920) made the first collection of ostracodes in strata now recognized as Newbury, at a site in the northwest corner of the Salem quadrangle, but was unable to find the locality again and did not report it until the Glen Mills find had been well publicized. On rediscovery by Toulmin, with others, of a 15-cm bed of ostracode-bearing limestone at the site described by Foerste, an additional collection was submitted to J. M. Berdan, who reported (1958, written commun. in Toulmin, 1964, p. A17) "leperditiids and other small ostracodes ... too poorly preserved to be identified." She concluded that the leperditiids were definitely Silurian or Devonian forms, probably Late Silurian or Early Devonian. N. P. Cuppels in 1964 discovered an ostracode fauna in the area north of Kent Corner where the calcareous mudstone member is principally exposed. Ostracodes commonly 2 to 10 mm and exceptionally as much as 18 mm in length are now recognized as characteristic of this highest stratigraphic element; if careful search is made they are found in practically every outcrop of the member. Berdan (1965, written commun. to Cuppels) appraised this fauna (U.S. Geol. Survey Colln. 7341-SD) as composed of large leperditiids referable to *Herrmannina* sp.

and at least two genera of smaller ostracodes too poorly preserved for further identification. She concluded that fossils of this aspect range from Middle Silurian to not younger than Early Devonian.

The similarity of the association of leperditiid and small ostracodes in the two collections provoked speculation that the Salem quadrangle locality and the Kent Corner locality might represent one zone in the Newbury Complex. If lithologic associations are given some weight, this possibility becomes less attractive. The Salem fossil site is now covered, but the exposures of the vicinity are dominated by sandy mudstone outcrops among which small isolated exposures of andesite are interspersed. These lithologies are not a part of the calcareous mudstone member near Kent Corner. Rather, the mudstone-andesite association resembles either the mudstones that contain the *Salopina* fauna or a section of sedimentary strata exposed along the Mill River, 600 m north of the Glen Mills locality and stratigraphically near the base of the porphyritic andesite member—or stratigraphically farther below the Kent Corner ostracode locality than the other fossil localities. As noted earlier, the Newbury of the Salem quadrangle may be a mélange of mudstone and andesite slivers from different parts of the section juxtaposed by faults. Thus, both paleontologically and lithologically, these ostracode-bearing elements provide less definitive age determinations than do the units dated by the *Salopina* fauna.

Since the later collections of the *Salopina* fauna of the Newbury were first assessed and deemed to be of Late Silurian to Early Devonian age, studies by J. M. Berdan and Anders Martinsson of ostracodes in strata along the Maine coastal belt have bracketed more narrowly the Silurian-Devonian boundary and further defined the sequence of units in fragmentary exposures of formations that are similar in provenance to the Newbury (Berdan, 1971; Brookins and others, 1973). In particular, ostracodes that have aspects more in common with the fauna of the Baltic-British area than with fauna of the Appalachian area of North America indicate that the Leighton Gray Shale Member of the Pembroke Formation is Pridoli in age and, furthermore, that the Silurian-Devonian boundary is at or somewhat above the contact between the Leighton and the overlying Hershey Red Shale Member of the Pembroke. Although the ostracodes most diagnostic of age in the Leighton have not yet been found in the Newbury, several of the Newbury ostracodes also occur in the Leighton and also are of European aspect. Moreover, a general consensus prevails that the other forms—especially the brachiopods—equate with the fauna typical of the Leighton Gray Shale Member. Thus, indirectly, ostracode zonation that defines the age of that member in the Eastport area also indicates that the paleontologic material mostly narrowly diagnostic for the Newbury may be wholly of Pridoli age. The porphyritic andesite member, then, represents the latest part of the Silurian. Because the leperditiids of the calcareous mudstone member might range into the Gedinnian, the possibility that higher strata of the Newbury are as young as Early Devonian cannot be ruled out.

Although physical features do not strongly support the possibility, it should be reiterated that a significant hiatus may be defined by the erosional break between the porphyritic andesite and banded rhyolite vitrophyre members. Strata above the contact differ in petrographic character from those below and thus perhaps mark quite separate eruptive episodes. No fossils have yet been found to bracket a maximum age for the members below this contact. The underlying strata must be about the same age as the fossiliferous rocks, but an older age, comparable to that based on fossil dating in similar successions to the northeast along the Maine coast, is still permissible.

ENVIRONMENTS OF ACCUMULATION

Depositional units of the Newbury Volcanic Complex range in type from lavas poured onto a land surface to subaqueous deposits of epiclastic sediments derived from volcanic rocks; spanning these extremes are breccias and tuffs that require interpretation both as to whether they were fragmented by pyroclastic or epiclastic processes and whether they were deposited above or below sea level.

Certain aspects are common to most of the stratigraphic elements in members below the siliceous siltstone member. Clastic constituents are wholly of volcanic origin. Even the more mature sediments totally lack quartz derived from basement or other terranes. Each layer within a member is grossly—and commonly in detail—monolithologic and only exceptionally includes detritus as foreign as that from the underlying layer. These features suggest that lavas and volcaniclastic strata were laid down rapidly, with little opportunity to incorporate previously formed detritus. But the lack of lateral change in coarseness of breccias or in notable thinning of layers and the lack of quaquaversal dips or other diagnostic primary structures argue against these being deposits piled up adjacent to source vents.

The rhyolitic tuffs and tuff breccias in the rhyolite tuff member at the bottom of the Newbury, those immediately above the basalt flow member, and those that locally constitute the basal 120 m of the banded rhyolite member all evolved as explosively comminuted, highly vesicular glassy ash dispersed as either ash falls or ash flows. The latter origin is favored, owing to the lack of sorting and of pronounced stratification that are expected characteristics of ash-fall deposits (Ross and Smith, 1961). Although platy pumice fragments of some localities are planar in disposition, they are not notably compacted and flattened to give the foliate structure typical of welded tuffs. Shards are little distorted, but an incipient welding is suggested by indications of plastic yield to provide accommodation of shards one to another. Uncommonly, local welding is clearly indicated by lune-shaped shards molded against larger clasts or by glass bubbles that are flattened and stretched. Apparently, the ash flows retained too little heat to promote conspicuous welding features.

The flow-banded rhyolite vitrophyre was laid down as a succession of viscous lavas, which solidified mostly as glass. Explosive episodes are recorded in the minor deposits of air-fall tuffs that locally intervene between flows.

Of more significance are features that suggest a hiatus before the porphyritic andesite member was laid down. As noted above, flow banding tends to be obliterated, quartz veinlets become abundant upward, and locally the banded rhyolite vitrophyre member is capped by a rubble of bleached vitrophyre. These and associated features suggest an erosional interval during which silicification occurred. The rhyolite conglomerates that occur in the overlying porphyritic andesite member geographically to the east of the rubble capping and 200 m above the base of the andesite might be explained as lateral equivalents of the capping except that they appear to be stratigraphically somewhat too high to be reasonably extrapolated to the particular exposure. In any event, the occurrence of well-rounded as well as angular rhyolite clasts in the andesite member indicates that part of the rhyolite was deeply covered by andesite while another part stood high enough to furnish detritus that was subjected to considerable abrasion—perhaps on a beach. It further demonstrates a notable discordance between members and indicates that lateral variations in this part of the formation can be expected over small distances.

The more mafic parts of the pile are more difficult to interpret, mostly because strata representing a variety of environments are juxtaposed and critical contacts between the diverse elements are rarely exposed.

A large proportion of the outcrops of the andesite(?) member lowest in the Newbury represent flows; however, the bulk of the member, which is not exposed, is very likely of andesitic mudstones. In contrast, flows of the overlying basalt member piled up subaerially with very little clastic residuum between flows.

The local lens of black aphanite (member 5, Table 1) may have been emplaced as mafic glassy flows or shallow sill-form intrusions.

The flows of the porphyritic andesite member are mainly subaerial deposits. The tuffs and breccias are largely of angular clasts aggregated without sorting or stratification, in massive tabular layers of appreciable thickness and of at least moderate lateral extent. Without further description, these might be regarded variously as pyroclastic rocks erupted subaerially and accumulated in either a subaerial or subaqueous setting or as pyroclastic rocks for which both eruption and deposition was subaqueous.

A few of the coarse monolithologic andesite breccias formed as crusts that crumbled and foundered in moving flows. Contrary to implications in past descriptions of the Newbury, breccias formed by Strombolian or lava-fountain eruptions—as would be evidenced in bread-crust bombs, spindle-shaped bombs, or pancake-shaped bombs accumulated as true agglomerates or spattercone agglutinates—are not apparent.

Subaqueous eruptions or flowage of lavas into water also are largely ruled out. Pillow lavas or their fragmented equivalents (Carlisle, 1963) are absent. A granular matrix, largely of crystalline lithic andesite fragments rather than scoria or other devitrified glassy materials, characterizes most of the breccias and tuffs. Even those breccias with the greatest amounts of vesiculated clasts lack the large proportions of vitroclastic materials ordinarily formed by quenching of lavas that flow into or are erupted into water. Diagnostically, massive breccias accumulated in water should be intercalated with finer grained volcanic sediments that exhibit small-scale structures indicative of turbidity-current deposits (Parsons, 1969). The finer textured layers associated with the andesite breccias in the Newbury mostly show planar stratification, are devoid of turbidity-current structures, and are sharply defined against, rather than mixed into, the breccia units.

Coupled with these considerations, the monolithologic character, lack of sorting, and absence of systematic lateral or vertical textural variation in many breccia layers seem to favor subaerial accumulation from pyroclastic flows (glowing-avalanche deposits) rather than from falls of aerially ejected pyroclastic material. The more heterolithologic breccias probably picked up andesite fragments of different aspect either during eruption or while in transit away from their vents. In the Newbury terrane, friction breccias that formed relatively nonviolently during slow extrusion of a dome and then spread as gravity flows could not be readily distinguished from explosively emitted pyroclastic deposits. The lack of features suggestive of domal extrusion or of taluslike wedges of breccia tends to negate this possibility.

Although most of the nonstratified andesitic breccias and tuff breccias are subaerial deposits, associations of certain of the finer grained massive volcaniclastic layers with water-laid materials do indicate marine accumulation. Andesite tuffs associated with the rhyolite conglomerates, which include discoid pebbles and cobbles locally shingled in a fashion suggestive of beach gravels, provide one example. In places, these conglomerates are sharply defined between andesite flows; elsewhere they are bounded above or below by massive layers of andesite tuff, which probably also are marine deposits. In these layers, 15 m or more in thickness, lapilli and smaller clasts—with scoria fragments greatly outnumbering lithic fragments—are set sparsely in a highly chloritic groundmass. Calcite, a cement common in marine volcaniclastic strata, is especially apparent as filling for the larger voids in the pumice, but calcite minutely disseminated through the groundmass perhaps constitutes a greater volume. The original constituents of this matrix are generally indeterminate, but in a few places are decipherable as fine shards, aligned with platy pumice clasts to provide a vague planar fabric that suggests bedding. Stratification is confirmed locally by lenses of abundant lapilli. Similar tuffs are associated with beds of andesite conglomerate and poorly stratified mudstones.

Tuffs in which shallow-water marine fossils are preserved are instructive. Brachiopod and pelecypod shells at the several sites in the porphyritic andesite member were much broken but, as far as can be determined from molds and casts, show little evidence of mechanical abrasion. Those still exposed at the Glen Mills locality lie within centimetres of a porphyritic flow and are in a crystal-lithic lapilli tuff. Much textural and mineral variety is apparent in the andesite of this tuff, but clasts that were once glassy are notably sparse. Stratification is imperceptible except as shown under the microscope by the vague alignment of plagioclase cleavage fragments and sparse platy lithic clasts. Clasts are little altered, and calcite is so sparse as to be easily overlooked. Thus, this tuff, probably of air-fall origin and certainly once saturated by marine water, gives little evidence of an aqueous environment. Calcite, a probable indicator of marine environment, is conspicuous as a filling of large vugs in the adjacent flow, and apparent as a cement in fine-grained tuffs that overlie the lapilli tuff.

A few andesite conglomerates are composed of subangular to well-rounded pebbles, cobbles, and boulders suspended in a sandy or granule-bearing mudstone that forms a large proportion of the bed. Occasional thin lenses of laminated sand or thin seams of fine silt may be the only indicators of internal stratification. The coarse clasts include a variety of andesitic materials and are altered irregularly throughout or are rimmed with alteration products. Apparently, these clasts were derived from different sources and variously exposed to weathering. These strata are interpreted as laharic deposits; some possibly are laterally transitional to water-laid conglomerates.

Distinctly stratified beds mark repeated episodes of shallow submergence while the pile of porphyritic andesite was accumulating. These strata, which range mostly from siltstone to coarse sandstone but locally include pebble conglomerate, are of ill-sorted, poorly rounded andesite detritus largely free of weathering products. Parallel stratification is dominant; small-scale, wispy cross-lamination is seen occasionally. These strata probably reflect rapid accumulation in shallow, fairly quiet waters that were at least in part marine, as demonstrated by faunal assemblages. Perhaps large proportions of the sediments were air-fall tuffs.

The massive mudstones that tend to take the place of the flows and volcaniclastic beds high in the member suggest that greater volumes of weathered detritus were formed and distributed onto marine mudflats as andesitic eruption diminished.

The three sedimentary members at the top of the Newbury appear to represent muds laid down in quiet water. The red mudstone member in its southernmost exposure obviously incorporates decomposed andesitic(?) detritus. Elsewhere all three members contain cleavage flakes of plagioclase that were plausibly derived from volcanic sources. According to J. M. Berdan (1973, oral commun.), the large leperditiids that characterize the calcareous mudstone member suggest an intertidal site of accumulation.

In summary, the Newbury obviously represents only part of the volcanic deposits that spread onto a shallow marine platform in this area. The basement, with detritus it might have supplied, was widely blanketed by volcanic deposits before the lowest of the members available for visible inspection was laid down. The lavas and explosively erupted ash and breccia flows that constitute most of the formation were laid down, episodically and subaerially, on surfaces of little relief maintained about at sea level while at least 4,500 m of the sequence accumulated. Lesser volumes of epiclastic detritus, supplemented by air-fall tuffs, accumulated as lagoon or mudflat deposits generally not subjected to the wave or current action of an open ocean. During submergence, exposed parts of the piled-up volcanic deposits possibly were too small to furnish much epiclastic material. The rhyolite lenses, emplaced by intrusion at a late stage, have no extrusive equivalents in the part of the pile that remains.

REGIONAL CORRELATIONS

The lithologic successions and the structural features to be considered in correlating Silurian-Devonian volcanic terranes of the New England coast have been summarized very well in a recent review by Gates (1969). Similarly, the data for paleontologic correlation have been reviewed by Berry and Boucot (1970). Details for the Newbury can be updated and added to these frameworks, but little refinement in regional correlations is provided by these data. First, we should recall that the Newbury succession as interpreted here is inverted and otherwise modified from the version that was available to Gates and Boucot for correlation. Furthermore, the "lower rhyolite member" reported by them is the micrographic rhyolite that intruded all the other members, and the uppermost sedimentary members were not recorded at the time of their reviews.

The Newbury Complex is the southernmost dated remnant of several remnants of Silurian-Devonian volcanic sequences that occur intermittently in a belt extending northeast along the coast of Maine and well into southern New Brunswick. Not only do the various volcanic terranes probably represent the accumulations about several eruptive centers, but they also represent accumulations slivered by numerous faults and apparently transported laterally along an even farther ranging system of strike faults—possible mostly wrench faults (Gates, 1969, p. 498–502; Webb, 1969). Thus, the remote possibility exists that offset segments can be matched, which would help elucidate the tectonic history of a large region. At present I can only point out areas where comparisons with the Newbury might be made when adequate data become available.

The nearest volcanic sequences to be considered as potential correlatives are the Lynn and Mattapan Volcanic Complexes, which crop out separately in east-trending belts along, respectively, the north and south borders of the Boston Basin, 21 to 27 km and 35 to 50 km south of the principal exposures of the Newbury (inset map, Fig. 1). Emerson (1917, p. 200) considered the Lynn Complex lithologically correlative to "the older and more felsitic part" of the Mattapan Complex, which is more widely but less continuously exposed in and near the southern part of the basin. The later flows of the Mattapan he supposed to be interstratified with lower members of the Roxbury Conglomerate, which in turn he assumed to be correlative with Pennsylvanian strata a bit farther south. Before the discovery of Silurian-Devonian fossils in the Newbury, the three complexes had been considered of the same age. Other workers (including Clapp, 1921, p. 29, and K. G. Bell, 1972, oral commun.), adducing marked lithologic similarities without citing kinship with specific units, have continued to postulate correlation of the Lynn and Newbury Volcanic Complexes. And with some misgivings, LaForge (1932, p. 29–35) proposed that the Mattapan was best restricted to volcanic rocks below the Roxbury Conglomerate and indicated that unspecified parts of the Mattapan were lithologically so similar that this complex too should continue to be regarded as correlative. The argument that the Mattapan, at least, should be regarded as younger than the Newbury has been revived occasionally (Billings, 1929, p. 106; Pollard, 1965).

With the Newbury succession now defined, comparisons of lithologic sequences in the Lynn and Mattapan might resolve the impasse. Unfortunately, the descriptions available for these formations are much too general for comparison. A cursory examination of thin sections, kindly provided by K. G. Bell, of greenish-gray to grayish-red strata from outcrops in the town of Lynn provided the only details available to me. Some of these specimens are of conspicuously flow-banded vitrophyre, very like the rock that makes up most of the rhyolite vitrophyre member of the Newbury. But most are lithic or crystal-lithic tuffs in which vitroclastic textures of the shardy groundmass and the structures in pumice and perlitic obsidian clasts are better preserved than in the Newbury rhyolite tuffs. Many specimens show only a hint of or no welding, but a fair number differ notably from the Newbury tuffs in exhibiting thorough welding of groundmass shards or in containing markedly compressed pumice clasts. Admittedly, in Newbury tuffs most nearly resembling these, the vitroclastic textures are almost wholly obliterated by devitrification, so perhaps greater similarity will be recognized with additional study. Phenocrysts of plagioclase are generally more abundant than in the Newbury rhyolites. Biotite, never seen in the Newbury, occurs sparsely in some specimens. Some specimens contain fragments of felty-textured andesite(?), a feature probably more representative of the lowest rhyolite tuff than other tuffs in the Newbury. The stratigraphic succession and relative volumes of the rock types represented in specimens is unknown. The Lynn rhyolites reported here are likely to be—but may not be—representative of a large part of the Lynn Complex. If determinable, the relationships of these rhyolites to elements of other reported lithologies will be resolved at localities west of the town of Lynn. In Lynn the volcanic pile is reported (LaForge, 1932, p. 31; K. G. Bell, 1973, oral commun.) to rest on an erosion surface deeply cut across an old granitic terrane. The latter may be correlative (K. G. Bell, 1973, oral commun.) with the Topsfield Granodiorite that is in fault contact with the Newbury. At present then, no features are known that would negate the reasonable speculation that the dominantly tuffaceous strata of the Lynn are lateral equivalents of either the rhyolite vitrophyre member or the lowest rhyolite tuff member in the Newbury; however, such correlation is merely permissive and yet to be proved.

Owing to lack of similar details for the Mattapan, Emerson's (1917) correlation of the basal felsites of the Mattapan with the Lynn rhyolites still cannot be confirmed.

To the northeast of the Newbury belt, the nearest correlative volcanic terranes are 210 to 260 km distant in the Penobscot Bay-Frenchmans Bay region of coastal Maine. The upper part of the Ames Knob Formation, which is the basal formation of the Silurian-Devonian sequence in southern Penobscot Bay, contains a recently reported fauna (Brookins and others, 1973, p. 1622) that is correlative with the Late Silurian (Pridoli) assemblage from the porphyritic andesite member of the Newbury. Lower strata of the Ames Knob include Silurian fossils older than any yet found in the Newbury. The Thorofare Andesite that overlies the Ames Knob is described (Smith and others, 1907; Gates, 1969) as dominated by pyroxene andesite—commonly with conspicuous plagioclase phenocrysts—that occurs as massive flows, flow-breccia, tuff-breccia, and coarse- and fine-grained tuffs. Outwardly then, if minor intercalations of porphyritic rhyolite and vitroclastic rhyolite tuffs are excepted, the 1,340-m-thick Thorofare has much in common with the porphyritic andesite member of the Newbury. The Vinalhaven Rhyolite, the next overlying and highest formation of the sequence, is composed of about 760 m of conspicuously banded flows, spherulitic flows and rhyolitic tuff and breccia, and is apparently rather similar to the banded rhyolite vitrophyre member of the Newbury. The maximum age of the Thorofare Andesite is defined as Pridoli, and parts of the andesite section were laid down in Pridoli time as shown by bracketing collections of fossils. But parts of the Thorofare contain diagnostic Early Devonian (Gedinnian) ostracodes that have been found only in the Hershey Red Shale Member of the Pembroke Formation and in the overlying Eastport Formation, farther northeast along the Maine coast (Berdan, 1971; Brookins and others, 1973). Though upper members of the Newbury might be as young as Gedinnian, none of the Newbury assemblages are known to range into that part of the Gedinnian represented in the Hershey Red Shale Member. Although Rb-Sr isotope dating suggests that the Vinalhaven Rhyolite could fall within the Pridoli-Gedinnian interval, the Vinalhaven overlies the Thorofare strata that contain the diagnostic Early Devonian ostracodes, and it too is apparently wholly younger than the Newbury. In summary, the Newbury is correlative with the upper part of the Ames Knob Formation and part of the Thorofare Andesite but has no apparent correlatives in other parts of the Silurian-Devonian sequence of southern Penobscot Bay.

On viewing the Newbury, D. B. Stewart, one of the most active recent students of the Penobscot Bay terranes, suggested (1971, oral commun.) that the banded rhyolite, the andesite flows, breccias, and tuffs, and the intrusive rhyolite may well have counterparts in the Castine Volcanics, which are exposed in northern parts of Penobscot Bay. Unfortunately, owing to lensing out of stratigraphic units and their interruption by faults and plutons, the succession of units cannot be established for the Castine (D. B. Stewart, 1974, written commun.). Rb-Sr data supplemented by collections of poorly preserved ostracodes (Brookins and others, 1973) do indicate that the Castine Volcanics are possible temporal equivalents of the Newbury. With somewhat less certainty, Rb-Sr data also suggest that the Cranberry Island Series of Shaler (1889), exposed some 30 km farther east, may be contemporaneous.

The Silurian-Devonian volcanic rocks of this region and those of eastern Massachusetts differ significantly in the geologic events that have affected each. The volcanic rocks of the Penobscot Bay area are in the chlorite grade of regional metamorphism; some are schistose (Brookins and others, 1973; D. B. Stewart, 1973, written commun.). Furthermore, the volcanic rocks are intensely cleaved along zones as much as 1 km wide adjacent to regional strike-slip faults. The Newbury and Lynn rocks (specific data are still needed for the Mattapan) lack a metamorphic overprint. The regional strike-slip faults that displace the Newbury are narrowly defined; in the adjacent rocks, cleavage related to faulting—indeed, any significant cleavage—is lacking. The volcanic rocks of Penobscot Bay have been intruded by plutons of diorite to granite composition; adjacent rocks were contact metamorphosed to as high as sillimanite grade through aureoles as much as 2 km wide. The Rb-Sr data from the volcanic rocks of Penobscot Bay and one of the plutons suggest the possibility that certain of the granites may be comagmatic with the volcanic rocks, but, for another example, a 30-m.y. interval between the volcanic rocks and a younger pluton is too long to support the hypothesis that all the Devonian granites are comagmatic (Brookins and others, 1973). Certain of the major faults that displace the volcanic rocks are bracketed as Devonian in age from their relationships to volcanic rocks and to the radiometrically dated plutons. The northeast-trending, regional faults of eastern Massachusetts (see Bell and Alvord, this volume)—some of which bound the Newbury and Lynn terranes—might be of like age and might be a part of the great fault belt which also displaced Silurian-Devonian volcanic rocks along the Maine coast. The Newbury and Lynn sequences have not been invaded by granitic plutons. Nor have plutons been recognized elsewhere that would put a limiting minimum age on the specific faults.

In the Machias-Eastport area, 100 to 170 km farther northeast along the Maine coast, another dominantly volcanic sequence is as much as 13,700 m thick and encompasses strata of a somewhat greater age span—Early Silurian to Early Devonian (Bastin and Williams, 1914; Gates, 1969). Under the guidance of Olcott Gates, whose recent work has done much to elucidate stratigraphic details that again demonstrate shallow marine accumulation, I was privileged to review representative exposures of the sequence in 1971. With the possible exception of a few basalts that are late in the sequence, the rocks of this terrane apparently have little petrographic similarity to those of the Newbury. As previously noted, the temporal definition of the Newbury is based on the equivalence of its fossils to those in the Leighton Gray Shale Member of the Pembroke Formation of the Eastport area. The Pembroke Formation, in particular, differs from the Newbury in being largely of water-laid sedimentary strata free of volcanic rock; the formations immediately above and below also have a greater content of nonvolcanic strata than the Newbury (see Gates, 1969, p. 492-495, for descriptions). If parts of the Machias-Eastport terrane were displaced laterally, they will most likely be found to the northeast in New Brunswick—as along Passamaquoddy Bay or still farther distant in the Long Reach region, where paleontologically dated equivalents have been noted (Gates, 1969, p. 496)—rather than to the southwest.

ACKNOWLEDGMENTS

For beneficial reviews of drafts of this paper, I thank G. W. Leo, J. W. Skehan, S. J., and D. B. Stewart. I am especially indebted to N. P. Cuppels for use of unpublished data on fossil localities and collections and to J. M. Berdan for guidance on Silurian-Devonian correlations. Other colleagues, credited in the text, generously shared information and ideas.

REFERENCES CITED

Bastin, E. S., and Williams, H. S., 1914, Description of the Eastport quadrangle, Maine: U.S. Geol. Survey Geol. Atlas, Folio 192, 15 p.

Bell, K. G., and Alvord, D. C., 1976, Pre-Silurian stratigraphy of northeastern Massachusetts, *in* Page, L. R., Contributions to the stratigraphy of New England: Geol. Soc. America Mem. 148, this volume.

Berdan, J. M., 1971, Silurian to Early Devonian ostracodes of European aspect from the Eastport quadrangle, Maine [abs.]: Geol. Soc. America Abs. with Programs, v. 3, no. 1, p. 18.

Berry, W.B.N., and Boucot, A. J., 1970, Correlation of the North American Silurian rocks: Geol. Soc. America Spec. Paper 102, 289 p.

Billings, Marland, 1929, Structural geology of the eastern part of the Boston Basin: Am. Jour. Sci., ser. 5, v. 104, p. 97-137.

Boucot, A. J., 1968, Silurian and Devonian of the northern Appalachians, *in* E-an Zen, and others, eds., Studies of Appalachian geology, northern and maritime: New York, Interscience Pubs., p. 83-94.

Brookins, D. G., Berdan, J. M., and Stewart, D. B., 1973, Isotopic and paleontologic evidence for correlating three volcanic sequences in the Maine coastal volcanic belt: Geol. Soc. America Bull., v. 84, p. 1619-1628.

Carlisle, Donald, 1963, Pillow breccias and their aquagene tuffs, Quadra Island, British Columbia: Jour. Geology, v. 71, p. 48-71.

Clapp, C. H., 1921, Geology of the igneous rocks of Essex County, Massachusetts: U.S. Geol. Survey Bull. 704, 132 p.
Cuppels, N. P., 1967, Geologic implications of undeformed fossils in the Newbury Formation of pre-Acadian age in northeastern Massachusetts [abs.]: Geol. Soc. America Spec. Paper 115, p. 258.
dos Santos, E. R., 1960, The Newbury volcanics [M.S. thesis]: Cambridge, Massachusetts Inst. Technology, 55 p.
Emerson, B. K., 1917, Geology of Massachusetts and Rhode Island: U.S. Geol. Survey Bull. 597, 289 p.
Foerste, A. F., 1920, Presence of Upper Silurian sandstone in Essex County, northeastern Massachusetts: Geol. Soc. America Bull., v. 31, p. 206-207.
Gates, Olcott, 1969, Lower Silurian-Lower Devonian volcanic rocks of New England coast and southern New Brunswick, *in* Kay, Marshall, ed., North Atlantic—Geology and continental drift, A symposium: Am. Assoc. Petroleum Geologists Mem. 12, p. 484-503.
Goddard, E. N., chairman, and others, 1948, Rock-color chart: Washington, D.C., Natl. Research Council.
LaForge, Laurence, 1932, Geology of the Boston area, Massachusetts: U.S. Geol. Survey Bull. 839, 105 p.
Parsons, W. H., 1969, Criteria for the recognition of volcanic breccias: Review, *in* Larsen, L. H., and others, eds., Igneous and metamorphic geology: Geol. Soc. America Mem. 115, p. 263-304.
Pollard, Melvin, 1965, Age, origin, and structure of the post-Cambrian Boston strata, Massachusetts: Geol. Soc. America Bull., v. 76, p. 1065-1068.
Ross, C. S., and Smith, R. L., 1961, Ash-flow tuffs, their origin, geologic relations, and identification: U.S. Geol. Survey Prof. Paper 366, 81 p.
Shaler, N. S., 1889, The geology of the island of Mt. Desert, Maine: U.S. Geol. Survey, 8th Ann. Rept., pt. 2, p. 987-1061.
Smith, G. O., Bastin, E. S., and Brown, C. W., 1907, Description of the Penobscot Bay quadrangle, Maine: U.S. Geol. Survey Geol. Atlas, Folio 149, 14 p.
Toulmin, Priestley, III, 1964, Bedrock geology of the Salem quadrangle and vicinity, Massachusetts: U.S. Geol. Survey Bull. 1163-A, p. A1-A79.
Webb, G. W., 1969, Paleozoic wrench faults in Canadian Appalachians, *in* Kay, Marshall, ed., North Atlantic—Geology and continental drift, A symposium: Am. Assoc. Petroleum Geologists Mem. 12, p. 754-786.
Zartman, R. E., and Naylor, R. S., 1972, Structural implications of some U-Th-Pb zircon isotopic ages of igneous rocks in eastern Massachusetts [abs.]: Geol. Soc. America Abs. with Programs, v. 4, no. 1, p. 55.

MANUSCRIPT RECEIVED BY THE SOCIETY JULY 24, 1975
REVISED MANUSCRIPT RECEIVED FEBRUARY 23, 1976
MANUSCRIPT ACCEPTED FEBRUARY 27, 1976

Printed in U.S.A.

Geological Society of America
Memoir 148

Pre-Silurian Stratigraphy of Northeastern Massachusetts

KENNETH G. BELL*
U.S. Geological Survey
Denver Federal Center
Denver, Colorado 80225

AND

DONALD C. ALVORD
U.S. Geological Survey
8422 Federal Building
125 South State Street
Salt Lake City, Utah 84111

ABSTRACT

Stratified rocks of northeastern Massachusetts that are known or considered to be of pre-Silurian age are (1) a sequence of nonfossiliferous marine metasedimentary and metavolcanic rocks at least 2,800 m thick, including the Westboro Formation (formerly known as the Westboro Quartzite) and the Middlesex Fells Volcanic Complex (new name), that probably are of Precambrian age; (2) a sequence of nonfossiliferous eugeosynclinal rocks having an aggregate thickness of about 18,800 m, which includes the Greenleaf Mountain Formation (new name), Burlington Formation (new name), Marlboro Formation, Shawsheen Gneiss (new name), Fish Brook Gneiss, Nashoba Formation, and Tadmuck Brook Schist (new name), all of undetermined age; and (3) some scattered remnants of unmetamorphosed, partly fossiliferous marine sedimentary rocks of Cambrian age.

The Westboro Formation, Middlesex Fells Volcanic Complex, and some older unnamed gneiss and quartzite strata were domed and locally subjected to minor folding during emplacement of the Dedham Granodiorite, which intrudes them but does not intrude any of the stratigraphically higher metasedimentary rocks. These intrusive and structural features, together with local omissions of the upper part of the above-named formations, suggest a significant unconformity. These stratified rocks are tentatively correlated with the Blackstone Series of Rhode Island.

*Present address: 725 Allison Street, Lakewood, Colorado 80215.

The eugeosynclinal rocks constitute a generally homoclinal sequence, except where parts of the section are locally cut out or are repeated by faults or are repeated by minor folds. This sequence consists of at least 80 percent metamorphosed volcanic wackes; units characterized by strata of metamorphosed pelitic and carbonate sediments are interspersed through it.

The stratified rocks of Cambrian age form pendants in plutonic rocks and possibly also outliers. The Weymouth Formation of Early Cambrian age and the Hoppin Slate that possibly is correlative with it consist mainly of marine shale and limy shale interspersed with minor thin limestone strata. The Braintree Argillite, conformably overlying the Weymouth Formation, consists of marine shale, silty shale, and siltstone. The remnants of Cambrian rocks do not have any contacts with other stratified rocks.

INTRODUCTION

Recent detailed study and mapping of bed rock in northeastern Massachusetts has produced significant new information about the stratigraphy, lithology, and structure of this area. Stratified rocks of known and inferred pre-Silurian age are discussed in this paper, with emphasis on their characteristic mineralogic and lithologic features, protoliths, and depositional environments. These rocks are described in two categories: (1) unmetamorphosed fossiliferous formations of Cambrian age situated within and south of the Boston basin (Figs. 1,[1] 2) and (2) a thick sequence of metamorphosed stratified rocks of inferred pre-Silurian age that lies north and northwest of the Boston basin (Fig. 1) and in which no fossils have been discovered.

The Cambrian age of the fossiliferous rocks has been long established. The ages of the metamorphosed nonfossiliferous rocks described in this paper are by necessity inferred from their structural relations, metamorphosed condition, relations to the intrusive rocks of the region, and lithologic characteristics. Different units of these metamorphosed stratified rocks are intruded by one or more of three major plutonic series of the region. Ages of the intrusive rocks that are based on isotopic determinations tend to be somewhat erratic and discordant as a consequence of an episode of pervasive hydrothermal alteration that has affected most of them (Zartman and others, 1970). Isotopic ages of the intrusive rocks cannot be used to establish conclusively the ages of the stratified rocks. The lithology of these metamorphosed rocks is similar to that of the oldest stratified rocks along the Maine coast northeastward from Casco Bay, where fossiliferous rocks of Ordovician or Silurian age are interlayered with or overlie rocks similar to those described in this paper; other localities of generally similar rocks are in eastern Connecticut and western Massachusetts.

The stratigraphy is not completely known because of insufficient exposures. Locally, parts of the stratified rocks are displaced or assimilated by plutonic rocks, and parts of the column are cut out by faults. Large segments of the terrane of northeastern Massachusetts are covered by glacial deposits and by swamps. Outcrops are scarce in belts underlain by schistose rocks, marble, calc-silicate rocks, and some types of argillite, quartzite, and metamorphosed shale. Thicknesses of units can only be approximated, because contacts are rarely exposed, all of the stratified rocks enclose locally variable amounts of intrusive rocks, and faults have cut out or have caused repetition of some units or parts of units. Measured

[1] Figure 1 is in pocket inside back cover.

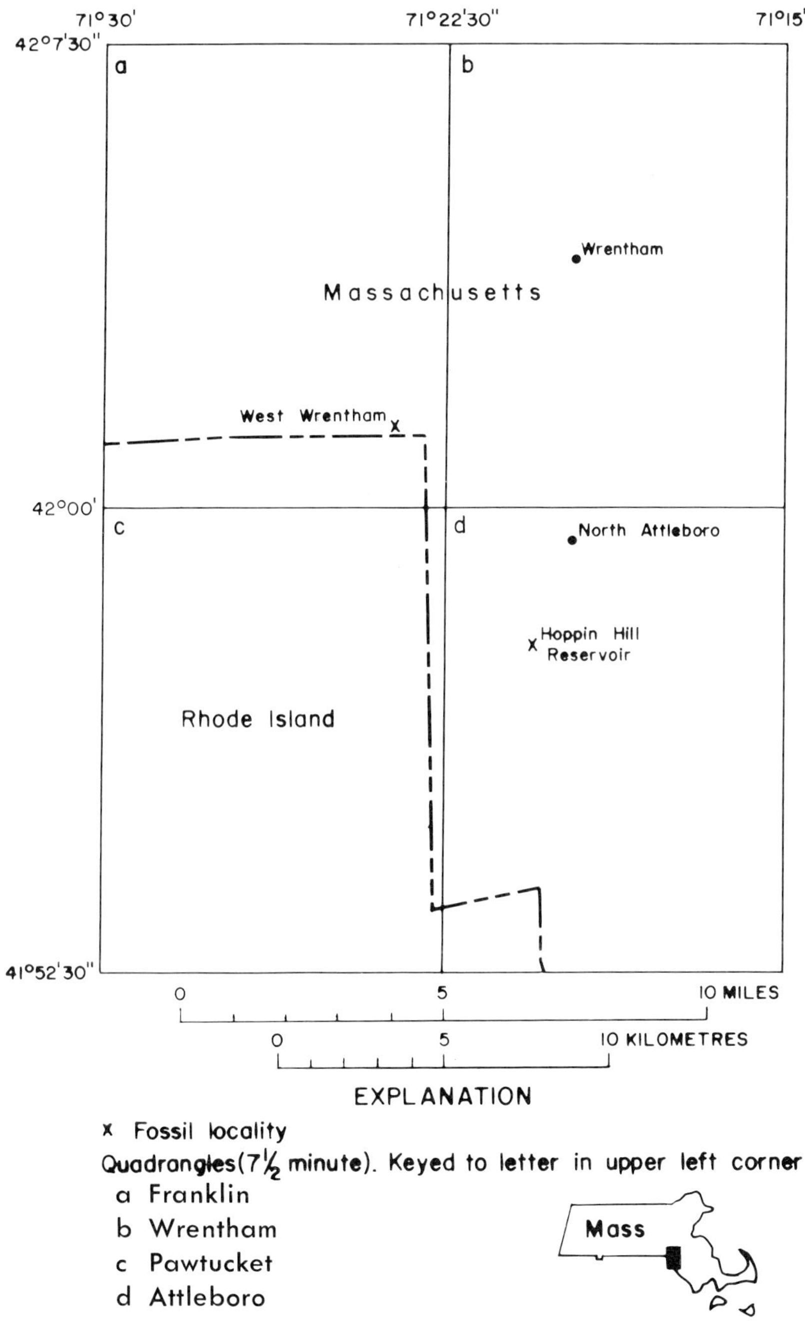

Figure 2. Localities of Cambrian fossils in southeastern Massachusetts.

sections can be obtained only for minor parts of the column and are not used in the descriptions that follow.

The stratigraphy presented in this paper is a summary of observations and unpublished mapping since 1965 by the following geologists: Andrew F. Shride (Georgetown quadrangle), William H. Dennen (Ipswich quadrangle), Norman P. Cuppels (Georgetown and Concord quadrangles), Donald C. Alvord (Billerica, Concord, Lowell, Westford, and Wilmington quadrangles), and Kenneth G. Bell (Lynn, Boston North, Lexington, Concord, Reading, Wilmington, Georgetown, and South Groveland quadrangles and reconnaissance mapping in other parts of the area). This work has been carried out under the supervision of Lincoln R. Page as part of a cooperative program of the State of Massachusetts Department of Public Works and the U.S. Geological Survey. Descriptions and comments concerning the fossiliferous rocks of Cambrian age have been obtained mainly from the literature, because many former exposures of these rocks have been obliterated during urban development.

SEDIMENTATION

The sedimentary and volcanic materials that formed the stratified rocks described herein were deposited mainly in a marine environment, probably near the southeast margin of a eugeosyncline. The sediments that formed the fossiliferous rocks of Cambrian age (Fig. 3) seem to have been deposited in a shallow, nearshore shelf environment. The sediments and volcanic materials that formed the lower units of the metamorphosed sequence, which include the Middlesex Fells Volcanic Complex and older units (Fig. 4), may have been deposited shortly before the eugeosyncline began to form or during its earliest stage. The stratified rocks younger than the Middlesex Fells Volcanic Complex (Fig. 4) seem to represent almost continuous deposition during eugeosynclinal development. Poorly preserved relict sedimentary structures in the latter rocks suggest east to west or southeast to northwest movement of volcanic detritus. Deposits formed almost wholly of epiclastic sediments are mostly very fine grained, and diagnostic sedimentary features indicating transport direction are not preserved. It is presumed that the deposits mostly had an initial gentle northwest dip, which increased as the eugeosyncline subsided and was filled.

STRUCTURAL FEATURES

The stratified rocks described herein mostly are truncated on the northwest and north by the Clinton-Newbury fault zone, one of the major structural features of the region. This zone is at least 2 km wide in the vicinity of the Wachusett-Marlborough Tunnel (Skehan, 1968, 1969, and this volume) and gradually narrows to a width of less than 100 m in the Newburyport West quadrangle (Andrew F. Shride, 1975, oral commun.). Our mapping along parts of this fault zone indicates that there is a very large fault (or faults) along its southeast and south margin. Many faults—some large, some small—lace the zone, bounding displaced masses of rock, many of which are identified with units on the northwest side of the zone. The Rattlesnake Hill fault zone described by Skehan (1968, 1969) is now considered by him to be the large fault bounding the southeast side of the Clinton-Newbury fault zone in the vicinity of the Wachusett-Marlborough Tunnel (1975, oral commun.). The line indicating the Clinton-Newbury fault zone in Figure

	Unit	Thickness	Lithology	Relict primary features	Protoliths
Fossiliferous formations of Cambrian age	Braintree Argillite	200+	Fine-grained slate(?) or shale. Low grade metamorphism, if any. Upper part of known section is dark gray to black nonfossiliferous. Lower part consists of greenish and greenish-gray slates(?) or shales interlayered with red slates(?) or shales. Bedding plane parting.	Bedding is well preserved. Unit contains a *Paradoxides* fauna. Remains of *trilobites* are flattened parallel to bedding planes.	Protolith was peletic sediment. Deposition was in a shallow, probably shelf, marine environment.
	Weymouth Formation	200+	Fine-grained red, brown, green, and black slates(?) or shales. Thin strata of white limestone in upper part. Low-grade metamorphism, if any. Some of the slate(?) or shale contains abundant calc-silicate nodules somewhat flattened parallel to the bedding, 2 to 5 centimetres diameter. Slate(?) or shale has bedding plane parting. Limestone is massive; layering, if any, is indistinct.	Bedding is well preserved. Limestone strata contains an *Olenellus* fauna. Calc-silicate nodules are a conspicuous feature of some gray and black shale beds.	Protolith of the slate(?) or shale was peletic sediment, possibly with minor carbonate component. The calc-silicate nodules are interpreted to be altered limy concretions. Limestone strata represent short episodes of carbonate sedimentation. Deposition was in a shallow, probably shelf, marine environment.

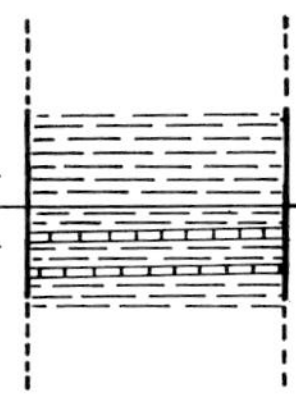

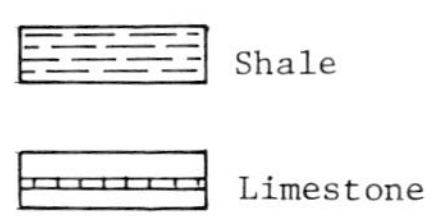

Figure 3. Fossiliferous formations of Cambrian age.

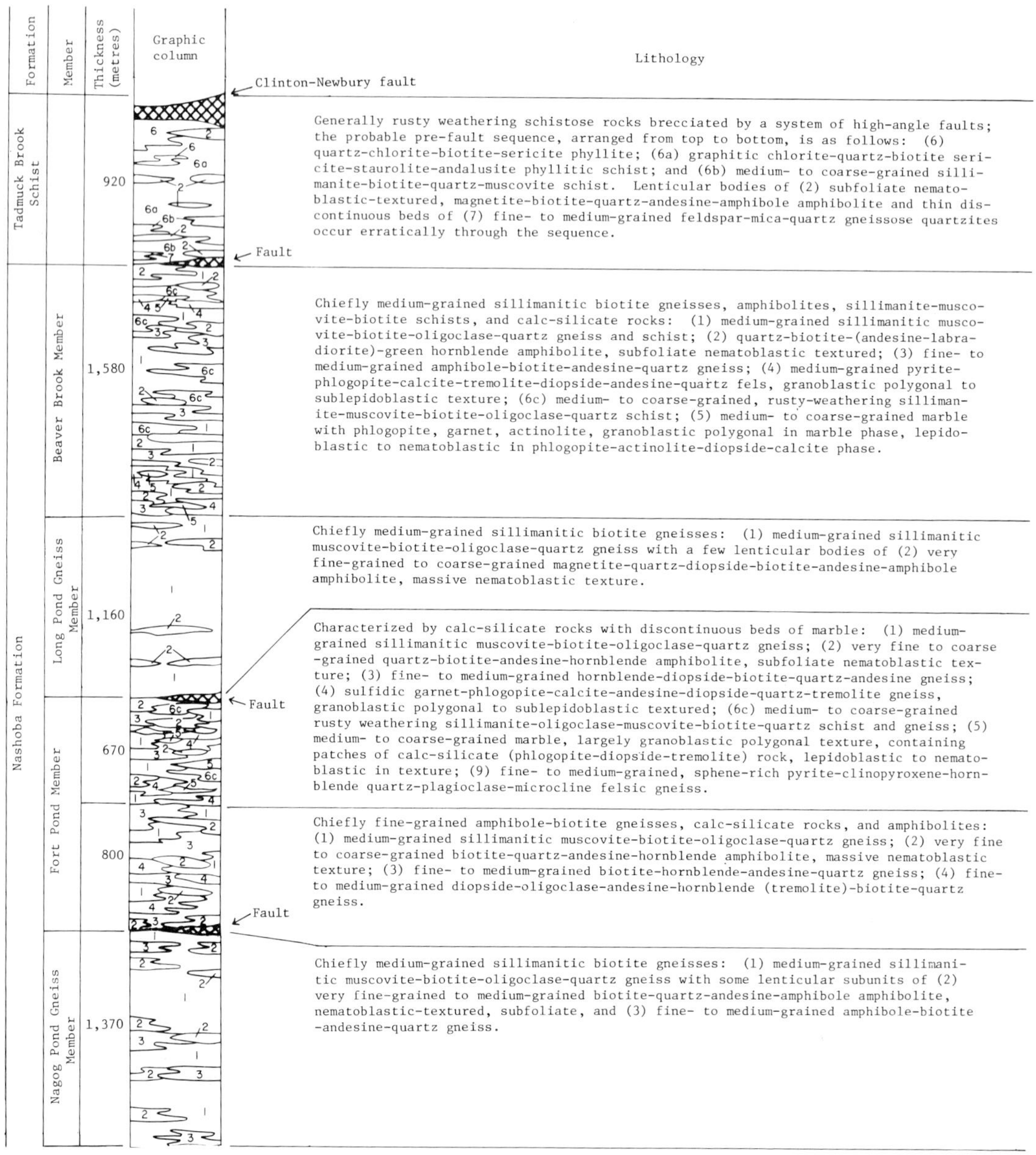

Figure 4. Generalized stratigraphic column of nonfossiliferous rocks of northeastern Massachusetts. Unit thicknesses are plotted to scale. Internal lithologies of the units are drawn schematically to suggest relative abundance and, to a lesser degree, position in the unit. Stratigraphic thicknesses shown are near maximum as determined from detailed mapping at 1:24,000 scale. Locations of sections are shown on Figure 1. Rocks are described megascopically. Minerals are listed in order of increasing abundance, but only essential minerals and characterizing accessory minerals are cited. The order of the minerals in rock names is descriptive of the cited lithosome, but variation from relative

Relict primary features	Protolith
(6), (6a) Thin laminae to thin beds having some rarely discernible low-angle cross laminae. (2) Thin laminae to very thin beds; parallel stratified to cross-laminated where fine-grained; massive lenticulate, thick to very thick beds where medium to coarse grained. (7) Laminae to thin lenticulate beds, low-angle cross-stratified, locally graded.	(6), (6a) (6b) Interstratified clay-mica and quartzose silt and shale, carbonaceous in part. (2) Andesitic to basaltic waterlain tuff and lava flows. (7) Fine-grained clayey and silty quartz sandstone.
(1) Generally poorly defined, discontinuous, lenticular, cross-stratified, laminated to medium beds. (2) Thin-bedded, distinct, even, planar, and parallel-stratified where fine grained; massive lenticulate thick to very thick beds where medium to coarse grained. (3) Generally distinct relict laminated to thin-bedded, parallel to very low-angle cross-stratified; locally beds are graded. (4) Uneven lenticular to horizontal stratified, laminated to medium bedded. (6c) Thin laminae to thin beds, low-angle cross stratified. (5) Bedding nebulitic, uneven, thin to medium.	(1) Volcaniclastic and epiclastic plagioclase-quartz sandstone interstratified with subsidiary clay-mica shale. (2) Andesitic to basaltic waterlain tuff and lava flows. (3) Volcaniclastic and epiclastic plagioclase-quartz sandstone and siltstone interlaminated with slightly calcareous clay-mica shale. (4) tuffaceous calcareous sandstone interbedded with siltstone and fine-grained plagioclase-quartz sandstone. (5) Clayey to sandy dolomitic limestone.
(1) Obscure to fairly distinct thin laminae to medium beds, discontinuous, lenticular, cross-stratified. (2) Very thin parallel beds to very thick lenticular beds.	(1) Volcaniclastic and epiclastic plagioclase-quartz sandstone interlaminated with subsidiary clay-mica shale. (2) Andesitic to basaltic waterlain tuff and submarine lava flows.
(1) Poorly defined laminae to medium thick, lenticular, discontinuous beds. (2) Thinly laminated to thick bedded, cross stratified to massive. (3) Generally distinct relict laminated to thin-bedded, parallel to low angle cross-stratified. (4) Laminated to medium bedded, lenticular discontinuous to horizontal stratified. (6c) Thinly laminated to thin bedded. (5) thin to medium bedded, uneven, generally obscure. (9) thinly laminated to very thin bedded, cross stratified.	(1) Volcaniclastic and epiclastic plagioclase-quartz sandstone interbedded with subsidiary clay-mica shale. (2) Andesitic to basaltic waterlain tuff and submarine lava flows. (3) Volcaniclastic and epiclastic plagioclase-quartz sandstone and siltstone interlaminated with slightly calcareous clay-mica shale. (4) Tuffaceous calcareous sandstone interbedded with volcaniclastic and epiclastic clayey siltstone and sandstone. (6c) Sulfidic shale interlaminated with siltstone and fine grained plagioclase-quartz sandstone. (5) Clayey to sandy dolomitic limestone. (9) Felsic tuffaceous siltstone or tuff.
(1) Obscure to rarely distinct thin laminae to medium beds, discontinuous, lenticular-bedded; cross-stratified. (2) laminated to thin-bedded, very low angle cross-stratified. (3) laminated to thin-bedded, parallel-stratified to very low angle cross-stratified, slabby weathering. (4) Thinly laminated to thin-bedded, very planar-stratified.	(1) Volcaniclastic and epiclastic plagioclase-quartz sandstone interstratified with subsidiary clay-mica silty shale. (2) Andesitic to basaltic waterlain tuff and submarine lava flows. (3) Volcaniclastic and epiclastic plagioclase-quartz sandstone and siltstone interlaminated with slightly calcareous clay-mica shale. (4) Tuffaceous calcareous sandstone interbedded with volcaniclastic and epiclastic clayey siltstone and sandstone.
Abundant syntectonic injection of Acton granite and associated pegmatites into the rocks of this unit has distorted and locally obliterated most primary features in many areas. (1) Obscure to distinct thin laminae to thin beds, fairly planar-stratified to discontinuous lenticular bedded, low to high angle cross-stratified. (2) Very distinct laminae to thin beds, very planar stratified with very low angle cross-stratification, to massive appearing. (3) slabby-weathering, generally distinct relict laminae to thin beds, planar stratified with some low angle cross lamination.	(1) Volcaniclastic and epiclastic plagioclase-quartz sandstone interlaminated with subsidiary shale. (2) Andesitic to basaltic waterlain tuff and lava flows. (3) Volcaniclastic and epiclastic plagioclase-quartz sandstone and siltstone interlaminated with slightly calcareous shale.

abundances is common. Grain size is determined by the modal dimensions of the framework minerals: quartz and plagioclase in quartzofeldspathic gneiss and fels; mica in schist; amphibole and plagioclase in amphibolite; and calcite and dolomite in marble. Class intervals used are as follows: very fine (less than 0.5 mm); fine (0.5 to 1 mm); medium (1 to 5 mm); coarse (5 mm to 1.5 cm). Bedding thickness classes are from Ingram (1954): thinly laminated (less than 0.3 mm); laminated (0.3 to 1.0 cm); very thin bedded (1 to 3 cm); thin bedded (3 to 10 cm); medium bedded (10 to 30 cm); thick bedded (30 to 100 cm); very thick bedded (greater than 1 m).

(Continued on following four pages)

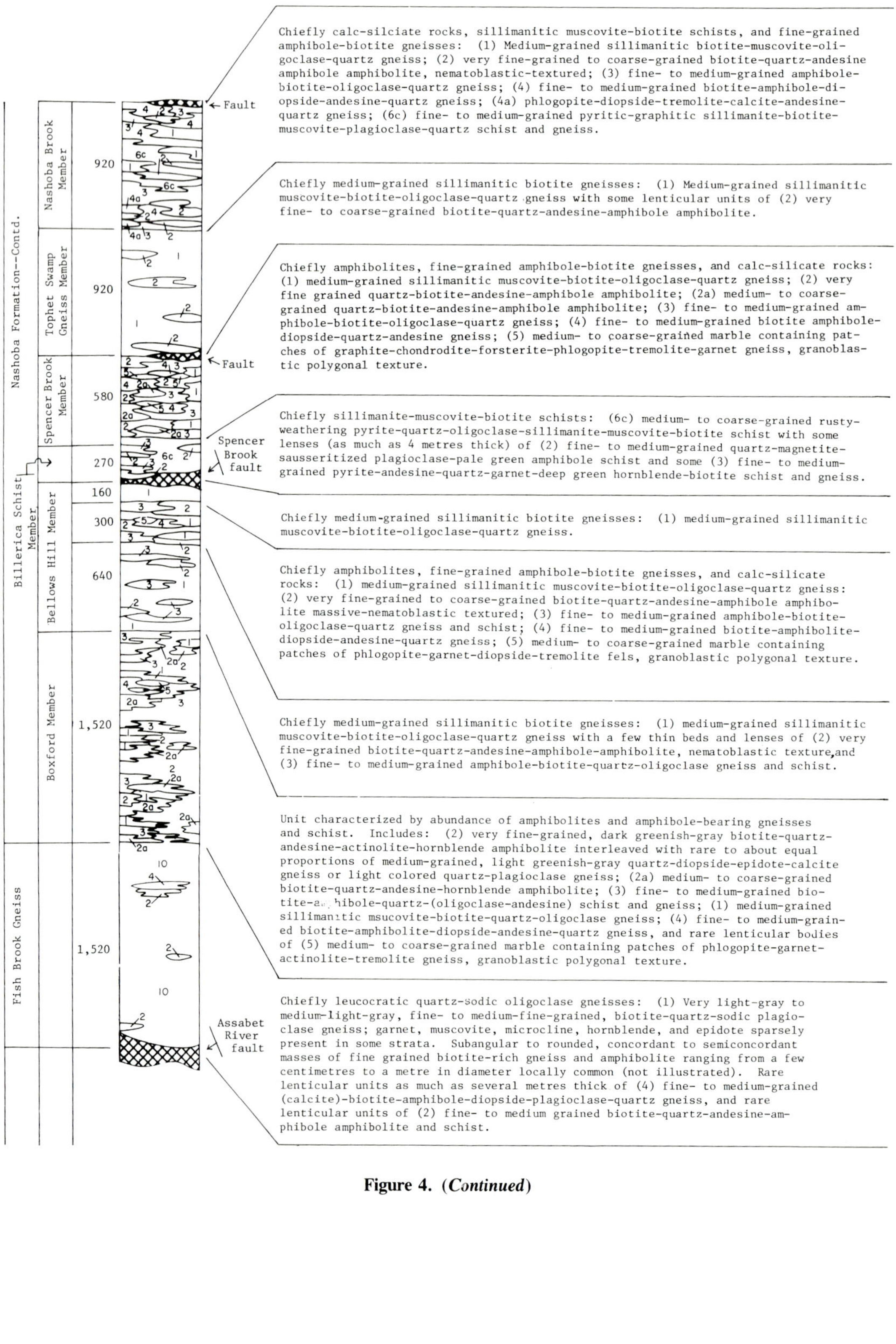

Figure 4. ***(Continued)***

(1) Obscure to distinct thin laminae to medium beds, unevenly discontinuous, lenticular banded, cross-stratified. (2) Platy weathering planar-stratified laminae and thin beds to massive. (3) platy weathering, generally distinct relict laminae to thin beds, wispy low angle cross-stratified, slabby weathering, obscurely uneven lenticular medium-bedded. (6c) Thin laminae to very thin beds, very low-angle cross-laminated.	(1) Volcaniclastic and epiclastic plagioclase-quartz sandstone interbedded with subsidiary clay. (2) Andesitic to basaltic waterlain tuff and lava flows. (3) volcaniclastic and epiclastic plagioclase-quartz sandstone and siltstone interlaminated with slightly calcareous clay-mica shale. (4)-(4a) Tuffaceous calcareous sandstone interbedded with volcaniclastic and epiclastic clayey siltstone, sandstone, and marl. (6c) Interstratified black sulfidic shale, quartzose siltstone, and minor plagioclase-quartz sandstone.
(1) Obscure to distinct laminae to medium beds, lenticular, discontinuous planar stratified and persistent, low angle cross stratified. (2) distinct relict laminae to thin beds, planar stratified, very low angle cross stratified to massive.	(1) Volcaniclastic and epiclastic plagioclase-quartz sandstone interlaminated with subsidiary clay-mica shale. (2) Andesitic to basaltic waterlain tuff and lava flows.
(1) Obscure to distinct laminae to medium beds, discontinuous, lenticular, cross-stratified. (2) Distinct thin laminae to thin beds, continuous planar with very low-angle cross laminations. (2a) Thin to thick irregular lenticular beds, massive appearing with nematoblastic subfoliate texture. (3) Distinct relict thin laminae to thin beds, continuous planar with low angle cross stratification. (4) Obscure uneven medium thick beds to planar-stratified laminae to thin beds, low angle cross-stratified. (5) thin- to medium-bedded, uneven, nebulitic.	(1) Volcaniclastic and epiclastic plagioclase-quartz sandstone interbedded with subsidiary clay-mica shale. (2) Andesitic to basaltic waterlain tuff. (2a) Intermediate to mafic submarine lava flows. (3) Volcaniclastic and epiclastic plagioclase-quartz sandstone and siltstone interlaminated with slightly calcareous clay-mica shale. (4) Tuffaceous calcareous sandstone interbedded with volcaniclastic and epiclastic clayey siltstone and sandstone. (5) Clayey to sandy dolomitic limestone and marl.
(6c) Thin laminae with considerable intrafolial injection of granite, inferred thinly laminated. (2) obscure to distinct relict planar-stratified laminae and thin beds. (3) thin laminae to thin beds, generally planar stratified, some low angle cross stratification and thin graded beds.	(6c) Black sulfidic shale interstratified with minor plagioclase-quartz. (2) Andesitic to basaltic waterlain tuff, submarine lava flows. (3) Tuffaceous slightly calcareous siltstone interlaminated with volcaniclastic and epiclastic clay-mica-quartz siltstone.
(1) Obscure to fairly distinct laminae and medium beds, discontinuous, lenticular, low angle cross-stratified.	(1) Volcaniclastic and epiclastic plagioclase-quartz sandstone interlaminated with subsidiary clay-mica shale.
(1) Obscure to fairly distinct laminae to medium beds, uneven lenticular discontinuous to moderately planar, low angle cross-stratified. (2) Distinct platy weathering planar-stratified laminae to thin beds having some low angle cross stratification, to thick lenticulate bedded. (3) distinctly relict laminated to thin-bedded, planar to low angle cross-stratified. (4) Obscure, uneven, medium thick beds to planar-stratified laminae and thin beds, low angle cross stratified. (5) Only faint indications of bedding.	(1) Volcaniclastic and epiclastic plagioclase-quartz sandstone interstratified with subsidiary clay-mica shale. (2) Andesitic to basaltic waterlain tuff and lava flows. (3) volcaniclastic and epiclastic plagioclase-quartz sandstone and siltstone interlaminated with slightly calcareous clay-mica shale. (4) Tuffaceous calcareous sandstone interbedded with volcaniclastic and epiclastic clayey siltstone and sandstone. (5) Argillaceous to quartz-sand dolomitic limestone.
(1) Obscure to fairly distinct laminae to medium beds, lenticular, discontinuous, low angle cross stratified. (2) laminated to thin-bedded, very low angle cross-stratified, locally massive. (3) distinct thin laminae to thin beds, low angle cross stratified to planar.	(1) Volcaniclastic and epiclastic plagioclase-quartz sandstone interlaminated with subsidiary clay-mica shale. (2) Andesitic to basaltic waterlain tuff. (3) Volcaniclastic and epiclastic plagioclase-quartz sandstone and siltstone interlaminated with slightly calcareous clay-mica shale.
(2) Thinly laminated to thin beds, planar to very low angle cross-stratified. (2a) Locally shows pillow, columnar, and flow structures. (3) Generally distinct planar laminae to thin relict beds, low angle cross stratified. (1) Obscure to fairly distinct uneven lenticular laminae to thin beds, low angle cross stratified. (4) Obscure, uneven, medium-thick beds to planar-stratified laminae and thin beds, low angle cross stratified. (5) Bedding indistinct, uneven lenticular, thin to medium.	(2) Largely basaltic waterlain tuff. (2a) Andesitic to basaltic submarine lava flows and hypabyssal sills. (3) Volcaniclastic and epiclastic plagioclase-quartz sandstone and siltstone interlaminated with slightly calcareous clay-mica shale. (1) Volcaniclastic and epiclastic plagioclase-quartz sandstone interlaminated with subsidiary clay-mica shale. (4) Tuffaceous calcareous sandstone interbedded with volcaniclastic and epiclastic clayey siltstone and sandstone. (5) Argillaceous to quartz-sand dolomitic limestone and marl.
(10) Thin laminae to thin beds, largely distinctively to obscurely ripple-bedded, amplitudes ranging from 1 to 10 centimeters; locally beds are subplanar and cross stratified, (4) Largely obscure thin laminae to thin beds, cross-stratified, (2) Bedding generally obscure, thinly laminated, locally ripple bedded.	(10) Fine- to medium-grained sodic plagioclase-quartz sandstone with a very little to as much as 10 percent clay-mica shale or mafic tuff concentrated chiefly in ripple laminae. (4) Tuffaceous calcareous sandstone interstratified with volcaniclastic and epiclastic clayey siltstone and sandstone. (2) Andesitic to basaltic waterlain tuff interlaminated with varying amounts of volcaniclastic and epiclastic siltstone.

Figure 4. (*Continued*)

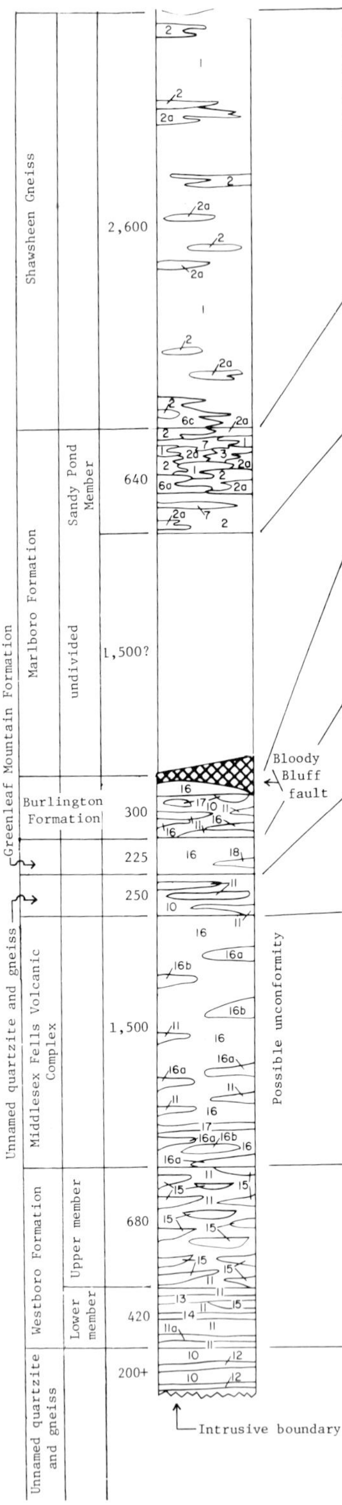

Largely medium grained muscovite-biotite-oligoclase-quartz gneisses: (1) medium-grained, locally sillimanitic, muscovite-biotite-oligoclase-quartz gneiss enclosing some lenticular bodies of (2) medium- to coarse-grained biotite-quartz andesine-amphibole amphibolite, nematoblastic- to subfoliate-textured and (2a) very fine-grained to fine-grained biotite-quartz-andesine amphibole amphibolite; and some (6c) rusty weathering sillimanite-muscovite-biotite-quartz schist locally near the base.

Largely complexly intertongued (2a) medium- to coarse-grained biotite-quartz-plagioclase-hornblende amphibolite, nematoblastic to subfoliate texture, and (2) very fine grained biotite-quartz-plagioclase-actinolitic amphibolite. Also present are (6a) rusty weathering andalusite-quartz-muscovite-biotite schist; (1) fine-to medium-grained biotite-muscovite-plagioclase-quartz gneiss; and (7) medium-grained muscovite-biotite-quartz plagioclase gneiss.

A complexly instratified sequence containing lenticular bodies of medium-grained plagioclase-quartz-mica schist, hornblende-biotite-plagioclase gneiss, calc-silicate-bearing gneisses, and rare lenses of marble and calc-silicate gneiss together with subsidiary (?) proportions of dark-greenish gray, thinly bedded very fine grained amphibolite and medium-grained massive lepidoblastic-textured amphibolite. Too poorly exposed and mapped to illustrate.

Includes: (10) fine-grained, light-colored, laminated biotite-quartz-feldspar gneiss; (11) fine-grained white to pale-gray quartzite; (16) fine-grained thinly laminated, dark-colored with fine white pin-stripe oligoclase-hornblende amphibolite; (17) lenticular bodies of metamorphosed boulder conglomerate, boulders are granite, diorite, or quartzite, matrix is fine- to medium-grained, gray biotite-quartz-feldspar gneiss. Metamorphic foliation parallels bedding.

(16) Fine-grained, thinly bedded or laminated dark gray to dark greenish-gray with fine whitish pin-stripe oligoclase-hornblende amphibolite, some laminae containing minor quartz and/or biotite; and (18) minor zones of thinly interlayered amphibolite and calc-silciate rock. Metamorphic foliation parallels bedding

Includes: (11) fine-grained white to pale gray quartzite, and (10) fine-grained laminated to thin-bedded hornblende-biotite-quartz-feldspar gneiss.

The mafic rocks all have been metamorphosed to amphibolite facies; locally they subsequently have been intensely chloritized and/or epidotized. Includes: (16) fine-grained, thinly bedded or laminated oligoclase-hornblende amphibolite, some laminae containing minor biotite; (16a) amphibolite having relict pillow and/or flow structures, (16b) amphibolite having relict features of coarse pyroclastic detritus; (17) carbonate and calc-silicate mineral zone, and (11) fine-grained, white to light-gray, generally rusty-weathering quartzite. Minor parts of (16) have fine, white pin-stripe. Metamorphic folication parallels bedding.

Includes: (11) fine- to medium-grained, white to light-gray quartzite, generally including less than 5 percent total feldspars and micas; (11a) fine-grained white or black-speckled feldspar-biotite-quartz schist; (13) fine- to medium-grained, light greenish-gray to light-green quartzose calc-silicate rocks, (14) fine-grained gray to black argillite and metamorphosed gray to black shale and limy shale, and (15) fine-grained light-gray to light-greenish-gray amphibole-feldspar gneiss. Amphiboles include pale-green hornblende and actinolite.; feldspar is mostly oligoclase; very minor potassium feldspar. Metamorphic foliation parallels bedding.

Includes: (10) fine- to medium-grained thinly bedded or laminated biotite-quartz-feldspar (K-feldspar and plagioclase) gneiss, which in localities of slightly higher metamorphic grade include minor quantities of garnet and amphiboles, and (12) fine- to medium-grained, white, pale gray, or pale-pink quartzite; some units contain minor quantities of biotite and feldspars. Metamorphic foliation parallels bedding.

Figure 4. (*Continued*)

(1) Primary thin laminae are largely obliterated by metamorphic recrystallization but very thin to medium beds are generally distinct, having lenticular discontinuous habit, low- to medium-angle cross stratification; many exposures are notably ripple-bedded, amplitudes of ripples ranging from 5 cm to more than 2 metre, (2) Generally massive, locally agglomeratic. (2a) Thinly laminated to thin bedded, planar to low angle cross stratified. (6a) Schistose mass with locally discernible thin crinkly laminae. Lower part of unit includes a few lenticular masses of agglomerate; outlines of subangular to angular fragments as much as 50 cm in maximum dimension are well preserved.	(1) Volcaniclastic and epiclastic sandstone interlaminated with subsidiary clay-mica shale. (2a) Andesitic to basaltic submarine lava flows and hypabyssal sills. (2) Basaltic waterlain tuff. (6a) Aluminous muds and silts.
(2a) Massive to uneven thick bedded. (2) Thinly laminated to thin bedded, planar to very low angle cross stratified. Primary structure in remaining lithologies (6a, 1, 7) too poorly known to describe.	(2a) Andesitic to mafic submarine lava flows and hypabyssal sills. (2) Basaltic waterlain tuff. (6a) Black sulfidic shale interstratified with quartzose siltstone. (1) and (7) Volcaniclastic and epiclastic plagioclase-quartz sandstone and siltstone interlaminated with subsidiary clay-mica shale. Depositional environment was entirely subaqueous and marine.
Poorly exposed and incompletely mapped.	Dominantly andesitic to basaltic waterlain tuff and submarine lava flows intermixed with subsidiary deposits of volcaniclastic and epiclastic sediments and very minor deposits of marine limestone and associated calcareous sediments.
Quartzite (11), fine-grained gneisses (10), and amphibolite (16) are thin bedded or laminated, interlayered. Obscure cross bedding and scour and fill structures. Conglomerate (23) has well preserved subangular to rounded pebbles, cobbles, and boulders.	(10) Probably epiclastic quartz-feldspar sediments grading to (11) quartz sand with minor feldspar; both deposited in a littoral to shallow marine environment. (16) Fine-grained mafic tuff, probably initially airborne, deposited subaqueously. (17) Boulders, cobbles, pebbles, and finer grained epiclastic sediments, deposited subaqueously.
(16) Distinct residual bedding in parts of the unit. Obscure and inconclusive cross bedding.	(16) Fine-grained mafic tuff, probably initially airborne; deposited subaqueously. (18) Intermittent minor deposition of carbonate sediment. Deposition was subaqueous in a marine environment.
Obscure bedding and cross bedding in the quartzites.	(10) Fine-grained epicalstic quartz-feldspar sediment and mafic tuff; probably initially airborne. (11) Quartz sand mixed with minor pelitic sediment, derived from sapropelized terrane. Deposition probably in a littoral to shallow marine environment.
Relict pillow and flow structures in (16a), relict clasts in (16b), Local ripple-bedding in upper part of (16). Obscure to well preserved cross-bedding in parts of (16b).	(16) Fine-grained mafic tuff, probably initially airborne. (16a) Mafic volcanic flows. (16b) fine- to coarse-grained mafic pyroclastic detritus. (17) Mixed mafic tuff and carbonate sedimentation. Deposition mostly, perhaps wholly, in a marine environment.
Some graded bedding and cross-bedding in (11), channels, scour and fill structures in (11) and (15).	(11) Thoroughly winnowed and sized quartz sand from a sapropelized terrane, beach and channel deposits. (11a) Quartz sand mixed with minor quantities of pelitic sediment. (14) mixed quartz silt, pelitic sediment, and carbonate sediment. (15) Fine-grained mafic tuff, probably initially airborne. Depositional environment seems to have fluctuated between littoral and shallow marine.
Obscure cross bedding, minor scour and fill structures.	(10) Fine-grained, winnowed subgraywacke sediments devoid of rock fragments. (12) Quartz sand from a sapropelized terrane. Deposition in a subaqueous, probably marine, environment.

Figure 4. ***(Continued)***

1 is drawn along the trace of the major fault or faults forming the southeastern and southern boundary of the zone. This line corresponds fairly closely to the boundary of gneisses and schists of undetermined age shown on the map of Emerson (1917).

The other major northeast-trending faults, in probable order of decreasing displacement, are the North Boundary fault, forming the present north boundary of the Boston basin; the Bloody Bluff fault; the Assabet River fault; and the Spencer Brook fault. The North Boundary fault has a substantial overthrust component, rocks on the northwest and north sides being thrust southeastward over the rocks of the Boston basin. The Mystic fault (Fig. 1) is transverse to the North Boundary and Bloody Bluff faults and to the Clinton-Newbury fault zone. It also seems to have a substantial overthrust component, rocks on the northwest side being thrust over those on the southeast side. The other faults mentioned seem to be high-angle fractures along which movement was predominantly strike slip.

The Clinton-Newbury, North Boundary, and Bloody Bluff faults are major structural features dividing the region into tectonic plates of different and distinct lithologies (Fig. 1). The magnitudes of displacement on the Clinton-Newbury fault zone are unknown. Structural trends and stratigraphic contacts on opposite sides of this fault zone are discordant. None of the stratigraphic units described herein can be correlated with any cropping out along the northwest and north sides of the fault zone throughout its trace in Massachusetts. The probable horizontal component of displacement is in the range of tens or possibly hundreds of kilometres, and the vertical component is in the range of thousands or tens of thousands of metres. The minimum horizontal component of displacement on the North Boundary fault, based on the distance between masses of Dedham Granodiorite enclosing xenoliths and pendants of the Westboro Formation (formerly known as the Westboro Quartzite) on opposite sides of the fault, is about 22 km; the probable vertical component of displacement is in the range of thousands or tens of thousands of metres. No lithologic units have been correlated across the Bloody Bluff fault; its horizontal component of displacement probably is measurable in kilometres or tens of kilometres, and the vertical component of displacement is measurable in thousands or tens of thousands of metres.

The tectonic plate bounded by the Clinton-Newbury and Bloody Bluff faults is disrupted by the Assabet River fault, the Spencer Brook fault, and several other lesser northeast-trending faults (not shown in Fig. 1). Displacements on these faults are much less than those on the Clinton-Newbury and Bloody Bluff faults but are sufficient to cause substantial disruptions of stratigraphic units. The Marlboro Formation, Shawsheen Gneiss, Fish Brook Gneiss, Nashoba Formation, and Tadmuck Brook Schist (Fig. 4) have an aggregate thickness of about 15,000 m in the Westford-Concord-Billerica quadrangle area. Local incomplete sections of these units result from displacements on zones of subparallel east-northeast-trending high-angle reverse faults and related transverse faults. Mapping along the Assabet River fault in the Concord and Billerica quadrangles shows a slicing out of about 900 m of strata from the Fish Brook Gneiss in a distance of 11 km. The Assabet River and Spencer Brook faults, together with several less extensive subparallel fault zones, have in a distance of 40 km progressively sliced out post-Marlboro strata and the igneous rocks intruding them, thus narrowing their outcrop belt gradually from an 18-km width in the Concord-Westford-Billerica quadrangle area to a 5.5-km width in the vicinity of the Wachusett-Marlboro Tunnel (Fig. 1); (see Skehan, this volume). Although the proportion of intrusive rock decreases southwestward and there may be some thinning of the stratigraphic units, the slicing out of both stratified and intrusive rocks by the many faults is a major factor in the diminution of width of the outcrop belt.

The tectonic plates are broken into structural blocks—that is, generally angular segments bounded on all sides by faults—by secondary faults transverse to the major faults shown in Figure 1. Displacements on the transverse faults (only the larger ones shown in Fig. 1) are small compared to those on the major faults, the horizontal component generally being less than 1 km and the vertical component generally being estimated at less than 2,000 m. Transverse faults are best developed on the plate bounded by the Bloody Bluff and the North Boundary faults and in the area to the east of the Mystic fault. Within structural blocks there is a multitude of minor faults, which, in places, have reduced the bed rock to virtual megabreccias. The faults disrupt all of the stratified and plutonic rocks within the area described. Evidence obtained from localities outside the area indicates that the disruptions and movements occurred during post-Pennsylvanian and pre-Cretaceous time (Wanless and Bell, 1976).

The stratified rocks of pre-Silurian age mostly form a homoclinal sequence. The older units, which include the Middlesex Fells Volcanic Complex and all stratified rocks underlying it (Fig. 4), were domed during intrusion of the Dedham Granodiorite and were subjected to minor drag folding along the flank of the pluton. These stratified rocks generally have a northeast strike and a northwest dip, but locally deviate from this trend because of drag folding and (or) tilting of structural blocks. The eugeosynclinal rocks overlying the Middlesex Fells Volcanic Complex (Fig. 4) have quite persistent northeast strikes and steep, predominantly northwest dips. This feature is attributed to the initial dip of the homoclinal sequence, plus subsequent tilting of the tectonic plate—of which these stratified rocks form a major part—to a very steep or high-angle attitude. Local deviations from the regional attitude exist where these rocks have been severely dilated or domed by emplacement of plutonic rocks; local steep dips to the southeast mostly are related to severe dilation and distortion by intrusive rock and in some places are related to minor folding or to rotation of structural blocks within the tectonic plate. Coarse ripple bedding and slump structures that were formed during sedimentation or prior to lithification are locally characteristic of some units (see Fig. 4). Chevron folds with amplitudes ranging from a few centimetres to a few metres are locally conspicuous features of some of the upper members of the Nashoba formation.

RELATIONS TO PLUTONIC ROCKS

Igneous rocks that are parts of at least three major plutonic series have intruded various units of the pre-Silurian stratified sequence. The plutonic series, from oldest to youngest, are (1) the Dedham Granodiorite and unnamed comagmatic rocks ranging from gabbro to granite, (2) a comagmatic series that includes the Andover Granite, diorites, aplites, and pegmatites, and (3) the Cape Ann plutonic series (of local usage), which includes Nahant Gabbro, Salem Gabbro-Diorite, Cape Ann Granite, and several minor facies (Bell and Dennen, 1972; Dennen, 1972); the Quincy Granite probably is a satellitic part of the Cape Ann plutonic series. Rocks of each plutonic series are restricted to individual tectonic plates or structural blocks.

Within the area covered by this discussion, the Dedham Granodiorite occurs in two structural blocks: one is bounded by the North Boundary, Mystic, and Walden Pond faults; and the other, including part of the coastline and some islands, is bounded on the south by the North Boundary fault and on the west by an unnamed transverse fault (Fig. 1). The Dedham Granodiorite was emplaced as part of a large pluton or a batholith that domed the intruded stratified rocks and partially stoped its way into them. Rocks of the Middlesex Fells Volcanic Complex, the Westboro Formation, and older unnamed stratified remnants (Fig. 4) are wrapped around the flank of the batholith and locally have been subjected to minor folding, probably as the intrusive rock was emplaced. The Dedham Granodiorite does not intrude any stratified rocks younger than the Middlesex Fells Volcanic Complex.

Rocks of the Cape Ann plutonic series (of local usage) are confined to the tectonic plate bounded by the North Boundary and Bloody Bluff faults and to the area east of the Mystic fault, with the exception of the structural blocks that include the Dedham Granodiorite. These rocks were emplaced as a large pluton. Along the margin of the pluton and projecting from its upper surface, there are several domed masses of intrusive rocks that were emplaced mainly by shouldering aside and doming the stratified rocks coincident with some stoping. Emplacement of these domes caused substantial small-scale faulting and some minor folding of the stratified rocks along their flanks. Also, sills of the intrusive rocks are wedged for short distances into some of the strongly foliated amphibolitic and schistose units.

The stratified rocks in the tectonic plate bounded by the Bloody Bluff fault, Mystic fault, and Clinton-Newbury fault zone are intruded by a plutonic series that includes the Andover Granite and comagmatic quartz diorite and diorite. These rocks were emplaced partly as plutons that shouldered aside the hosts and partly as sills and dikes, some quite large. No rocks of any other plutonic series have been conclusively identified in this plutonic plate.

METAMORPHISM

The fossiliferous formations of Cambrian age are virtually unmetamorphosed and do not exhibit any of the metamorphic features of the great sequence of metamorphosed nonfossiliferous rocks. The Cambrian formations are hornfelsed adjacent to intrusive contacts with plutonic rocks. Parts of them are somewhat chloritized and epidotized; this feature is attributed to local hydrothermal alteration.

All the rocks of the nonfossiliferous sequence have been subjected to regional metamorphism. The metamorphic foliation characteristically parallels the layering and bedding features of all units. Foliation crosscutting the bedding is of local occurrence only, being observed cutting across drag folds near intrusive contacts with plutonic rocks. The metamorphism is considered to be predominantly thermal, caused by depth of burial and locally enhanced by additional heat from injections of plutonic rocks.

The major faults have cut across and displaced metamorphic isograds; consequently, each tectonic plate is characterized by metamorphic facies different from those of the plates with which they are juxtaposed. Post-Silurian rocks of the Boston basin that are truncated by the North Boundary fault are virtually unmetamorphosed. Rocks in the plate bounded by the North Boundary and Bloody Bluff faults and in the plate east of the Mystic fault (Fig. 1) have been subjected to low-grade metamorphism that produced biotite and hornblende; garnet occurs in a few strata and always within a few centimetres of intrusive contacts with plutonic rocks. Rocks in the plate bounded by the Bloody Bluff fault and the Clinton-Newbury fault zone are metamorphosed to higher grade facies. Sillimanite-bearing strata are dispersed through all of the formations, except possibly the Fish Brook Gneiss, that are exposed in the plate. There seems to be no sillimanite isograd crossing the plate, because sillimanitic units are cut off both to the northeast and southwest by faults and plutonic rocks. Garnet is a minor metamorphic constituent of some strata, occurring most abundantly in the Nashoba Formation. Rocks lying northwest and north of the Clinton-Newbury fault zone, and which are truncated by it, exhibit low-grade metamorphic facies and are sillimanitic only near a few intrusive contacts with plutonic rocks.

The age of metamorphism cannot be conclusively established from field observations in eastern Massachusetts. It generally is inferred to be Acadian, mainly because stratified rocks of Early Devonian age of the Merrimack synclinorium were metamorphosed during this orogeny. All stratified units above the Middlesex Fells Volcanic Complex (that is, the eugeosynclinal rocks) seem to have been subjected to a single episode of metamorphism, which may have been Acadian. The Middlesex Fells Volcanic Complex and older stratified rocks may have been subjected to an earlier metamorphism. Local garnetiferous contact zones adjacent to Dedham Granodiorite suggest metamorphism coincident with emplacement of the granodiorite. This episode has been assigned to late Precambrian time (Fairbairn and others, 1967). Also, the unmetamorphosed condition of the Cambrian formations implies that (1) they were separated from the nonfossiliferous rocks at the time of metamorphism by a considerable distance in either the horizontal or the vertical sense, or possibly both, or (2) the metamorphism of stratified rocks intruded by the Dedham Granodiorite occurred in Precambrian time.

The Dedham Granodiorite is the only plutonic rock showing evidence of possible regional metamorphism. Most of it has been subjected to a pervasive cataclasis that moderately fractured quartz and feldspar grains. If this feature was caused by dynamic metamorphism, it did not attain a grade sufficient to reconstitute the mineralogy. The Andover Granite and comagmatic diorites are locally foliated parallel to the walls of the plutons. This feature is attributed to flowage. Rocks of the Cape Ann plutonic series (of local usage) are devoid of metamorphic features.

STRATIFIED ROCKS OF CAMBRIAN AGE

Unmetamorphosed, fossiliferous rocks of Cambrian age crop out in four localities of eastern Massachusetts. The Weymouth Formation, having an *Olenellus* fauna of Early Cambrian age (Burr, 1900; Grabau, 1900; Foerste, 1889) crops out in the vicinity of Weymouth in the Weymouth quadrangle and at Nahant in the Lynn quadrangle. The Braintree Argillite, having a *Paradoxides* fauna of Middle Cambrian age (Rogers, 1856; Grabau, 1900), overlies the Weymouth Formation in the Weymouth locality. The Hoppin Slate and other unnamed strata crop out in the vicinity of Hoppin Hill Reservoir in the Attleboro quadrangle; the Hoppin Slate has a fauna described as Early Cambrian by early investigators (Shaler and Foerste, 1888), this being the basis for correlation with the Weymouth Formation on the Cambrian Correlation Chart (Howell and others, 1944). Studies by Shaw (1950) indicate that it has a somewhat older Early Cambrian *Obolella* fauna. Small outcrops having an Early Cambrian fauna, situated about 0.8 km south of West Wrenthem Village in the Wrentham quadrangle, have been described briefly by Foerste (in Shaler and others, 1899, p. 393) and Emerson (1917, p. 36).

Outcrops of the fossiliferous rocks in the Weymouth locality have been mostly obliterated during urban development, so the following description is taken from Crosby (1900, p. 307):

The data, and especially the paleontologic data, are still inadequate for a definite determination of the general succession of the Cambrian strata of the Blue Hills Complex, but the following sequence seems to accord best with the facts now in hand, and may be stated provisionally:
be stated provisionally:

Lower Cambrian (Olenellus) zone:

1. Siliceous series (quartzite, siliceous slates, etc.), not exposed.
2. Argillaceous and calcareous series. Reddish-brown to reddish and brownish black slates, often banded or interstratified with green slates and characterized throughout by calcareous segregations (lenticular concretions or clay stones) and more or less continuous calcareous layers, which pass upward into thin beds of white crystalline limestone.

Middle Cambrian (Paradoxides) zone:

1. Greenish to greenish gray slates, holding the Paradoxides fauna and often banded or interstratified with reddish brown slate, and changing gradually upward to
2. Dark gray to black (carbonaceous) slates which have a great thickness, but have as yet afforded no fossils.

The Middle Cambrian beds, both the dark and the light series, are frequently pyritiferous; and they are also occasionally slightly calcareous. The succession is believed to be conformable throughout, and the Lower Cambrian red beds appear to pass gradually upward into the Middle Cambrian green beds.

A section of the Weymouth Formation, about 12 m thick, that is intruded by the Nahant Gabbro is exposed in cliffs at the most easterly point of Nahant. Beds of fossiliferous white limestone, 15 to 30 cm thick, are exposed during intervals of low tide. The limestone is interlayered with and overlain by reddish-brown to black hornfelsed shale, mostly as beds a few centimetres thick. Much of the shale is characterized by conspicuous, closely spaced nodules of greenish calc-silicate minerals that are somewhat flattened parallel to the bedding. The maximum dimension of the nodules ranges from 2 to 5 cm. These nodules are considered to be thermally metamorphosed limy concretions.

Woodworth (in Shaler and others, 1899, p. 105), in describing the North Attleboro locality, stated, "The Olenellus fauna occurs in little-altered, red, calcareous shales and slates at this latter place in close proximity to granite (hornblendic granitite)." Emerson (1917, p. 36-37) further described these localities: "The Lower Cambrian strata, in the Narragansett Basin, which are mapped under the name Hoppin slate, are exposed at two localities. One of these, from which the formation is named, is at Hoppin Hill, North Attleboro. . . . The other locality is in Wrentham, half a mile south of West Wrentham village, just north of the Rhode Island boundary. . . . The rocks consist chiefly of red shale or slate, with layers and nodules of white limestone, overlying greenish shale or slate, beneath which is a basal white quartzite. They are not greatly altered. The fossils are found mainly in the limestone and red slate. . . . In both localities the beds dip steeply toward the neighboring granite." Emerson's (1917) map shows the nearest granitic rock of both localities to be Dedham Granodiorite.

The relations of the Cambrian rocks at the North Attleboro locality are of particular interest, because attempts have been made to use them, with differing conclusions, as a basis for an inferred age of the Dedham Granodiorite and of the nonfossiliferous, metamorphosed stratified rocks it intrudes. The diversity of observations and conclusions are summarized here because of the importance of this subject in determining the chronologic sequence of the rocks.

Woodworth (in Shaler and others, 1899, p. 105) assigned an Algonkian age to the nonfossiliferous stratified rocks, because they have been subjected to one more episode of metamorphism than have the fossiliferous rocks. He did not mention contact relations and only concluded that "the relation of the granitic intrusives to the pre-Cambrian on the one hand and to the Cambrian on the other is simply to show that the granite is younger than the former, and that the sedimentary rocks are of different ages." Warren and Powers (1914, p. 459-460) described the position of outcrops of granite with relation to outcrops of red shale, limestone, and sandstone along the east side of Hoppin Hill and on the bed of the reservoir; they concluded that although the contact was not observed, it is "apparently due to intrusion and not to faulting." They also stated, "Many years ago excavations were made on the west side of Hoppin Hill which showed the granite in fault contact with the sediments." The conclusion of Warren and Powers (1914, p. 460) is the basis for Emerson's (1917, p. 37) statement, "The relation of the two rocks was long in doubt, but that the granite is younger now seems certain."

Dowse (1950, p. 97) described the contact of granite and sandstone observed in a shallow excavation on the east side of Hoppin Hill and concluded that "the sedimentary contact of basal white quartzite against the granite is definitely proved and the possibility of a fault contact or intrusive contact is clearly eliminated." Shaw (1950, p. 578) stated, "The core of Hoppin Hill is a granite correlated with the Dedham granite. At the base of the Hoppin slate lies a massive partly crossbedded white quartzite, dipping steeply away from the granite. No apophyses of the granite were seen in the quartzite, but neither were recognizable pebbles of granite noted in the quartzite." Skehan (1969, p. 798) stated, "At Hoppin Hill, the Lower Cambrian strata overlie an erosion surface of granodiorite . . . for which an Sr-Rb whole-rock age of 540 ± 26 m.y. was determined by Fairbairn and others (1965)." Fairbairn and others (1967) reported a revised age of 514 ± 17 m.y. for this granodiorite.

The observations cited above do not conclusively establish the relationship between the Hoppin Slate and the Dedham Granodiorite. Inasmuch as the granodiorite was subjected to a long period of erosion after intruding the metamorphosed Westboro Formation and Middlesex Fells Volcanic Complex, it seems most likely that the unmetamorphosed Cambrian strata rest unconformably on a basement of the older rocks.

The thickness of the formations of Cambrian age is unknown. It probably is measurable in hundreds of metres. Crosby (1900, p. 308), in describing the Weymouth locality, stated that it is very great, being thousands of feet (probably 1,000 m or more), but this estimate cannot be verified, because many exposures that he examined have been obliterated.

The remnants of formations of Cambrian age are isolated from all other stratified rocks of the region by masses of intrusive rocks and by major faults. It has not been possible to determine their position in the stratigraphic column relative to that of the other units described in this paper.

The protoliths of the Hoppin Slate and Weymouth Formation were predominantly pelitic. Minor quantities of carbonate sediment were deposited intermittently with the pelitic sediments, and, at least locally, there were short intervals of almost wholly carbonate sedimentation. The protoliths of the Braintree Argillite varied from pelitic to silty. The sedimentation occurred in a marine shelf environment that periodically supported an abundant fauna. The presence of pyritiferous shales devoid of fossils in the lower part of the sequence may be indicative of periods of anaerobic conditions.

NONFOSSILIFEROUS STRATIFIED ROCKS

The stratigraphic column of the nonfossiliferous metamorphosed rocks was compiled from information collected in the area between the north border of the

Boston basin and the Clinton-Newbury fault (Fig. 1). Inasmuch as this area is within an imbricate fault zone (Bell, 1968), it is possible that units or parts of units that can be seen elsewhere may be cut out by faults or may be concealed under overthrust sheets and, consequently, do not appear in the compiled column. Descriptions presented in the text and in Figure 4 are characteristic of the rocks as they exist along the traverses shown in Figure 1; these descriptions are supplemented with information from other localities.

All of these stratified rocks enclose dikes, sills, veins, and small plutons. The proportion of intrusive rock varies from locality to locality, ranging from 15 to 80 percent of the volume. Estimated thicknesses of units presented in Figure 4 do not include intrusive rock. The figures presented are the maximum observed for each unit along the traverses shown in Figure 1.

Stratified Rocks Intruded by the Dedham Granodiorite

The oldest rocks of eastern Massachusetts are, from oldest to youngest, some remnants of generally fine grained quartzites and gneisses, the Westboro Formation, and the Middlesex Fells Volcanic Complex. These rocks are best and most completely exposed in the north part of the Boston North quadrangle. Some of these rocks, mainly parts of the Westboro Formation and the Middlesex Fells Volcanic Complex, can be traced intermittently from near the northeast coast of Massachusetts southwestward into Rhode Island, where they were designated the "Blackstone Series" by Woodworth (in Shaler and others, 1899, p. 104–109). They are considered to be correlatives of the Blackstone Series as described by Woodworth (in Shaler and others, 1899, p. 104–109), Quinn and others (1949), and Quinn (1971). The sequence of lithologies is the same throughout the belt in which these rocks are found, consisting of a lower part composed predominantly of fine-grained quartzite and gneiss, a middle part having a large proportion of quartzite, and an upper part composed predominantly of mafic metavolcanic rocks.

These stratified rocks are unique in that the Dedham Granodiorite, representing the oldest major plutonic series of the region, intrudes them and does not intrude any of the younger metamorphosed stratified rocks of the region. Also, the upper part of the older stratified rocks is missing in some localities where no evidence of faulting on a scale that could cause such omissions has been found. For example, in the southwest part of the Lexington quadrangle, the entire Middlesex Fells Volcanic Complex and the upper part of the Westboro Formation are missing. These features suggest that the upper surface of these older stratified rocks may represent a significant unconformity.

Unnamed Stratified Remnants. The oldest of the stratified rocks intruded by the Dedham Granodiorite are known only as xenoliths and pendants. These rocks have not been formally named, because the remnants are small and scattered, the various lithologies cannot be traced from one outcrop area to another, and features delineating the units are lacking.

In the north-central part of the Boston North quadrangle, on a hill about 1 km in diameter at the north side of Stevens Pond, there is exposed lit-par-lit interlayering of Dedham Granodiorite and very fine grained, grayish biotite-quartz-feldspar gneiss. The estimated thickness of the section represented by the slablike xenoliths of gneiss, which range from 1 cm to 5 m in thickness and which are exposed on the hilltop, is about 150 m. The attitude of the gneiss is concordant with that of the Westboro Formation exposed about 0.5 km to the north.

South of the lit-par-lit gneiss and presumably stratigraphically below it, there are fine-grained quartzites, feldspathic quartzites, and other gneissic rocks represented only by widely scattered xenoliths in the granodiorite. These xenoliths are too few in number to permit compilation of a stratigraphic section of them. The distribution of remnants seems to indicate interlayering of quartzite and gneiss.

The quartzite and feldspathic quartzite remnants are mostly equigranular and massive and are nearly devoid of recognizable bedding or layering features. Their original color was whitish or very pale gray; many of them are now pale pink because of pervasive hydrothermal alteration. The gneisses are light- to medium-gray on fresh surfaces; weathered surfaces are generally shades of brown. The biotite occurs as minute flakes, mostly lying nearly parallel to planes of lamination. The finely laminar structure is inconspicuous on fresh surfaces but commonly is clearly indicated by finely fluted surfaces of weathered rock. Nothing is known of the floor upon which this sequence of stratified remnants was deposited. No contact with the overlying Westboro Formation has been seen, although concordant attitudes suggest a conformable contact.

The sediments from which these unnamed remnant rocks were formed were derived from at least two sources. The greater volume of sediment, as determined from the amount of outcrop that can be seen, was fine-grained detritus of probable submarine volcanic origin. Rock fragments, if present at all, constituted a very minor proportion of it. The quartzite layers were formed mainly from sand such as would be derived from a sapropelized terrestrial terrane and which was cleaned and sized on beaches. Sedimentation seems to have occurred in a shallow, nearshore marine environment. The abundance of quartzite layers suggests continuous transportation of beach sands to the depositional site; deposition of the sand occasionally was overwhelmed by rapid deposition of volcaniclastic detritus.

Westboro Formation. The Westboro Formation, formerly known as the Westboro Quartzite, can be traced intermittently from the northeast corner of the Boston North quadrangle southwestward to the northwestern part of Rhode Island. A belt of similar and probably correlative rocks crops out in a belt extending from the central part of the Marlboro quadrangle southwestward to the northeastern part of Connecticut. Faults and bodies of intrusive rocks interrupt the continuity of the outcrops. The Westboro Formation is best exposed in the north-central part of the Boston North quadrangle, where it is about 1,100 m thick. It is divided into a lower member about 420 m thick and an upper member about 680 m thick.

The name of the Westboro Quartzite is here changed, because a large proportion of it consists of rocks other than quartzite. Most of the natural exposures of rocks in both the upper and lower members are massive quartzite, and this feature gives an erroneous impression of the character of the formation. Many excavations made during the course of construction projects since about 1950—such as construction of interstate highways, other freeways and roads, foundations for large buildings, trenches for utilities, and cuts and grading for Bentley College campus in the southwest part of the Lexington quadrangle—have exposed considerable amounts of argillite, metamorphosed shale and metamorphosed limy shale that do not have slaty cleavage, calc-silicate rocks, and micaceous quartzite interlayered between massive quartzite strata, particularly in the lower member. Several varieties of these rock types commonly occur within a section of strata 30 m thick or less.

Some of the quartzites of eastern Massachusetts that have previously been mapped as parts of the Westboro Formation (Emerson, 1917; LaForge, 1932) are now recognized as being parts of other formations. Quartzite strata above the Middlesex Fells Volcanic Complex are here included in a poorly exposed, unnamed unit and in the Burlington Formation (Fig. 4). Such separation was predicted by Emerson (1917, p. 30–31).

Probably less than one-half of the lower member is massive quartzite, and this rock constitutes all of the outcrops in many localities. Disruptions by minor faults and intrusive rocks further hinder determination of the stratigraphy. Near the base of the lower member, there is an interval of bedded quartzite and minor micaceous quartzite about 60 m thick. It includes a zone of reddish-brown to chocolate-colored quartzite 3 to 5 m thick that is an easily recognizable marker unit. The color is caused by minute flakes of reddish-brown biotite uniformly dispersed through the rock. In the middle part of the member, massive white to pale-gray quartzite occurs mainly as lenticular bodies that are dispersed both laterally and vertically. These bodies range from 100 m to 1.5 km in length and have maximum thicknesses ranging from 5 to 60 m. They cannot be traced continuously along any horizon. The distribution and lenticular character of these bodies suggest that they are channel fillings. The upper 100 to 150 m of the member consist of interlayered quartzite, argillite, metamorphosed shale, and quartz-bearing calc-silicate strata ranging from a few centimetres to as much as 3 m in thickness. All of the quartzite is fine grained and equigranular.

The upper member consists of interlayered quartzite and gneiss. Beds or layers of both types of rock tend to be lenticular, having maximum thicknesses that range from a few centimetres to as much as 15 m. Channels and scour-and-fill structures are characteristic features. The proportion of gneiss in the lower part of this member is approximately 35 percent, and it increases to approximately 75 percent at the top of the member. The interlayering and gross sedimentary features of both types of rock are readily seen on outcrops. Internal sedimentary features such as lamination, cross-bedding, and graded bedding within beds of both quartzite and gneiss are obscure.

The quartzite and gneiss are so fine grained that their mineral compositions can be determined only under the microscope. The quartzite is white to pale gray and even grained and has a saccharoidal texture. Most of it is composed of 90 percent or more quartz, with minor quantities of feldspars, amphiboles, and micas. The gneiss is composed mainly of feldspar and pale-green hornblende. Some gneiss layers contain minor quantities of quartz, either dispersed or as tiny streaks and lenses, which indicates an intermingling of two sediments.

Gross bedding and layering in the Westboro Formation is excellently displayed wherever excavations have been made through sections of interlayered rock types and also on outcrops of the upper interlayered quartzite-gneiss member. Outcrops of the lower member are predominantly massive quartzite in which indicators of stratification are generally limited to thin, discontinuous micaceous streaks and obscure graded bedding. Some of the quartzite lenses in the lower member contain abundant cylindrical structures about 3 mm in diameter, which are suggestive of filled worm burrows, but which may have been formed in some other manner.

The contact between the Westboro Formation and the overlying Middlesex Fells Volcanic Complex is exposed at two places in the Boston North quadrangle. The best exposure is about 0.8 km east of Griswold Pond in the area cleared for a power line on the south slope of a hill (lat 42°28′52″N, long 71°2′45″W). The other exposure is near the west shore of Spot Pond at the south side of Deer Hill (lat 42°27′5″N, long 71°27′5″W). The basal stratum of the metavolcanic complex at these localities is basaltic pillow lava metamorphosed to amphibolite. The change from quartzite and gneiss to mafic metavolcanic rock is abrupt. In minor detail metavolcanic rock locally truncates small scour-and-fill structures at the top of the quartzite-gneiss member. Locally, small fragments of quartzite and gneiss are included in the basal part of the metabasalt at the contact. There is no conclusive evidence of a significant unconformity; therefore, the two formations are considered conformable. The lower contact of the Westboro Formation is not known to be exposed. It is defined as the transition from fine-grained, laminated gneissic rocks to interlayered fine-grained quartzite, metamorphosed mudstone, and limy mudstone.

The rocks composing the Westboro Formation were formed from several varieties of sediments. The quartzite was formed from a highly degraded and winnowed clastic sediment that most probably was beach sand. Argillites and metamorphosed shales were formed from very fine grained silts and muds. The calc-silicate and limy rocks indicate intermittent comingling of carbonate and other sediments in a subaqueous environment. The gneissic strata of the upper member were formed from fine-grained, initially airborne tuff that became mixed with variable proportions of other sediments near or at depositional sites.

The depositional environment of the Westboro Formation seems to have fluctuated between shallow marine and littoral. Fine-grained, well-graded quartz sand was the most abundant sediment and was deposited intermittently throughout the formation, indicating a rather persistent source. Intervals of well-stratified rocks, particularly in the lower member, represent subaqueous deposition under quiescent conditions. Intervals of strata characterized by abundant scour-and-fill structures and lenticular channel deposits indicate at least partial deposition from flowing water, either shallow marine currents or terrestrial streams and their distributaries flowing across a littoral zone. Quartzite in the lower member seems to form increasingly continuous strata toward the southwest end of the outcrop belts, possibly indicating a gradual change from channel deposits to beach deposits. The tuffaceous constituent of the upper member probably was transported from a distant source; coarse pyroclastic deposits and flows have not been found in the Westboro Formation.

Middlesex Fells Volcanic Complex. The thick unit of mafic metavolcanic rocks that overlies the Westboro Formation in eastern Massachusetts is here named the "Middlesex Fells Volcanic Complex." Exposures of these rocks occur intermittently in a belt extending from the coastline in the Ipswich quadrangle southwestward to the northern part of Rhode Island. These rocks attain a maximum thickness of about 1,500 m, being best exposed and most completely preserved in the northern part of the Boston North quadrangle. We selected the area including the northern part of the Middlesex Fells Reservation and extending slightly northeast to within about 1.6 km of the north edge of the Boston North quadrangle as the type locality of the Middlesex Fells Volcanic Complex.

The Middlesex Fells Volcanic Complex formerly was included in the Marlboro Formation as described by Emerson (1917) and LaForge (1932). Recent mapping in northeastern Massachusetts has shown that these metavolcanic rocks are older than the unit that crops out in the type locality of the Marlboro Formation as described by Emerson (1917, p. 25) and are separated from it by other stratified rocks (Fig. 4). This separation of units was anticipated by Emerson (1917, p. 31), who commented, "The terrane mapped as the Marlboro Formation in northeastern Massachusetts thus corresponds fairly well in general character to the Marlboro Formation of central Massachusetts and of Rhode Island, but it is notably more complex, includes much more rock of igneous origin, and may represent a longer period of time. . . . It is therefore possible that the formation, at least in northeastern Massachusetts, may be eventually broken up into several formations."

The Middlesex Fells Volcanic Complex is correlated with the mafic upper part of the Blackstone Series of Rhode Island as described by Woodworth (in Shaler and others, 1899), Quinn and others (1949), and Quinn (1971). The complex is at least 1,500 m thick in the Georgetown quadrangle, but it is poorly exposed in this locality. It has about the same thickness in the type locality in the Boston

North quadrangle, where there are many large outcrops and artificial exposures of it. Outcrops of metamorphosed amygdaloidal basalt and fine-grained tuff of the middle part of the complex are abundant in the southeast part of the Concord quadrangle. Elsewhere, remnants of the complex occur as pendants and xenoliths in intrusive rocks and form parts of structural fault blocks.

The Middlesex Fells Volcanic Complex is a typical volcanic pile, composed of lentils of metamorphosed flow rocks, coarse pyroclastic deposits, and fine-grained tuffs. Many of the deposits within it cannot be traced from one outcrop area to another. Fine-grained metamorphosed tuffs seem to have the greatest continuity, but they lack diagnostic features that would permit identification from one outcrop area to another. Lentils of metamorphosed flow rocks constitute about 20 percent of the complex and are distributed throughout it, but they are most abundant in the lower half; those within the lower 300 m commonly show well-preserved pillow structures. Lentils of metamorphosed coarse pyroclastic detritus are common but do not seem to have continuity. Lentils and strata of fine-grained metamorphosed tuff probably constitute about 60 percent of the complex and are most abundant in its upper half. Lentils of light-gray or white, rusty-weathering quartzite occur sparsely throughout the complex but are most abundant in the lower half; they are interpreted to be channel fillings. A zone at least 15 m thick that is rich in calc-silicate and carbonate minerals is present near the base of the complex in the Boston North quadrangle; its exact stratigraphic position cannot be determined, because intrusive rocks have displaced part of the section in this locality. There are laminated, fine-grained gneiss units formed from volcaniclastic sediments in the upper 300 m of the complex; locally these gneiss units are ripple bedded.

Most of the rocks of the Middlesex Fells Volcanic Complex are fine grained, are dark gray to black, and have a metamorphic foliation that parallels the bedding or layering. Locally, they are dark greenish gray because of epidotization and retrogressive hydrothermal alteration of pyroxenes, amphiboles, and biotite to chlorite. Most of these rocks are now composed mainly of hornblende and plagioclase. Oligoclase is the most abundant plagioclase, followed by andesine. Small pods and irregular veins of epidote are common.

The metamorphosed tuff lentils are now mostly thinly laminated amphibolite lentils. Some of those near the top of the complex are conspicuously white pinstriped. Those in the lower and middle parts of the complex are generally darker and have inconspicuous lamination or are seemingly massive. The proportion of hornblende varies considerably from lentil to lentil, ranging from 35 to 80 percent. Biotite is a minor constituent of some lentils and is a major constitutent of a few schistose lentils.

The metamorphosed flows and coarse pyroclastic sediments do not have a thinly laminated structure; outlines of fragmental detritus, pillows, amygdules, and flow banding are readily discernible, except in localities of intense retrogressive alteration. Amygdules generally consist of chlorite and epidote; some are partly composed of oligoclase or carbonate minerals. There are a few lentils of light-gray or reddish-gray dacite in the upper part of the complex that are slightly altered, if at all, by metamorphism; all of those observed are less than 5 m thick.

These metavolcanic rocks represent an important episode of mafic volcanism. The presence of pillow structures, ripple bedding, and a calc-silicate–carbonate mineral zone indicates that much, or perhaps all, of the complex was deposited subaqueously, probably in a marine environment. Some of the lentils of flow rock that do not have pillows may have been deposited subaerially; if so, features that would indicate oxidized surfaces of flows have been obliterated during metamorphism. The tuffaceous sediments that formed the finely laminated amphibolites initially may have been airborne, but they seem to have been mostly, and perhaps entirely, deposited subaqueously.

Age. There is no direct evidence that indicates the age of the stratified rocks intruded by the Dedham Granodiorite. Geologists who have described these rocks in greatest detail have assigned a Precambrian age to them (Crosby, 1880; Shaler, 1888, p. 15–18; Woodworth in Shaler and others, 1899, p. 105; Warren and Powers, 1914, p. 445; Emerson, 1917, p. 27; LaForge, 1932, p. 17). The Precambrian age is based mainly on the observation that these rocks are metamorphosed, whereas nearby fossiliferous rocks of Cambrian age are virtually unmetamorphosed (Woodworth, in Shaler and others, 1899, p. 105).

The age of the Dedham Granodiorite has not been established on the basis of its field relations with other rocks. Ages that have been assigned to it are late Precambrian (Billings, 1929, p. 103), Ordovician or Silurian (Chute, 1940; LaForge, 1932, p. 22), and Devonian (Emerson, 1917; Page, 1968). During its early history the granodiorite batholith was subjected to a long period of erosion; the cover and much of the upper part of the intrusive rock were removed, and stratified rocks were deposited unconformably on it. The contact of the granodiorite and the fossiliferous rocks of Cambrian age at the Hoppin Hill Reservoir locality has been reported as an unconformity (Dowse, 1950; Skehan, 1969). The Lynn Volcanic Complex and the Mattapan Volcanic Complex lie unconformably on the eroded surface of the Dedham Granodiorite and the stratified rocks it intrudes. Compositional and lithologic similarities of these unmetamorphosed volcanic rocks with the fossil-bearing Newbury Volcanic Complex (Shride, this volume) suggest that they are of Late Silurian or Early Devonian age. These relations, combined with the unmetamorphosed condition of the younger stratified rocks, suggest that the Dedham Granodiorite is of Precambrian age.

Most of the Dedham Granodiorite has been subjected to pervasive hydrothermal alteration; it is within the area of Permian radiometric age disturbance of New England (Zartman and others, 1970). Zartman and Naylor (1972; 1975, oral commun.) have determined a Precambrian U-Th-Pb zircon isotopic age for a sample of the granodiorite that shows no visible effects of alteration. The assignment of a probable Precambrian age to the stratified rocks intruded by the Dedham Granodiorite seems to agree best with the field relations.

Stratified Rocks between the Middlesex Fells Volcanic Complex and the Bloody Bluff Fault

Most of the stratified rocks between the top of the Middlesex Fells Volcanic Complex and the Bloody Bluff fault are poorly exposed and can be described only in general terms. These rocks are divided into three units: (1) unnamed schist, gneiss, and quartzite; (2) Greenleaf Mountain Formation; and (3) Burlington Formation. These rocks are not intruded by the Dedham Granodiorite, and no folding that might be associated with the emplacement of the granodiorite has been found in them. They have the same general regional trends, attitudes, and facings as the stratified rocks northwest of the Bloody Bluff fault, but no unit has been correlated across the fault.

Unnamed Gneiss and Quartzite. Xenoliths and small pendants of mostly light-colored, fine-grained quartzite, feldspar-quartz gneiss, biotite-feldspar-quartz gneiss, and quartz-hornblende-feldspar gneiss are sparsely distributed in rocks of the Cape Ann plutonic series (of local usage) in a belt about 1,500 m wide along the northwest side of the Middlesex Fells Volcanic Complex in the Lexington and Concord quadrangles. The thickness of section represented probably does not exceed 250 m and may be a fraction of this amount. No contact with the underlying Middlesex Fells Volcanic Complex or the overlying Greenleaf Mountain Formation has been observed. These rocks are not named, because stratigraphic details are too incomplete.

The quartzite remnants are impure, containing substantial proportions of feldspars, much of which is highly altered, and small proportions of micas and opaque minerals. These rocks are generally fine grained but are inequigranular; quartz grains especially vary in size within a ratio of at least 10:1. Bedding or layering features in them are obscure. A weak metamorphic foliation can be seen in thin sections. These quartzite rocks differ conspicuously from the equigranular, almost pure quartz rocks of the Westboro Formation. They probably were formed from poorly graded epiclastic sediments derived from a source such as a saprolite.

The gneissic remnants are fine grained, somewhat inequigranular, distinctly foliated by metamorphism, and mostly made of alternating laminae or thin layers of light and dark minerals. The quartz content ranges from 10 to 30 percent. Part of the feldspathic content is so thoroughly sericitized that the species cannot be determined; the remainder is clear, unaltered oligoclase and andesine. Hornblende is more prevalent than biotite.

The sediments that formed these rocks probably were deposited subaqueously in a marine environment. The scattered exposures seem to represent interstratified deposits of epiclastic sediment and initially airborne tuff; some strata have compositions that indicate mingling of the two kinds of sediment.

Greenleaf Mountain Formation. A fine-grained, very thinly laminated dark greenish-gray to dark gray rock consisting mainly of amphibolite crops out abundantly in an arcuate belt south of the Bloody Bluff fault. This rock is here named the "Greenleaf Mountain Formation." It has a maximum thickness of about 225 m in the vicinity of Greenleaf Mountain—selected as the type locality—in the northern part of the Lexington quadrangle. The Greenleaf Mountain Formation is separated from the Bloody Bluff fault by the lens-shaped remnant of the Burlington Formation described below and is cut off by this fault and minor subsidiary faults in the central part of the Reading quadrangle and near the western border of the Lexington quadrangle (Fig. 1). The formation is segmented in several places by masses of intrusive rock.

The amphibolite is composed mainly of hornblende and oligoclase; some beds or laminae contain minor quantities of biotite, and some contain a very small quantity of quartz. Much of the amphibolite has a closely spaced, fine whitish pinstripe caused by planar segregation of oligoclase parallel to the bedding. Minor parts of the formation consist of alternating layers of amphibolite and pale green calc-silicate rock that are mostly less than 1 cm thick. On the south side of Greenleaf Mountain, there is considerable lit-par-lit interlayering of amphibolite and protoclastic intrusive rock.

The Greenleaf Mountain Formation is included in the Marlboro Formation on the maps of Emerson (1917) and LaForge (1932). Possible correlation with the Sandy Pond Amphibolite Member of the Marlboro Formation as redefined in this paper has been considered but not accepted, because the sequence of lithologies above and below the two thick amphibolite units cannot be matched. Zones or beds of carbonate rocks such as occur in the upper part of the lower member of the Marlboro Formation have not been found in the stratified remnants underlying the Greenleaf Mountain Formation. The rusty-weathering muscovite- and sillimanite-bearing schist at the base of the Shawsheen Gneiss, which seems to immediately overlie the Sandy Pond Amphibolite Member of the Marlboro Formation, has not been found in the Burlington Formation overlying the Greenleaf Mountain Formation.

Contacts of the Greenleaf Mountain Formation with adjacent formations are not exposed. The contacts are defined as being at the change from fine-grained dark amphibolite to a different lithology.

The amphibolite of the Greenleaf Mountain Formation was formed mainly from fine-grained tuff that was deposited in a subaqueous, presumably marine, environment. The tuff may have been initially airborne. The calc-silicate parts represent periodic intervals of minor carbonate deposition.

Burlington Formation. The stratified rocks overlying the Greenleaf Mountain Formation are poorly exposed in a lens-shaped area about 4.8 km long and about 1,000 m in maximum width in the northwestern part of the Lexington quadrangle. These rocks lie along the southeast side of the Bloody Bluff fault and are truncated by the fault. They occur as xenoliths and pendants in rocks of the Cape Ann plutonic series (of local usage). Rock types represented are quartzite, quartz-feldspar gneiss, mica-quartz-feldspar gneiss, amphibolite, metamorphosed arkose, and metamorphosed conglomerate. Their thickness probably does not exceed 300 m. This section of stratified rocks is here named the "Burlington Formation." The northwest slope of Greenleaf Mountain in the northern part of the Lexington quadrangle is selected as the type locality.

The quartzites, gneisses, and amphibolites of the Burlington Formation are fine-grained, randomly interlayered strata ranging from a few centimetres to several metres in thickness. The quartzites tend to be impure, containing substantial amounts of feldspars and micas, and grade into gneisses. The sediments that formed these rocks were deposited subaqueously, probably in a shallow-marine to littoral environment. Some of the sediments, especially the quartz-rich sands, may have been derived from a sapropelized terrestrial terrane and were winnowed and sized on beaches before final deposition. The amphibolites represent eruptions of fine-grained mafic tuff that may have been initially airborne. The various gneissic rocks were formed either from mixtures of mafic tuff and terrestrially derived beach sand or from volcaniclastic detritus.

The metamorphosed conglomerates seem to be small lenticular deposits. One of these deposits contains smooth rounded boulders as much as 60 cm in diameter; their forms indicate transport by large, fast-flowing streams. Another deposit contains smaller subangular clasts, mostly of quartzite, but also a few of diorite. A third deposit consists of subrounded pebbles and cobbles of granitic rock in an arkosic matrix. These deposits do not seem to represent a single horizon and are interpreted to be channel fillings. Their exact position is obscured by masses of intrusive rock, but they seem to lie near or possibly at the top of the Burlington Formation. If so, these conglomerates may indicate an unconformity at this horizon and are not actually part of the Burlington Formation.

Age. The sequence consisting of the unnamed stratified remnants, the Greenleaf Mountain Formation, and the Burlington Formation seemingly rests unconformably on an eroded surface of the Middlesex Fells Volcanic Complex. This relationship fixes the position of these units in the stratigraphic section. These units are intruded by the Cape Ann plutonic series (of local usage). On the basis of these relations, permissible ages for these stratified rocks range from late Precambrian to possibly Ordovician.

Stratified Rocks between the Bloody Bluff Fault and Clinton-Newbury Fault Zone

A steeply dipping homoclinal sequence of metasedimentary rock strata, with an aggregate thickness of 18,070 m is isolated within the tectonic plate bordered by the Bloody Bluff fault on the south, the Mystic fault on the east, and the

Clinton-Newbury fault zone on the north (Fig. 1). The metasedimentary rocks within this tectonic plate are shown on the preliminary geologic map of Massachusetts and Rhode Island (Emerson, 1917) as the Marlboro Formation of Algonkian (?) age, gneisses and schists of undetermined age, and the Brimfield Schist and Worcester Phyllite, with its Harvard Conglomerate Lentil, of Carboniferous age.

Hansen (1956) named the "gneisses and schists of undetermined age" of Emerson (1917) the "Nashoba Formation." Hansen also interpreted the rocks of the sillimanite-grade Brimfield Schist and the chlorite-grade Worcester Phyllite, both assigned to the Carboniferous by Emerson (1917), as equivalent and separate metamorphic facies of the same strata. These rocks, together with the Harvard Conglomerate Lentil of the Worcester and a unit of interbedded quartzite and phyllite (which Hansen named the "Vaughn Hills Member"), were defined and mapped by Hansen as the Worcester Formation of Carboniferous age. Castle (1965) assigned the name "Boxford Formation" and a probable pre-Silurian age to some predominantly mafic metasedimentary rocks in the South Groveland quadrangle and the name "Fish Brook Gneiss" to some felsic gneissic rocks in the same quadrangle, which he interpreted to be either intrusive rock or an ancient "dome" gneiss. The Marlboro Formation is redefined in this paper and is divided into an unnamed lower member and the newly named upper Sandy Pond Amphibolite Member.

The basal 2,600 m of the Nashoba Formation of Hansen are separated and defined as the newly named Shawsheen Gneiss. During the course of our mapping, the stratigraphically adjacent Fish Brook Gneiss and Boxford Formation of Castle (1965) were traced from their type localities in the South Groveland quadrangle southwestward into the Billerica, Concord, and Maynard quadrangles, where the Fish Brook Gneiss is underlain by the lowermost part of the Nashoba Formation of Hansen (1956) (the Shawsheen Gneiss as defined herein) and where the Boxford includes the belt of amphibolite mapped by Hansen to the southeast of East Acton village in the Maynard quadrangle. The upper 10,890 m of the Nashoba Formation as described by Hansen (1956) are redefined under that name at formational rank. Underlain by the Fish Brook Gneiss, the redefined Nashoba Formation is divided into ten members, including the following: at its base, the Boxford Formation of Castle (1965), reduced in rank to member, and, in ascending order, the Bellows Hill Member, Billerica Schist Member, Spencer Brook Member, Tophet Swamp Gneiss Member, Nashoba Brook Member, Nagog Pond Gneiss Member, Fort Pond Member, Long Pond Gneiss Member, and Beaver Brook Member.

Perhaps the Shawsheen Gneiss and Fish Brook Gneiss should be ranked as members of the Nashoba Formation; there seems to be no valid reason why they should not be so ranked. Many of the textural and compositional features of the Shawsheen Gneiss and the gneissic members of the Nashoba Formation are generally similar. The Fish Brook Gneiss has some textural features in common with the other gneissic units, but because it has a rather unusually distinctive lithology, it is described here as a unit of formational rank. This choice necessitates assigning formational rank to the underlying Shawsheen Gneiss.

The name "Gospel Hill Gneiss" was suggested by Hansen (1956, p. 39) for a large mass of rock in the Hudson and Maynard quadrangles that he "regarded as a granitized product of the Nashoba and Marlboro formations." This rock is part of a sill-like body or tongue that extends from a large mass of Andover Granite in the Wilmington quadrangle southwestward through the Lexington, Concord, Hudson, and Maynard quadrangles. It was described and mapped as a part of the Andover Granite by Emerson (1917, p. 220), an interpretation with which we agree. This granite encloses many xenoliths and possibly some pendants of metamorphosed rocks having foliation conformable to the regional trend. We identified the metamorphosed remnants as parts of the Shawsheen Gneiss and Sandy Pond Amphibolite Member of the Marlboro Formation.

The newly named Tadmuck Brook Schist overlies the Nashoba Formation and includes the chiefly aluminous strata earlier mapped as Brimfield Schist by Emerson (1917) and as the Worcester Formation by Hansen (1956), but it does not include the Harvard Conglomerate Lentil of the Worcester Phyllite as mapped by Emerson (1917) and the Vaughn Hills Member of the Worcester Formation (Hansen, 1956). The Harvard Conglomerate Lentil and the Vaughn Hills Member, within the area outlined in this paper (Fig. 1), most likely are fault blocks of Merrimack Group rocks juxtaposed with the Tadmuck Brook Schist as a result of Clinton-Newbury faulting. Because of evident structural isolation from any rocks in the Worcester area that might be validly considered either Brimfield Schist or Worcester Phyllite, it is concluded that these names cannot have valid application in the area outlined in this paper (Fig. 1).

Marlboro Formation. The name "Marlboro Formation" is here redefined as the stratigraphic interval between the Bloody Bluff fault and the base of the Shawsheen Gneiss, which includes the rocks at the type locality in the town of Marlboro, described by Emerson (1917, p. 25-26). The formation is subdivided into two members. The lower 1,500 m, of heterogenous makeup and poorly exposed, is left unnamed (see Skehan, this volume). The upper 640 m, made up mostly of bedded and massive amphibolites, is here named the "Sandy Pond Amphibolite Member"; its type locality is designated as the northeast-trending rocky ridges at the south and east sides of Sandy Pond in the central part of the Concord quadrangle. The formation is truncated by the Bloody Bluff fault in the eastern part of the Concord quadrangle.

The name "Marlboro Formation" originally was used by Emerson and Perry (1907) to designate the upper mafic part of the Blackstone Series of Rhode Island as described by Woodworth (in Shaler and others, 1899). Later, Emerson (1917, p. 25) redefined or extended the name to include several areas of predominantly mafic metavolcanic rocks in eastern Massachusetts and selected the large ledge north of Main Street in Marlboro, Massachusetts, as the type locality. Emerson also commented (p. 31) that the rocks in the various areas have notable dissimilarities, and he predicted a possible eventual breakup into several formations. Subsequent mapping has verified this prediction. Quinn and others (1949) abandoned the use of the name "Marlboro" for the upper mafic part of the Blackstone Series and named this unit "Hunting Hill Greenstone." Our mapping in northeastern Massachusetts has resulted in separating rocks indicated to be Marlboro Formation on the maps of Emerson (1917) and LaForge (1932) into the following units: (1) Middlesex Fells Volcanic Complex, (2) Greenleaf Mountain Formation, (3) Marlboro Formation as redefined herein, and (4) Boxford Member of the Nashoba Formation.

The lower unnamed member is made up of easily weathered and eroded rock types, and much of its outcrop area is covered with thick glacial deposits. The thickness is unknown, because the basal part is everywhere cut off by the Bloody Bluff fault. The estimated thickness of 1,500 m for this member represents the maximum thickness between the fault and the base of the Sandy Pond Amphibolite Member in the western part of the Concord quadrangle. The lower member consists, in unknown proportions, of the following lithologies: medium-grained plagioclase-quartz-biotite-muscovite schist; amphibole-biotite-plagioclase-quartz gneiss; calcareous calc-silicate gneiss; marble; thin-bedded, very fine grained amphibolite; and medium-grained lepidoblastic amphibolite.

The rocks of the Sandy Pond Amphibolite Member are fairly well exposed on a discontinuous ridge that separates the Shawsheen Gneiss on the north from the lower part of the Marlboro on the south. The lower one-third of the Sandy Pond Amphibolite Member is made up almost entirely of dark greenish-gray, very fine grained, thinly laminated to thin-bedded, planar stratified amphibolite, within which occur a few lenticular bodies of medium-grained, massive lepidoblastic amphibolite and, very rarely, thin beds and laminae of pale brownish-gray, fine-grained impure quartzite. The upper two-thirds of the member consists predominantly of thin-bedded amphibolite and massive amphibolite complexly interstratified with lenticular bodies of fine- to medium-grained biotite-muscovite-quartz-plagioclase gneiss, felsic muscovite-biotite-quartz-plagioclase gneiss, rusty-weathering biotite-rich to muscovite-rich plagioclase-quartz schist locally containing some staurolite and andalusite, and a few thin beds of very light brownish-gray, fine-grained, impure quartzite. The thinly bedded amphibolites enclose discontinuous thin lenses of scapolite-plagioclase-clinozoisite-calcite-quartz-epidote gneiss. Some amphibolite beds contain notable quantities of biotite, garnet, or magnetite.

The sediments that formed the lower part of the Marlboro Formation were mostly mafic tuff and fine-grained volcaniclastic and epiclastic sediments. Lentils or beds formed from each type of sediment are distributed throughout the lower part of the formation, but most of it seems to have been composed of variable mixtures of these sediments. There are one or more zones of carbonate sedimentation (see Skehan, this volume). Deposition was in a subaqueous, presumably marine environment. The Sandy Pond Amphibolite Member was formed from andesitic to basaltic water-laid tuff and submarine lava flows interlayered with very minor deposits of carbonate sediments and fine-grained epiclastic or volcaniclastic sediments.

Shawsheen Gneiss. The basal 2,600 m of the Nashoba Formation, as described by Hansen (1956), is here separated from that formation and is named the "Shawsheen Gneiss." The formation has been traced from the north-central part of the Reading quadrangle southwestward into the Hudson quadrangle, except for gaps where it has been displaced by plutonic rocks. Typical lithologies of the formation are well exposed in a belt along the northwest side of the Shawsheen River in the Billerica and Wilmington quadrangles, chosen as a type locality, and in the Harold Parker State Forest in the Reading quadrangle, also chosen as a type locality. The top of the formation is nowhere exposed; it is cut out by the Assabet River fault from the Wilmington quadrangle southwestward to the vicinity of the Wachusett-Marlboro Tunnel, where the remaining part of the formation may be less than 400 m thick (see Skehan, this volume). An actual contact of the Shawsheen Gneiss with the Sandy Pond Amphibolite Member of the Marlboro Formation is not exposed, but the stratigraphic section seems to be virtually complete; large quantities of intrusive rocks have been injected into the strongly foliated metamorphic rocks in the zone that includes the contact. The basal strata of the Shawsheen Gneiss in part have been cut out by the Bloody Bluff fault or other minor faults and in part have been stoped out by intrusive rocks from the Lexington quadrangle northeastward.

The Shawsheen Gneiss consists chiefly of varieties of medium-gray, inequigranular but generally medium grained, unevenly layered muscovite-biotite-quartz-oligoclase gneisses (some layers are sillimanitic to sillimanite-rich) that are irregularly interstratified with a few lenticular layers of amphibolite. The basal strata, some 50 to 250 m thick, sparsely exposed from the vicinity of the Wachusett-Marlboro Tunnel northeastward into the Concord quadrangle, consist of rusty-weathering sillimanite-muscovite-biotite-quartz schist. Within a thickness of about 500 m above the rusty-weathering schist, there are lenticular masses that are agglomeratic in appearance and in which outlines of subangular to angular clasts as much as 50 cm in diameter are well preserved. Many exposures of the gneissic rocks are conspicuously ripple bedded; amplitudes of ripples may be as much as 1 m but generally are less than 50 cm. Relict structures and the composition of the Shawsheen Gneiss closely resemble the dominant biotite gneiss members of the Nashoba—for example, the Long Pond Gneiss Member (Fig. 4).

The protoliths of the gneissic strata of the Shawsheen Gneiss were mostly wackes resulting from submarine effusion of volcanic ejecta of intermediate composition. These materials were subjected to water reworking over moderate distances, during which process small amounts of generally aluminous terrigenous epiclastic silts and clays were transported into the area and accumulated as thin, discontinuous laminae dispersed through the wacke beds. Periods of exceptionally rapid volcanic effusion, or perhaps deposition close to the volcanic source, resulted in only slightly water-worked deposits and agglomeratic deposits. The fine-grained, thinly laminated and the massive amphibolite strata interspersed throughout the formation represent deposition of mafic tuff or, possibly in some instances, mafic flows. The rusty-weathering sillimanitic schist at the base of the formation represents deposition, during a volcanically quiescent period, of clays, muds, and silts derived from a deeply weathered terrane.

Fish Brook Gneiss. The Fish Brook Gneiss has been mapped from the north-central part of the Georgetown quadrangle southwestward to the eastern edge of the Maynard quadrangle, except for gaps where it locally has been displaced by plutonic rocks. It is cut off by faults at the northeast and southwest ends of its map area. The Fish Brook Gneiss everywhere is in fault contact with the Shawsheen Gneiss or is separated from it by masses of intrusive rocks. Features such as intervals of ripple-bedded strata, interspersed thin lenticular layers of amphibolite, and a dominantly volcaniclastic character are common to both formations and suggest a gradational sequence. The two formations probably are conformable. The Fish Brook Gneiss is conformable with the overlying Boxford Member of the Nashoba Formation; rocks of the two lithologies can be seen interstratified in the zone of contact.

The Fish Brook Gneiss consists chiefly of very light gray to light gray, fine- to medium-grained biotite-quartz-plagioclase gneiss. It characteristically weathers to pale yellow, which is caused by oxidation of dispersed, extremely fine grained pyrite. Bedding, characteristically faint but locally distinct, ranges in form from distinctive undulate ripple laminae to sets of fairly even planar-stratified thin beds that are cross-stratified. The amplitudes of ripples may be as much as 10 cm but generally are less than 3 cm. Thin lenticular layers of amphibolite and biotite-hornblende-plagioclase schist and gneiss are sparsely distributed throughout the formation. Locally, subangular to rounded masses of amphibolite or biotite-hornblende-gneiss as much as 1 m in diameter are sparsely and erratically distributed in the dominant fine-grained light-colored gneiss. One stratum, about 10 m thick, of medium-dark greenish-gray to brownish-gray diopside-rich calc-silicate gneiss has been observed in the upper part of the formation in the Billerica quadrangle.

The paucity of mafic minerals—generally less than 10 percent and for much of the formation less than 5 percent—in the dominant lithology of the Fish Brook Gneiss, compared to 20 percent or more of mafic minerals in nearly all of the rocks of the other formations, makes this unit the most lithologically distinctive in the metasedimentary sequence of the area. Biotite is the principal mafic constituent. It is most abundant in the upper and lower parts of the formation. Hornblende is a very minor constituent, and many of the samples examined are devoid of this mineral. Plagioclase, which generally is oligoclase and albite, constitutes 50 to 75 percent of the dominant gneissic facies. Oligoclase is most abundant in the upper and lower parts of the formation. Much of the dominant gneissic facies of the middle part of the formation, exposed in the South Groveland quadrangle, is biotite-quartz-albite rock. Garnet is a very minor constituent of some strata.

The protoliths of the Fish Brook Gneiss were predominantly volcanic wackes of felsic to intermediate composition. The preponderance of albite, accounting for more than 95 percent of the feldspars in parts of the formation, suggests that some of the feldspathic constituent initially may have been calcium-rich zeolite, which by base exchange in an ocean-water environment was converted to sodium zeolite. During metamorphism the zeolite became feldspar. The thin mafic strata represent influxes of mafic tuff that was probably initially airborne from a distant source.

Nashoba Formation. The Nashoba Formation is a heterogeneous conformable sequence of complexly interstratified metasedimentary and metavolcanic rocks. It consists of about 50 percent lenticularly bedded, medium-grained biotite-rich gneisses; 15 percent evenly layered, platy stratified, fine- to medium-grained, commonly diopsidic, amphibole- and biotite-rich gneisses and schists; 15 percent thinly bedded to massive amphibolites; 10 percent sillimanite-rich mica schists; and 10 percent weakly foliated granoblastic diopsidic calc-silicate-bearing rocks qualifying as gneiss or schist depending on the presence or absence of appreciable amounts of biotite. Thin lenticular units of marble, laminated to thinly bedded quartzite strata, and other varieties of rock probably constitute 1 percent or less of the sequence. The aggregate thickness of these rocks is about 10,890 m.

The Nashoba Formation in the Billerica, Concord, and Westford quadrangles is divided into ten locally mappable lithostratigraphic members, principally on the basis of differentiating intervals of strata made up mostly of biotite-rich gneisses from intervals composed of a more heterogeneous variety of rock types. The biotite-rich gneiss intervals, which contain subordinate proportions of amphibolites and little or no other rocks, are herein designated, from oldest to youngest, the Tophet Swamp Gneiss Member, the Nagog Pond Gneiss Member, and the Long Pond Gneiss Member (Fig. 4). The more complex intervals of strata—distinguished by having, in addition to biotite-bearing gneisses and amphibolites, notable proportions of calc-silicate rocks, amphibole-biotite gneisses, and, less commonly, sillimanitic mica schist or marble—are herein designated the Boxford Member, the Bellows Hill Member, the Billerica Schist Member, the Spencer Brook Member, the Nashoba Brook Member, the Fort Pond Member, and the Beaver Brook Member (Fig. 4).

About 50 percent of the Nashoba Formation consists of varieties of tough, medium-gray, inequigranular but chiefly medium grained, unevenly banded or layered biotite-bearing gneisses. Most of these varieties of gneiss are composed of about 1 percent or less sillimanite, 2 to 5 percent muscovite, 20 to 25 percent biotite, 30 to 40 percent calcic oligoclase, and 35 to 50 percent quartz. Garnet and K-feldspar either are absent or constitute no more than 2 percent of most of the gneiss, but locally either of these minerals may constitute as much as 5 percent of the rock; their distribution is erratic and neither characterizes mappable units. Where K-feldspar amounts to more than 2 percent of these gneisses, it occurs mainly as large crystals in close proximity to or within migmatite zones; such K-feldspar seems to be exogenic. Sillimanite is distributed widely throughout the biotite-bearing gneiss, mostly as fine-grained clusters and isolated crystals interleaved with biotite and muscovite and as scattered thin needles in quartz grains. Locally, it occurs intergrown with fine-grained quartz and magnetite in almost cylindrical pods; in a few places, sillimanite that is pseudomorphic after kyanite or andalusite has been found. Inasmuch as the metamorphic grade of the Nashoba changes laterally, the abundance of some of these minerals may increase or decrease within a particular unit.

In areas unaffected by large-scale faulting or injection of intrusive rocks, the rocks of the Nashoba Formation are virtually undeformed, and original sedimentary structures of the protoliths generally are fairly well preserved. Small-scale faulting subparallel to the schistosity and stratification locally disrupts the continuity of the relict stratification in such a way as to form a false cross-stratification that might suggest an incorrect facing. Metamorphic foliation is parallel to nearly all compositional layering, showing a penetrative habit only locally, as in the crests of small drag folds, where it is clearly axial plane.

Relict sedimentary structures can be observed at most exposures of the biotite-bearing gneisses. These structures are mainly varieties of current bedding that locally retain fine details of internal cross laminae and incomplete ripple marks, but examples of graded beds, slump structures, and calcareous concretions (now calc-silicate lenses) are also encountered. A common banding or layering of the biotite-bearing gneisses consists of light-gray, discontinuous wispy granulose subfoliate layers of oligoclase, quartz, and a little mica that alternate with similar, but generally thinner, discontinuous medium to dark gray highly foliate layers of biotite and subsidiary muscovite, sillimanite, plagioclase, and quartz. Other minerals such as amphiboles and garnets, where present, favor the schistose layers. Most layers range from a featheredge to 1 or 2 cm in thickness and from a few centimetres to a few metres in lateral persistence.

The remaining 50 percent of the Nashoba Formation consists principally of evenly and planar stratified fine- to medium-grained amphibole- and biotite-rich gneisses and schists, amphibolites, sillimanite-quartz-mica schists, diopsidic calc-silicate rocks, marble, and a very small amount of quartzite (Fig. 4). Locally, thin beds and lenses of these rocks are interleaved with one another and with the biotite-bearing gneiss in almost any proportion or order of sequence, but most of them occur as large, irregularly intertongued stratiform bodies of nearly uniform lithologic character that typically range from a few metres to as much as 200 m in thickness and may be as much as 2 km in length. Margins of these bodies are exposed in only a few places. Contacts with adjacent and similar bodies of different lithologic character or with enclosing biotite-bearing gneisses are conformable and may be intertongued. Contacts range from sharp and abrupt to diffuse and gradational.

Relict primary structures found in the Nashoba rocks are indicative of deposition in a subaqueous environment. The mineralogic compositions and textures indicate that protoliths of the amphibolite and biotite-rich gneiss were derived from the same or a similar provenance and underwent similar transport and deposition as the protoliths of the Shawsheen Gneiss. The presence of micas and minor amounts of calc-silicate minerals in nearly all subunits of the formation indicate that deposition of fine-grained terrigenous sediments and carbonate sediments may have been maintained at a fairly uniform rate. The planar stratified, thin-bedded, fine-grained diopsidic amphibole- and biotite-rich gneiss and schist units represent periods during which deposition of the coarser grained volcaniclastic sediments was at a minimum, and thin, evenly bedded, slightly magnesian and calcareous volcanic wackes were formed. During periods of minimum or no volcanic activity, calcareous muds and sands interstratified with thin beds of dolomitic limestone were deposited. Such deposition also may have coincided with shoaling of the depositional area because of immediately preceding rapid deposition of volcaniclastic sediments or possibly because of tectonic movement. The sulfidic sillimanite-muscovite-biotite schist units represent mainly terrigenous aluminous clay-rich sediments eroded from a deeply weathered terrane and deposited under euxinic conditions.

Tadmuck Brook Schist. The uppermost unit considered to be of pre-Silurian age is here named the "Tadmuck Brook Schist." It consists chiefly of aluminous schistose rocks interstratified with subsidiary amphibolite and sparse thin beds of fine-grained impure quartzite (Fig. 4). Sills, lenses, veins, and pods of granite, pegmatite, aplite-granophyre, and quartz are distributed erratically through the unit. The aluminous rocks exist in three phases: (1) rusty-weathering, medium-dark to silvery gray, medium- to coarse-grained, sillimanite-biotite-quartz-muscovite schist; (2) graphitic chlorite-quartz-biotite-sericite-muscovite-staurolite-andalusite phyllitic schist; and (3) some quartz-chlorite-biotite-sericite phyllite. Pyrite, magnetite, albite-oligoclase, sphene, zircon, and apatite are ubiquitous in these rocks as accessories; tourmaline and K-feldspar are present locally. Alteration of biotite and staurolite to chlorite and andalusite to sericite is common.

Most of the fine detail of sedimentary structure of the Tadmuck Brook Schist has been eliminated by recrystallization and tectonic shear, but the depositional structure is locally preserved by thin beds of quartzite, amphibolite, and other compositional banding such as layers rich in staurolite and andalusite crystals, all of which are parallel to the foliation.

Lenticular bodies of dark greenish-gray, fine- to coarse-grained amphibolite composed of quartz, andesine, and a dark green amphibole occur sparsely and irregularly throughout the aluminous rock sequence. Brown biotite, titaniferous magnetite, sphene, pyrite, apatite, and zircon are ubiquitous as accessory minerals in the amphibolites; garnet, calcite, and veinlets of scapolite are notable at some localities. Chlorite formed from ferromagnesian minerals and saussurite minerals formed from plagioclase are sparse to abundant. Exposures of amphibolite within the unit are fairly abundant, but the thickness of rock exposed seldom exceeds 10 m, and none of these bodies can be traced laterally to a second outcrop. Most of the amphibolite is medium grained and has a subfoliate-fibroblastic texture within massive lenticulate, thick to very thick beds; the rock of a few exposures of amphibolite is fine grained, thinly laminated to thin-bedded, and cross-stratified.

The uppermost strata of the Tadmuck Brook Schist are truncated on the north by the main strand of the Clinton-Newbury fault zone (Fig. 1) and are in fault contact with either metasedimentary rocks of the Merrimack Group or intrusive rocks belonging principally to the Ayer or Chelmsford Granite (of local usage). The upper limit of the Tadmuck Brook Schist is arcuate, extending from the vicinity of Lawrence southwestward about 64 km to the vicinity of Shrewsbury, and is formed by the Clinton-Newbury fault zone and other smaller faults that slice down into the Nashoba Formation at both ends of the arcuate belt. Many minor faults, evidently related to the Clinton-Newbury fault, lace through the Tadmuck Brook Schist, cutting it at high angles, and contribute to the discontinuity and irregular distribution of its rock types. At some localities the lowermost beds of the Tadmuck Brook Schist seem to be intertongued with beds of quartzofeldspathic gneiss and amphibolite of the Nashoba Formation; at other localities they seem to be in fault contact.

The protoliths of the Tadmuck Brook Schist were mainly terrigenous sediments that resulted from the erosion of a deeply weathered and sapropelized terrane. These sediments were deposited at moderate depth under generally euxinic marine conditions as laminae and thin beds of slightly sulfidic aluminous clay (illite-kaolinite) alternating with laminae of quartzose silty clay. The clay accumulation was interrupted by deposition of a few thin beds of fine-grained quartzose sand and wacke sand and was several times overwhelmed by andesitic to basaltic submarine lava flows accompanied by minor deposits of water-laid tuff.

Age. The stratified rocks here considered to be part of a eugeosynclinal sequence—that is, all rocks stratigraphically above the Middlesex Fells Volcanic Complex—are devoid of any features or relations that might provide direct evidence of their age. These rocks are not known to be fossiliferous, and they have no sedimentary contacts with rocks that are fossiliferous.

Degree of metamorphism does not provide any conclusive evidence of the age of the eugeosynclinal rocks. They have been subjected to regional metamorphism, whereas the fossiliferous rocks of Cambrian and of Silurian and Devonian age of eastern Massachusetts are unmetamorphosed. The fossiliferous rocks of Cambrian age may be younger than the metamorphism that has altered the other stratified rocks in their vicinity, which are included in the Westboro Formation and Middlesex Fells Volcanic Complex. Fairbairn and others (1967) have suggested, on the basis of isotopic data, a Precambrian age for this metamorphism. On the other hand, the Cambrian rocks may have been outside the zone of metamorphism and may have been moved close to metamorphosed rocks by displacements on faults. No evidence suggests a correlation of the eugeosynclinal rocks and the Middlesex Fells Volcanic Complex. If the rocks of the eugeosynclinal sequence are correlative with the older rocks of the Merrimack synclinorium of central New England, then their metamorphism could be as recent as Early Devonian. The fossiliferous rocks of Silurian and Devonian age in the Newbury basin of eastern Massachusetts probably have never been buried deep enough to be regionally metamorphosed (see Shride, this volume).

Isotopic ages of plutonic rocks that intrude the eugeosynclinal sequence, such as the Andover Granite and Sharpners Pond Tonalite, are given as Devonian or older (Zartman and others, 1970). Even though these isotopic ages may differ somewhat from the age of intrusion because of the widespread thermal activity that has altered isotopic ratios of rocks of the region, the general designation of pre-Silurian age for the stratified eugeosynclinal host rocks is consistent.

CHANGES OF STRATIGRAPHIC NOMENCLATURE

Changes of stratigraphic nomenclature introduced in the foregoing discussion and terms not used in this report are here summarized.[2]

Names Redefined

Stratigraphic units we have redefined on the basis of field work and petrographic studies are the following:

Westboro Formation (formerly known as the Westboro Quartzite). Discussed in the preceding text and Appendix 1.

Marlboro Formation. Discussed in the preceding text and Appendix 1.

Boxford Member. The Boxford Formation is redefined as the Boxford Member of the Nashoba Formation.

Nashoba Formation. The Nashoba Formation, redefined and subdivided into ten newly named members, is discussed in the preceding text and Appendix 1.

[2] The supporting evidence for these changes is presented in Appendix 1 which is on microfiche in pocket inside back cover.

New Names Introduced

New names of stratigraphic units are introduced in the preceding discussion. Details of unit boundaries, predominant lithologies, and type localities are presented in Appendix 1.

New formation names are as follows: Middlesex Fells Volcanic Complex, Greenleaf Mountain Formation, Burlington Formation, Shawsheen Gneiss, and Tadmuck Brook Schist.

The Nashoba Formation is subdivided into the following newly named members: Bellows Hill Member, Billerica Schist Member, Spencer Brook Member, Tophet Swamp Gneiss Member, Nashoba Brook Member, Nagog Pond Gneiss Member, Fort Pond Member, Long Pond Gneiss Member, and Beaver Brook Member.

Names Not Used in This Report

Cherry Brook Formation. The name "Cherry Brook Formation" has been used by Nelson (1974) to designate some scattered remnants of metavolcanic rocks that crop out in the south-central part of the Concord quadrangle and in the Natick quadrangle. These stratified metavolcanic rocks are a small segment of the Middlesex Fells Volcanic Complex as described in this report, and therefore the term "Cherry Brook Formation" is not used in this paper.

Stonybrook Quartzite. The name "Stonybrook Quartzite" was given by Hobbs (1899, p. 110) to quartzite that crops out in the southeast part of the Concord quadrangle. These strata are indicated to be part of the Westboro Quartzite (Westboro Formation as described herein) on the map of Emerson (1917). They are in the lower member of this formation. Inasmuch as these strata are part of the Westboro and this name has been applied to them since the publication of Emerson's map, the name "Stonybrook Quartzite" is not used.

Lincoln Slate. The term "Lincoln Slate" was used by Hobbs (1899, p. 110–111) to designate "a micaceous slate, containing lenticular masses of an impure cherty limestone" in the central part of the Concord quadrangle. The thickness of this so-called slate is stated to be "probably over 5,000 feet." The locality mentioned by Hobbs is within the outcrop belt of the lower undivided part of the Marlboro Formation as redefined in this report. We have examined outcrops of several lithologies within the belt (Fig. 4) but have not seen any rock that properly can be called a slate. The term "Lincoln Slate" therefore is not used.

Kendall Green Slate. The name "Kendall Green Slate" was used by Hobbs (1899, p. 110) to designate a "schistose slate, using the term in its widest sense." Hobbs described this unit as conformably overlying the Stonybrook Quartzite [part of the Westboro Quartzite as mapped by Emerson (1917) and accepted in this paper] in a "small area west of Waltham" and having a thickness "estimated at probably over 5,000 feet." Detailed mapping in this locality by Bell reveals that the only rocks in this locality that conformably overlie quartzite are poorly exposed, metamorphosed limy shale, micaceous quartzite, and calc-silicate rocks that are interstratified with massive quartzite beds of the lower member of the Westboro Formation. Hobbs (p. 110) also described the "slate" as consisting of "hornblende, quartz and many other constituents." This description suggests that he observed part of the upper member of the Westboro Formation, outcrops of which probably have been obliterated during urban development. In this locality the Westboro Formation is separated from overlying mafic metavolcanic rocks of the Middlesex Fells Volcanic Complex by faults and masses of intrusive rock. Inasmuch as the

Kendall Green Slate of Hobbs (1899) seems to be part of the Westboro Formation, the former term is not used in this paper.

Kendal Green Formation. The term "Kendall Green Slate" of Hobbs (1899, p. 10) was redefined as Kendal Green Formation by Nelson (1974) and applied to rocks described as very fine grained felsic tuff. Detailed studies of these rocks have led Bell to conclude that they are not of tuffaceous origin but are facies of the protoclastic intrusive rocks of the Cape Ann plutonic series (of local usage). These protoclastic rocks intrude the earlier facies of the plutonic series and form veins, sills, and dikes in metamorphosed stratified rocks in the wall of the batholith. Bodies of the protoclastic rock range from veins 1 cm or less in width and 1 m or less in length to pods or lenticular masses as much as 50 m wide and 200 m or more long. All of the protoclastic rock has well-developed flow foliation. It is not metamorphosed. Those bodies of protoclastic rock not enclosed by other plutonic rocks do not occupy a specific stratigraphic horizon; they have diverse attitudes. LaForge (1932) gave the name "Woburn Formation" to bodies of the protoclastic rocks in the Lexington quadrangle. If these rocks are to have a formal name, Woburn Formation has precedence over Kendal Green Formation. The protoclastic rocks will be fully described and illustrated in a separate report.

Woburn Formation. The rocks described under the name "Woburn Formation" are identified as protoclastic intrusive rocks of the Cape Ann plutonic series (of local usage) by Bell. For this reason they are not included in the stratigraphic section and the name "Woburn Formation" is not used in this paper.

Gospel Hill Gneiss. The name "Gospel Hill Gneiss" was suggested by Hansen (1956, p. 39) for a large mass of rock in the Hudson and Maynard quadrangles that he "regarded as a granitized product of the Nashoba and Marlboro Formations." This rock is part of a sill-like body or tongue that extends from a large mass of Andover Granite in the Wilmington quadrangle southwestward through the Lexington, Concord, Hudson, and Maynard quadrangles. It was described and mapped as a part of the Andover Granite by Emerson (1917, p. 220), this being an interpretation with which we are in agreement. This granite encloses many xenoliths, and possibly some pendants, of metamorphic rocks having foliation conformable to the regional trend. The metamorphic rocks are mostly rusty-weathering schist and biotitic gneiss of the Shawsheen Gneiss as described in this paper. Locally the granite intrudes or tongues into and encloses xenoliths of the Sandy Pond Amphibolite Member of the Marlboro Formation. Inasmuch as these rocks are identified as parts of the Andover Granite, Marlboro Formation, and Shawsheen Gneiss, the name "Gospel Hill Gneiss" is not used in this paper.

SUMMARY

Remnants of unmetamorphosed fossiliferous rocks of Cambrian age are sparsely distributed in eastern Massachusetts and have no contacts with any other stratified rocks. The Cambrian rocks were formed from silts, muds, and minor carbonate sediments deposited in a shallow quiescent marine environment.

The nonfossiliferous stratified rocks are regionally metamorphosed. At least two sequences are represented. The sediments and volcanic deposits that formed the older sequence were deposited in a littoral to shallow-marine environment. The younger stratified rocks form a homoclinal eugeosynclinal sequence.

The older nonfossiliferous sequence—represented by some unnamed remnants, the Westboro Formation, and the Middlesex Fells Volcanic Complex—is intruded by the Dedham Granodiorite; the younger eugeosynclinal sequence is not intruded by this granodiorite. This relationship, combined with local omissions of the upper part of the older sequence that cannot be attributed to displacements on faults or displacements by plutonic rocks and which may be due to an ancient erosional episode, suggests a possible significant unconformity.

Materials that formed the nonfossiliferous rocks were derived from three sources: (1) volcaniclastic detritus and laval flows; (2) muds, clays, and sand eroded from a deeply weathered and sapropelized land; and (3) marine carbonate sediments. At least 80 percent of the material was of volcanic origin.

Deposition of the eugeosynclinal sequence tended to be somewhat cyclic. During periods of minimum volcanic activity, mainly pelitic and carbonate sediments were deposited and are now represented by calcic and dolomitic marble, calc-silicate rock, and micaceous schist units. These rocks are interlayered with amphibolite and biotite-hornblende schist formed from mafic tuff and flows. Periods of maximum volcanic activity are represented by thick units of biotite-quartz-plagioclase gneiss formed from volcanic wacke and amphibolite rocks formed from mafic tuff.

The volcaniclastic protoliths included both mafic and salic types. The mafic types were predominantly fine grained tuffs, some of which may have been initially airborne; mafic flow rocks occur in minor amounts throughout the sequence. The felsic types were mainly medium- to coarse-grained wackes.

The present feldspar constituents of the felsic units are mainly plagioclases. Potassium feldspar is a very minor constituent, even of quartz-rich gneisses.

The sediments and volcanic materials that formed the eugeosynclinal sequence were predominantly, and possibly entirely, deposited in a subaqueous, presumably marine environment. This environment was subjected, at least locally, to episodes of considerable turbulence.

Possible ages of the nonfossiliferous rocks are inferred from indirect envidence. The older sequence, intruded by the Dedham Granodiorite, is probably of Precambrian age. The eugeosynclinal sequence is older than the Andover Granite that intrudes it; this relationship is consistent with a pre-Silurian age.

REFERENCES CITED

Bell, K. G., 1968, Faults in eastern Massachusetts: Geol. Soc. America, Abs. for 1967, Spec. Paper 115, p. 250.

Bell, K. G., and Dennen, W. H., 1972, Plutonic series in the Cape Ann area: Geol. Soc. America Abs. with Programs, v. 4, no. 1, p. 2.

Billings, M., 1929, Structural geology of the eastern part of the Boston basin: Am. Jour. Sci., v. 18, 5th ser., p. 93-137.

Burr, H. T., 1900, A new Lower Cambrian fauna from eastern Massachusetts: Am. Geologist, v. 25, p. 41-50.

Castle, R. O., 1965, Gneissic rocks in the South Groveland quadrangle, Essex County, Massachusetts, *in* Geological Survey research 1965: U.S. Geol. Survey Prof. Paper 525-C, p. 81-86.

Chute, N. E., 1940, Preliminary report on the geology of the Blue Hills quadrangle, Massachusetts: Massachusetts Dept. Public Works Bull. 1, 52 p.

Crosby, W. O., 1880, Contributions to the geology of eastern Massachusetts: Boston Soc. Nat. History Occasional Papers 3, 286 p.

——1900, Geology of the Boston basin, pt. 3, The Blue Hills complex: Boston Soc. Nat. History Occasional Papers 4, v. 1, p. 289-563.

Dennen, W. H., 1972, Correlations of igneous rocks by chemical signatures of minerals: Geol. Soc. America Abs. with Programs, v. 4, no. 1, p. 13.

Dowse, A. M., 1950, New evidence for the Cambrian contact at Hoppin Hill, North Attleboro, Massachusetts: Am. Jour. Sci., v. 248, p. 95-99.

Emerson, B. K., 1917, Geology of Massachusetts and Rhode Island: U.S. Geol. Survey Bull. 597, 289 p.

Emerson, B. K., and Perry, J. H., 1907, The green schists and associated granites and porphyries of Rhode Island: U.S. Geol. Survey Bull. 311, 74 p.

Fairbairn, H. W., Moorbath, S., Ramo, A. O., Pinson, W. H., Jr., and Hurley, P. M., 1965, Rb-Sr age of granitic rocks of southeastern Massachusetts and the age of the Lower Cambrian at Hoppin Hill, *in* Variations in isotopic abundances of strontium, calcium, and argon and related topics: Massachusetts Inst. Technology, Dept. Geology and Geophysics, U.S. Atomic Energy Comm. Contract AT (30-1)-1381, M.I.T.-1381-13, 13th Ann. Prog. Rept. for 1965, p. 3-10.

——1967, Rb-Sr age of granitic rocks of southeastern Massachusetts and the age of the Lower Cambrian at Hoppin Hill: Earth and Planetary Sci. Letters, v. 2, p. 321-328.

Foerste, A. F., 1889, The paleontological horizon of the limestone at Nahant, Massachusetts: Boston Soc. Nat. History Proc., v. 24, p. 261-263.

Grabau, A. W., 1900, Paleontology of the Cambrian terranes of the Boston Basin, pt. 3, The Blue Hills complex: Boston Soc. Nat. History Occasional Papers 4, v. 1, p. 601-694.

Hansen, W. R., 1956, Geology and mineral resources of the Hudson and Maynard quadrangles, Massachusetts: U.S. Geol. Survey Bull. 1038, 104 p.

Hobbs, W. E., 1899, Some new fossils from eastern Massachusetts: Am. Geologist, v. 23, p. 109-114.

Howell, B. F., and others, 1944, Correlation of the Cambrian formations of North America [chart no. 1]: Geol. Soc. America Bull., v. 55, p. 993-1002.

Ingram, R. L., 1954, Terminology for the thickness of stratification and parting units in sedimentary rocks: Geol. Soc. America Bull., v. 65, p. 937-938.

LaForge, Laurence, 1932, Geology of the Boston area, Massachusetts: U.S. Geol. Survey Bull. 839, 105 p.

Nelson, A. E., 1974, Changes in nomenclature of upper Precambrian to lower Paleozoic (?) formations in the Natick quadrangle, eastern Massachusetts, and their tentative correlations with rocks in Rhode Island and Connecticut: U.S. Geol. Survey Bull. 1395-E, 14 p.

Page, L. R., 1968, Devonian plutonic rocks in New England, *in* Zen, E-an, and others, eds., Studies of Appalachian geology, northern and maritime: New York and London, Interscience Pub., p. 371-383.

Quinn, A. W., 1971, Bedrock geology of Rhode Island: U.S. Geol. Survey Bull. 1295, 68 p.

Quinn, A. W., Ray, R. G., and Seymour, W. L., 1949, Bedrock geology, Pawtucket quadrangle, Rhode Island-Massachusetts: U.S. Geol. Survey Geol. Quad. Map GQ-[1].

Rogers, W. B., 1856, On trilobites from Braintree and on the geologic relations of the district: Boston Soc. Nat. History Proc., v. 6, p. 27-29.

Shaler, N. S., 1888, On the geology of the Cambrian district of Bristol County, Massachusetts: Harvard Coll. Mus. Comp. Zoology Bull. 16 (Geological ser. no. 2), p. 13-26.

Shaler, N. S., and Foerste, A. F., 1888, Preliminary description of North Attleboro fossils: Harvard Coll. Mus. Comp. Zoology Bull. 16 (Geological ser. no. 2), p. 27-41.

Shaler, N. S., Woodworth, J. B., and Foerste, A. F., 1899, Geology of the Narragansett Basin: U.S. Geol. Survey Mon. 33, 402 p.

Shaw, A. B., 1950, A revision of several Early Cambrian trilobites from eastern Massachusetts: Jour. Paleontology, v. 24, p. 577-590.

Shride, A. F., 1976, Stratigraphy and Correlation of the Newbury Volcanic Complex, northeastern Massachusetts, *in* Page, L. R., ed., Contributions to the stratigraphy of New England: Geol. Soc. America Mem. 148 (this volume).

Skehan, J. W., 1968, Fracture tectonics of southeastern New England as illustrated by the Wauchusett-Marlborough Tunnel, east-central Massachusetts, *in* Zen, E-an, and others, eds., Studies of Appalachian geology, northern and maritime: New York and London, Interscience Pub., p. 371-383.

——1969, Tectonic framework of southern New England and eastern New York, *in* Kay,

M., ed., North Atlantic—Geology and continental drift, a symposium: Am. Assoc. Petroleum Geologists Mem. 12, p. 793-814.

Skehan, J. W., S. J., and Abu-Moustafa, A. A., 1976, Stratigraphic analysis of rocks exposed in the Wachusett-Marlborough Tunnel, east-central Massachusetts, *in* Page, L. R., ed., Contributions to the stratigraphy of New England: Geol. Soc. America Mem. 148 (this volume).

Wanless, H. R., and Bell, K. G., 1976, New England, *in* McKee, E. D., Crosby, E. J., and others, eds., Paleotectonic investigations of the Pennsylvanian System in the United States, Pt. 1: U.S. Geol. Survey Prof. Paper 853 (in press).

Warren, C. H., and Powers, S., 1914, Geology of the Diamond Hill-Cumberland District in Rhode Island-Massachusetts: Geol. Soc. America Bull, v. 25, p. 435-476.

Zartman, R. E., and Naylor, R. S., 1972, Structural implications of some U-Th-Pb zircon isotopic ages of igneous rocks in eastern Massachusetts: Geol. Soc. America Abs. with Programs, v. 4, no. 1, p. 54-55.

Zartman, R. E., Hurley, P. M., Krueger, H. W., and Giletti, B. J., 1970, A Permian disturbance of K-Ar radiometric ages in New England, its occurrence and cause: Geol. Soc. America Bull., v. 81, p. 3359-3373.

MANUSCRIPT RECEIVED BY THE SOCIETY MARCH 4, 1975
REVISED MANUSCRIPT RECEIVED NOVEMBER 18, 1975
MANUSCRIPT ACCEPTED DECEMBER 8, 1975

Printed in U.S.A.

Geological Society of America
Memoir 148

Stratigraphic Analysis of Rocks Exposed in the Wachusett-Marlborough Tunnel, East-Central Massachusetts

James W. Skehan, S.J.
Boston College, Weston Observatory
Weston, Massachusetts 02193
and
Department of Geology and Geophysics
Boston College
Chestnut Hill, Massachusetts 02167

AND

A. Adel Abu-moustafa
Department of Earth Sciences
Salem State College
Salem, Massachusetts 01970

ABSTRACT

An unusually complete section of several important Paleozoic(?) formations was exposed in the 12,806-m-long Wachusett-Marlborough Tunnel in the Clinton, Shrewsbury, and Marlboro quadrangles of east-central Massachusetts. The Marlboro Formation, the stratigraphically lowest formation cut by the tunnel, is at least 2,705-m-thick and has been divided into 31 members composed of distinctive sequences of eight lithologic types or facies referred to herein as lithologic units. Several of these units have recognizable subdivisions.

The overlying Nashoba Formation, at least 3,740 m thick, has been divided into 30 members composed of distinctive sequences of seven of the same eight lithologic units that form the members of the Marlboro Formation. Some of the lithologic unit subdivisions that are recognized in the Marlboro also occur in the Nashoba, but others do not. Additional lithologic unit subdivisions are found only in the Nashoba.

The layered rocks above the Nashoba Formation are divided into 15 unnamed formations or members composed of distinctive sequences of 12 lithologic units. Five of these lithologic units are recognized in the Nashoba and Marlboro Formations, but the other seven lithologic units are restricted to the unnamed formations or members.

Of the total measured stratigraphic thickness of 7,834 m, the Marlboro Formation makes up 34.5 percent; the Nashoba Formation, 47.7 percent; and the unnamed formations or members, 17.7 percent.

The northwestern part of the stratigraphic succession is in the chlorite and biotite zones of metamorphism, whereas the remainder of the succession ranges from the sillimanite-potassium feldspar to garnet zones or their mafic rock equivalents.

The originally sedimentary and volcanic section has been thickened by the intrusion of 2,521 m of igneous rocks. Three plutonic bodies (the Ayer Granodiorite, the Rattlesnake Hill muscovite granite, and the Andover Granite), having a total thickness of 1,422 m measured normal to the enclosing strata, account for 56 percent of the thickening.

The Marlboro Formation has been thickened 14.1 percent (444 m) by intrusive rocks; the Nashoba Formation, 17.9 percent (816 m); and the unnamed formations or members, 47.6 percent (1,260 m). The Andover Granite cut by the tunnel is intruded near the transitional boundary of the Nashoba with the Marlboro Formation and encloses rocks of Nashoba affinities. The unnamed formations have been thickened chiefly as a result of the intrusion of the Ayer Granodiorite and the Rattlesnake Hill muscovite granite plutons. Eighteen members of the unnamed formations or members compose a section 956 m thick which is devoid of intrusive rocks.

The protoliths of the layered rocks in the Wachusett-Marlborough Tunnel section were a succession of sedimentary and volcanic accumulations formed in environments that ranged from pelagic to shallow marine to subaerial. The protoliths of the Marlboro Formation were dominantly basaltic pyroclastic rocks and lava flows with interlayered quartz-rich and calcium carbonate-rich sedimentary rocks and limestone. The protoliths of the coticule (a garnet-quartz rock) and epicule (an epidote- and (or) hornblende-quartz rock) and quartzite, closely associated with calc-silicate rocks and amphibolite, were probably manganese-, iron-, and aluminum-rich sandstone or chert. These sedimentary rocks had well-developed beds that had been deposited as part of a volcanic sequence probably in a pelagic to shallow marine environment. The protoliths of the Nashoba Formation were probably predominantly sedimentary rocks whose clastic particles were derived from the weathering and erosion of volcanic rocks, including ultramafic and mafic to felsic flows, pyroclastic rocks, sandstone, limestone, and sandy lime mudstone. The Nashoba Formation is also made up in part of rocks whose protoliths were the same as those of the most common rocks of the Marlboro Formation.

The protoliths of the lower unnamed formations and members were probably fine-grained strandline deposits that graded upward into interbedded quartz-rich shale and high aluminum, sulfide-rich euxinic shale that formed offshore in an enclosed, noncirculating basin. These sedimentary rocks graded upward into a mixture of sandy lime mudstone, shaly sandstone, and volcanogenic sedimentary rocks, which were succeeded by more interbedded sandy lime mudstone and sandstone. These were overlain by a unique hornblende-actinolite ultramafic, pyroclastic volcanic rock, which was in turn succeeded by a thin, felsic, volcanogenic sequence.

The northwestward-dipping stratigraphic section is folded and cut by numerous faults. The axial planes of the folds commonly are overturned to the southeast; this suggests that the upper beds have been transported eastward relative to the strata below. Although premetamorphic faults are present, the dominant faults are postmetamorphic, northwestward-dipping thrust and reverse faults. Northeastward-striking, steeply dipping normal faults, mostly formed later than the thrust faults, are cut by northwestward-striking faults and by faults of other orientations as well; these are typically marked by quartz, calcite, and barite veins.

INTRODUCTION

A detailed, systematically measured stratigraphic section (Tables 1 through 4)[1] was developed for the 12,806-m-long Wachusett-Marlborough[2] Tunnel from data collected during detailed underground mapping from 1960 to 1962 (Skehan, unpub. maps). The locations and attitudes of faults, relative movement directions of faulted blocks, and igneous intrusions of relatively large size were determined (Tables 5 and 6). The presence of folds, which caused duplication of beds, and the more important kinds of mineralization features were noted (Tables 5 and 6).

Because the tunnel stations were surveyed in terms of feet by the Metropolitan District Commission (MDC) engineers, all measurements in the subsurface were recorded in the English system of measurement and later were recalculated in the standard metric notation. Since all station locations given in official records of the MDC are distances measured according to the English system, they will be so referred to in this paper.

Measurements of lithologic boundaries and structural features were made in the tunnel with a 100-ft tape that was drawn taut between each MDC station. Thicknesses were determined by measurement of breadth of outcrop on cross sections (scale, 1 in. = 20 ft) of the tunnel; the thicknesses then were corrected for both strike and dip. Thicknesses of intrusive bodies were calculated (1) from the attitude of the enclosing strata, (2) by measurements normal to the foliation of the intrusive rock, or (3) by measurements normal to the walls of the intrusive body. An average thickness of folded beds was computed that represents the average of the measurement taken normal to the beds on the flank and the measurement taken normal to the beds along the axial plane of the fold.

Thin-section petrographic control (Table 7) for all representative specimens (collected by Skehan) of major lithologic units of the stratigraphic sequence was provided by Abu-moustafa (1969) except for certain specimens from the Marlboro Formation that were studied by Assem Badawy (1973, written commun.).

RELATED STUDIES

The Wachusett-Marlborough Tunnel (Fig. 1) is an 8-mi-long aqueduct with a 16-ft interior diameter. The tunnel was driven through a generally westward-dipping, homoclinal sequence (Skehan, unpub. maps and 1967, 1968b, 1969b) near Worcester and Framingham, Massachussetts (Fig. 2). Subsequent to the reconnaissance reports by Hitchcock in 1832 and 1841, Crosby (1899a, 1899b; Fig. 2) recognized that parts of eastern Massachusetts are cut by westward-dipping thrust faults along the Nashua River valley near Clinton, along the northwestern margin of the surface outcrop of the Clinton-Newbury fault zone (Skehan, 1968b, 1969a; Fig. 1).

Emerson (1917) mapped the bedrock exposures of the area as part of the larger study of Massachusetts and Rhode Island. Throughout most of the area, his map shows the Clinton-Newbury fault zone as an uninterpreted, pronounced structural discontinuity.

K. G. Bell and other members of the U. S. Geological Survey (1967, oral commun.; this volume), traced the major fault and its branches, first recognized in the tunnel, from Clinton to the vicinity of Newbury, Massachussetts. Recent interpretations by Barosh (1972) and Bell (1972) of an aeromagnetic map of the region (Zietz

[1] Tables 1, 2, 3, 4, 5, 6, 7, and 9 are on microfiche in pocket inside back cover.

[2] The spelling "Marlborough" is used for the township and the tunnel; "Marlboro" is used for the city in which the type section of the Marlboro Formation is exposed.

and others, 1972) indicate that the Clinton-Newbury fault zone and the Wachusett-Marlborough fault block are geophysically two of the most easily recognized features in eastern Massachusetts; the rocks involved in them are readily distinguished from the weakly magnetic rocks of the Merrimack group northwest of the Clinton-Newbury fault zone.

Preliminary stratigraphic interpretations of the tunnel and later studies of certain structural elements from shaft A to shaft B were made by Skehan (unpub. maps and 1967, 1968b, 1969b). In addition to Abu-moustafa's (1969) petrologic and geochemical research, Legarde (1967) and Lacroix (1968) carried out surface field mapping and related laboratory studies on the Nashoba and Marlboro Formations, respectively. Fischer (1964) and Gore (1967) made magnetometer surveys on the surface along the line of the tunnel. Hansen (1956) made a study of the bedrock and glacial features of the adjacent Hudson and Maynard quadrangles (Fig. 2). R. Z. Gore (in prep.) studied the Ayer Granodiorite of eastern Massachusetts.

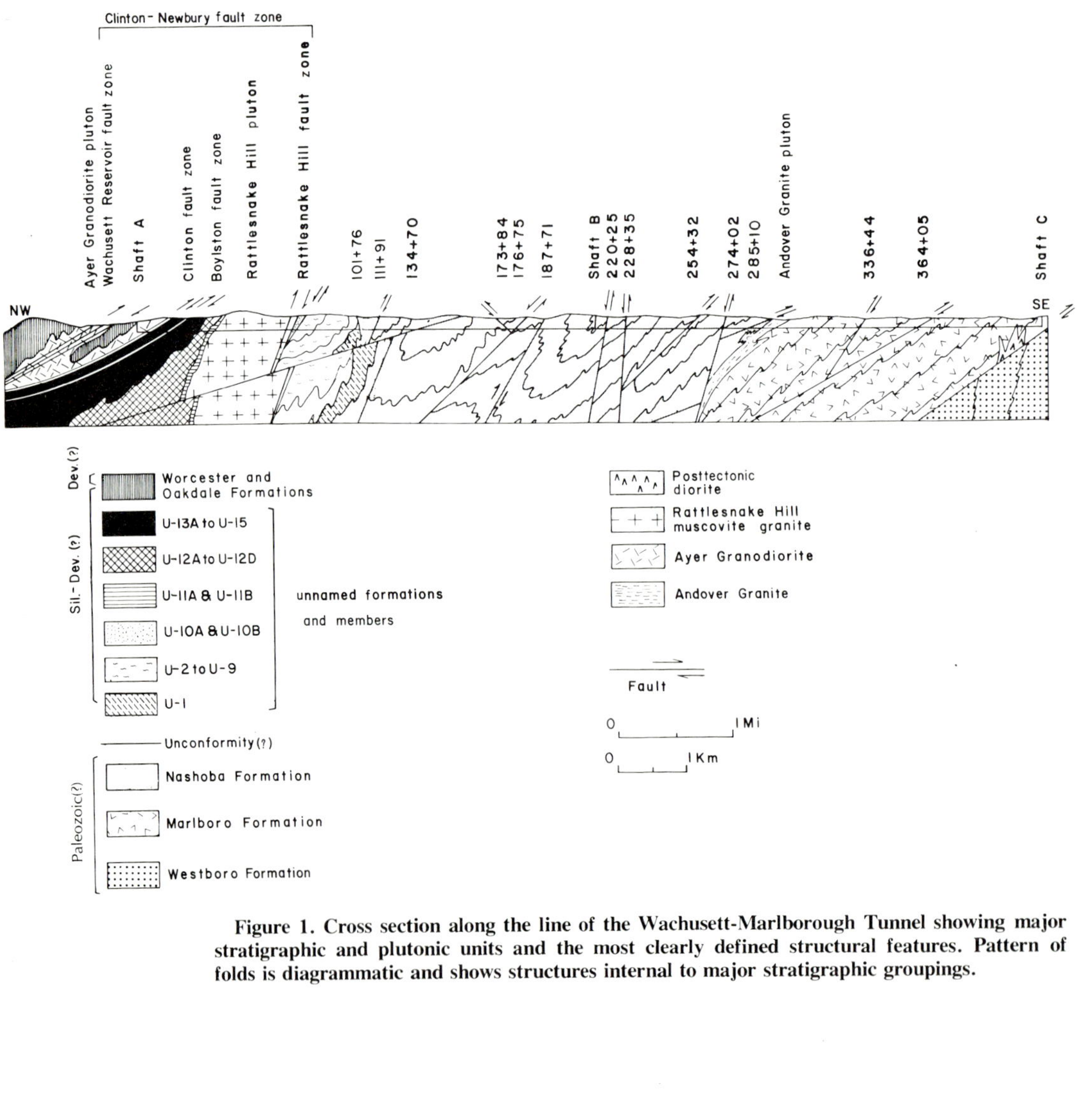

Figure 1. Cross section along the line of the Wachusett-Marlborough Tunnel showing major stratigraphic and plutonic units and the most clearly defined structural features. Pattern of folds is diagrammatic and shows structures internal to major stratigraphic groupings.

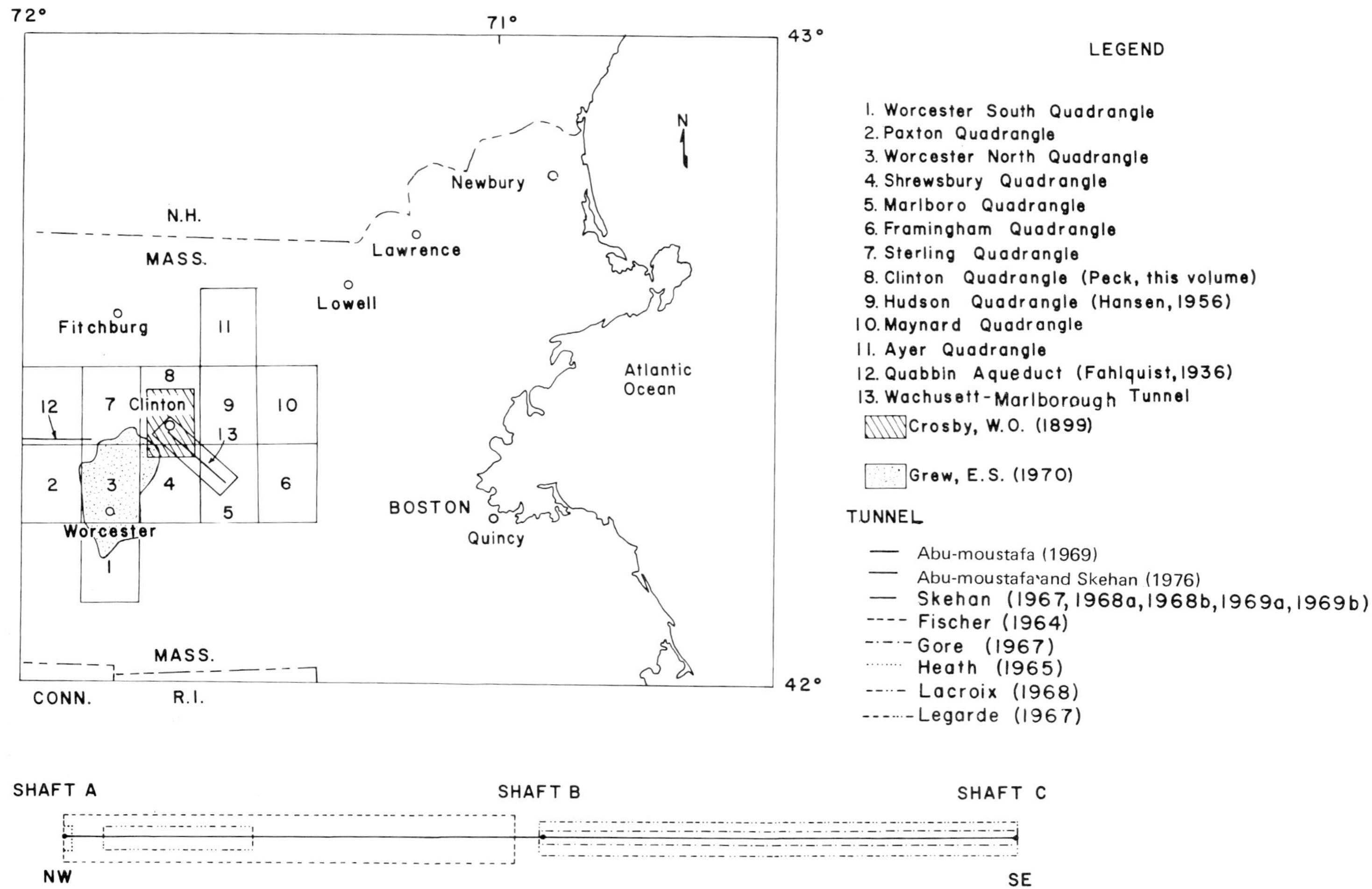

Figure 2. Regional geographic setting and sources of published and unpublished data in the vicinity of the Wachusett-Marlborough Tunnel.

GENERAL STRATIGRAPHY

The stratigraphic section exposed in the tunnel is 7,834 m thick and generally dips westward. The section consists of intensely faulted, homoclinally folded strata, parts of which are retrograded to the chlorite and biotite zones and parts of which are metamorphosed by prograde metamorphism to the garnet and to sillimanite-potassium feldspar zones and thickened by abundant intrusive igneous bodies. The stratigraphic succession is made up of 15 lithologic units that occur in varying combinations and that can be identified in the field. Consequently, it has been possible to make three major stratigraphic divisions: the Marlboro Formation, the Nashoba Formation, and the unnamed formations or members. The relative amounts of the 15 lithologic units form the basis for further differentiation into mappable members.

In addition to having been transformed by regional metamorphism, the stratigraphic succession has been altered locally to varying degrees by shearing along faults and by hydrothermal metamorphism that chloritized and sericitized certain minerals and produced many veins. Postmetamorphic faults served as channels for the hydrothermal solutions.

Lithologic Units

The entire stratigraphic succession is made up of fifteen distinct lithologic units, some of which have recognizable mineral compositions that vary and thus are useful in establishing further subdivisions. Table 7 gives the modes of representative rocks in each of these units. Table 8 summarizes their importance in each of the members of the formations. These units, their inferred protoliths, and (where possible) their environments of deposition are presented below.

Units 1, 1A, and 1B. Gneiss and Schist. The description of unit 1 applies only to the Marlboro Formation, whereas that of units 1A and 1B applies only to the Nashoba Formation.

Unit 1. Biotite-rich Gneiss and Schist. Well- to poorly foliated, gneissic to schistose, equigranular to porphyroblastic, biotite-rich (>20 percent), feldspar-quartz gneiss and schist with variable amounts of garnet, muscovite, and opaque minerals. Porphyroblasts are mostly garnet, feldspar, or both. Sillimanite was not observed in the Marlboro Formation where it is exposed in the tunnel. Unit 1, where subdivided, is further differentiated according to the mineral composition of the porphyroblasts as follows: unit 1_1—microcline; unit 1_2—plagioclase; unit 1_3—microcline and plagioclase; and unit 1_4—garnet (>5 percent and <5-cm diam).

Unit 1A. Porphyroblastic, Biotite-Feldspar-Quartz Gneiss and Schist. Dominantly gneissic, in places schistose. Mostly well-layered; porphyroblastic; biotite-rich (20 to 60 percent); variable amounts of garnet, sillimanite, muscovite, and as yet undifferentiated opaque minerals. The porphyroblasts are microcline, plagioclase, and garnet. Unit 1A, where subdivided, is further differentiated according to the mineral content of the porphyroblasts as follows: unit $1A_1$—microcline; unit $1A_2$—plagioclase; unit $1A_3$—microcline and plagioclase; and unit $1A_4$—garnet (>5 percent and <5-cm diam).

Unit 1B. Nonporphyroblastic, Biotite-Feldspar-Quartz Gneiss and Schist. Dominantly gneissic, in places schistose. Mostly well-layered to equigranular; biotite-rich (20 to 60 percent); variable amounts of garnet, sillimanite, muscovite, and opaque minerals.

Protoliths. The protoliths of units 1, 1A, and 1B were dominantly mafic sedimentary rocks that were derived in part from the basaltic volcanic pile and in part from felsic volcanic rocks. These rocks (Table 1) contained substantial quartz and feldspar and minerals rich in potassium and iron. The high potassium content suggests that the protolith may have been either iron-rich claystone that adsorbed potassium, mafic volcanic rocks that contained potassium zeolites, or possibly a mixture of mafic components with potassium-rich keratophyres.

Unit 2. Quartz-Feldspar-Biotite Gneiss and Schist. Well-foliated to poorly foliated; gneissic to schistose; equigranular to porphyroblastic. Less biotite (<20 percent) and more quartz and feldspar than unit 1. Variable amounts of chlorite, muscovite, garnet, and opaque minerals. Unit 2, where subdivided, is further differentiated according to the mineral composition of the porphyroblasts as follows: unit 2_1—sillimanite and (or) muscovite (sillimanite noted in Nashoba Formation only); unit 2_2—feldspar, chiefly plagioclase; unit 2_3—feldspar, chiefly microcline; unit 2_4—garnet.

Unit 2 contains more calcium and silica and less potassium, iron, and magnesium than unit 1. The protoliths may have been glass fragments partly leached so that iron was not concentrated and less silica and calcium were removed; in part, the protoliths may have been felsic to intermediate volcanic flows and volcanogenic detritus.

Unit 3. Biotite-Quartz Granulite, Calc-Silicate Granulite, and Marble. Occur in separate beds or interlayered in zones with or without layers of other units. Wide variation in mineral composition.

Unit 3-1. Brown Biotite-Quartz or Quartz-Feldspar Granulite. Commonly thin-bedded; garnet-rich; associated with biotite- and garnet-rich gneiss of unit 1A or with calc-silicate granulite of unit 3-2.

Unit 3-2. Calc-Silicate Granulite. Greenish-gray to brown; thin- to thick-bedded; fine- to coarse-grained; in part, porphyroblastic. Calcite-, plagioclase-, and quartz-bearing rocks with variable amounts of diopside, graphite, serpentine, epidote, biotite, garnet, clinozoisite, pyrophyllite, and talc.

Unit 3-3. Marble. Mostly calcite; white, gray, green, or pink; well-bedded, banded, or massive; in places may contain quartz, biotite, clinozoisite, grossularite, diopside, forsterite, muscovite, graphite, and rhodochrosite.

Protoliths. The protoliths of the rocks of unit 3 and its subdivisions must have been characterized by abundant calcium and magnesium. In the Marlboro Formation (Table 3), these rocks are closely associated with metabasaltic rocks of unit 4 as well as with the less abundant units 1 and 2. In the Nashoba Formation, which contains a smaller proportion of unit 4 and more of units 1 and 2, the association is similar.

These relationships suggest that the sea water of the volcanic environment was enriched in calcium and magnesium. Thus, the protoliths of unit 3-1, enriched to some degree in potassium, were volcanogenic sedimentary rock that contained precipitates of calcium and magnesium. The protoliths of unit 3-2 contained less potassium than those of unit 3-1 but were equally rich in calcium, magnesium, and silica. The protoliths of unit 3-3 were chiefly calcite limestone or dolostone with variable but lesser amounts of silica, iron, and aluminum.

Unit 4. Hornblende Gneiss, Schist, and Amphibolite. Massive to poorly foliated amphibolite. Well-layered (relict bedding), mostly nonporphyroblastic gneiss and schist. Variable amounts of biotite, quartz, epidote, garnet, muscovite, sericite, chlorite, opaque minerals, and clinozoisite.

Unit 4 is interbedded with all other units and is closely associated with units 3-2, 3-3, and 7, and, in the Marlboro Formation, also with unit 8. Unit 4, where subdivided, is further differentiated according to the rock type as follows: unit 4_1—gneiss; unit 4_2—schist; and unit 4_3—amphibolite.

The protoliths of unit 4_3 were mafic lava flows and pyroclastic rocks, whereas

TABLE 8. AGGREGATE THICKNESSES (IN METRES) OF LITHOLOGIC UNITS IN

Lithologic Units	Members												
	Unnamed members or formations												
	U-1	U-2	U-3	U-4	U-5	U-6	U-7	U-8	U-9	U-10A	U-10B	U-11A	U-11B
3	0.61	0.45				17.07	0.15			33.22			
4	8.23												
5	53.64	10.21	2.13	2.59	32.92	50.59	31.24	12.34	4.11				
6											4.57		
9													
10													
11													
12													
13												4.27	14.02
14									56.69				
15		52.73	54.26	41.30	55.78	45.26	29.26	45.57					
Total	62.48	63.39	56.39	43.89	88.70	112.92	60.65	57.91	60.80	33.22	4.57	4.27	14.02

	Members of														
	N-1	N-2	N-3	N-4	N-5	N-6	N-7	N-8	N-9	N-10	N-11	N-12	N-13	N-14	N-15
1			60.65	147.83	14.33	52.73	166.73	42.67	383.13	61.42	300.84		36.42	3.96	28.96
2				84.12	37.18		7.62		2.74	15.08		27.43	8.53		
3		0.60					0.30		10.06	2.74			1.07	2.74	42.36
4		11.57				83.82	13.41	14.63	1.83	34.14		49.37		10.68	
5							12.19		2.13	0.61					22.40
6	36.58					15.55			38.10		71.63				
7										2.44				0.30	1.07
Total	36.58	12.17	60.65	231.95	51.51	152.10	200.25	57.30	437.99	116.43	372.47	76.80	46.02	17.68	94.79

	Members of															
	M-1	M-2	M-3	M-4	M-5	M-6	M-7	M-8	M-9	M-10	M-11	M-12	M-13	M-14	M-15	M-16
1	18.60				7.32					74.06	113.69			12.19		
2	50.90	34.14			51.20								3.66	8.23		
3	18.29	7.62				14.02	4.27	3.35	15.31	19.20	4.57	20.42	12.19	1.83	3.96	
4	14.94	92.05	2.89	25.00	152.70	84.13	95.11	25.91	8.30	5.18	33.53	11.89	79.70	60.04	44.81	46.63
5	9.76		12.80	2.44			6.71		0.76			26.52	9.45		36.27	3.96
6															2.74	20.73
7													0.15		3.05	
8															41.45	3.05
Total	112.49	133.81	15.69	27.44	211.22	98.15	106.09	29.26	24.37	98.44	151.79	58.83	105.15	82.29	132.28	74.37

the protoliths of units 4_1 and 4_2 contained in addition quartz and feldspar sandstone, a variety of clay minerals, and sandy lime mudstone. Unit 4 rocks make up about 50 percent of the Marlboro Formation but only about 8.5 percent of the Nashoba Formation.

Unit 5. Quartzite and Micaceous Quartzite. Quartz-rich rock subdivided according to mica content and the presence or absence of foliation.

Unit 5-1. Quartzite. Faintly bedded, "varve"-like; faintly foliated to massive; pink to brown, gray to black; fine- to coarse-grained, in part "pebbly" and feldspar-rich; vitreous or granular.

Unit 5-2. Micaceous Quartzite. Faintly foliated, brown biotite and (or) muscovite quartzite.

Protoliths. The protoliths of unit 5 consisted of quartz sandstone, quartz-feldspar sandstone, pebble-bearing sandstone, chert- and clay-bearing quartz sandstone and (or) chert. Unit 5 rocks make up about 6 percent of the Marlboro Formation,

THE STRATIGRAPHIC DIVISIONS, WACHUSETT-MARLBOROUGH TUNNEL

Members												Total
Unnamed members or formations												
U-12A	U-12B	U-12C	U-12D	U-13A	U-13B	U-13C	U-13D	U-13E	U-13F	U-14	U-15	
2.75												54.25
												8.23
2.74	22.56		83.52									308.59
		8.23				0.91	4.57			11.58		29.86
											7.62	7.62
					106.98	181.66	46.94	96.62	48.16		0.61	480.97
					87.48							87.48
				12.80								12.80
												18.29
												56.69
												324.16
5.49	22.56	8.23	83.52	12.80	194.46	182.57	51.51	96.62	48.16	11.58	8.23	1,388.94

Nashoba Formation

N-16	N-17	N-18	N-19	N-20	N-21	N-22	N-23	N-24	N-25	N-26	N-27	N-28	N-29	N-30	
37.49	23.47	79.25	52.73		223.12	7.62	16.45	2.13	52.73	163.06	68.27	53.04	19.81		2098.84
27.43	81.99			31.39	156.36			2.13		107.29	10.67	39.62			639.58
	27.43		71.02	21.34	0.61		0.61	1.83	6.10	0.61	2.44	2.13	70.41	1.92	266.32
18.90						12.80	7.47	22.87	19.81		25.60			10.36	337.26
		1.22			0.61								7.62		46.78
	9.14	25.90			45.72					2.44		0.91		99.88	345.85
							1.07	0.30	0.30						5.48
83.82	142.03	106.37	123.75	52.73	426.42	20.42	25.60	29.26	78.94	273.40	106.98	95.70	97.84	112.16	3,740.11

Marlboro Formation

M-17	M-18	M-19	M-20	M-21	M-22	M-23	M-24	M-25	M-26	M-27	M-28	M-29	M-30	M-31	
			3.66	17.68	5.79										252.99
1.83															149.96
17.53	12.19	5.18	49.38	98.45	87.63	42.37	9.75				1.52				449.03
75.44	3.96	30.49		7.92	2.44	39.62	67.67	113.23	26.37	64.00	35.67	19.81	42.06	33.84	1,345.33
	3.05	12.80		3.05	17.06				4.88	0.61	1.52	14.18	5.79	7.62	179.23
1.83	5.49	0.30	25.30		12.19					29.87	29.57			4.57	132.59
0.68	0.61	0.30			0.08	33.53	10.67	16.91	2.59	3.05	2.44	4.73	0.61	0.61	80.01
0.08	0.61	0.61	5.48		0.08	36.57	4.88	3.05	9.14	3.05	7.01			0.30	115.36
97.39	25.91	49.68	83.82	127.10	125.27	152.09	92.97	133.19	42.98	100.58	77.73	38.72	48.46	46.94	2,704.50

about 1 percent of the Nashoba Formation, and about 22 percent of the unnamed formations and members.

The protoliths of the greater part of unit 5 rocks in the Marlboro Formation probably were quartz sandstone and possibly chert. The protoliths of the unit 5 rocks in the Nashoba Formation probably contained more quartz sandstone and, because of the volcanic environment, less chert. The protoliths of the thicker sections of unit 5 that occur in the unnamed formations or members (that is, in U-1, U-6, and U-7)[3] were strand or nearshore deposits, whereas the thinner sections of unit 5 (and those sections in which clay is more abundant, for example, in U-15) were deposits in a deeper basin.

Unit 6. Muscovite- and (or) Chlorite-rich Schist and Gneiss. Quartz- and plagio-

[3]To avoid confusion, note explicitly that U-1, U-2, . . . U-15 mean the various unnamed formations or members, whereas unit 1, unit 2, . . . unit 15 refer to the lithologic units that make up the tunnel section.

clase-rich schist and gneiss with muscovite and (or) chlorite, with or without biotite. Probably variants of units 1A, 1B, and 2 but are more commonly schistose than gneissose. Chlorite may be the result of hydrothermal alteration. Opaque minerals may be as much as 12 percent; sillimanite and garnet may be abundant; feldspar and garnet porphyroblasts occur in places.

The protoliths of unit 6 rocks were moderately aluminum-rich claystone bodies that may have been formed of detritus from hydrothermally altered felsic and mafic volcanic piles. Such rocks make up nearly 9 percent of the Nashoba Formation but are a minor component of the Marlboro and unnamed formations and members. The protoliths may have accumulated on the submarine flanks of island volcanoes or in offshore basins near mafic and felsic volcanic centers.

Unit 7. Garnet Coticule.[4] Quartz and garnet in thin beds and lenses; very fine grained to cherty or medium-grained; pink- to flesh-colored. Associated with garnet-microcline-hornblende-quartz-clinozoisite gneiss and plagioclase-biotite gneiss of units 1A, 2, 3, 4, and 5 and, in the Marlboro Formation, also with unit 8.

The protolith of unit 7 was a layered manganiferous iron formation consisting of cherty iron- and manganese-rich carbonate and fine-grained clastic sedimentary rocks. Although it makes up only a small fraction of the Marlboro Formation, unit 7 is an important and easily recognized marker in field mapping; it occurs only in the upper half of the part of the Marlboro Formation that is exposed in the tunnel and sparsely as thin beds in the Nashoba. The environment of deposition was probably the submerged flanks of volcanoes, chiefly mafic in composition, that furnished small amounts of iron-manganese clay to the area of sand deposition or chert precipitation.

Unit 8. Epidote Epicule.[5] Quartz, epidote, and hornblende in thin beds and lenses; very fine grained to cherty or medium-grained. Associated with unit 7 coticule rock and rocks of units 4 and 5. Epicule is a distinctive fine-grained, pistachio-green, flinty rock that is closely associated with the pink garnet coticule beds and that contains coarse-grained, fascicular hornblende crystals.

The protoliths of unit 8 were layered, cherty iron- and calcium-rich carbonate and fine-grained clastic sedimentary rocks. Unit 8 occurs only in the upper part of the Marlboro Formation that is exposed in the tunnel and constitutes an important, easily recognized marker in field mapping.

Unit 9. Biotite-Quartz-Feldspar Schist. Gray, foliated and layered, fine-grained biotite-quartz-feldspar schist whose protolith was a light-colored volcanogenic rock.

Unit 10. Hornblende-Actinolite. Blocky to massive, mottled-black to green, layered to massive, coarse- to fine-grained hornblende-actinolite with minor amounts of oligoclase-andesine, chlorite, epidote, calcite, and sericite. Unit 10 contains little feldspar and therefore is of ultramafic composition.

The protoliths of unit 10 were probably dark-colored, volcaniclastic rocks. Other possibilities include massive to layered igneous intrusive rocks or a combination of layered volcaniclastic rocks that alternated with sills of similar composition. These may have been sea-floor volcanic extrusive rocks of unusual composition and (or) intrusive rocks either derived directly from the lower crust or mantle

[4] Defined by Renard (1878) as thin spessartine-rich layers in metamorphic rocks from the Ardennes, Belgium (from Latin *cos, cotis*, which means whetstone).

[5] Epicule is a new term that refers to rocks whose physical appearance and texture are similar to garnet coticule except that the latter are buff to pink, whereas epicule is light to dark green. Garnet coticule rocks in the Marlboro Formation are closely associated with epidote and hornblende epicule.

or by magmatic differentiation of basaltic magma.

Unit 11. Hornblende-Actinolite. The mineral composition of unit 11 is the same as that of unit 10, but unit 11 is uniformly much finer grained and has a poorly developed layering. The protoliths and environments of deposition were probably essentially the same as for unit 10; only slight changes could account for the finer grained character and the poorly developed layering.

Unit 12. Interbedded Hornblendite and Quartz-Feldspar Gneiss. Light- to dark-gray to black; fine- to coarse-grained; layered with alternating bands of epidote, feldspar, and hornblende. The protoliths of unit 12 consisted of thin, interbedded volcanic ash or lava flows of alternating felsic and mafic composition.

Unit 13. Feldspathic Muscovite-Quartz Schist. Leached, coarse-grained, feldspar-bearing muscovite-quartz schist. Weathers to rust color. Basal 1.22 m contains lenses or "pebbly" masses of light-blue to purple quartz. The protoliths of unit 13 were either clay-bearing, felsic, volcanogenic sedimentary rocks or hydrothermally altered, felsic volcanic rocks.

Unit 14. Quartz-Muscovite-Biotite Phyllite. Uniformly foliated, dark gray to black quartz-muscovite-biotite phyllite. Weathers rusty to yellowish on exposure to air and water; contains thin quartz lenses; encloses quartzite beds of unit 5-1. The protoliths were largely euxinic, black mudstone beds that enclosed thin quartz and quartz-feldspar sandstone and (or) chert beds. These may represent nearshore or shoreline deposits derived from a felsic volcanic pile or from precipitation of silica gel.

Unit 15. Graphite-Andalusite-Sillimanite-Quartz Schist. Dark-gray to dark-brown and black, coarse-grained, porphyroblastic graphite-muscovite-biotite-staurolite-andalusite-sillimanite-quartz schist. Contains sillimanite after andalusite in crystals as much as 45 cm long. Disseminated sulfide minerals are present in notable amounts throughout but are especially abundant from tunnel stations 88+00 to 94+15 (see Table 1; station locations in hundreds of feet plus feet, measured from northwest end of tunnel); weathers rusty to yellowish on exposure to air and water.

The protoliths of unit 15 were much the same as those of unit 14 except that unit 15 either began with more aluminum or was recrystallized at a higher grade of metamorphism than unit 14. In contrast to unit 14, unit 15 is interbedded with more abundant and thicker quartz and quartz-feldspar sandstone and (or) chert (unit 5-1) with minor sandy lime mudstone and related sedimentary rocks (units 3-1 and 3-2). The stratigraphic section in which unit 15 is present contains unit 5-1 in variable amounts. In unnamed formations or members U-2, U-3, and U-4, unit 5-1 makes up a small fraction of the section, whereas in members U-6 and U-7, unit 5-1 forms a larger part of the section than does unit 15 (see Table 8).

VARIATIONS IN THICKNESS

The thickness of the stratigraphic succession has been affected by the intrusion of igneous plutons as well as by smaller concordant and discordant intrusive igneous rocks. The intrusive igneous rocks have an aggregate thickness of 2,521 m, but their distribution throughout the stratigraphic succession is far from uniform. The accuracy of estimates of the thickness of the stratigraphic succession also depends in part on recognizing the amount of repetition of the rock units caused by folding as well as on recognizing, if possible, where the folded, layered rocks have been tectonically thinned or thickened. Folds in the tunnel section appear to be of two kinds: (1) those formed during the development of the Merrimack synclinorium, whose eastern flank is crossed by the tunnel, and (2) those formed by drag along faults.

Detailed stratigraphic analysis suggests that there has been no major repetition of the section by folding, at least on a scale recognizable within this eight-mile-long section (Tables 1 through 6). Small-scale repetition of beds was recognized locally from station 115+00 to station 125+00.

Although variations in thickness that were caused by intrusion of igneous rocks and by folding are calculated with some degree of confidence, the accurate determination of stratigraphic thickness depends to an even greater extent on an accurate assessment of how much section has been repeated or eliminated by faulting. Table 6 records the orientation of faults, their relative movement direction and other related features that indicate the extent to which the stratigraphic section has been disrupted. These data are presented so that workers on the surface may use them as a basis for a more complete regional structural and stratigraphic analysis. Figure 1 shows a few of the faults and fault zones that may be among those most important regionally in computing the thickness of the stratigraphic section before tectonism.

The recognition of unconformities and disconformities of regional or even of local importance also affects the estimates of the thickness of the stratigraphic section before deformation and erosion.

INTRUSIVE IGNEOUS ROCKS

A detailed petrographic study of the intrusive igneous rocks of the tunnel has not yet been undertaken. Nevertheless, on the basis of field observations and preliminary petrographic data, these rocks are subdivided into several series as follows:

Oliverian Plutonic Series

In several places in the tunnel, it has been possible to recognize pretectonic sills, small pegmatite bodies, and quartz veins that show cataclastic textures and structures. Such features are characteristic of pretectonic intrusive igneous rock series (Page, 1968, p. 374–375). On the assumption that the layered rocks of the tunnel are of Paleozoic age, such intrusive igneous rocks are assigned provisionally to the Oliverian Plutonic Series (designated PS-1 in the tables). The ultramafic rocks, however, may be part of the Highlandcroft Plutonic Series. These pretectonic rocks make up a relatively small part of the intrusive igneous rocks that thicken the stratigraphic succession.

Representatives of rocks of the Oliverian Plutonic Series are noted in Table 5. They are recognized chiefly in the Marlboro Formation and tend to be best displayed in the lower part of the section (as in members M-4 and M-5), where the coarse-grained, originally gneissoid buff-pink granite has a sugary, blastoporphyritic texture. The Oliverian Plutonic Series has the following recognizable divisions:

A. Banded Gray Granite. The primary flow banding is crossed by a later-formed metamorphic foliation that was also imposed on the enclosing rocks.

B. Banded Gray Pegmatite. Primary flow bands of biotite and microcline grade from gray to buff and into layers rich in garnet and blue quartz. The bands have been granulated and are crossed by the foliation of the enclosing country rocks.

C. Banded Buff to Pink Pegmatite. The banding of the buff to pink microcline-rich pegmatite is sheared and crossed by a later-formed foliation.

D. Ultramafic Rocks. Concordant, thin, serpentine- and talc-rich lenses. These may belong to the earlier Highlandcroft Plutonic Series. Formerly, these rocks were interpreted to belong to the New Hampshire Plutonic Series (below; designated PS-1D in the tables).

New Hampshire Plutonic Series

Most of the intrusive igneous rocks of the tunnel are syntectonic and early posttectonic, as evidenced by their internal structures and their relationships to the enclosing metamorphosed layered rocks. These rocks were intruded and crystallized under orogenic stresses. On the assumption that the layered rocks of the tunnel are of Paleozoic age, the syntectonic intrusive igneous rocks would have been intruded and crystallized under Acadian orogenic stresses and are therefore provisionally assigned to the New Hampshire Plutonic Series (designated PS-2 in the tables). Page (1968, p. 377) presented criteria that allow the recognition and separation of syntectonic members from early posttectonic members of this series. Representatives of this series in the tunnel are described in Table 5. Although elsewhere in New England the New Hampshire Plutonic Series does not typically include pink granite and pink pegmatite (Page, 1968, p. 375), such rocks are common in the tunnel section and, on the assumption that they are of Paleozoic age, thus extend the variations found in the New Hampshire Plutonic Series (Table 5). The New Hampshire Plutonic Series has the following recognizable divisions (A through G, below, are syntectonic; J through O are early posttectonic):

A. Gray Granite to Granodiorite. Strikingly gneissoid; foliation concordant with that of wall rocks; dominantly porphyritic but smaller bodies nonporphyritic. Potassium feldspar dominantly buff to pink, but may be gray (Table 5). Individual crystals may be gray at surface and buff to flesh-pink in subsurface in Ayer Granodiorite. Protoclastic structures were produced by stresses developed during intrusion. Granite may grade into gray aplite. Two plutons belong to this series: the Ayer Granodiorite and the Andover Granite.

The Andover Granite (Table 5; sta. 290+55 to 298+72) is porphyritic, foliated, quartz-potassium feldspar-plagioclase-biotite-muscovite granite. It is gray, strikingly gneissoid, and dominantly concordant with the foliation and in part with the bedding of the enclosing rocks.

The Ayer Granodiorite (PS-2A) is of two varieties: (1) greenish, granulated, foliated, porphyritic rock and (2) gray, granulated, foliated, more compact, less porphyritic rock.

B. Fine-Grained Microcline Granite. Buff to flesh-pink; foliated parallel to walls when it contains biotite; chiefly concordant but locally crosscuts enclosing rocks.

C. Fine- to Medium-Grained, Biotite-Quartz-Microcline Granite. Dominantly nonporphyritic; buff to flesh-pink but may be gray locally; foliated parallel to walls; chiefly concordant but locally crosscuts enclosing rocks; may grade into aplite.

D. Fine- to Medium-Grained, Biotite-Quartz-Microcline Pegmatite. Dominantly buff to flesh-pink, locally gray; foliated parallel to walls; chiefly concordant, but locally crosscuts enclosing rocks. D_1 consists of related concordant quartz veins that follow folds.

E. "Pebbly" Gray Pegmatite. Foliated; with rounded, coarse-grained microcline crystals; chiefly concordant.

F. Gray Biotite Pegmatite. Fine- to medium-grained biotite-quartz pegmatite; well-foliated; with relatively euhedral gray microcline; where biotite is absent, a rare occurence, pegmatite is white. Chiefly concordant.

G. Brick-Red Pegmatite. Fine- to medium-grained; commonly garnetiferous and biotite-rich; closely associated with amphibolite (unit 4) or biotite schist; microcline is brick-red probably as a result of incorporation of iron during intrusion.

H and I. Ultramafic Rocks. Probably belong to the Oliverian Plutonic Series (PS-1D in the tables).

J. Muscovite Granite. Gray to buff muscovite-biotite-quartz-plagioclase granite; generally equigranular but may be foliated near contact with enclosing rocks; when buff potassium feldspar is absent, rock is blue gray.

K. Fine- to Medium-Grained, Biotite-Quartz-Microcline Pegmatite. Dominantly buff to flesh-pink; foliated near contact with enclosing rocks; equigranular away from contact; may in part be concordant but locally may crosscut enclosing rocks.

L. Gray Granite. Gray, fine- to medium-grained biotite granite; commonly crosscuts foliation of country rock or follows fault; chiefly lacks foliation but may be somewhat foliated near contact with wall rock; may grade into flesh-pink microcline granite or gray aplite.

M. Gray Pegmatite. Fine- to medium-grained biotite-microcline-quartz pegmatite; chiefly crosscuts enclosing rock.

N. Fine-Grained Pink Granite. Microcrystalline granite without biotite; commonly crosscuts foliation and bedding.

O. Gray to Buff Aplite. May be a finer grained phase of the gray granite or fine- to medium-grained, biotite-quartz-microcline pegmatite.

Upper Devonian Plutonic Series

Page (1968, p. 380–381) described a third plutonic series in New England, which does not fit the criteria for the Oliverian or New Hampshire Plutonic Series or the White Mountain Plutonic-Volcanic Series. These rocks are herein designated as the Upper Devonian plutonic series. Rocks of this late posttectonic series may be exposed in the tunnel. The series includes dark-gray biotite diorite, buff to fine-grained biotite monzonite-diorite, gray aplitic granite, and aplite dikes. Page (1968, p. 381) indicated that one of the distinctive features of the Upper Devonian plutonic series is that it is a source of metallic ores. It is probable that the relatively abundant veins containing metallic ore minerals of the tunnel are related to the intrusive rocks of this series rather than to the White Mountain Plutonic-Volcanic Series, representatives of which have as yet not been recognized in the tunnel. The Upper Devonian plutonic series (PS-3 in the tables) has the following recognizable divisions:

A. Biotite Diorite. Dark-gray; equigranular; medium-grained. Crosscuts foliation of country rock by stoping.

B. Fine-Grained Biotite Monzodiorite. Buff to dark-gray; equigranular; fine-grained. Crosscuts foliation of country rock by stoping, located from station 402+22 to station 404+74.

C. Gray Aplitic Granite. Very fine grained.

D. Aplite Dikes. Gray to white, buff to light-green. Dikes are intrusive into fine-grained biotite monzodiorite (B, above; PS-3B in tables) and are prominent between stations 404+40 and 404+50; the monzodiorite is bleached within 1 cm of the contact with the dikes.

E. Veins. Calcite, rhodochrosite, quartz (including rose and purple quartz), sulfide minerals, and related minerals.

F. Pegmatite. Buff-colored; crosscuts the fine-grained biotite monzodiorite (B, above) at station 404+28.

Post-Paleozoic Plutonic Series

Post-Paleozoic basaltic dikes are recorded at several locations in the tunnel. Intrusion of basalt must have taken place during an episode of dilatation that followed the deformation responsible for the reverse fault motions and associated fractures, such as must have been dominant during the folding. The presence of sheared muscovite granite in the most prominent of these reverse faults, in turn cut by basalt, suggests that episodes of compression and dilatation have developed more than once. Although petrographic data are not available, these dikes are considered to be tholeiitic and possibly related to the Medford Diabase.

STRUCTURE

Folds

The 7,834-m-thick section through which the tunnel was driven is part of the eastern limb of the Merrimack synclinorium whose folds are overturned to the east. The data in Tables 1 to 4 shows that there is no large-scale repetition of the stratigraphic section. Locally, as at station 115+00 to station 125+00, some small-scale repetition of beds by folding was observed.

In addition to folds associated with the formation of the Merrimack synclinorium, there are folds formed by movements on faults. Numerous examples are noted in Table 6. In many instances, the relative motions of the blocks on either side of the fault may be deduced from the attitude of the associated folds on the assumption that such folds were produced by drag along the fault.

Several postmetamorphic faults—chiefly oriented parallel to the axial planes of associated folds, such as the faults that dip southeastward between stations 172+00 and 180+00—are unrelated to the folds that developed during formation of the Merrimack synclinorium. We interpret such discordant folds or refolded folds as being more or less locally developed within fault-bounded blocks.

Faults

Many faults and other structures are of importance in reconstructing an accurate measured stratigraphic section for the rocks exposed in the Wachusett-Marlborough tunnel. All faults that have disrupted the stratigraphic sequence to any potentially important degree are listed in Table 6. Faults whose displacement is less than the distance of their exposure in the tunnel generally have not been listed in Table 6.

This information should prove useful to geologists mapping on the surface where faults are not observable directly. On the basis of data collected in the subsurface only, it is commonly not possible to evaluate definitely the relative importance of various faults in producing variations in the thickness of the original stratigraphic succession. Nevertheless, in the process of subsurface mapping and in the analysis of the fault data, certain features may be recognized that suggest that specific faults and fault zones may be more important than others in reconstructing the regional stratigraphy. Those faults that are now considered of greatest importance to regional stratigraphic studies are shown in Figure 1.

Thrust and Reverse Faults. On the basis of data shown in Table 6, it is clear that the two fault zones at station 5+12 to station 7+56 and at station 11+23 to station 13+56 are so intensely deformed and so pervasively fractured that they must be of regional significance. The fault zone at station 5+12 to station 7+56 strikes N40°E and dips 35°NW; it is referred to as the Clinton-Newbury fault zone (Skehan, 1968b). On the surface, it was traced northeastward to near Lowell and Newbury by U.S. Geological Survey geologists (K. G. Bell, 1967, oral commun.; L. R. Page, 1967, oral commun.; Bell and Alvord, this volume). Other nearby dislocation zones, such as the Boylston fault zone (sta. 11+23 to sta. 13+56) and other unnamed zones in which closely spaced faults are clustered may be subsidiary branches of the Clinton-Newbury fault zone.

The Rattlesnake Hill fault zone (sta. 69+07 to sta. 75+18; Fig. 1, Table 6) is characterized by fine-grained, light-gray, well-laminated, intensely sheared muscovite granite. This rock has conspicuous slip surfaces separated by 1- to 2-mm-thick layers of very finely banded ultramylonite that includes finely divided muscovite and forms well-defined drag folds. These folds plunge 25°, N25°W and, together with slickensides on slip and fault surfaces that plunge 65°, N25°W throughout this zone, indicate northeastward tectonic movement. The approximate orientation of the fault zone as calculated from tunnel exposures is N50°E, 65°NW. This fault zone may be an important branch of the Clinton-Newbury fault zone and possibly forms its southeastern margin.

Throughout the tunnel, the other dominant faults appear to be northeastward-striking, north-westward-dipping thrust and high-angle reverse faults. Thrust faults are especially well developed in the northwestern part of the tunnel in the Clinton-Newbury fault zone and also in the southeastern part of the tunnel that is relatively near the Bloody Bluff fault zone or one of its branches (Fig. 1).

The high-angle reverse faults typically strike northeastward and dip northwestward. Such faults are more numerous than thrust faults. Crosscutting relationships with other faults indicate that the thrust and reverse faults are early postmetamorphic features produced by eastward-directed compressive forces or by westward-directed underthrusting of blocks lying to the east.

Normal Faults. The normal faults for the most part were formed later than the thrust or reverse faults; for example, the normal faults near station 217+19 to station 222+73 cut the reverse faults. Some of the normal faults contain mylonite or ultramylonite and cut early posttectonic granite bodies. The dominant orientation of normal faults is northeastward with steep northwestward or southeastward dips. A few normal faults are gently dipping, as the one at station 176+75 (Fig. 1) that truncates a reverse fault at station 176+94 (Table 6).

Northwestward-Striking Postmetamorphic Faults. At several locations, the tunnel intercepts distinctive, northwestward-striking, steeply dipping faults that typically contain veins of quartz, calcite, some barite, sulfide minerals, and other metallic minerals not yet studied in detail. In some places, the veins are composed of quartz crystals and dogtooth spar in open vugs. These faults appear to be spaced several hundred feet apart and have a somewhat variable strike and dip. These are the youngest faults in the tunnel and were produced by extension, possibly during the latter part of the Mesozoic Era.

One such fault at station 134+70 to station 135+15 is oriented N35°W, 65°SW. It is a reverse fault with a left-lateral component; the fault zone is filled with very finely banded mylonite or ultramylonite and some calcite. Branching off this fault are numerous smaller faults (whose displacements are measured in centimeters) and joints that are filled with calcite and quartz.

The fault at station 134+70 to station 135+15 may connect with the fault at station 141+47, which is oriented N50°W, 65° to 85°SW and is also left-lateral and filled with a calcite vein as much as 1.22 m thick. The fault is associated with pyrrhotite-quartz breccia. Offset on the fault has brought together two different sequences exposed on either side of the tunnel, respectively (Table 2). It is not yet possible to determine the amount of offset on this and other faults having similar characteristics. Cuppels (1964) mapped similar faults in the Concord Quadrangle that have not appreciably offset the Bloody Bluff fault.

Another example of this type of left-lateral fault occurred from station 375+00 to station 376+00; the fault is oriented N40°W, 10°SW and contains a calcite vein.

DETAILED STRATIGRAPHY OF FORMATIONS

The measured thicknesses, location, and variations from the general description of the individual lithologic units are presented in Tables 1, 2, and 3. The location listed in these tables for each bed, group of beds, or interlayered sequence of beds of variable composition refers specifically to the location on the southwest wall of the tunnel at a height of 91 cm above the invert (low point of the cylindrical tunnel). The remarks under Variations in Thickness and under Intrusive Igneous Rocks (above) provide a description of those features over and above the least common denominator given in the general lithologic description. These descriptions should enable the field geologist more readily to correlate specific distinctive beds in this measured section with their counterparts exposed on the surface. Moreover, information is provided in the remarks mentioned above that should be sufficient to permit the reconstruction of the exact measured stratigraphic position of any given bed or lithologic unit. The length of the tunnel interval that corresponds to any given lithologic unit or sequence of units was deliberately restricted so that, using the data of the several tables, one could reconstruct an approximate map and cross section. This was not possible here on the scale of the original mapping because of limitations of space.

Marlboro Formation

Emerson (1917, p. 25) proposed the name "Marlboro" for a sequence of rocks found in northeastern Massachusetts and adjacent Rhode Island. He described the formation as consisting of biotite, hornblende, and epidote-chlorite schists; actinolite quartzite; steatite with intercalated quartzite; conglomerate; and limestone beds. A typical locality exposing the Marlboro Formation was mentioned by Emerson (1917, p. 25) to be a wall-like outcrop located on the north side of Main Street in the town of Marlboro, Massachusetts. Here Emerson described the outcrop as composed of thin layers (8 to 23 cm thick) of conglomerate interbedded with black biotite schist.

Hansen (1956, p. 8) mapped the Marlboro Formation in the Maynard quadrangle as two separate belts. He described the formation as consisting predominantly of amphibole schist with five beds of quartzite and marble. Hansen also indicated that possibly a large part of the stratigraphic sequence of the Marlboro contains muscovite-biotite schist, but that the two-mica schist was not well exposed within the Maynard quadrangle.

Table 8 provides a synoptic view of the lithologic makeup of the 31 members of that part of the Marlboro Formation exposed in the tunnel as well as their variations. Table 3 is complementary to Table 8 in that it provides the systematic stratigraphic data that allow compilation of the aggregate thickness of each lithologic type in each of the members. Table 8 is thus a partial analysis of Table 3.

After the various rock types were mapped in the field, petrographic analysis of representative samples (Table 7) provided the basis for classifying the rock types of the entire Marlboro Formation into eight lithologic units. Some of these units were further subdivided on the basis of features discussed under General Stratigraphy (above).

Approximately 50 percent of the Marlboro Formation is made up of rocks of unit 4. Unit 4 rocks are present in all members except M-20. The relative proportions of hornblende schist and gneiss and of equigranular amphibolite, however, are variable from member to member. Table 8 provides the data for comparing the relative amounts of unit 4 rocks, or any other lithologic unit in each member, to the total thickness of the member. Table 3 provides data on the interlayering of the various lithologic facies.

The calcareous rocks of unit 3 make up a prominent part of the Marlboro Formation, 449 m out of a total of 2,705 m. In the lower 24 members, unit 3 rocks are present in variable amounts in all members except M-3, M-4, M-5, and M-16. In the uppermost members, unit 3 rocks are absent except for a total of 1.52 m in M-28.

The biotite-feldspar-quartz rocks of units 1 and 2 aggregate the next largest thickness but are restricted to nine members, namely, M-1, M-2, M-5, M-10, M-11, M-14, M-20, M-21 and M-22; M-10 and M-11 have the greatest thicknesses of such rocks.

The quartz rocks of unit 5 make up a thickness of 179 m and are dispersed throughout the formation. Such rocks are absent from eleven members but are consistently present in M-26 through M-31.

The schist of unit 6 aggregates 133 m but is absent from M-1 through M-14; it is consistently present in M-15 through M-20 and sporadically present in M-22, M-27, M-28, and M-31.

The part of the Marlboro Formation exposed in the tunnel forms two major distinctive stratigraphic divisions on the basis of the presence or absence of units 7 and 8—coticule and epicule rock—even though in terms of total thickness they make up only 195 m. The lower twelve members, M-1 through M-12, contain neither coticule nor epicule beds. The upper members for the most part contain variable amounts of each, but both are missing in M-14 and M-21. Unit 7 is absent and unit 8 is present in M-16 and M-20, whereas unit 8 is absent and unit 7 is present in M-29 and M-30. In the Marlboro Formation, by contrast with the Nashoba Formation, units 7 and 8 rocks are associated with those of unit 1 and (or) unit 2 only in M-13, M-17, M-20, and M-22.

Environment of Deposition. The probable protoliths of the several lithologic units (see General Stratigraphy, above) and their repetition in differing patterns (Table 8) throughout the stratigraphic column allow certain conclusions to be reached concerning the environment of deposition of the Marlboro Formation.

The Marlboro Formation exposed by the tunnel is fundamentally part of a basaltic volcanic pile that contains lesser amounts of felsic volcanic rocks (unit 2) and that developed in a marine environment, as suggested by the interlayering of carbonate rocks (unit 3), sandstone, and chert (units 5, 7, and 8). The rocks of unit 1, part of unit 2, and unit 6 represent detritus derived from weathering of mafic as well as felsic volcanic terranes that were interlayered with lava flows and pyroclastic rocks on the flank of the volcanic pile. The Marlboro Formation probably represents the deposition about a submarine volcano or volcanic chain in a shallow to medium depth of water. This situation would have been suitable for deposition that resulted in the carbonate rocks of unit 3 and sandstone, siltstone, and possibly chert of units 5, 7, and 8.

Nashoba Formation

The Nashoba Formation (Tables 2 and 8) was originally named the Bolton Gneiss (Perry and Emerson, 1903, p. 79) after the town of Bolton, Massachusetts, where good exposures are found. The name Bolton Gneiss was later abandoned (Emerson, 1917, p. 81) because it had been previously used for a formation located in Connecticut.

The formation remained unnamed until Hansen (1956, p. 32) proposed the name Nashoba Formation after the good exposures located along Nashoba Brook in the Maynard and Westford quadrangles. The Nashoba Formation as described by Hansen consists of the following lithologic units: predominantly biotite gneiss with lesser amounts of biotite schist, interbedded layers of amphibolite schist and hornblende gneiss with quartzite beds that have to a large degree been feldspathized, and a few beds of marble. The Nashoba Formation also contains an abundance of pegmatite and quartz veins. Hansen also described a series of granitized rocks within the Nashoba for which he proposed the name Gospel Hill Gneiss. These rocks are now considered to be part of a large mass of Andover Granite (Bell and Alvord, 1976).

Althouth the pegmatite veins are absent from N-1, N-2, and N-30, they are represented in all other members. Unit 1 is represented in all members except N-12 and N-20 and may be thought of as a variation of unit 2. Unit 2 is difficult to distinguish from unit 1 where its biotite content is relatively high but is sporadically represented, commonly in substantial thickness, throughout the section. It is present in N-12 and N-20 where unit 1 is not represented.

The end members of unit 1 are extremely biotite-rich feldspar-quartz rocks, whereas end members of unit 2 may be relatively depleted in biotite and enriched in quartz and feldspar. Intermediate members of unit 1 grade into rocks of unit 2. Together they constitute the most abundant lithologic units in the Marlboro Formation and are typically highly magnetic because of their enrichment in magnetite, pyrrhotite, and (or) chalcopyrite.

The muscovite- and chlorite-rich schist and gneiss of unit 6, with an aggregate thickness of 346 m, make up the next most abundant rock type. Rocks of unit 6 make up all of N-1 and most of N-30 (the base and top of the Nashoba Formation as it is exposed in the tunnel), and are sporadically represented throughout the rest of the section. Unit 6 rocks are present in, but are usually less than 10 to 15 percent of the total thickness of, eight other members. Unit 6 rocks are about 20 percent of N-11. The rocks of unit 6 may be thought of as variants of units 1 and 2 but are distinguished by different micaceous minerals. These three lithologic units had similar environments of deposition that differed only in the provenance of the mica-forming components.

The 266-m-thick calcareous rocks of unit 3 are the next most abundant rock group (Table 8) and are commonly associated with unit 4 rocks. Unit 3 rocks are present in 19 of the members and are absent from N-1, N-3 through N-6, N-8, N-11, N-12, N-16, N-18, and N-22. Unit 4 rocks are present in 15 members and absent from N-1, N-3, N-4, N-5, N-11, N-13, N-15, N-17 through N-21, N-26, N-28, and N-29. Where unit 3 and unit 4 rocks are present in the same formation, they tend to be interlayered.

In the Nashoba Formation, the quartz rocks of unit 5 are 47 m thick and are sparsely represented in N-7, N-9, and N-10. They are associated with unit 3 and unit 4 rocks and, in N-10, also with coticule beds. The protoliths of these rocks were probably sandstone and possibly in part chert. The unit 5 rocks in N-15, N-18, N-21, and N-29 are associated with unit 3 rocks except in N-18, where the overlying or underlying members contain unit 4 rocks.

Coticule rock of unit 7 is a distinctive rock type in both the Nashoba and the Marlboro Formation. Its maximum total thickness is 2.4 m (in N-10) in the lowest members of the Nashoba. Unit 7 rocks are present also in N-14, N-15, and N-23 through N-25. Coticule is most consistently associated with rocks of units 3 and 4 in the Marlboro, but in the Nashoba, it is also associated consistently with unit 1, with or without unit 2 rocks.

Environment of Deposition. The probable protoliths of the several lithologic units discussed under General Stratigraphy (above) and the patterns of sequences (Table 8) in the stratigraphic column (Table 2) allow conclusions to be drawn concerning the environment of deposition of the Nashoba Formation.

Units 1, 2, and 6 make up 3,084 m of the total 3,740 m of the Nashoba Formation. Their protoliths were composed of quartz- and feldspar-rich rocks. Many of these rocks contained abundant clay minerals of variable composition; some were aluminum-rich. This implies an environment close to a source of detritus derived from weathering of mafic and felsic volcanic and plutonic rocks. In the lower part the Nashoba Formation, rocks of units 1 and 2 and (or) unit 6 may make up the entire member, as in N-1 and N-3 through N-5, or may be interlayered with lesser amounts of rocks of unit 4, with or without small amounts of unit 3 and unit 5, as in N-6 through N-10 and N-14.

The presence of abundant clay minerals implies a rich source of hydrothermally altered or weathered felsic to mafic feldspar-rich rocks. Such rocks are accumulated in a marine environment. The presence of high-aluminum clay minerals in units 1, 2, and 7, coupled with the presence of carbonate deposits in unit 3 (Table 8), suggests a relatively nearby land source of hydrothermally altered felsic rocks or lateritic soils.

The lithologic units of the Marlboro and Nashoba Formations are similar but are built up of different successions and relative thicknesses. This suggests that the Nashoba Formation was formed largely of altered volcanogenic detritus and volcaniclastic debris interlayered with lesser amounts of basaltic pyroclastic rocks and flows (unit 4), related carbonate rocks (unit 3), and sandstone and possibly chert (units 5 and 7) on the lower flank of a volcano or seamount.

Unnamed Formations and Members

Table 8 gives the lithologic composition of the fifteen unnamed formations and members and their variations. These rocks form the stratigraphically upper part of the tunnel section. Table 1 is complementary to Table 8 in providing the systematic stratigraphic data for compiling the aggregate thickness of each lithologic unit in each formation or member. Lithologic units 9 to 15 are compositionally quite different from those that make up the underlying formations (Nashoba and Marlboro). Lithologic units 3, 4, 5, and 6, which occur in the underlying formations, are found interlayered with lithologic units 9 to 15.

The abrupt change from the character of sedimentation and environments represented by the Nashoba and Marlboro Formations to those of the unnamed formations and members suggests that the contact described at station 101+72 is an unconformity.

U-1 through U-9 (see footnote 3) comprise a sequence of related rocks. U-1 consists dominantly of finely layered quartzite (unit 5) with lesser amounts of interbedded calc-silicate rocks and amphibolite. Unit 5 rocks are probably derived from standline sand deposits. The thickness of unit 5 rocks in U-1 (54 m) is greater than the total of such rocks in the Nashoba Formation and about 75 percent greater than the thickest section of such rocks in the Marlboro Formation (namely, 36 m in M-15). The protoliths of the unit 5 rocks in the Marlboro and Nashoba Formations were probably deposited farther offshore and may include recrystallized chert.

The rocks of U-2 through U-9 contain variable amounts of lithologic units 5, 14, and 15. The protoliths of the unit 15 dark schist that contains sulfide minerals, sillimanite, andalusite, and muscovite were derived from high-aluminum, euxinic clay deposits. Unit 15 rocks occur in greater, but variable, abundance in U-8 and U-2 through U-5 than in U-6 and U-7, which contain slightly greater thicknesses of quartz-rich material. The lack of sillimanite-andalusite in the phyllite of U-9 is probably the result of an originally low aluminum content but may have been caused by metamorphic conditions that were different from those that affected U-1 through U-8.

U-10A and U-10B make up the stratigraphic section between the top of U-9 and station 76+42 and the southeastern intrusive contact of the Rattlesnake Hill pluton. U-10A consists of a 33-m-thick distinctive sequence of calc-silicate granulite and brown biotite-quartz granulite. U-10B is feldspar-chlorite schist that is less than 5m thick. U-10A is unique in that nowhere else in the succession is there such a thick sequence formed only of interlayered unit 3 rocks. Comparable thicknesses of calc-silicate granulite are found in N-15, N-19, N-29, M-20, M-21, M-22, and M-23, but there the unit 3 rocks are interlayered with other lithologic units.

The environment of deposition of U-10A was one in which sandy and feldspathic lime mud was being deposited alternately with quartz, carbonate, and feldspar sand and clay enriched in magnesium, iron, and potassium. U-10B represents quartz and feldspar sand that contained iron-rich clay. U-11A, 4.27 m thick, is a feldspar-muscovite-quartz shist that occurs as a xenolith near the southeastern margin of the Rattlesnake Hill pluton in the Rattlesnake Hill fault zone. U-11B, a total of 14.02 m thick, comprises a series of schist septa nearly identical in appearance to U-11A but the two are separated from each other by muscovite granite sills near the northwestern margin of the Rattlesnake Hill pluton. The protoliths of U-11B were either clay-bearing, felsic volcanogenic sedimentary rocks of hydrothermally altered volcanic rocks.

The rocks of U-12A through U-12D form a succession. U-12A consists of an interbedded, 6-m-thick sequence of thin-bedded unit 3 calc-silicate rocks and quartzite; this sequence is succeeded by 23 m of unit 5 quartzite that makes up U-12B. U-12C is a 8-m-thick gray muscovite schist, overlain by U-12D, which consists of 84 m of quartzite. The environment of deposition of these well-banded, quartzite layers was probably a littoral or nearshore area. The protolith of the muscovite schist was a sandy claystone that originated farther offshore.

U-13A through U-13F is a very distinctive sequence of rocks. U-13A is a 13-m-thick sequence of interlayered, interbedded hornblende gneiss and quartz-feldspar gneiss described as lithologic unit 12. This grades into an unusual sequence of blocky, massive to fine-grained, mottled black to gray, layered to massive hornblende-actinolite (U-13B through U-13F, 573 m thick), which may be either an intrusive or a pyroclastic rock. Related rocks exposed at Carville Basin near shaft A are clearly of two types, intrusive and pyroclastic rocks. Banding or layering in parts of the sequence may be the result of crude layering developed in pyroclastic accumulations or of flow banding. The sequence, if pyroclastic, may also enclose sills of comparable composition; however, the contact relationships are not readily recognized. We think that these are pyroclastic rocks of unusual ultramafic composition. They may be ridge-type explosive volcanic rocks that possibly derived their layering in the same way that the first-formed pyroclastic rocks of Surtsey were developed (Sheridan, 1971).

Unit 14, only 12 m thick, is a chlorite-actinolite-calcite gneiss. Its protoliths were derived from an iron-rich clay and sandy lime mud.

U-15 abuts the southeastern margin of the Ayer Granodiorite and consists of 61 cm of hornblende (unit 10) enclosed in 7.62 m of biotite-quartz-feldspar schist (unit 9). Its protoliths were presumably felsic volcanogenic rocks.

Table 9 (see footnote 1) compares these fifteen units with units mapped by Grew (1970) and Peck (1976).

CONCLUSION

The 8-mi-long Wachusett-Marlborough Tunnel exposes an unusually complete section of Paleozoic(?) formations. The protoliths of the Marlboro Formation were dominantly basaltic pyroclastic rocks and lava flows with interlayered volcanogenic sedimentary rocks, carbonate and sandy lime mudstone, sandstone, and chert. The overlying Nashoba Formation was formed basically from the same rock types, but in different proportions. It was mainly a variety of micaceous volcanogenic sedimentary rocks with lesser hornblende-rich rocks and somewhat lesser amounts of carbonate rocks, lime mudstone, and sandstone. The unnamed formations and members in the uppermost part of the stratigraphic sequence consist largely of different lithologic units that are interbedded with lithologic units found also in the lower part of the Marlboro and Nashoba Formations. These uppermost rocks are separated from the underlying formations by a slight angular unconformity. The rocks of the unnamed formations and members were derived from fine-grained strand deposits that grade upward into interbedded quartz-rich rocks and aluminum- and sulfide-rich euxinic shale formed as offshore deposits in an enclosed, noncirculating basin. These sedimentary rocks grade upward into a mixture of sandy lime mudstone, shaly sandstone, and volcanogenic sedimentary rocks. Next are more interbedded sandy lime mudstone and sandstone layers, perhaps succeeded by pyroclastic volcanic rocks that are now a thick sequence of unique hornblende-actinolite rocks. These in turn are succeeded by a thin volcaniclastic sequence.

Detailed stratigraphic data as well as data providing controls for evaluating variations in thickness of the stratigraphic sections are presented in the several tables. Thus, the regional geology may be reconstructed by using the pertinent details for correlation with features on the surface.

ACKNOWLEDGMENTS

M. P. Billings initially suggested that this investigation be undertaken and made all the necessary arrangements with the Metropolitan District Commission (MDC) for carrying out the tunnel study. We also acknowledge Billings's help in the field and laboratory. We are grateful for the insights, discussions, and suggestions of L. R. Page and H. R. Dixon, who spent may weekends examining the tunnel section in detail with one of us (Skehan). We thank L. R. Page for his many helpful suggestions and painstaking editing that have materially improved the manuscript. M. A. Gheith and C. W. Wolfe assisted during the petrographic study of the tunnel specimens. The petrographic study was partially supported by National Science Foundation Grant GP 4969.

Walter W. Correia and Donald L. Guernsey of the Starnd Laboratory of

Massachusetts Institute of Technology allowed us the use of the spectrographic laboratory and gave us valuable assistance and advice. W. H. Dennen provided us with the use of the Cabot Laboratory of M.I.T. for the preparation of samples for spectrochemical analysis.

We also acknowledge helpful discussions with numerous colleagues but especially with D. Alvord, A. Badawy, P. J. Barosh, K. G. Bell, R. W. Bromery, N. P. Cupples, W. L. Fischer, R. Z. Gore, E. S. Grew, R. H. Jahns, M. B. Kane, A. V. LaCroix, C. N. Legarde III, R. J. Martin III, D. L. O'Leary, M. H. Pease, J. H. Peck, and G. Snyder. Many of Skehan's former students assisted with the field work, especially F. Accetta. Help was given by E. B. Murphy, T. J. Nadowski, L. J. Serino, and J. J. Vertic of the Construction Division of MDC and by R. M. Comeau, P. deCaprariis, M. Dewey, T. H. Hastie, N. Hellier, J. J. Gallagher, J. Moran, G. Saulnier, R. J. Smith, and M. L. Gannon.

We are indebted to C. W. Barton, L. Mc. Ferrigan, P. P. Gremillion, R. A. Olson, K. M. Phinney, and C. E. Porter for help in assembling the data and in preparation of the manuscript.

REFERENCES CITED

Abu-moustafa, A. A., 1969, Petrography and geochemistry of the Nashoba Formation of the Wachusett-Marlborough Tunnel, Massachusetts [Ph.D. thesis]: Boston, Mass., Boston Univ., 223 p.

Abu-moustafa, A. A., and Skehan, James W., S. J., 1976, Petrography and geochemistry of the Marlboro Formation, Wachusett-Marlborough Tunnel, east-central Massachusetts, *in* Lyons, P. C., and Brownlow, Arthur, ed., Contributions to the geology of New England: Geol. Soc. America Memoir 146, p. 31-70.

Barosh, P. J., 1972, Structural geology of Rhode Island and southern Massachusetts as revealed by aeromagnetic data: Geol. Soc. America Abs. with Programs, v. 4, p. 444-445.

Bell, K. G., 1972, Interpretation of the aeromagnetic map of northeastern Massachusetts: Geol. Soc. America Abs. with Programs, v. 4, p. 448-449.

Bell, K. G., and Alvord, D. C., 1976, Pre-Silurian stratigraphy of northeastern Massachusetts, *in* Page, L. R., ed., Contributions to New England stratigraphy: Geol. Soc. America Memoir 148, this volume.

Crosby, W. O., 1899a, Geology of the Wachusett Dam and Wachusett Aqueduct Tunnel of the Metropolitan Water Works in the vicinity of Clinton, Massachusetts: Technology Quart., v. 12, p. 68-69.

——1899b, Geological history of the Nashua Valley during the Tertiary and Quaternary Periods: Technology Quart., v. 12, p. 288-324.

Cuppels, N. P., 1964, The Marlboro Formation in the Concord quadrangle, *in* Skehan, James W., S. J., ed., New England Intercoll. Geol. Conf., Guidebook, 56th Ann. Mtg., Boston: p. 81-89.

Emerson, B. K., 1917, Geology of Massachusetts and Rhode Island: U.S. Geol. Survey Bull. 597, 289 p.

Fahlquist, F. E., 1936, Geology of region in which Quabbin Reservoir is located: Ann. Rept. Metropolitan District Water Supply Comm., Boston, Mass., App., p. 1-46.

Fischer, W. L., 1964, A magnetic study of the Wachusett-Marlborough Tunnel area [M.S. thesis]: Chestnut Hill, Mass., Boston College, 74 p.

Gore, R. Z., 1967, An evaluation of the magnetometer in metamorphic terrain [M.S. thesis]: Chestnut Hill, Mass., Boston College, 69 p.

Grew, E. S., 1970, Geology of the Pennsylvanian and pre-Pensylvanian rocks of the Worcester area, Massachusetts [Ph.D. thesis]: Cambridge, Mass., Harvard Univ. 263 p.

Hansen, W. R., 1956, Geology and mineral resources of the Hudson and Maynard quadrangles, Massachusetts: U.S. Geol. Survey Bull. 1038, 104 p.

Heath, M.M.M., 1965, Geochronology of the Ayer Granite in the Wachusett-Marlborough Tunnel, Clinton, Massachusetts, *in* Variations in isotopic abundance of strontium, calcium, and argon and related topics, U.S. Atomic Energy Comm. Contract AT (30-1)-1381: Cambridge, Massachusetts Inst. Tech. Dept. Geology and Geophysics, p. 15-19.

Hitchcock, E., 1832, Report on the geology of Massachusetts: Am. Jour. Sci., v. 22, p. 1-70.

——1841, Final report on the geology of Massachusetts, Vol. I and II: Amherst, Mass., Amherst College, vol. I., 299 p.; Northampton, Mass., Smith College, vol. II, 530 p.

Lacroix, A. V., 1968, Structure and contact relationships of the Marlboro Formation, Marlboro, Massachusetts [M.S. thesis]: Chestnut Hill, Mass., Boston College, 83 p.

Legarde, C. N., III, 1967, The petrography, structural geology and tectonic history of a portion of the Wachusett-Marlborough Tunnel area, Massachusetts [M.S. thesis]: Chestnut Hill, Mass., Boston College, 100 p.

Page, L. R., 1968, Devonian plutonic rocks in New England, *in* Zen, E., and others, eds., Studies of Appalachian geology, northern and maritime: New York, John Wiley & Sons, Inc., p. 371-383.

Peck, J. H., 1976, Silurian and Devonian stratigraphy in Clinton quandrangle, central Massachusetts, *in* Page, L. R., ed., Contributions to New England stratigraphy: Geol. Soc. America Memoir 148, this volume.

Perry, J. H., and Emerson, B. K., 1903, The geology of Worcester, Massachusetts, Worcester, Mass.: Worcester Nat. Hist. Soc., 166 p.

Renard, A., 1878, Sur la structure et la composition mineralogique du coticule: Acad. Royale Belgique, Mem. Sav. Etr. tome XLI.

Sheridan, M. F., 1971, Textural analysis of Surtsey tephra: A preliminary report: Surtsey Research Soc. Prog. Rept., no. 6, p. 150-151.

Skehan, James W., S. J., 1967, Geology of the Wachusett-Marlborough Tunnel, east-central Massachusetts, A preliminary report, *in* Farquhar, O., ed., Economic geology in Massachusetts—Conf., Amherst, Proc.: Amherst, Massachusetts Univ. Graduate School, p. 237-244.

——1968a, Evidence of easterly tectonic transport in the Appalachians of eastern Massachusetts: Geol. Soc. America Spec. Paper 115, p. 292.

——1968b, Fracture tectonics of southeastern New England as illustrated by the Wachusett-Marlborough Tunnel, east-central Massachusetts, *in* Zen, E., and others, eds., Studies of Appalachian geology, northern and maritime: New York, John Wiley & Sons, Inc., p. 281-290.

——1969a, Recognition of faults of regional importance by mapping of joints: Geol. Soc. America Spec. Paper 121, p. 376-377.

——1969b, Tectonic framework of southern New England and eastern New York, *in* Kay, Marshall, ed., North Atlantic geology and continental drift: Am. Assoc. Petroleum Geologists Mem. 12, p. 793-814.

Zietz, Isidore, Gilbert, F., and Kirby, J. R., compilers, 1972, Aeromagnetic map of New England: U.S. Geol. Survey Open-File Map, scale 1:250,000.

MANUSCRIPT RECEIVED BY THE SOCIETY APRIL 18, 1974
REVISED MANUSCRIPT RECEIVED FEBRUARY 18, 1975
MANUSCRIPT ACCEPTED MARCH 14, 1975

Printed in U.S.A.

Geological Society of America
Memoir 148

Silurian and Devonian Stratigraphy in the Clinton Quadrangle, Central Massachusetts

JOHN H. PECK*
U.S. Geological Survey
80 Broad Street
Boston, Massachusetts 02110

ABSTRACT

Metasedimentary rocks of probable Silurian to Devonian age underlie about three-fourths of the Clinton quadrangle. Metamorphosed marine mudstone, siltstone, fine-grained graywacke, and limy siltstone form two thick sequences. Primary structures preserved within the rocks include compositional layering, laminations, tabular and lenticular beds, low-angle cross-stratification, and graded beds. The thicker sequence (4,570 m to 6,100 m) is divided into four units (in ascending order): (1) white to light-gray, very fine grained quartzite interlayered with minor thin beds of gray phyllite; (2) alternating very thin laminae of dark-gray to greenish-gray phyllite and light-brown ankeritic metasiltstone; (3) dark-gray carbonaceous slate and phyllite and minor interbedded metagraywacke; and (4) dark-gray phyllite with abundant chiastolite porphyroblasts and many beds of light-gray metagraywacke. Intrusive igneous rocks form the base of this sequence, and a fault at the top separates it from the other sequence. The second, thinner sequence (1,830 m) consists chiefly of thin-bedded biotitic feldspathic quartzite, greenish- to purplish-gray calc-silicate rock, and thin layers of biotite schist. These rocks are probably older than the rocks of the first sequence. The rocks of the second sequence previously were included in the Oakdale Quartzite, those of the first sequence in the Oakdale Quartzite and the Worcester Formation of central Massachusetts. They may be broadly correlative with parts of the Gonic, Rindgemere, Towow, Berwick and Eliot Formations and possibly with part of the Kittery Quartzite of southern New Hampshire and southwestern Maine.

*Present address: Stone and Webster Engineering Corp., P.O. Box 2325, Boston, Massachusetts 02107.

INTRODUCTION

The Clinton quadrangle is in east-central Massachusetts about 59.5 km west of Boston (Fig. 1). Mapping of the quadrangle is part of a cooperative program between the U.S. Geological Survey and the Massachusetts Department of Public Works. The general distribution of the major rock units is shown in Figure 1.

Rock units of probable Silurian to Devonian age are designated by number rather than name because the formal stratigraphic nomenclature is in a state of flux, and final designations have yet to be proposed. The numbering system is based on geographic distribution in the quadrangle, rather than on age relations. Emerson (1917) mapped rocks of unit 5, unit 1, and part of unit 2 as Oakdale Quartzite, part of units 2 and 4 and all of unit 3 as Worcester Phyllite, and part of unit 4 as chiastolite schist facies of the Worcester. Both the Oakdale Quartzite and the Worcester Phyllite were considered to be Carboniferous in age.

No fossils have been found in any of the units under consideration, and the Silurian(?)-Devonian age designation is based on regional correlation. The rocks are on strike with similar metasedimentary rocks in southeastern New Hampshire considered by Billings (1956, p. 9) to be Silurian(?) and Devonian (see also Sriramadas, 1966; Freedman, 1950; Sundeen, 1971). The units originally were marine silts and clays and strongly resemble fossil-bearing rocks in the vicinity of Skowhegan and Waterville, Maine (Ludman, this volume; Osberg, 1968; Osberg and others, 1968).

Radiometric age determinations on the Ayer Granite which intrudes unit 1 do not support the age based on correlation. The lead-alpha age of zircon from the Ayer in the Clinton quadrangle ranges from 520 ± 60 m.y. to 410 ± 50 m.y. (Zartman and others, 1965, p. D5). The U-Th-Pb isotopic age of zircon from the Ayer Granite is about 435 ± 10 m.y., according to recent work by Zartman (1971, written commun.). The apparent discrepancy between the radiometric age and the age based on regional correlation has not been resolved.

The major structures in the quadrangle are the Clinton-Newbury thrust fault zone (Skehan, 1968) on the east side, a homoclinal sequence on the east limb of the Merrimack synclinorium, and a normal fault in the northwest part of the quadrangle. The Clinton-Newbury zone separates high-grade (sillimanite-muscovite) metasedimentary and metavolcanic rocks of pre-Silurian(?) age from low- and medium-grade (chlorite to andalusite) Silurian(?)-Devonian(?) rocks of the Merrimack synclinorium.

A generalized columnar section of the Silurian(?)-Devonian(?) rock units is shown in Figure 2. Units 1 through 4 constitute an apparently conformable sequence that trends generally north and dips west. Primary sedimentary features indicate that the sequence is part of a west-facing homocline. Unit 5 is separated from unit 4 by a high-angle normal fault of regional extent (Novotny, 1961, p. D48). Rocks of unit 5 strike northeasterly and dip moderately to the southeast. Emerson (1917) mapped the rocks of unit 5, unit 1, and presumably part of unit 2 as the Oakdale Quartzite overlain by the Worcester Phyllite (most of unit 2 and units 3 and 4). Novotny (1961) concluded that the rocks of unit 5 (Oakdale) are older than units 1 through 4. Unit 5 is probably older than unit 1, but until more detailed mapping is completed on both sides of the fault north and west of the Clinton quadrangle, the exact stratigraphic relation of unit 5 to the other sequence remains obscure.

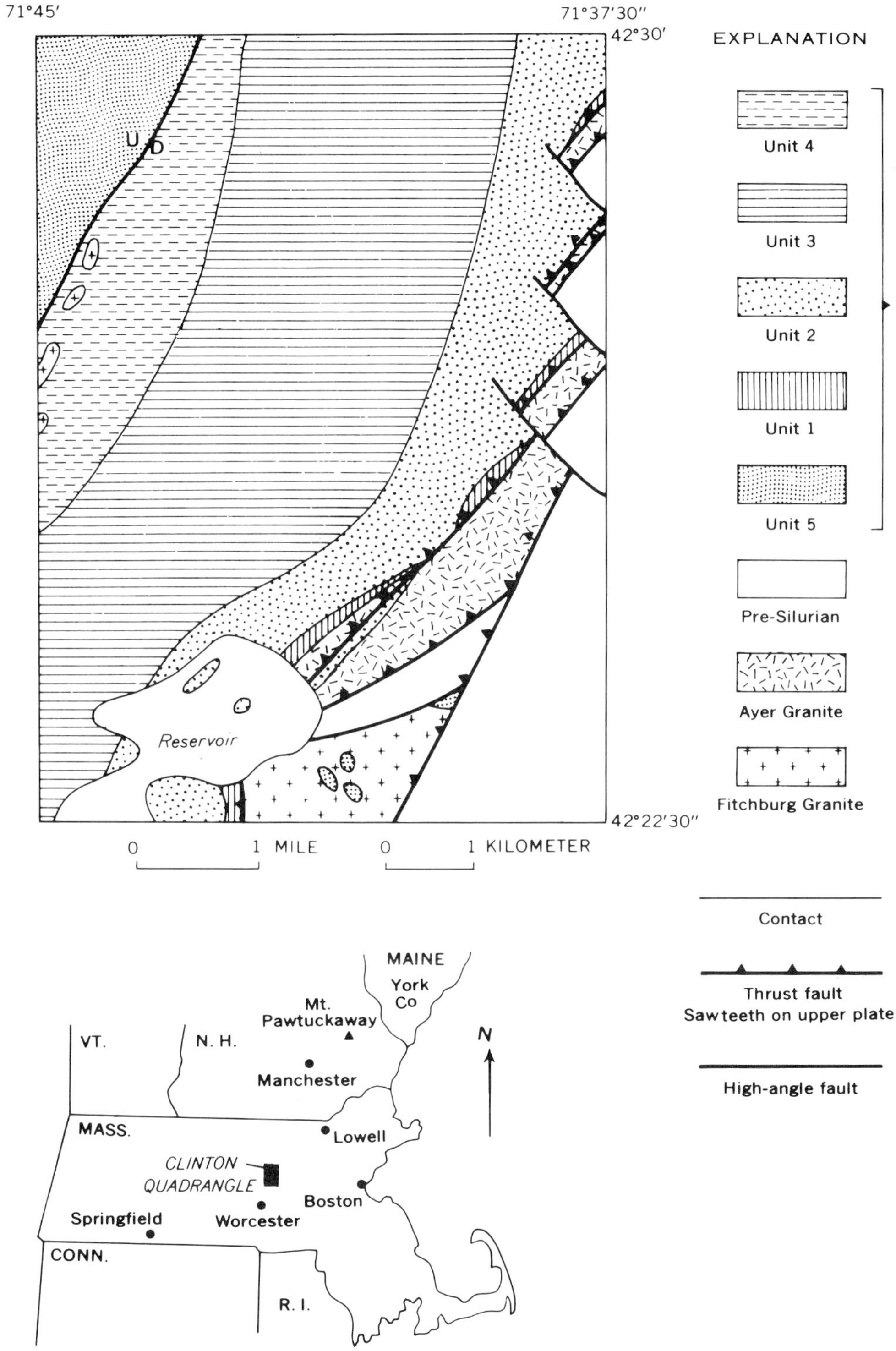

Figure 1. Index map and generalized geologic map of the Clinton quadrangle, Massachusetts.

UNIT	THICKNESS IN METRES	SECTION	PRESENT LITHOLOGY	ORIGINAL LITHOLOGY	PRIMARY STRUCTURES	ENVIRONMENT OF DEPOSITION
4	1220–1830	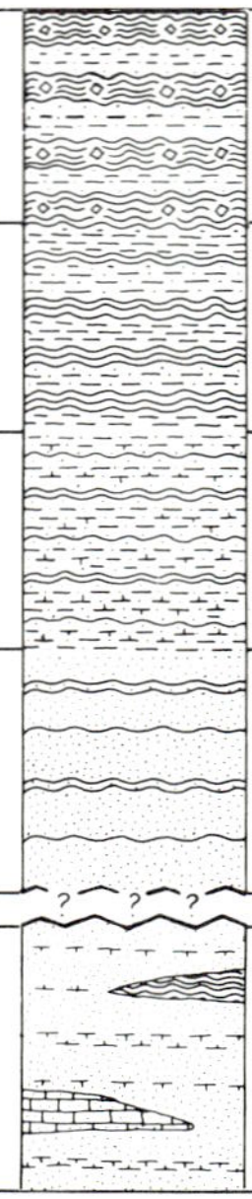	Gray chiastolite-bearing phyllite; interbedded metagraywacke.	Carbonaceous mud-stone and graywacke.	Tabular beds 1.25 cm to 1.25 m thick; graded bedding; low-angle crosslaminations in graywacke.	Low-energy deep marine basin with periodic currents and restricted circulation.
3	1830–2440		Dark-gray slate and phyllite; minor meta-graywacke.	Carbonaceous mud-stone and minor graywacke.	Tabular beds 0.6 cm to 0.6 m thick; graded beds; thin laminations; crosslaminations in graywacke.	Low-energy deep marine basin; reducing conditions; turbidity currents.
2	1220–2130		Laminated metasiltstone and greenish phyllite; calcareous metasiltstone.	Thinly laminated siltstone, shale, and calcareous siltstone.	Crossbeds, thin parallel laminations; tabular beds 0.3 cm to 0.3 m thick; cyclic banding.	Low-energy deep marine basin; with few currents; cyclic deposition.
1	0–91		Light-gray quartzite; some interbedded gray phyllite.	Quartzose siltstone and silty mudstone.	Tabular beds 5 cm to 1.5 m thick; thin laminations in phyllite.	Winnowed shelf area at moderate depth; muds accumulating adjacent to silt shoals.
5	1830+		Biotitic, feldspathic quartzite; calc-silicate beds, minor biotite schist.	Arkosic siltstone and fine sandstone; limy siltstone and minor shale.	Tabular and lenticular beds up to 7.6 cm thick; crossbeds, rare ripple marks, thin laminations.	Moderately deep marine with some currents; reworked volcanic debris, carbonate.

Figure 2. Generalized columnar section of Silurian(?) and Devonian(?) rocks in the Clinton quadrangle, Massachusetts.

DESCRIPTION OF ROCK UNITS

Unit 5

Unit 5 consists of about 1,830 m of thin-bedded feldspathic quartzite with interbedded calc-silicate rock and subordinate beds and lenses of biotite schist. The feldspathic quartzite is mostly a very fine grained, tan to brown equigranular rock containing abundant reddish-brown biotite disseminated through the groundmass. Chlorite and calcite are locally common. Feldspar makes up as much as 35 percent of the rock. Greenish-gray calc-silicate beds are thin and commonly lenticular. Thin biotite schist layers are interleaved throughout the unit, but are most abundant in the lower part near the west border of the quadrangle. Most schist beds contain staurolite and garnet. Bedding in unit 5 is well defined, trends generally northeastward, and has a moderate to steep southeast dip.

Originally the feldspathic quartzite was feldspathic or arkosic silt and fine sand. Possibly these sediments were derived from fresh volcanic terrane because the sediment type is very immature. Many layers have cross-bedding and some are scoured. Thin lenticular beds and laminae are also common and are probably indicative of current deposition, as are the ripple marks or structures resembling ripple marks in a few beds. Graded beds are rare. The calc-silicate beds indicate periodic deposition of limy sediments within the mostly arkosic deposits, whereas schist layers represent influxes of clayey material.

Unit 5 was probably deposited in a moderately deep marine environment far enough offshore to have been beyond sites of very coarse clastic accumulations, but shallow enough to experience active currents. The environment was one of moderately high energy; deposition was rapid and very little reworking took place. The location was probably a fairly active basin adjacent to volcanic terrane.

Unit 5 crops out poorly within the Clinton quadrangle, and no subdivision has been attempted. The top and bottom stratigraphic boundaries of the unit are not present within the quadrangle. It is separated from unit 4 by a steeply dipping fault.

Unit 1

Unit 1 is chiefly light gray quartzite. The quartzite is very finely crystalline, thin to thick tabular bedded, resistant, well sorted, and clean. The quartzite contains dark-gray phyllite and lenticular pods, and the phyllite also is present as thicker beds and sequences of beds above and below the main body of quartzite. The quartzite probably represents metamorphosed silt-shoal deposits formed by currents winnowing out clays that were then deposited as muds surrounding the large lenses of clean silt. The depositional environment was below wave base in restricted or protected areas where very low velocity currents or oscillatory motions could effectively winnow clay from silt.

As mapped, unit 1 ranges from zero to 91 m in thickness in the quadrangle, but averages about 46 m thick in most places. Parts of the variation in thickness may be due to lenticular deposition originally, but most of the variation is due to faulting or igneous intrusion. Unit 1 grades upward into unit 2.

Unit 2

Unit 2 is composed chiefly of interlayered ankeritic metasiltstone and phyllite but has many thin beds of calcareous metasiltstone. Thin laminae of alternating lithologies are the most conspicuous and characteristic feature of the unit (Fig. 3). The metasiltstone is brownish gray to light gray, fine grained, mostly well sorted, and consists dominantly of quartz but has some feldspar. Ankeritic spots and ankerite-rich laminae are abundant within the silty beds and on weathering impart a limonitic stain and distinctive spotted brown appearance to the rock. The phyllite is a dark greenish-gray, medium gray, and locally light greenish-gray, very fine grained rock composed mostly of quartz, sericite, and chlorite. The calcareous metasiltstone is similar in appearance and composition to the other metasiltstone except that calcite makes up as much as 15 percent of the rock. Natural outcrops of unit 2 are characterized by the distinct brown weathering of the silty layers and greenish to grayish tints in the phyllite layers. Fresh rock is mostly gray. Crinkle folding (Fig. 3) is characteristic of most outcrops of unit 2.

The bedding throughout this unit shows cyclic deposition of silt-clay couplets in a slightly carbonate-producing environment. In thin section, some of the couplets are poorly graded from silt at the base to clay at the top, but most couplets seem to be separate laminae of different composition with sharp boundaries. The persistence of laminations no more than a few millimeters thick for many meters along strike indicates that most of this unit was laid down in a marine environment of very low energy. The cyclic deposition was probably either seasonal or controlled by some similar periodic process.

The thickness range of unit 2 within the Clinton quadrangle is 1,220 m to 2,130

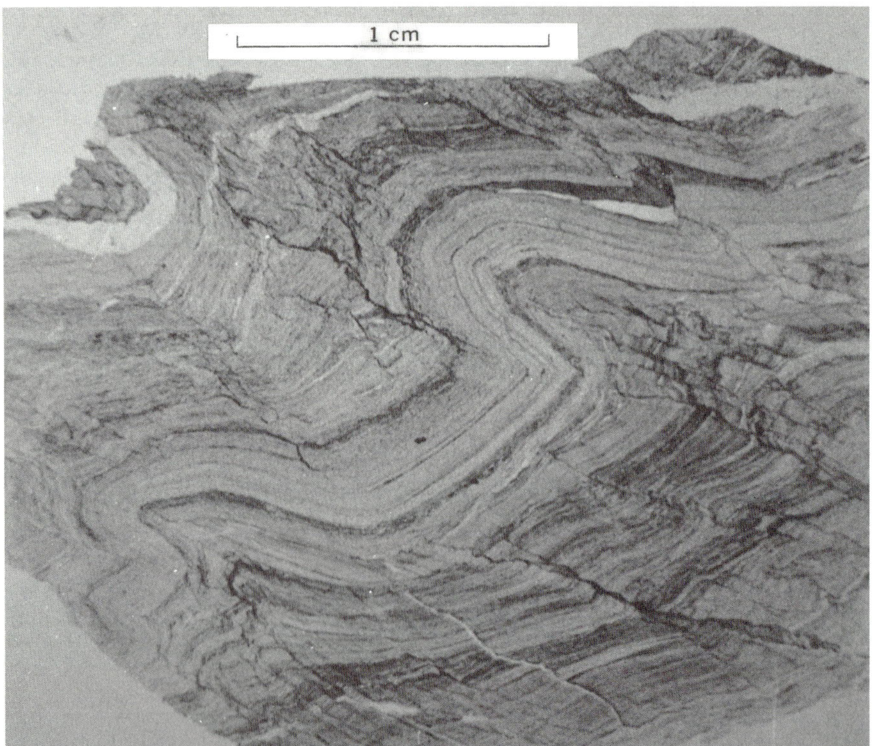

Figure 3. Photograph of thin section showing characteristic crinkled laminations in rocks of unit 2. Light-colored layers are quartzose metasiltstone.

m (Fig. 2). The contact between unit 2 and unit 3 is not exposed in the quadrangle but is assumed to be conformable because bedding is parallel and topping sense is the same on both sides of the contact. The contact is exposed in the Shirley quadrangle about 1.6 km north of the Clinton quadrangle boundary. There the rock types characteristic of the two units are interlayered through a section greater than 20 m thick.

Unit 3

Unit 3 is dominantly dark-gray slate and phyllite and light-gray metagraywacke. In contrast to unit 2, the rocks of unit 3 are carbonaceous and have few iron-bearing minerals. The slate and phyllite consist mostly of quartz, sericite, and black carbonaceous dust. The metagraywacke is mostly silt-sized quartz, feldspar, sericite, biotite, and carbonaceous material. Thin layers of graphite are present in some places. Tongue-shaped metasiltstone bodies as much as 7.6 m thick are found in unit 3 near the southwestern corner of the quadrangle.

The slate and phyllite are mostly in thick-bedding sets that internally contain laminations. Bedding in the slate is obscure in many places. Graded beds are abundant throughout the unit (Fig. 4), and sequences of graded beds alternate with sequences of more homogeneous slate or phyllite. Most graded beds have metagraywacke at the base, but some beds of metagraywacke are not graded. Primary cross-stratification (Fig. 5) is often present in the metagraywacke. Rare tongue-shaped metasiltstone bodies show little or no internal primary structures and may be submarine slump deposits. They range from about 1 m to as much as 7.6 m in thickness and wedge out rapidly in an updip direction.

The environment of deposition of unit 3 was probably one of relatively low energy in deep marine water that was receiving silt and clay rich in organic material. The general carbonaceous aspect and local graphite layers indicate a reducing environment. Graded beds suggest that much of the sediment was deposited in the basin by turbidity currents. Cross-laminations in metagraywacke, thinning of graded beds eastward, and the tongue shape of some metasiltstones all indicate that the basin was being filled from the west.

Unit 3 is probably 1,830 m to 2,440 m thick in the Clinton quadrangle. Several folds within the unit and wide areas of no outcrop make thickness determinations approximate only. Unit 3 grades upward into unit 4 by an increase in the thickness of metagraywacke beds and a decrease in the amount of slate. The change is gradual through about 90 m to 120 m, and the contact is placed at the lowest occurrence of metagraywacke beds more than .60 m thick.

Unit 4

Unit 4 is composed mostly of thin beds of dark-gray laminated phyllite containing abundant chiastolite or andalusite porphyroblasts and many interlayered thin to thick beds of silt-sized light-gray metagraywacke. Carbonaceous dustlike material is nearly everywhere present in the phyllite and metagraywacke but is more abundant in the phyllite. The chiastolite porphyroblasts are mostly euhedral to subhedral, commonly .635 cm to 1.27 cm across, and about 7.35 cm long. They are randomly oriented with respect to bedding and foliation except that in beds grading from metagraywacke at the base to phyllite at the top the porphyroblasts are restricted to the upper part. The phyllite is not so well laminated as the phyllite in unit 3 because of disruption of the laminae by the chiastolite. The phyllite beds are mostly less than 15.24 cm thick. Most beds of metagraywacke are cross-laminated

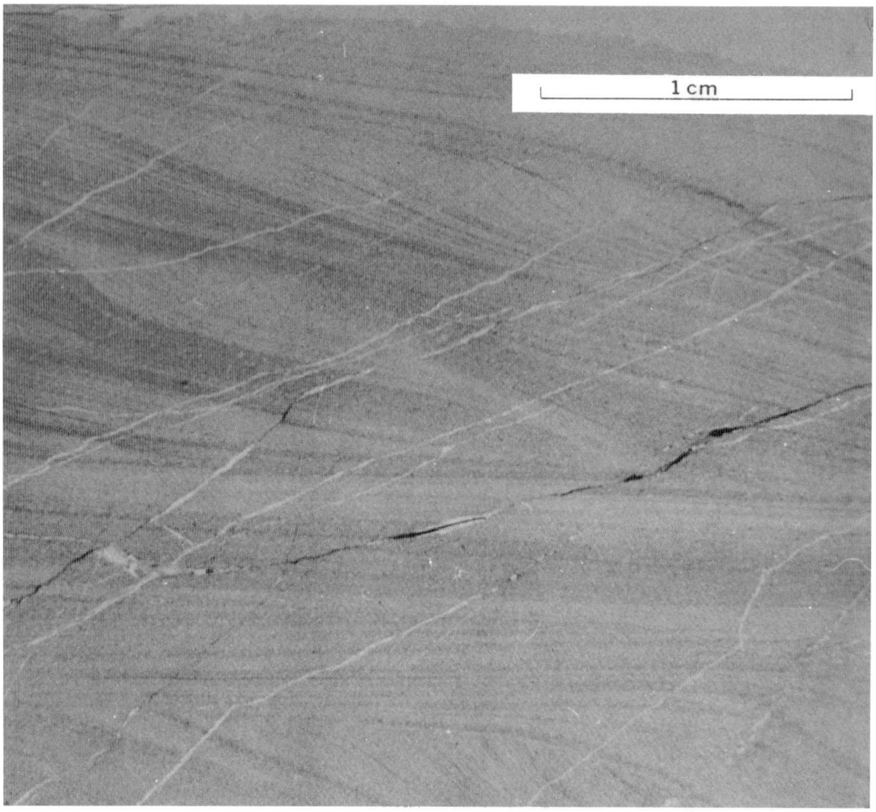

Figure 4. Photograph of thin section showing graded beds in fine-grained rocks of unit 3. Light bands consist of fine silt-sized quartz and feldspar grains; darker bands consist of quartz and sericite grains and carbonaceous dust. Tops to the right.

and range from about 7.35 cm to .30 m in thickness, but some beds are as much as 12.1 m thick. Tabular graded beds are very common in unit 4 and commonly contain cross-laminations in the lower or middle parts.

Unit 4 is distinguished from unit 3 on the basis of more and thicker beds of metagraywacke and the abundance of chiastolite or andalusite porphyroblasts. The units are, however, gradational, and the contact is difficult to locate precisely. In the Clinton quadrangle, chiastolite appears in the rock sequence at or near the boundary between unit 3 and unit 4. The chiastolite is a contact metamorphic mineral resulting from the intrusion of unit 4 by the Fitchburg Granite. In areas of good exposures, the metamorphic boundary (chiastolite-no chiastolite) is seen to cut across the bedding at a low angle. However, within the quadrangle at least, the presence of chiastolite is a useful field guide for recognition of unit 4 rocks in areas of poor exposures.

The environment of deposition of unit 4 was much the same as that of unit 3 except that the basin was somewhat shallower and energy may have been somewhat higher. Turbidity currents were the primary depositional agent, bringing in slightly coarser sediment than that of unit 3. The pelitic material was rich in aluminous minerals but poor in iron; hence there are relatively small amounts of chlorite, biotite, or garnet and an abundance of chiastolite or andalusite in the metamorphosed rock. The cross-laminations in the metagraywacke indicate a western source of sediment. Unit 4 has a maximum thickness of about 1,830 m within the quadrangle. The upper boundary is a fault.

POSSIBLE CORRELATIONS

The rocks in the Clinton quadrangle are on strike with rock units in the Worcester area mapped recently by Grew (1970). The Pennsylvanian rocks at the Worcester "coal mine" occupy a wedge along a fault and are younger than all other metasedimentary rocks in the Worcester area except possibly some small bodies of metaconglomerate (Grew, 1970, 1971). Graphitic coaly material in fault gouge was recently found in borings made at Memorial Hospital in Worcester, approximately 2,740 m southwest of the "coal mine" (D. F. Goodell, 1972, oral commun.). The presence of the coaly material in the borings substantiates Grew's conclusion that the Pennsylvanian coal-bearing rocks are only in fault slivers in the Worcester area. No coal-bearing or fossiliferous rock has been found in the Clinton quadrangle, and none of the rock units there can be correlated with the Pennsylvanian Worcester Formation as redefined by Grew (1970). A tentative correlation of the Clinton units with those of the Worcester area is made in Figure 6.

Correlation of the Clinton units with rocks southwest of the Worcester area on a unit-for-unit basis cannot be made at this time because of unresolved structural

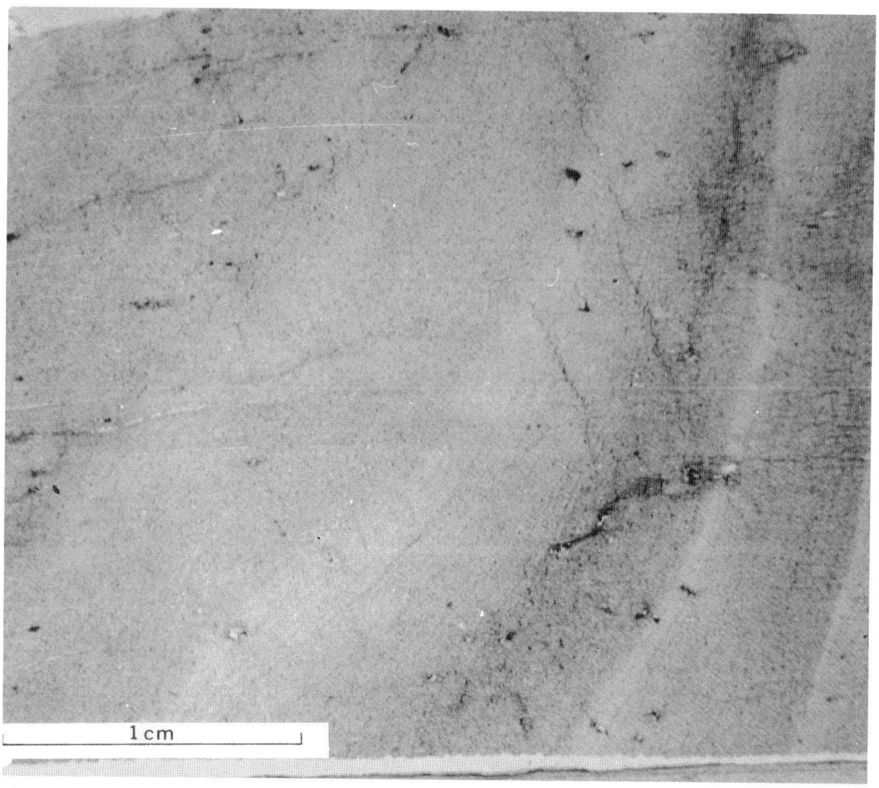

Figure 5. Photograph of thin section showing cross-laminated metagraywacke from unit 3. Laminations are outlined by concentrations of carbonaceous dust. Currents moved from left to right and changed direction slightly as sedimentation progressed.

Age	CLINTON QUADRANGLE	WORCESTER AREA Grew (1970)	WORCESTER AND CLINTON AREAS Emerson (1917)
Pennsylvanian		Worcester Formation Harvard Conglomerate	Worcester Phyllite
Silurian (?) and Devonian (?)	fault	unconformity	
	Unit 4	Holden Formation (part)	Chiastolite schist facies of Worcester Phyllite
	Unit 3		Worcester Phyllite
Silurian (?)	Unit 2	Oakdale Formation (part)	Oakdale Quartzite Worcester Phyllite
	Unit 1	Tower Hill Quartzite Member of Boylston Formation	Oakdale Quartzite (part)
	?		
		Boylston Formation	Boylston Schist Worcester Phyllite
	?		
	Unit 5	Oakdale Formation (part)	Oakdale Quartzite Paxton Quartz Schist
	?		
	?	Boylston Formation	Boylston Schist Brimfield Schist
Pre - Silurian (?)	Pre - Silurian rocks	Nashoba Formation	Gneisses and schists of undetermined age

Figure 6. Tentative correlation chart of units in the Clinton quadrangle and the Worcester area, Massachusetts. Query in Clinton quadrangle column indicates relation uncertain. Relative age of the Worcester Formation and the Harvard Conglomerate is uncertain (Grew, 1970).

problems. By projection along strike, the units in the Clinton area may correlate with the lower part of the stratigraphic section described by Peper and Pease (this volume) in northeastern Connecticut.

The Clinton units are also on strike with rocks included in the Merrimack Group [Silurian(?)] of Billings (1956) in southeastern New Hampshire, but detailed correlation on a unit-for-unit basis does not seem possible now. Sriramadas (1966) mapped some of the rocks east of the Massabesic Gneiss (Fitchburg pluton) in the Manchester quadrangle, New Hampshire, as Littleton Formation which he considered to be of Devonian age. Freedman (1950) also mapped a more extensive

Littleton Formation in the Mount Pawtuckaway quadrangle, New Hampshire. Sundeen (1971) believed that much of Freedman's Littleton belonged in the Merrimack Group. Some of the Clinton rocks may be correlative with Sriramadas' and Freedman's Littleton Formation, but it is still unclear whether the rocks east of the Massabesic Gneiss of Sriramadas in New Hampshire are Devonian.

It seems possible then to correlate the Clinton rocks by way of the Merrimack Group and Littleton(?) rocks in southeastern New Hampshire with the section in York County, Maine (Hussey, 1962, 1968). Lithologically, units 5, 1, and 2 of the Clinton quadrangle could fit in the Kittery, Eliot, and Berwick part of Hussey's section (1962, p. 19-26), but not on a unit equivalent basis. Units 3 and 4 of the Clinton area most closely resemble the Gonic, Rindgemere, and Towow Formations of the York County section (Hussey, 1962, p. 26-31; 1968, p. 294-297).

All the above correlations except possibly those in Figure 6 are questionable to some degree because there are no fossils in any of the rocks discussed. The Silurian(?)-Devonian(?) age designation is mostly based on projection along strike with rocks tentatively correlated by further projection and lithic similarity to fossil-bearing rocks in the Waterville, Maine, area (see Billings, 1956, p. 99-105; Osberg, 1968). The radiometric age data from the Clinton quadrangle, mentioned above argue for a Late Ordovician or older age rather than a Silurian-Devonian age for the Clinton rocks. The apparent discrepancy between the radiometric age data and the regional correlations remains to be worked out.

REFERENCES CITED

Billings, M. P., 1956, The geology of New Hampshire, Part II, Bedrock geology: New Hampshire Plan. Devel. Comm., 203 p.

Emerson, B. K., 1917, Geology of Massachusetts and Rhode Island: U.S. Geol. Survey Bull. 597, 289 p.

Freedman, Jacob, 1950, Stratigraphy and structure of the Mount Pawtuckaway quadrangle, southeastern New Hampshire: Geol. Soc. America Bull., v. 61, p. 449-492.

Grew, E. S., 1970, Geology of the Pennsylvanian and pre-Pennsylvanian rocks of the Worcester area, Massachusetts [Ph. D. thesis]: Cambridge, Harvard Univ., 263 p.

——1971, Stratigraphy of the Worcester area of Massachusetts: Geol. Soc. America Abs. with Programs, v. 3, p. 33-34.

Hussey, A. M., 2d, 1962, The geology of southern York County, Maine; Maine Geol. Survey Spec. Geol. Studies Ser., no. 4, 67 p.

——1968, Stratigraphy and structure of southwestern Maine, *in* Zen, E-an, White, W. S., Hadley, J. B., and Thompson, J. B., Jr., eds., Studies of Appalachian geology, northern and maritime: New York, Interscience Pubs. Inc., p. 291-301.

Ludman, Allan, 1976, Fossil-based stratigraphy in the Merrimack synclinorium, central Maine, *in* Page, L. R., ed., Contributions to the stratigraphy of New England: Geol. Soc. America Mem. 148 (this volume).

Novotny, R. F., 1961, A regional fault in east-central Massachusetts and southern New Hampshire: U.S. Geol. Survey Prof. Paper 424-D, p. D48-D49.

Osberg, P. H., 1968, Stratigraphy, structural geology, and metamorphism of the Waterville-Vassalboro area, Maine: Maine Geol. Survey Bull. 20, 64 p.

Osberg, P. H., Moench, R. H., and Warner, Jeffrey, 1968, Stratigraphy of the Merrimack synclinorium in west-central Maine, *in* Zen, E-an, White, W. S., Hadley, J. B., and Thompson, J. B., Jr., eds., Studies of Appalachian geology, northern and maritime: New York, Interscience Pubs. Inc., p. 241-253.

Peper, J. D., and Pease, M. H., Jr., 1976, Summary of stratigraphy in the Brimfield area, Connecticut and Massachusetts, *in* Page, L. R., ed., Contributions to the stratigraphy of New England: Geol. Soc. America Mem. 148 (this volume).

Skehan, J. W., 1968, Fracture tectonics of southeastern New England as illustrated by the Wachusett-Marlborough Tunnel, east-central Massachusetts, *in* Zen, E-an, White, W. S., Hadley, J. B., and Thompson, J. B., Jr., eds. Studies of Appalachian geology, northern and maritime: New York, Interscience Pubs. Inc., p. 281–290.

Sriramadas, Aluru, 1966, Geology of the Manchester quadrangle, New Hampshire: New Hampshire Dept. Resources and Econ. Devel. Bull. 2, 78 p.

Sundeen, D. A., 1971, The bedrock geology of the Haverhill fifteen-minute quadrangle, New Hampshire: New Hampshire Dept. Resources and Econ. Devel. Bull. 5, 125 p.

Zartman, Robert, Snyder, George, Stern, T. W., Marvin, R. F., and Buckman, R. C., 1965, Implications of new radiometric ages in eastern Connecticut and Massachusetts: U.S. Geol. Survey Prof. Paper 525-D, p. D1–D10.

MANUSCRIPT RECEIVED BY THE SOCIETY APRIL 18, 1974
REVISED MANUSCRIPT RECEIVED FEBRUARY 18, 1975
MANUSCRIPT ACCEPTED MARCH 14, 1975

Printed in U.S.A.

Geological Society of America
Memoir 148

Summary of Stratigraphy in the Brimfield Area, Connecticut and Massachusetts

JOHN D. PEPER
U.S. Geological Survey
National Center (925)
Reston, Virginia 22092

AND

M. H. PEASE, JR.
U.S. Geological Survey
80 Broad Street
Boston, Massachusetts 02110

ABSTRACT

A thick, lithologically varied sequence of high-grade metamorphic rocks, pre-Acadian in age, has been delineated in the Southbridge, Wales, Westford, Eastford, Stafford Springs, and Monson quadrangles of south-central Massachusetts and adjacent Connecticut. The sequence consists largely of metamorphosed aluminous mudstone and siliceous siltstone. Metamorphosed limy mudstone, euxinic mudstone, fine-grained volcaniclastic debris, and lava form marker units that may extend tens of kilometres laterally. The sequence is divided into four formations, each containing many lenticular units. Three formations in the southeastern part of the area are separated by thrust faults and apparently top westward in an imbricate homoclinal sequence. The fourth and youngest formation is in isoclinal synclines in a faulted block that parallels the eastern border of the Monson Gneiss. The observed rock sequences contain rocks similar in gross lithology to sequences present in east-central Massachusetts and southern New Hampshire.

INTRODUCTION

Since 1964, a lithologically varied sequence of high-grade metamorphic rocks has been mapped in northeast-central Connecticut and adjacent Massachusetts (Fig. 1) by workers of the U.S. Geological Survey as part of cooperative mapping programs with the Massachusetts Department of Public Works and the Connecticut State Geological and Natural History Survey. These rocks are in a fault-bound wedge east of the Bronson Hill anticlinorium and are shown as Brimfield Schist and Hebron Gneiss on preliminary geologic maps of Connecticut by Rodgers and others (1956) and Dixon and Lundgren (1968), and as Brimfield Schist and Paxton Quartz Schist on Emerson's (1917) geologic map of Massachusetts and Rhode Island. Our paper briefly summarizes the metamorphic rock sequence in the Brimfield area, giving particular attention to protoliths (original rock types). The Brimfield area, as herein defined, comprises parts of the Southbridge, Monson, Stafford Springs, South Coventry, Spring Hill, and Hampton quadrangles, and the Eastford, Westford, and Wales quadrangles—all in Connecticut and (or) Massachusetts (Fig. 1).

The original materials (protoliths) from which these high-grade metamorphic rocks

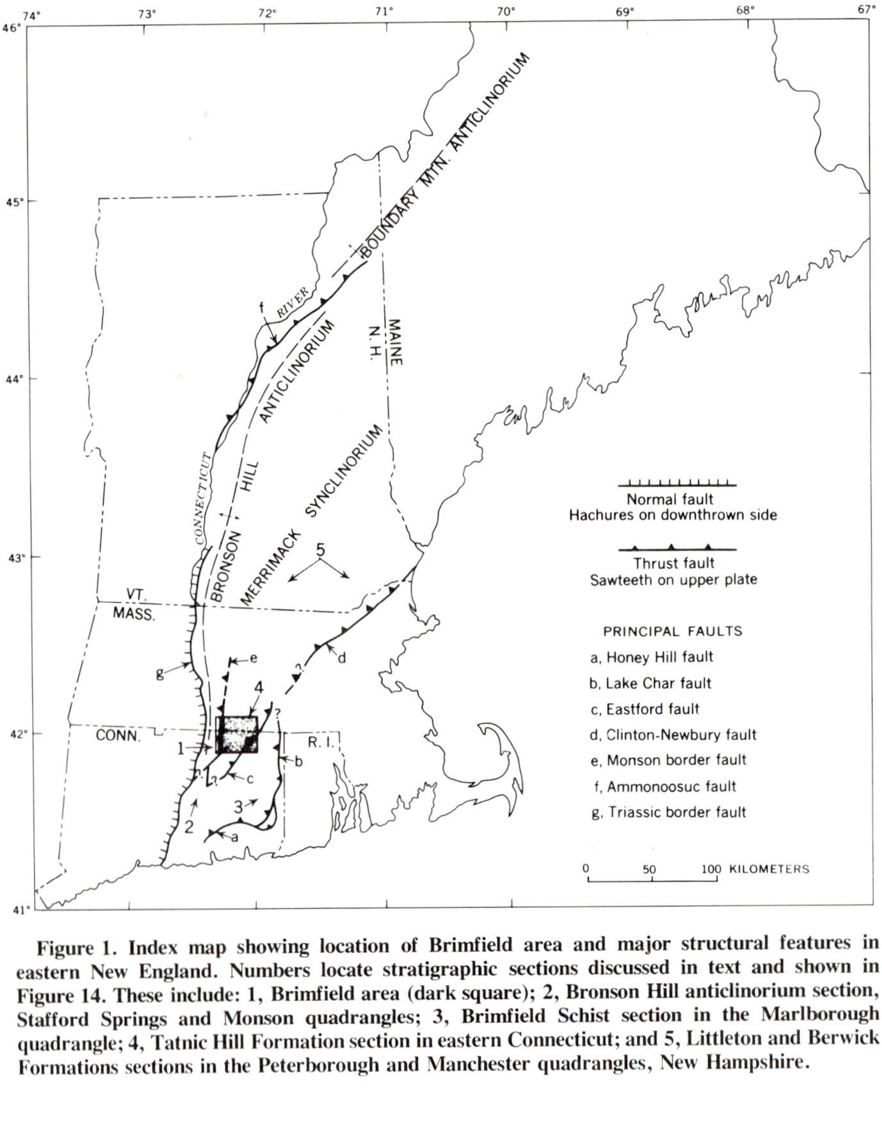

Figure 1. Index map showing location of Brimfield area and major structural features in eastern New England. Numbers locate stratigraphic sections discussed in text and shown in Figure 14. These include: 1, Brimfield area (dark square); 2, Bronson Hill anticlinorium section, Stafford Springs and Monson quadrangles; 3, Brimfield Schist section in the Marlborough quadrangle; 4, Tatnic Hill Formation section in eastern Connecticut; and 5, Littleton and Berwick Formations sections in the Peterborough and Manchester quadrangles, New Hampshire.

were formed can be ascertained only approximately. Many examples of relict gross sedimentary structures such as bedding (including cross-bedding and graded bedding) are still observable; however, original fine textures and structures have been obscured or obliterated by recrystallization and deformation. Differences in protolith composition are reflected in variations in metamorphic mineralogy among rock types. The presence or absence of relict bedding, types of relict bedding, and metamorphic mineralogy are noted in tabular form in the stratigraphic columns (Figs. 3, 8), along with the type of protolith we infer from these data.

STRUCTURAL SETTING AND METAMORPHISM

The geologic map and composite column (Fig. 2) show the broad features of the geology of the Brimfield area. Rocks of the area dip westward in a largely homoclinal sequence and are in a fault wedge bounded on the west by the Monson border fault and on the east by the Eastford fault. Four units, A through D, are recognized from east to west; the composite column shows their structural sequence and generalized subdivision.

Numerous small high-angle cross faults and large and small high- and low-angle near-strike faults complicate the stratigraphic succession and affect reported thicknesses. Gneiss along major and minor faults is crushed and sheared (Pease and Peper, 1968). Sills and lenticular bodies of gneissoid pegmatite are present throughout the section. These thin to thick (4 cm to 10 m, rarely as thick as 100 m) bodies locally make up 25 percent of the exposures, and they increase the apparent thicknesses of the units by an unknown amount.

Unit A crops out in the southeastern part of the area between the Eastford fault and the Black Pond and Keach Pond faults. Rocks of the unit dip northwest at moderate angles. Most of the rocks face northwest and probably become younger to the northwest. A few minor folds, however, are indicated by local reversals in facing direction.

Unit B crops out in a westward-dipping, homoclinal sequence between the Black Pond and Keach Pond faults and the Kinney Pond fault. The sequence is internally faulted and is corrugated by asymmetric small-scale and minor folds that show west-over-east movement sense and plunge shallowly northward. Except on the east limbs of recognized folds, rocks of the unit face west, and the sequence is inferred to become younger to the west.

The Black Pond fault separates units A and B in the southeastern part of the area, but apparently dies out to the northeast in the Southbridge quadrangle, where the contact between these two units is gradational. Unit A corresponds approximately to the uppermost part of Emerson's (1917) Paxton Quartz Schist, and unit B to the lowermost part of his Brimfield Schist. The Keach Pond fault is also the trace of a metamorphic isograd that marks the southeastern boundary of sillimanite-bearing rocks in which prograde muscovite is absent. The trace of the Kinney Pond fault, which marks the top of unit B, bends westward in the South Coventry quadrangle and intersects the Monson border fault in the Rockville quadrangle. Strata of units A and B also are warped around to a westward strike beneath the Kinney Pond fault in the South Coventry quadrangle, where they dip northward in gentle open folds.

Unit C is west and north of the Kinney Pond fault. Rocks of the unit dip steeply west and, for the most part, face west. Subunit C_5 includes Emerson's type locality of the Brimfield Schist at the town of Brimfield in the Wales quadrangle. Unit C is broken near midsequence by the Rock Meadow fault. The relative ages

Figure 2. Generalized geologic map and composite column of Brimfield area.

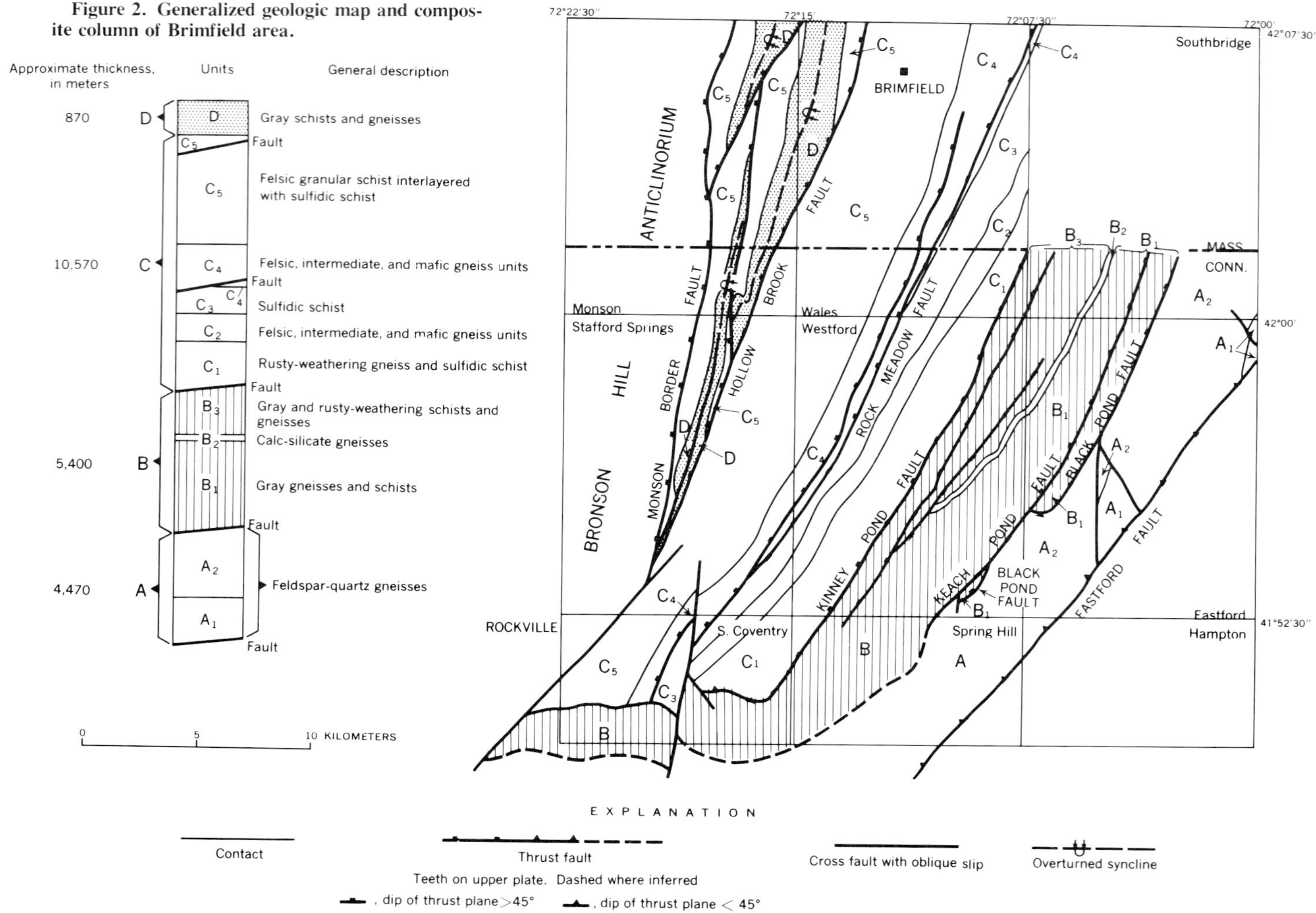

of rocks on either side of this fault cannot be firmly established. Subunits C_4 through C_5 are, however, considered to be younger than subunits C_1 through C_3, because the rocks appear to face west and there is no apparent repetition of section on opposite sides of the fault. unit C is also broken near the top of the sequence by the Hollow Brook fault. The Hollow Brook fault dies out northward in the Warren quadrangle.

Unit D is stratigraphically younger than unit C. In the fault wedge between the Hollow Brook and Monson border faults, unit D is in gradational contact with subunit C_5 in two isoclinal synclines that are overturned to the east. The axial trace of the easternmost syncline has been offset and repeated by a north-trending fault, and its axial trace is cut out southward along another north-trending fault.

INTERNAL SUBDIVISION AND PETROGRAPHY OF UNITS A AND B

The internal subdivision and salient petrographic features of units A and B are summarized in Figures 2 and 3. Unit A consists chiefly of light- to medium-gray quartz-feldspar-biotite gneiss and schist and is divided into two subunits. Subunit A_1 contains abundant, uniform, evenly and thickly layered (2 cm to 2 m), medium- to coarse-grained, gray quartz-feldspar-biotite gneiss. This gneiss characteristically contains subhedral megacrysts (0.5 to 3 cm) of potassium feldspar and polymineralic megacrysts of potassium feldspar, quartz, and plagioclase; it is interlayered in thick lenses with subordinate amounts of medium-grained quartz-feldspar-biotite gneiss, amphibolite, and minor quartz schist.

Subunit A_2 is heterogeneous (Figs. 3 and 4). It consists mostly of gray quartz-feldspar-biotite schist and layered light- and dark-gray, quartz-feldspar-biotite gneiss. It includes thinly layered, dark-gray, biotite-rich granular schist, thinly layered amphibole-, clinopyroxene-, and epidote-bearing gneiss, and layered and massive amphibolite. Thin layers of amphibolite are interlayered with the other gneisses in subordinate amounts forming thin (30 m) marker units at two stratigraphic horizons. Also, near the top of subunit A_2 are layers of gray aluminous schist containing muscovite and minor sillimanite. The schist occurs in discontinuous thin to thick (2 cm to 7 m) lenses.

Rocks of unit A are inferred to record an interval of marine deposition of fine to coarse sediment that was chiefly derived from volcanic materials (rhyodacitic and andesitic water-laid tuff and water-worked clastic debris). Influxes of subordinate, more aluminous epiclastic debris occurred near the end of this depositional interval.

Unit B (Figs. 3, 5 through 7) is divided into three subunits. A thick lower subunit (B_1) composed largely of brownish-gray garnetiferous gneiss and schist is separated by a thin calc-silicate gneiss subunit (B_2) from a thick upper subunit (B_3) that is predominantly rusty-weathering feldspathic sillimanite gneiss. Figures 6 and 7 show calc-silicate-bearing sequences in subunit B_1; the lithology shown in Figure 7 is characteristic of subunit B_2. Figure 5 shows a common lithology in subunit B_3.

In sharp contrast to unit A, unit B records an interval of deposition marked by accumulation of pelitic sediment and epiclastic debris. The early part of this depositional interval, recorded by rocks of subunit B_1, was marked chiefly by accumulations of calcareous to slightly calcareous semipelitic and pelitic sediment (garnetiferous gneiss and schist), and by minor lenticular accumulation of euxinic pelitic mud (sulfidic schist) and volcaniclastic sand and mud of intermediate and mafic composition (layered gneiss and amphibolite). Subunit B_2 represents an abundant influx of calcareous silt. Deposition of euxinic and pelitic mud alternated with calcareous mud in the earlier part of the interval recorded by subunit B_3; the calcareous mud decreases in abundance upward. Near the end of the interval, epiclastic sand and mud accumulated, in part alternating with euxinic sediment. The close of the interval is marked by the accumulation of more extensively reworked quartz-rich and locally calcareous deposits (fine- to medium-grained quartzose and calcic-plagioclase-bearing biotite gneiss that is locally cross-bedded and graded).

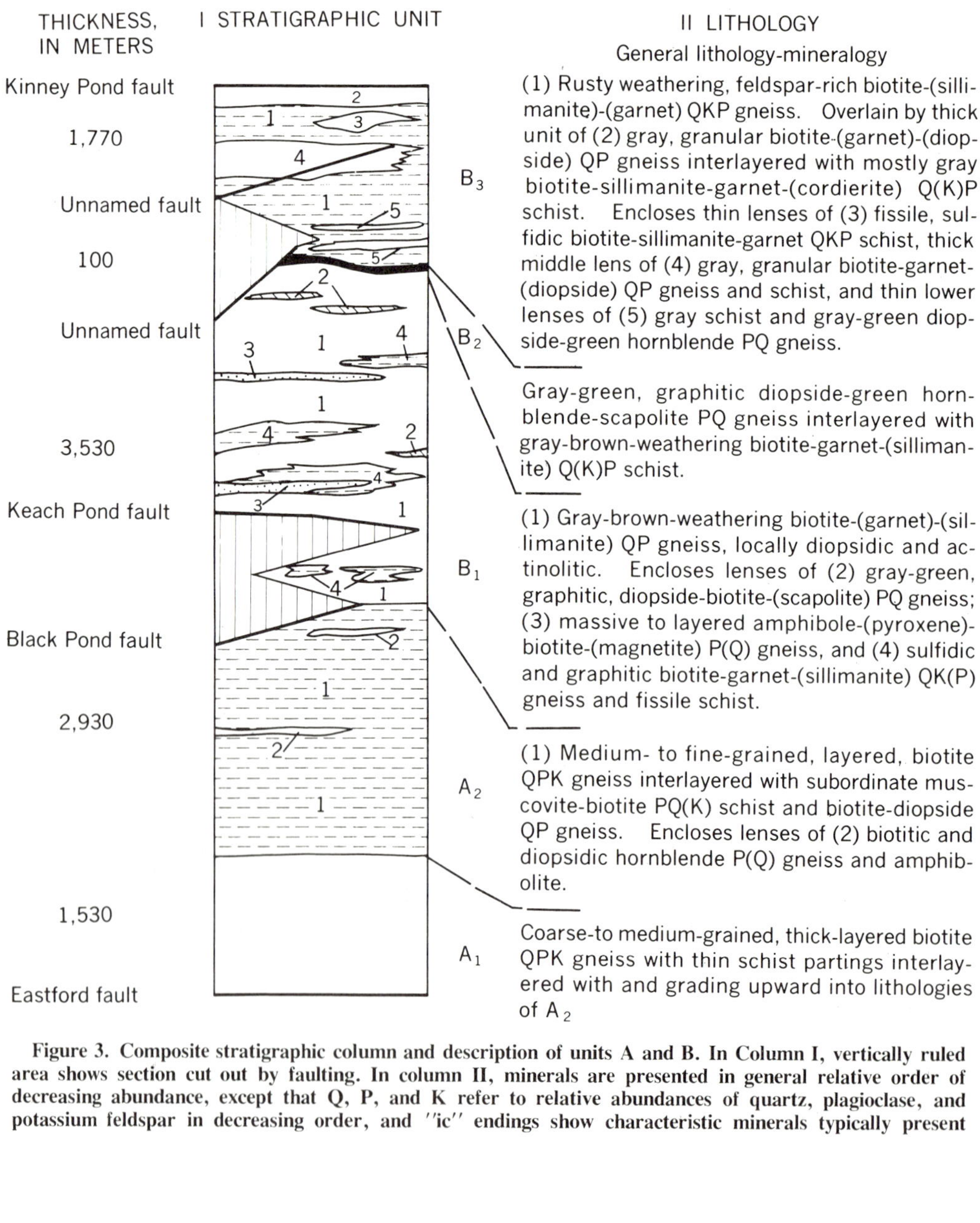

Figure 3. Composite stratigraphic column and description of units A and B. In Column I, vertically ruled area shows section cut out by faulting. In column II, minerals are presented in general relative order of decreasing abundance, except that Q, P, and K refer to relative abundances of quartz, plagioclase, and potassium feldspar in decreasing order, and "ic" endings show characteristic minerals typically present

INTERNAL SUBDIVISION AND PETROGRAPHY OF UNITS C AND D

The internal subdivision and salient petrographic features of units C and D are shown in Figures 2 and 8. Unit C is divided into five subunits. Subunits C_1, and C_3, and C_5 are typically rusty-weathered in outcrop; subunits C_2 and C_4 are nonrusty.

The rusty-weathering subunits form thick sequences of well-layered biotite gneiss containing sillimanite, garnet, and cordierite interlayered with variable thicknesses

III LITHOLOGY Relict primary features	IV INFERRED PROTOLITH	
(1) Thin to thick bedded, rare cross stratification. (2) Very thin to thick bedded; cross stratification; a few graded beds. (3) Mostly very thin bedded to medium bedded. (4) Very thin to thick bedded, thin to medium graded beds; local cross stratification. (5) Very thin to medium bedded, local cross stratification.	Q,P P,Q P Q,P C,Q,P	(1) Mudstone, shale. (2) Shale, slightly calcareous siltstone. (3) Black sulfidic shale. (4) Slightly calcareous siltstone, shale. (5) Calcareous siltstone, mudstone, shale.
Thin to medium bedded; laminations in calc-silicate gneiss.	P,C,Q	Shale, calcareous siltstone, mudstone.
(1) Very thin to thick bedded, local cross stratification. (2) Thin to medium bedded. (3) Laminated to thick bedded, cross stratified, possible lava pillows. (4) Very thin to medium bedded laminations.	Q,P C Q,M P,Q	(1) Slightly calcareous and noncalcareous shale and mudstone. (2) Calcareous silstone, sandstone. (3) Volcaniclastic sandstones and siltstones, in part quartz-poor, dacite lava. (4) Black, sulfidic shale, mudstone.
(1) Laminated to massive; medium bedded, lenticular beds; mostly very thin bedded, local cross stratification. (2) Very thin to thin bedded, lenticular beds.	Q,C,P Q,C,P	(1) Calcareous and noncalcareous siltstone, volcaniclastic in part, mudstone and shale. (2) Volcaniclastic sandstone, siltstone, sandy shale, calcareous siltstone.
Very thin to very thick beds, lenticular beds, cross stratification.	Q	Volcaniclastic rocks.

in amounts less than 5 percent. Minerals in parentheses might not be present in every hand sample of the unit. In column III, bedding thickness is from McKee and Weir (1953) as modified by Ingram (1954). In column IV, letters preceding protolith names show major constituents in decreasing order of abundance: P, pelitic; Q, quartzo-feldspathic; M, mafic; C, calcareous (Fyfe and others, 1958, p. 200).

Figure 4. Coarse-grained megacryst gneiss (below arrow) gradationally overlain by thinly layered light- to medium-gray quartz-plagioclase - biotite - (hornblende) gneiss. Lower part of subunit A_2. Scale in inches.

of fissile graphitic and sulfidic garnetiferous sillimanite schist (Fig. 9). Thin amphibole- and pyroxene-bearing gneiss layers, 50 to 150 m thick, form lenticular marker units traceable for distances ranging from a thousand meters to several kilometers. Subunit C_1 commonly contains thin layers (10 to 25 cm) of dark-gray, fine-grained, quartz-rich diopside gneiss. Similar graphitic calc-silicate gneiss is present in subunit C_3. Distinctive white sillimanite knots (Fig. 10), 0.5 to 1.0 cm in diameter, are present locally in the schist layers of subunit C_3. Subunit C_5 is distinguished from subunits C_1 and C_3 by its greater percentage of well-layered biotite gneiss. In the upper part of subunit C_5, fine- to medium-grained granular quartz-feldspar-biotite gneiss is interlayered on a scale of 2 to 20 cm with slightly

Figure 5. Light-gray, medium-grained, granular quartz - plagioclase-biotite - garnet-(diopside) gneiss (metasiltstone, calcareous in part) of subunit B_3. Note that carbonate-bearing layers weather into negative relief. Scale in inches.

Figure 6. Interlayered granular and schistose biotite- and diopside-bearing quartz-plagioclase gneiss in subunit B_1. Relatively sharp break at base of light-gray granular gneiss (arrow) and grading of granular gneiss into overlying medium-gray, coarser grained, more schistose gneiss that is richer in biotite suggest that these two layers are components of a relict graded bed. Tops up. Scale in inches.

less amounts of rusty-weathering fissile schist and schistose gneiss (Fig. 11).

The nonrusty gneiss subunits of unit C are composed mostly of dark- and light-gray biotite gneiss layers which locally contain amphibole and pyroxene (Fig. 12). Within these nonrusty subunits, quartz-feldspar-rich gneiss is interlayered with quartz-poor biotite gneiss. The nonrusty gneiss is generally thin- to thick-layered, but massive gneiss is also present.

Belts of gneiss tens to hundreds of meters thick and tens of kilometers long have been mapped within subunits C_2 and C_4. These mapped gneiss belts (not shown in Fig. 2) are intertongued and interlayered with each other and consist of varied proportions of distinctive gneiss types, including amphibolite, banded

Figure 7. Laminated diopside- and scapolite-bearing plagioclase (quartz) gneiss (calcareous metasiltstone) in subunit B_1. Scale in inches.

amphibolite, calc-silicate gneiss with diopside and scapolite, hornblende-bearing gneiss, and garnet-quartz gneiss. Subordinate amounts of rusty-weathering gneiss and fissile sulfidic schist are also present.

Unit C represents the accumulation of a great thickness of epiclastic debris derived from an eroding land surface and deposited in a subsiding marine basin. Volcaniclastic debris was the main component for subunits C_2 and C_4 (Fig. 8).

The record begins with accumulation of silt and sand in a euxinic environment, followed by deposition of noneuxinic epiclastic mud. A significant change in the

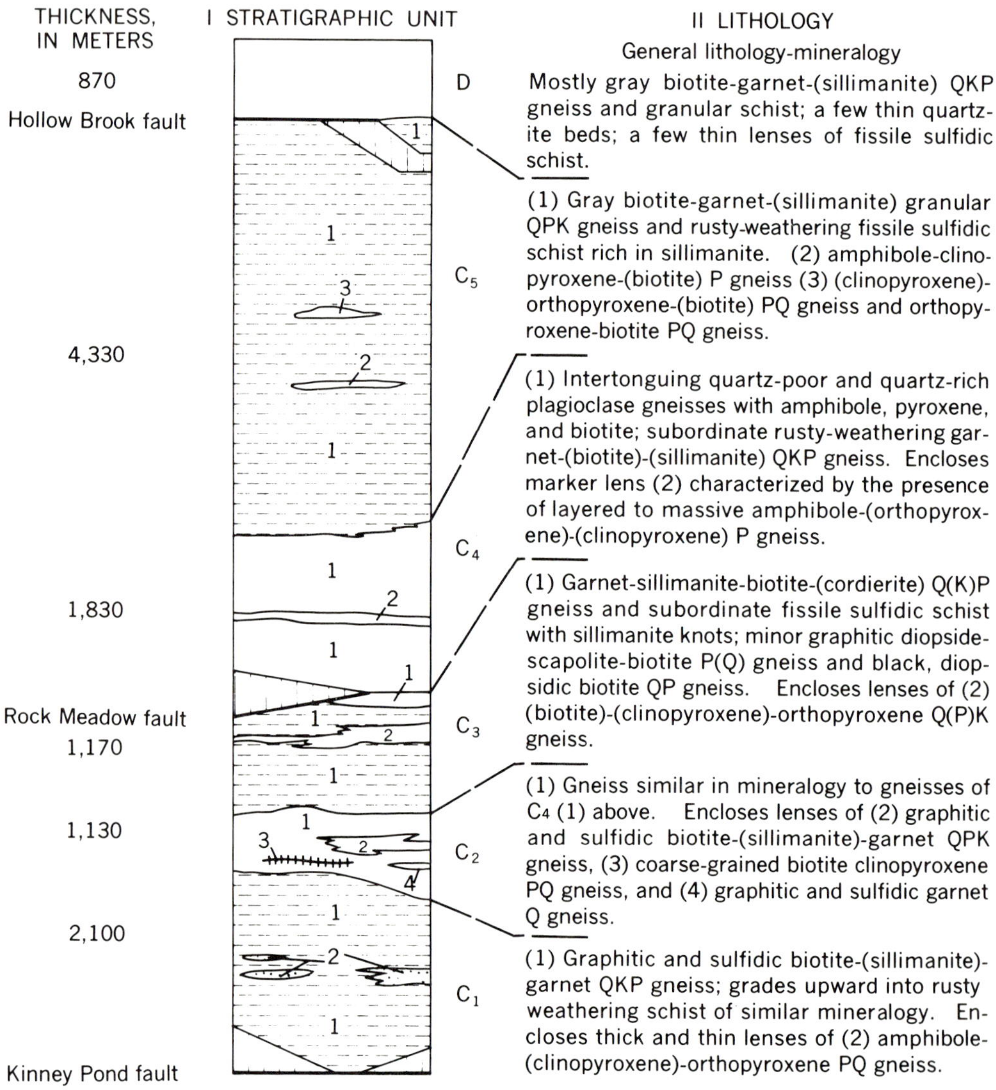

Figure 8. Composite stratigraphic column and descriptions of units C and D. In column I, vertically ruled areas show section cut out by faulting. In column II, minerals are presented in general relative order of decreasing abundance, except that Q, P, and K refer to relative abundances of quartz, plagioclase, and potassium feldspar in decreasing order, and "ic" endings show characteristic minerals typically present

nature of materials deposited in the basin took place when large amounts of volcanic debris, which formed the volcaniclastic sandstone and siltstone of subunit C_2, were introduced. This interval of volcaniclastic deposition was followed by return to deposition in a euxinic environment, in which volcaniclastic sand represented only a minor component near the end of the interval. The second significant volcanic episode (Fig. 8, subunit C_4) is represented by thick deposits of volcaniclastic material and included what may represent a thin lava flow. Thin layers of epiclastic debris represent brief periods of volcanic quiescence. During the last half of the unit C

III LITHOLOGY Relict primary features	IV INFERRED PROTOLITH
Very thin to thick bedded; characteristically cyclically bedded; some graded beds.	P,Q Mudstone; siltstone; sandstone; black, sulfidic shale.
(1) Very thin to medium bedded; a few graded beds. (2 and 3) Thin to medium bedded; a few cross-stratified beds.	Q,P (1) Siltstone; black sulfidic shale. M,Q,C (2 and 3) Volcaniclastic siltstone; calcareous siltstone.
(1) Very thin to medium bedded; some cross-laminated beds. (2) Thin-bedded, laminated; and massive.	Q,C,B (1) Volcaniclastic siltstones, tuffs, both calcareous in part; black shale. Q,M,P (2) Intermediate to mafic tuffs; lava flow or hypabyssal sill; mudstone.
(1) Very thin to medium bedded (gneiss); laminated, very thin bedded (schist); a few graded beds. (2) Laminated thin beds; massive.	P,Q,C (1) Mudstone, black shale, calcareous siltstone. Q,M,C (2) Volcaniclastic siltstone, tuffs, lava, calcareous siltstone.
(1) Very thin to medium bedded, some cross-stratified beds. (2) Thin to medium bedded, laminated; few graded beds. (3) Massive. (4) Thin bedded.	Q,C,M (1) Volcaniclastic siltstone, calcareous in part; tuffs. P,C (2) Black shale, calcareous siltstone. M,Q (3) Mafic lava, tuffs. F (4) Iron-formation.
(1) Mostly thin bedded, laminated; a few graded or cross-stratified beds. (2) Thin to medium bedded, some cross-stratified beds; massive.	P,Q,C, (1) Black shale; mudstone; calcareous siltstone grading up into mudstone. Q,M (2) Volcaniclastic siltstone and subordinate lava.

in amounts less than 5 percent. In column III, bedding thickness is from McKee and Weir (1953) as modified by Ingram (1954). In column IV, letters preceding protolith names show major constituents in decreasing order of abundance; P, pelitic; Q, quartzofeldspathic; M, mafic; C, calcareous; F, ferruginous (Fyfe and others, 1958, p. 200).

Figure 9. Laminated fissile sulfidic schist of subunit C_5. Pegmatitic material (light-gray layers under ruler) is in pelitic beds. Scale in inches.

interval, a great thickness of alternating euxinic and noneuxinic mud, quartz-rich silt, and fine sand was deposited, with very minor additions of volcaniclastic material and calcareous mud. Graded beds and crossbeds in this thick, well-layered sequence probably represent turbidite sedimentation in relatively deep water.

Unit D consists mostly of well-layered to massive, gray, granular gneiss and schist. Interlayering of quartz-feldspar granular gneiss and sillimanite schist on a scale of 2 to 10 cm gives many outcrops of the unit a characteristically ribbed appearance (Fig. 13). Local lenses of brownish-gray sulfidic sillimanite schist and thin beds of gray quartzite are also present in the unit. Rocks in the lower 100 to 150 m of unit D typically have simply and complexly graded beds and some cross-bedding. Rocks of unit D record an accumulation of extensively reworked epiclastic debris, marked by only sparse local accumulation of euxinic mud.

STRATIGRAPHIC CORRELATION

No fossils have been found in, and no radiometric dates have been assigned to the highly metamorphosed stratified rocks of the Brimfield area. In addition, correlation of these rocks with the stratified rocks of other areas of New England is difficult because they are separated from rocks to the southeast and west by the Eastford and Monson border faults, respectively, and to the north they strike into a broad region in central Massachusetts where no detailed mapping has been done.

Although these uncertainties preclude any positive correlation at this time, we maintain that a long-range correlation with the thick sequence of rocks mapped in New Hampshire as Littleton Formation (Billings, 1956) is more tenable than correlations with much thinner sequences closer at hand.

Unit A is equivalent to the upper part of Emerson's Paxton Quartz Schist; most of units B, C, and D were mapped by Emerson as Brimfield Schist. Emerson (1917) assigned a Carboniferous age to these two formations of the basis of fossils of Pennsylvanian age found at the Worcester coat mine. Recent work suggests, however, that the rocks surrounding the coal mine are probably older than the fossiliferous rocks (Grew and others, 1970, p. 120–124).

A summary of the prevailing stratigraphic and structural interpretation of the geology of eastern Connecticut by Dixon and Lundgren (1968) correlated rocks in several other areas in eastern Connecticut with the Brimfield Schist of Emerson. Inherent in their interpretation was the assumption that the sequence in the Brimfield area is inverted. Our more recent detailed mapping in the Brimfield area, however, has shown that although these other rocks are similar to rocks present in the Brimfield area, facing directions observed in rocks of the Brimfield area indicate that the sequence is upright. Furthermore, thicknesses of these other rock sequences do not approach the order of magnitude of the 20 km of section exposed in the Brimfield area.

Figure 14 contrasts the stratigraphic section in the Brimfield area (Fig. 14, col. 4) with the Brimfield Schist of Snyder (1970) in the Marlborough quadrangle (Fig. 14, col. 2) and with the sequence in eastern Connecticut (Fig. 14, col. 3) that includes the Putnam Gneiss of Snyder (1961) and Sclar (1958) and the Tatnic Hill Formation of Dixon (1964). It is re-emphasized here that the sequence in the Brimfield area may have been thickened an undetermined amount within mapped units by repetition on major and minor faults or by minor folds. Although the precise amount of thickening is impossible to ascertain, the lack of repetition of mapped units suggests to us that the original thickness of the sequence in the Brimfield area

Figure 10. Laminated schistose sillimanite-megacryst gneiss in subunit C_3. Note folds confined to pelitic zone between unfolded micaceous and quartz-feldspar-rich layers. Scale in inches.

Figure 11. Sillimanite gneiss in upper part of subunit C_5. Note sharp break (top of scale container) at base of light-gray granular gneiss with internal laminations. Granular gneiss (metasiltstone) grades into overlying coarse schistose layer (metashale), suggesting beds are upright. Scale in inches.

was relatively large (10 to 15 km or more) compared with the thin sequence on the Bronson Hill anticlinorium to the west. The Brimfield Schist in the Marlborough quadrangle (Snyder, 1970) is about one-tenth the apparent thickness of the sequence in the Brimfield area.

The Tatnic Hill Formation has its greatest outcrop width in the Norwich and Jewett City quadrangles. Here it contains several subunits in its lower part (Bates Pond lentil and graphite schist phase of Snyder, 1961) not present along strike to the northeast and has an apparent thickness of as much as 7.3 km. However, it has been suggested that the formation may have been considerably thickened tectonically in this area (Lundgren and Ebblin, 1972, p. 2789 and Fig. 2a). This thickened Tatnic Hill section is only half as thick as the sequence in the Brimfield area. Neither in the Marlborough area nor in the Norwich area can details of the stratigraphy be matched with those in the Brimfield area.

Part of unit C and unit D in the Brimfield area have been equated with the Partridge and Littleton Formations on the Bronson Hill anticlinorium (Fig. 14, col. 2) by some workers in northern Massachusetts (Thompson and others, 1968, p. 206-207, 214). This correlation is based on a stratigraphic interpretation, in which rusty-weathering schist and gneiss is Partridge Formation and gray-weathering schist that overlies it is Littleton Formation, and on a structural interpretation involving large-amplitude nappes folded about gneiss domes. Our recent mapping, however, has shown that the Brimfield area rocks are considerably more diverse and the sequence many times thicker than the proposed equivalent sequence in the Great Hill syncline of the Bronson Hill anticlinorium west of the Monson border fault.

The Brimfield area rocks strike north and northeast across central Massachusetts toward the thick sequence of geosynclinal rocks in the Merrimack synclinorium, mapped by Billings (1956) in New Hampshire as the Littleton Formation. Greene's (1970) publication on the geology of the Peterborough quadrangle in southern New

Hampshire has considerably strengthened this long-range correlation. His work provides a synthesis of the stratigraphic succession of Silurian and Devonian rocks in the Merrimack synclinorium (Greene, 1970, p. 15-16, 52-53) and establishes a sequence of gray- and rusty-weathering schist and volcanic rocks (Fig. 14, col. 5) that compares well in gross lithology, but again not in detail, with that in the Brimfield area.

Greene, using a trigonometric summation of strikes and dips collected on cross-strike traverses, estimated a thickness of about 7.3 km for the Littleton Formation in the Peterborough quadrangle alone. He would add to this 1.8 km of additional section from the Monadnock quadrangle to obtain an estimate of the general thickness of the Littleton Formation in southwestern New Hampshire (Greene, 1970, p. 40). An additional 5.5 km of Silurian and Devonian rocks below the Littleton Formation in the Manchester quadrangle (Sriramadas, 1966) would yield a general thickness of at least 15 km for the Silurian and Devonian sequence in southern New Hampshire. Thus, our estimated thickness of 20 km for the deformed rock sequence in the Brimfield area is more comparable to the thickness of the Merrimack synclinorium sequence in New Hampshire than to the sequence on the Bronson Hill anticlinorium.

The thick, steeply west-dipping, highly faulted sequence of high-grade metamorphic rocks exposed in the Brimfield area appears to represent the southernmost extent of the thick geosynclinal sequence that forms the Merrimack synclinorium and extends northeastward through New Hampshire and southern Maine. This tightly appressed sequence is known to fan out to include a progressively wider area of exposure northeastward. Many of the closely spaced near-strike faults common to the Brimfield area appear to die out in central Massachusetts. A cross-section (Callaghan, 1931) of the rocks exposed in a water tunnel that cuts across the strike some 24 km north-northeast of the Brimfield area shows the rocks to be in broad, gently dipping open folds.

Callaghan attempted to match the stratigraphic section in the tunnel with Emerson's

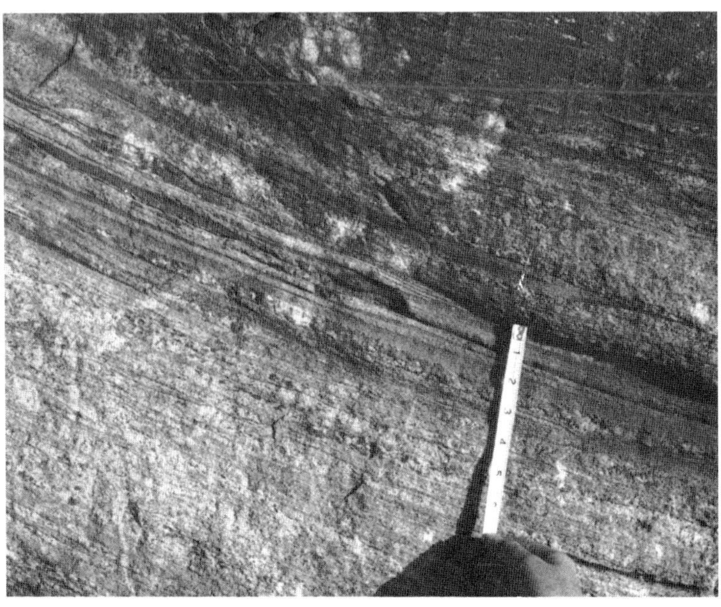

Figure 12. Thinly layered, light- and dark-gray, biotite- and hornblende-bearing plagioclase (quartz) gneisses of subunit C_2 show probable relict cross-stratification in metamorphosed water-worked tuffs. Scale in inches.

Figure 13. Interlayered granular biotite gneiss (metasiltstone) and biotite-garnet-sillimanite schist (coarsely recrystallized metapelite) of unit D. Note sharp truncation of biotite-rich (dark-gray) *lamina* in metapelite bed (x) by sharp discontinuity at base of graded bed A. Biotite content (darker shades of gray) and schistosity increase upward gradationally in bed A. Granular gneiss forming base of graded bed B has relatively sharp contact with underlying coarse schist and enclosed pegmatite stringer (y) and grades into overlying coarse schist (z).

section farther east but was unable to make positive correlation. Much of the tunnel section is generally on strike with rocks in the Brimfield area, and the lithology and thickness of stratigraphic units within the tunnel section do compare with stratigraphic units in the Brimfield area. Detailed areal geology is needed, however, before it will be possible to determine where the tunnel section fits in the Brimfield area sequence.

Finally, reconnaissance investigations in the Peterborough quadrangle suggest that a tentative general correlation can be made, allowing for long-distance facies changes and loss or addition of sections by faulting, between the informal units of Greene (1970) and the Brimfield area sequence. The Souhegan Member of the Littleton Formation, which characteristically contains persistent calc-silicate-bearing gneiss, is equated with unit B, and the rusty-weathering schist and gneiss interlayered with gray biotite gneiss and amphibolite gneiss of the Peterborough Member are equated with unit C. Rusty-weathering quartzite similar to the Francestown Member is also present in the upper part of a unit C correlative in the Warren quadrangle. Gray- to rusty-weathering aluminous schist similar to the Crotched Mountain Member is abundant in the upper part of unit C and in unit D. Correlation of rocks southeast of the Massabesic Gneiss is even less certain, but conceivably the Berwick Formation (Sriramadas, 1966) is equivalent to the lower part of unit B and perhaps to unit A (see Peck, this volume).

ACKNOWLEDGMENTS

In addition to our own mapping of the Eastford, Westford, Stafford Springs, and Monson quadrangles, this paper is a snythesis of unpublished mapping and of observations of several U.S. Geological Survey geologists, including Victor M. Seiders (Wales quadrangle), George E. Moore, Jr. (Southbridge quadrangle), and John S. Pomeroy (Warren quadrangle—north of Wales quadrangle). Professor Janet M. Aitken of the University of Connecticut provided and discussed with the writers basic geologic data in the field, in the Spring Hill and South Coventry quadrangles.

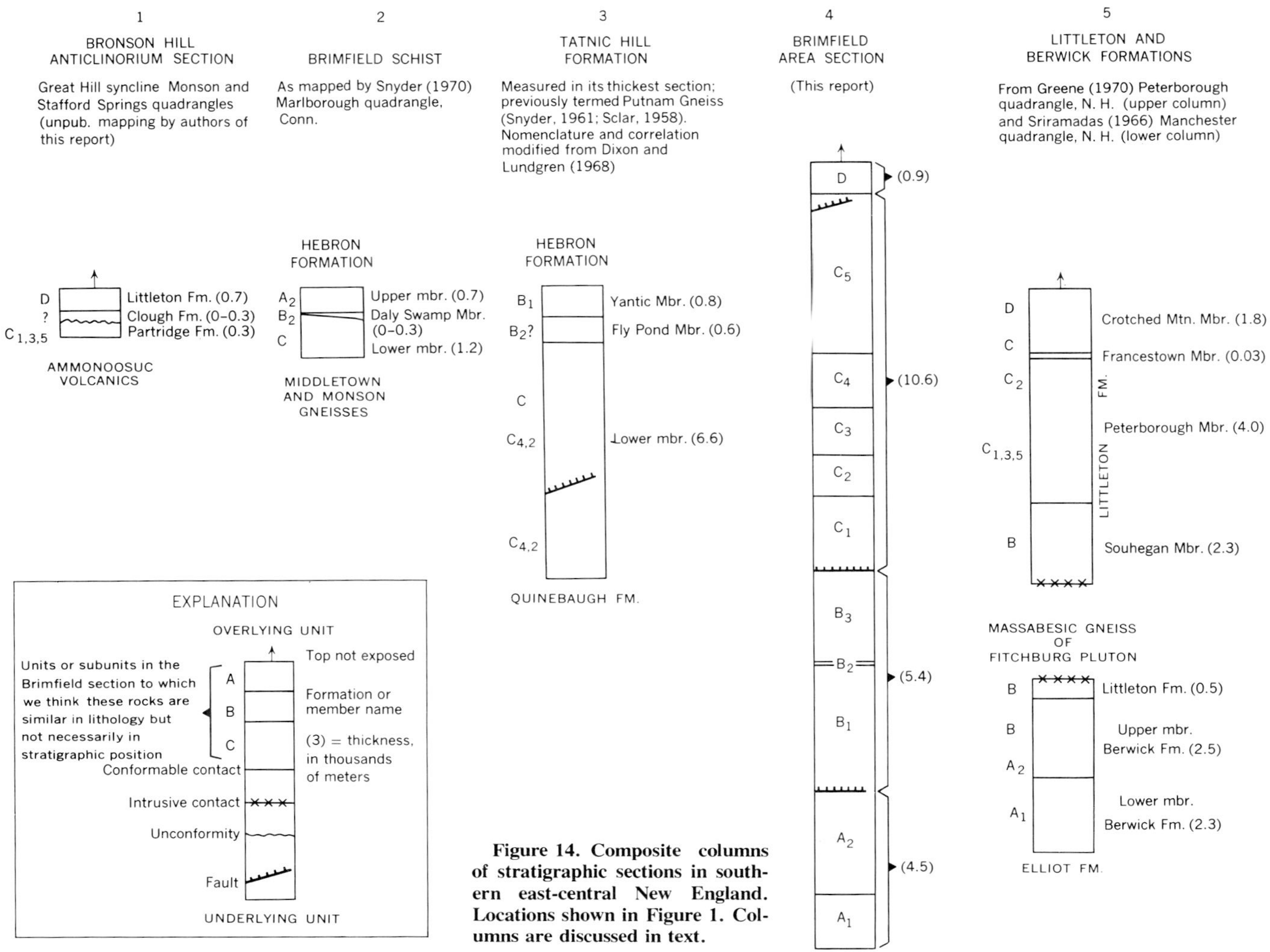

Figure 14. Composite columns of stratigraphic sections in southern east-central New England. Locations shown in Figure 1. Columns are discussed in text.

REFERENCES CITED

Billings, M. P., 1956, The geology of New Hampshire, Pt. II, Bedrock geology: Concord, New Hampshire State Planning and Devel. Comm., 203 p.

Callaghan, Eugene, 1931, A contribution to the structural geology of central Massachusetts: New York Acad. Sci. Annals, v. 33, p. 27-75.

Dixon, H. R., 1964, The Putnam Group of eastern Connecticut: U.S. Geol. Survey Bull. 1194-C, 12 p.

Dixon, H. R., and Lundgren, L. W., Jr., 1968, Structure of eastern Connecticut, *in* Zen, E-an, and others, eds., Studies of Appalachian geology, northern and maritime: New York, Interscience Pub., p. 219-229.

Emerson, B. K., 1917, Geology of Massachusetts and Rhode Island: U.S. Geol. Survey Bull. 597, 289 p.

Fyfe, W. S., Turner, F. J., and Verhoogen, John, 1958, Metamorphic reactions and metamorphic facies: Geol. Soc. America Mem. 73, 259 p.

Greene, R. C., 1970, The geology of the Peterborough quadrangle, New Hampshire: New Hampshire Dept. Resources and Econ. Devel. Bull. 4, 88 p.

Grew, E. S., Mamay, S. H., and Barghoorn, E. S., 1970, Age of plant fossils from the Worcester coal mine, Worcester, Massachusetts: Am. Journ. Sci., v. 268, p. 113-126.

Ingram, R. L., 1954, Terminology for the thickness of stratification and parting units in sedimentary rocks: Geol. Soc. America Bull., v. 65, no. 9, p. 937-938. [Discussion of paper by McKee and Weir, 1953.]

Lundgren, L. W., Jr., and Ebblin, Claude, 1972, Honey Hill fault in eastern Connecticut: Regional relations: Geol. Soc. America Bull., v. 83, no. 9, p. 2773-2794.

McKee, E. D., and Weir, G. W., 1953, Terminology for stratification and cross-stratification in sedimentary rocks: Geol. Soc. America Bull., v. 64, no. 4, p. 381-389.

Pease, M. H., Jr., and Peper, J. D., 1968, The Brimfield(?) and Paxton(?) Formations in northeastern Connecticut, Trip F-5: Connecticut State Geol. and Nat. History Survey Guidebook 2, 18 p.

Peck, J. H., 1976, Silurian and Devonian stratigraphy in the Clinton quadrangle, central Massachusetts, *in* Page, L. R., ed., Contributions to the stratigraphy of New England: Geol. Soc. America Mem. 148 (this volume).

Rodgers, John, and others, compilers, 1956, Preliminary geological map of Connecticut: Hartford, Connecticut State Geol. and Nat. History Survey, scale 1:253,440.

Sclar, C. B., 1958, The Preston gabbro and the associated metamorphic gneisses, New London County, Connecticut: Connecticut Geol. and Nat. History Survey Bull., no. 88, 136 p.

Snyder, G. L., 1961, Bedrock geology of the Norwich quadrangle, Connecticut: U.S. Geol. Survey Geol. Quad. Map GQ-144, scale 1:24,000.

——1970, Bedrock geologic and magnetic maps of the Marlborough quadrangle, east-central Connecticut: U.S. Geol. Survey Geol. Quad. Map GQ-791, scale 1:24,000.

Sriramadas, Alura, 1966, Geology of the Manchester quadrangle, New Hampshire: New Hampshire Dept. Resources and Econ. Devel. Bull. 2, 78 p.

Thompson, J. B., Jr., Robinson, Peter, Clifford, T. N., and Trask, N. J., Jr., 1968, Nappes and gneiss domes in west-central New England, *in* Zen, E-an, and others, eds., Studies of Appalachian geology, northern and maritime: New York, Interscience Pubs., p. 203-218.

MANUSCRIPT RECEIVED BY THE SOCIETY APRIL 18, 1974
REVISED MANUSCRIPT RECEIVED NOVEMBER 25, 1974
MANUSCRIPT ACCEPTED JANUARY 3, 1975

Printed in U.S.A.

Geological Society of America
Memoir 148

Pre-Silurian Stratigraphy of the New London Area, Southeastern Connecticut

Richard Goldsmith
U.S. Geological Survey
Reston, Virginia 22092

A section comparable in general aspect to that presented by Bell (1971) for northeastern Massachusetts is present in the New London area, southeastern Connecticut (Fig. 1). Disregarding a fossiliferous sequence, which is possibly in a different structural block, Bell described a 12,800-m-thick metamorphosed sequence of interbedded volcanic, volcaniclastic, and sedimentary rocks overlying a 1,520-m-thick sequence of metamorphosed quartzose and calcareous sedimentary rocks containing some volcaniclastic material. In the New London area, an upper, primarily volcanic sequence as much as 3,100 m thick (composed of amphibolite, hornblende and biotite gneisses, and some schist) overlies a primarily sedimentary section at least 1,860 m thick (consisting of quartzite, quartz-rich schist, pelitic and feldspathic schists, and some calc-silicate rocks, gray biotite gneiss, and amphibolite). The upper sequence in the New London area consists largely of metamorphosed volcanic and volcaniclastic rocks and related meta-intrusive rocks as well as small amounts of sedimentary rocks; the lower sequence primarily consists of metamorphosed quartzose and calcareous sedimentary rocks with lesser amounts of intercalated volcanic and volcaniclastic material. The two sequences have been complexly folded and are intercalated with granitic gneiss. Primary sedimentary features useful in determining tops of beds are rarely seen. Individual units show appreciable differences in thickness in different areas. Both sequences are pre-Silurian in age, and most probably the rocks range in age from late Precambrian or Cambrian to Ordovician.

Accurate estimates of thicknesses in the New London area are difficult to achieve because of the complex structure and numerous granitic intrusions. Sections have been measured in favorable localities, and thicknesses, except for the granite bodies, have been calculated (Table 1). Differences in thickness of sections are considered to be due to deposition and tectonism. The rocks in the New London area show a higher grade of metamorphism than comparable rocks in northeastern Massachusetts, and unit-by-unit comparisons cannot be made because of facies changes.

The lowermost part of the New London sequence is the Plainfield Formation, which consists of quartzite, quartz-mica schist, micaceous and garnetiferous gneisses,

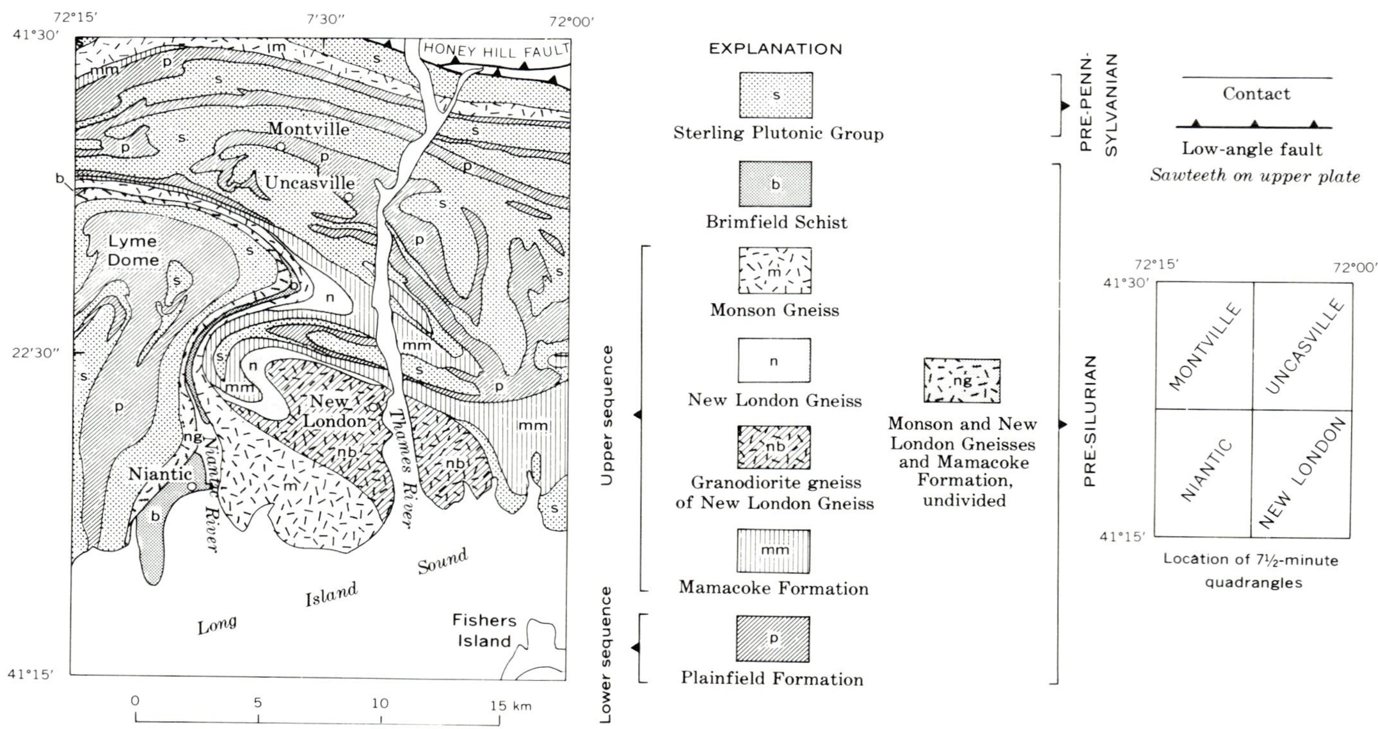

Figure 1. Generalized geology of the New London area, southeastern Connecticut.

and calc-silicate schist, gneiss, and quartzite. Two primarily quartzitic intervals are separated by a primarily nonquartzitic interval.

The lower part of the Plainfield Formation consists of thick units of white quartzite in beds 15 cm to 1 m thick containing sillimanitic partings and interbedded mica schist and mica gneiss. These rocks grade downward into sillimanitic and nonsillimanitic mica schist, gneiss, and micaceous feldspathic quartzite; these lower units, in their highest metamorphic grade, contain quartz-sillimanite nodules. Minor calcareous layers are present as well as some gray feldspathic and quartzose gneisses. Many of the rocks in this lower part are highly feldspathic. However, the base of the Plainfield Formation has not been recognized in the New London area. The lowermost rocks should possibly be considered as a separate formation beneath the Plainfield.

The middle part of the Plainfield Formation consists of a gray biotite-quartz-feldspar gneiss containing hornblende and diopside, a distinctive calc-silicate quartzite, and associated calc-silicate gneiss, amphibolite, garnetiferous schist, sugary-textured sillimanite-bearing biotite-feldspar-quartz gneiss, and thin beds of white-to-gray quartzite. Quartzite is less prominent in this part of the formation. In the area northeast of New London, the upper and middle parts of the formation are not clearly separable.

The upper part of the Plainfield Formation contains an upper thin-bedded gray quartzite, which shows thin micaceous partings. Interlayered are subordinate amounts of mica schist and mica gneiss, which may locally appear above the highest quartzite beds. The beds are primarily 1 to 7 cm thick but may be as much as 30 cm thick. The schist of the uppermost part is typically dark and biotitic and contains little muscovite or sillimanite. In the lower part of the upper quartzite section are thin-bedded, impure metasandstone units, beds (60 cm to 1 m thick) of white quartzite, and layers of white to light-green diopside-bearing calc-silicate gneiss.

The purer and thicker quartzite sequences of the Plainfield Formation grade laterally into impure quartzite and quartzose schist. The recurring lenticular outcrop pattern of the purer quartzite suggests original linear trends prior to folding. It is possible, but unlikely, that the upper quartzitic section is equivalent, because of folding, to the lower quartzitic section. If this were true, the middle member would be a unit below the quartzite, or in an interpretation favoring more complicated structure, it would be the equivalent to the upper part of the Mamacoke Formation.

TABLE 1. APPROXIMATE MAXIMUM THICKNESS OF ROCK UNITS IN THE NEW LONDON AREA, CONNECTICUT

Sequence	Rock unit	Maximum thickness (m)
Upper	Monson Gneiss	1,130
	New London Gneiss	610
	Mamacoke Formation	
	Upper	360
	Lower	1,000
Total		3,100
Lower	Plainfield Formation	
	Upper	260
	Middle	500
	Lower	1,100+
Total		1,860+

Sparse and somewhat questionable information about the direction of the top of the formation suggests that the section was not folded in this manner.

The upper part of the sequence includes the Monson Gneiss, New London Gneiss, and Mamacoke Formation, which are shown in Figure 1. These rocks consist largely of mafic to intermediate and felsic metavolcanic rocks. Mafic metavolcanic rocks are most abundant toward the top of the section, as in the Monson and New London Gneisses, whereas felsic and intermediate rocks are more abundant below, as in the Mamacoke Formation. Amphibolite units within and overlying the upper part of the Monson Gneiss in the northern part of the area are equivalent in part to the Quinebaug Formation in the Plainfield and Danielson areas to the north, as mapped by Dixon (1965, 1968). Included with the metavolcanic part of the section are massive gneiss units with features indicating flowage, such as the granodiorite gneiss of the New London Gneiss and massive phases of the Monson Gneiss in the Niantic area, which are interpreted as intrusive rocks penecontemporaneous with the volcanic rocks. In the lower part of the section, layered gray biotite gneiss of the Mamacoke Formation appears to interfinger with the underlying quartzitic rocks.

The lower and thicker part of the Mamacoke Formation consists of gray biotite-quartz-feldspar gneiss, which is generally rich in plagioclase but locally contains appreciable amounts of potassium feldspar forming granitic layers. The gneiss is indistinctly to distinctly layered, and the layers contain different amounts of biotite. The biotite within a given layer is characteristically, but not everywhere, present as evenly distributed individual flakes. Hornblende-bearing layers and layers of amphibolite occur in subordinate amounts. Northwest and west of New London, the Mamacoke contains a few thin quartzite beds, but these are lacking in the immediate New London area.

The upper part of the Mamacoke Formation consists of a layered sequence with a thin white quartzite unit locally at the top. Below the quartzite, the first relatively consistent unit is a coarse-grained, inequigranular, swirled, and somewhat blocky amphibolite, which in turn is underlain by a series of dark, even-textured layers of biotite-plagioclase gneiss, some of which contain abundant garnet with or without sillimanite. Below the dark gneiss is a light-colored sugary-textured biotite-quartz-feldspar gneiss with quartz-sillimanite nodules, either interfolded or interlayered with greenish calc-silicate gneiss and schist. The different layers in this sequence are not everywhere of the same thickness, and one or more may be missing locally. Lenses of nonlayered amphibolite are rare.

The New London Gneiss can be separated into a layered facies, a massive facies, and an aegerine-augite granite gneiss. The layered facies forms the upper and lower parts of the unit and consists of alternating layers of light-colored biotite-quartz-plagioclase gneiss (dacitic or granodioritic in composition) and amphibolite. In some places, the layering is sharp and uniform in outline and texture; in others, it is irregular. Likewise, in some places the layers are commonly less than 30 cm thick, but in other places the amphibolite or the granodioritic gneiss may be several meters thick. In general, the bottom part of the New London Formation is less distinctly layered than the upper part.

The massive facies is a granodiorite gneiss that is uniform in texture, grain size, and color and has shiny black biotite plates and conspicuous magnetite grains. The gneiss contains elongate inclusions and locally rotated blocks of amphibolite and shows evidence of flowage.

The Monson Gneiss in the New London area is a lenticularly layered, locally massive gray hornblende-biotite-quartz-plagioclase gneiss and, in places, a biotite-quartz-microcline-plagioclase gneiss containing scattered lenses of amphibolite.

Mafic minerals tend to be concentrated in clots or streaks. The Monson is relatively uniform in composition and texture throughout the area. On the east side of the Niantic River, however, the upper part of the Monson Gneiss consists of thin layers, as much as 7 cm thick, ranging from rocks rich in hornblende to rocks poor in mafic minerals. The Monson Gneiss in the northern part of the area is finer grained than it is in the south, and layers of amphibolite, 60 to 120 cm thick, are fairly abundant immediately below the Honey Hill fault.

REFERENCES CITED

Bell, K. G., 1971, Pre-Silurian stratigraphy of northeastern Massachusetts: Geol. Soc. America Abs. with Programs, v. 3, no. 1, p. 17.

Dixon, H. R., 1965, Bedrock geologic map of the Plainfield quadrangle, Windham and New London Counties, Connecticut: U.S. Geol. Survey Geol. Quad. Map GQ-481.

——1968, Bedrock geologic map of the Danielson quadrangle, Windham County, Connecticut: U.S. Geol. Survey Geol. Quad. Map GQ-696.

MANUSCRIPT RECEIVED BY THE SOCIETY APRIL 18, 1974
REVISED MANUSCRIPT RECEIVED DECEMBER 9, 1974
MANUSCRIPT ACCEPTED JANUARY 3, 1975

Printed in U.S.A.

Geological Society of America
Memoir 148

Summary Discussion of Stratigraphy of Eastern Massachusetts and Connecticut

H. Roberta Dixon
U.S. Geological Survey
Federal Center
Denver, Colorado 80225

ABSTRACT

Recent geologic work in eastern Massachusetts and Connecticut has demonstrated that the rocks of the area are broken into a series of fault blocks. From east to west, the faults bounding these blocks are the North Boston Basin fault (in Massachusetts and possibly extending south into Rhode Island), the Bloody Bluff (Massachusetts) and the Honey Hill-Lake Char (Connecticut) fault system, and the Clinton-Newbury fault (Massachusetts). All of these faults, except possibly the North Boston Basin, trend southwest toward a common area in southern Massachusetts that has not yet been mapped in detail, so the relationship between them has not yet been established. Although early work in the area lumped various rock units across the faults, the modern work indicates that distinct groups of rocks occur within each fault block, and repetition of units across the major faults has not been demonstrated.

The ages of the rocks in the area are not yet established. Fossil-bearing rocks from Cambrian to Pennsylvanian in age occur in fault-bounded blocks; their relationship to the metamorphosed rocks cannot be determined.

DISCUSSION

Within eastern Massachusetts and Connecticut are several major fault blocks, in many of which the stratigraphic succession of probable Paleozoic rock units has by now been fairly well studied. The boundaries of most of the fault blocks are well defined, although some critical gaps still exist. Correlation of rock units within the blocks is reasonably good, but correlation across the faults is, for the most part, not well established. Figure 1 illustrates the fault boundaries as known and the relationships that can be established or suggested across and between them, as well as some of the gaps and unresolved problems.

Most lower Paleozoic rocks in eastern Massachusetts and Connecticut have been metamorphosed to a medium to high grade. Low-grade to unmetamorphosed rocks, some of which are fossiliferous, occur in fault-bounded blocks in northeastern Massachusetts. In and south of the Boston basin, for example, there are isolated outcrops of fossiliferous rocks of Cambrian age (Bell and Alvord, this volume), and in the Newbury area, fossiliferous sedimentary rocks occur in a volcanic complex of Silurian-Devonian age (Shride, this volume). The relationship of these rocks to the surrounding more highly metamorphosed rocks is not known, because the nature and amount of displacement on most of the faults is not well established. The North Boston Basin fault has been mapped as a thrust fault (Billings, 1929; Billings and Rahm, 1966), but its trace is well known only north and west of Boston; its continuation to the southwest has not been mapped, but it may trend into Rhode Island, as suggested by Skehan (1969), and the Cambrian rocks thus would be confined to the fault block southeast of it.

North and west of the North Boston Basin fault and east of the Bloody Bluff fault is a sequence of metasedimentary and metavolcanic rocks that is generally similar to the sequence of rocks in southeastern Connecticut south and east of the Honey Hill-Lake Char fault, as described by Goldsmith (1966, and this volume). In both areas the oldest rocks are a series of clastic quartz-rich rocks—the Westboro Formation of Bell and Alvord (this volume) and the Plainfield Formation of Goldsmith (1966, and this volume). These quartz-rich rocks generally include a lower and an upper quartzite with a middle unit of quartz-rich schist, calc-silicate gneiss, and assorted lithologies. The Plainfield Formation has been traced from southern Connecticut north across eastern Connecticut into Massachusetts. Southern Massachusetts has not yet been mapped in detail, but Emerson (1917) showed isolated patches of Westboro quartzite in the granitic rocks that occupy most of the area between Connecticut and northeastern Massachusetts; presumably, these form a connecting link between the quartzose rocks of the two areas. The metamorphic grade of these rocks decreases from southwest to northeast; in southern Connecticut they are upper sillimanite grade (Goldsmith, 1967; Lundgren, 1966a), in northeastern Connecticut they are staurolite-kyanite grade (Dixon, 1974), and in the Boston area they are apparently no more than biotite grade (Bell and Alvord, this volume).

Overlying the quartz-rich rocks of both areas are interlayered metasedimentary and metavolcanic rocks. Although it would seem likely that these metavolcanic rocks are of comparable age, there are differences between the rocks of the two areas that make direct correlation difficult. In general, the metavolcanic rocks of Massachusetts are mafic (andesitic to basaltic; Bell and Alvord, this volume), and most of the interlayered metasedimentary rocks occur near the top of the sequence. In contrast, in Connecticut the metasedimentary rocks are restricted to the Mamacoke Formation at the base of the metavolcanic complex (Goldsmith, 1966, and this volume), and none have been identified higher in the sequence. The metavolcanic rocks in Connecticut are dacitic (granodiorite of the New London Gneiss) to andesitic (hornblende-biotite-quartz-oligoclase gneiss of the Monson Gneiss (Goldsmith, 1966, and this volume). In eastern Connecticut and southern Massachusetts, the metavolcanic rocks are apparently cut out by the Lake Char fault, so it is not possible to map a connection between the rocks of southern Connecticut and those of the Boston area.

In both Massachusetts and Connecticut, this sequence of clastic quartz-rich rocks overlain by metavolcanic rocks is cut off at the top by a major fault zone. In Massachusetts it is the Bloody Bluff fault and in Connecticut the Honey Hill-Lake Char fault. Both fault zones are characterized by thick belts of cataclastic rocks. In Connecticut, however, the cataclastic rocks are thickest on the north and west sides of the fault (Dixon and Lundgren, 1968), whereas in Massachusetts the thickest zone of cataclasis is east of the Bloody Bluff fault (R. O. Castle, 1972, written commun.). Thus the connection between these two fault zones is not yet clear.

West of the Bloody Bluff fault in Massachusetts and north and west of the Honey Hill-Lake Char fault in Connecticut are thick units of primarily metavolcanic rocks overlain by predominantly metasedimentary rocks. These include the Marlboro Formation and associated rocks and the Nashoba Formation as redefined by Bell and Alvord (this volume) in Massachusetts and the Quinebaug and Tatnic Hill Formations of Connecticut. In the past it has been assumed that the Marlboro correlated with the Quinebaug, and the Nashoba with the Tatnic Hill (Emerson, 1917, p. 79; Dixon, 1964). As detailed mapping progresses, however, these correlations are less certain. Certainly there are similar lithologies, including amphibolite, hornblende gneiss, sillimanite gneiss, calc-silicate gneiss, and marble in the two areas. The Quinebaug Formation, which consists of a variety of interlayered biotite and hornblende gneiss and amphibolite and lesser amounts of pelitic gneiss and calc-silicate gneiss, is bounded on the bottom and in part, if not entirely, on the top by faults, and it is cut out along these faults in southern Massachusetts. It could be correlative with the Marlboro Formation, with one of the amphibole gneiss units within the Nashoba, of Bell and Alvord (this volume), or possibly with none of the Massachusetts rocks. The Tatnic Hill Formation may be equivalent to a part of the Nashoba (although which part is uncertain) or to none of the Nashoba. As described by Bell and Alvord (this volume), in the Nashoba the various lithologies listed above occur as interlayered lenses, and the individual members are distinguished by the most prominent lithology. In the Tatnic Hill Formation, on the other hand, there is a distinct sequence of lithologic units mappable across Connecticut and into southern Massachusetts, where the lower part is cut out by the Lake Char fault. The upper part is apparently continuous with rocks mapped by Emerson (1917) as Boylston Schist and the upper (or western) part of his "gneisses and schists of unknown age." These rocks were mapped by Grew (1973) as units Sa and Sb; his Sb would be equivalent to Emerson's (1917) Boylston Schist and to the micaceous schist at the top of the Tatnic Hill Formation, his Sbl to the calc-silicate gneiss below the upper schist, and his Sa to the pelitic gneiss that is the upper part of the lower member of the Tatnic Hill. These rocks are west of the Clinton-Newbury fault as projected southwest by Bell and Alvord (this volume) and would thus be separated from the rest of the Nashoba by that fault. This is the interpretation shown in Figure 1, with the implication that the Tatnic Hill Formation (pS s) west of the fault may not be equivalent to the Nashoba (pS s) east of the fault.

The Tatnic Hill Formation has been correlated (Dixon and Lundgren, 1968) with a part of the Brimfield sequence (units A to C) of Peper and Pease (this volume). This correlation is based on structural interpretations and general lithologic similarities. The Brimfield Schist of Snyder (1970) in the Marlborough quadrangle and of Lundgren and others (1971) in the Moodus quadrangle has been traced south into the west and south limb of the folded Chester-Hunts Brook syncline; the Tatnic Hill Formation has been traced southwest from the type area into the east limb of this fold (Dixon and Lundgren, 1968; Lundgren, 1966b). Thus, the Tatnic Hill Formation and the Brimfield Schist are considered to be correlative units in opposite limbs of a syncline that, north of the Honey Hill fault, is overturned to the east with the Brimfield Schist in the overturned limb. Although specific details of the lithologies of units A to C of the Brimfield sequence of Peper and Pease (this volume) cannot be matched with details of the Tatnic Hill Formation, considering the subdivisions of the two units as a whole, there is a good general match of the lithologies in a reverse order. Thus the top of the Tatnic Hill Formation is similar in gross lithology to the bottom of the Brimfield sequence (unit A of

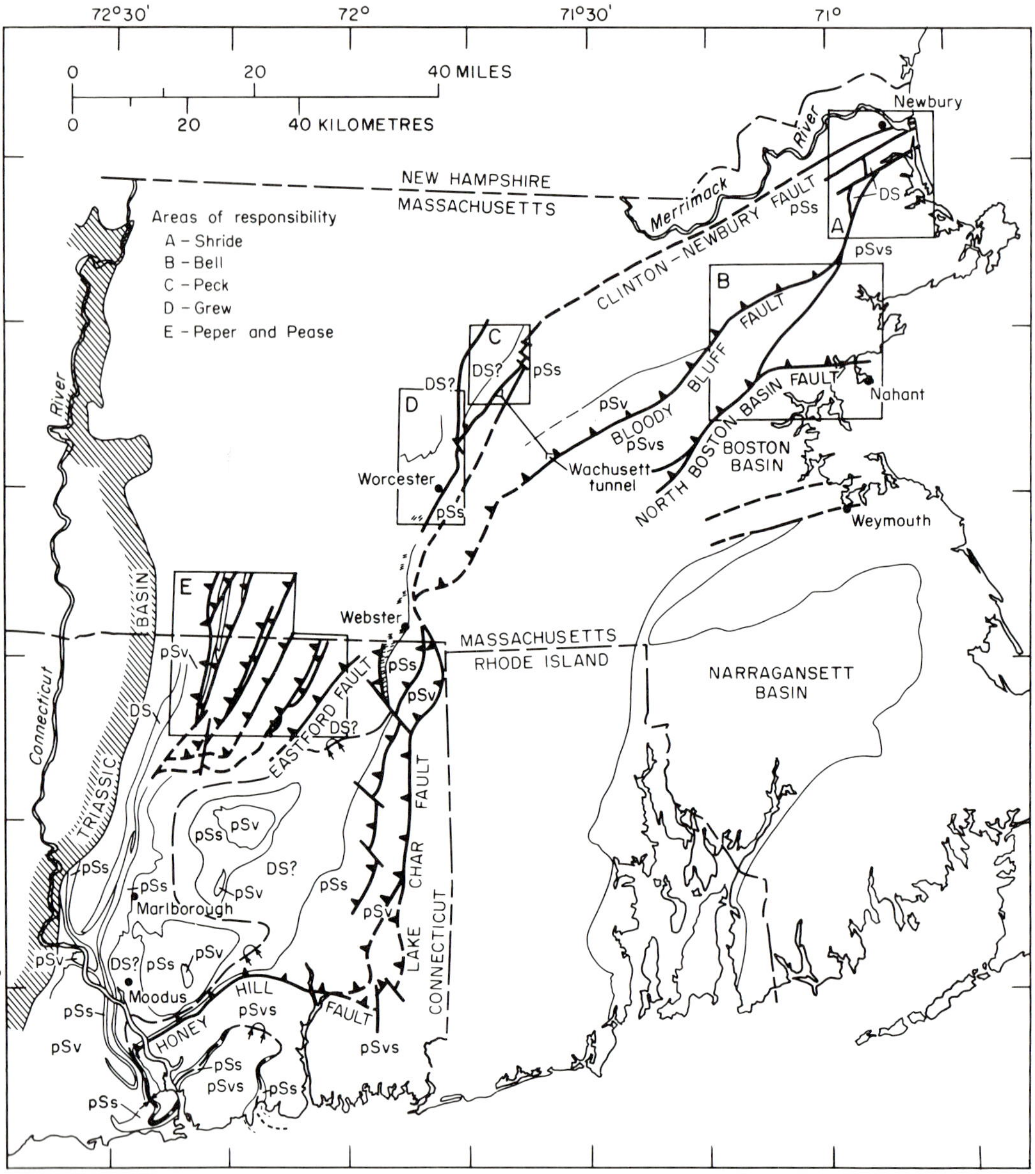

DS	Devonian and Silurian metasedimentary and volcanic rocks
DS?	Possible Devonian and Silurian metasedimentary rocks; possibly pre-Silurian
pSs	Pre-Silurian metasedimentary rocks
pSv	Pre-Silurian metavolcanic rocks
pSvs	Older pre-Silurian metavolcanic and metasedimentary rocks

Observed exposures of an apparently continuous belt of Ayer type granite (of Emerson, 1917) limiting fault extensions. Other granitic rocks abundant but not shown

Contact

Thrust fault or high-angle reverse fault--Sawteeth on upper plate

High-angle fault or dip of fault surface not known--Dashed where mapping not complete

Approximate trace of folded axial surface of Hunts Brook-Chester syncline

Border of Triassic basin

Peper and Pease, this volume); the middle parts of the two sequences are generally comparable; and the bottom of the Tatnic Hill is similar to unit C near the top of the Brimfield sequence. In addition, the metamorphic grades of the two units are similar but in a reverse order. The lower part of the Tatnic Hill is sillimanite-potassium feldspar metamorphic grade, most of the rest of the unit is sillimanite-muscovite grade, and only the uppermost part of the upper member is staurolite grade. In the Brimfield sequence, units A and part of B are sillimanite-muscovite grade, and the rest of unit B and unit C are sillimanite-potassium feldspar grade (Peper and Pease, this volume). Although apparently none of the Brimfield sequence is as low as staurolite grade, it overlies staurolite-grade rocks. Thus in the Brimfield sequence, high-grade metamorphic rocks structurally overlie lower grade rocks.

Peper and Pease (this volume) do not accept this interpretation and do not believe the Brimfield sequence in northern Connecticut and Massachusetts is overturned. Citing gross differences in thickness and their inability to match details of lithologies, they apparently reject the continuity of the Brimfield sequence in their area and the Brimfield Schist of Snyder (1970) in the Marlborough quadrangle, but would separate the rocks of the two areas by a fault. This fault was not mapped in either the Marlborough quadrangle (Snyder, 1970) or the Rockville quadrangle (Aitkin, 1955) to the north, nor is there an abrupt structural break suggesting a fault in either quadrangle. Therefore, the nature and location of this fault, which would separate rocks of similar lithology that trend into each other, is still uncertain.

The complex of dominantly metasedimentary and minor metavolcanic rocks of eastern Massachusetts and Connecticut is overlain by a sequence of generally lower grade metasedimentary rocks in which the metamorphic grade decreases toward the northeast. In Connecticut these rocks are mostly staurolite grade, and in central Massachusetts they are locally as low as chlorite grade (Grew, 1973). In Massachusetts these rocks are separated from the underlying Nashoba Formation by the Clinton-Newbury fault, but the southward extension of this fault is still open to conjecture, because near Worcester the fault breaks into several branches (Grew, 1973), and in Connecticut the contact of these rocks with the Tatnic Hill Formation appears to be a normal sedimentary contact. There is no agreement on the grouping and correlation of these upper metasedimentary rocks, but in general they can be divided into three units. Lowermost is a lenticular quartzite—unit Sc of Grew (1973) and unit 1 of Peck (this volume). This quartzite has not been observed south of the Worcester area. Overlying the quartzite is a unit consisting of thinly layered metasiltstone, commonly interlayered with calc-silicate rocks. This includes Grew's (1973) unit DSd, Peck's (this volume) unit 2, and the Hebron Formation in Connecticut. Grew (1973) would also include Peck's unit 5 (this volume) as part of this middle unit, although Peck (this volume) suggests that his unit 5 might be older than his units 1 to 4. The youngest unit consists of slate, phyllite, or schist, depending on the metamorphic grade, and it includes Grew's (1973) unit De, Peck's units 3 and 4 (this volume), and possibly the Scotland Schist in Connecticut. Unit D of the Brimfield sequence of Peper and Pease (this volume) may also correlate with these phyllitic rocks but probably does not.

The ages of the rocks in eastern Connecticut and Massachusetts only locally can be determined directly by fossils. The fossil-bearing rocks are, moreover, separated from the high-grade metamorphic nonfossiliferous rocks by faults, so

Figure 1. Generalized geologic map of parts of eastern Massachusetts and Connecticut. Compiled from Grew (1973); Bell and Alvord, Peck, Peper and Pease, and Shride (all this volume); compilation map of eastern Connecticut (Dixon, 1968, Pl. 1); and Bell (personal commun.).

they are of little value in dating most of the rocks in the area. Therefore, ages must be determined by other methods, primarily correlations and radiometric determinations, and for some units the results are conflicting. There is apparently general agreement among current workers in Massachusetts that the complex of metasedimentary and metavolcanic rocks east of the Clinton-Newbury fault is pre-Silurian in age. Except for fossil-bearing rocks (separated by faults from the metamorphic rocks) of Cambrian age at isolated outcrops, as at Nahant and Weymouth, Massachusetts (Bell and Alvord, this volume), there is little direct evidence on which to base this assignment. The rocks east of the Bloody Bluff fault have been cut by alkalic intrusive rocks that have been dated as 450 ± 25 m.y. old by Zartman and Marvin (1971). In addition, the clastic quartz-rich rocks have been intruded by the Northbridge Granite Gneiss of Emerson (1917), dated as 569 ± 4 m.y. old by Fairbairn and others (1967). These numbers suggest an Ordovician or older age for the rocks east of the Bloody Bluff fault but give no indication of the age of the rocks between the Bloody Bluff and Clinton-Newbury faults. In Connecticut the Tatnic Hill and Quinebaug Formations have been tentatively dated as pre-Silurian on the basis of correlation with pre-Silurian rocks in the Bronson Hill anticlinorium (see Dixon and Lundgren, 1968).

There is no general agreement on the age of the metasedimentary sequence west of the Clinton-Newbury fault and east of the Bronson Hill anticlinorium located immediately east of the Triassic basin in Massachusetts. These rocks are shown as SD? in Figure 1, as they have been assigned a Silurian to Devonian age by Grew (1973), Peck (this volume), and Skehan (this volume). The age assignment is based on correlation of these rocks with fossil-bearing rocks in Maine. The correlation with rocks in Maine is, however, a long-distance one, across a large area where detailed mapping has not been done, and thus it must be considered tentative. The age, moreover, is contradicted by recent radiometric work. As mentioned by Peck (this volume) and Grew (1973), analyses of zircon from the Ayer Granite by Zartman and Naylor (1972) give an age of 435 to 470 m.y. A slightly younger age of 410 m.y. was determined by Zartman (1974, personal commun.; Zartman and others, 1965) for the Canterbury Gneiss of Connecticut, by both whole-rock Rb-Sr analyses and U-Pb isotopes from zircons. The Canterbury Gneiss cuts the Hebron Formation, which is correlated with rocks cut by the Ayer Granite in Massachusetts; the Ayer Granite probably also cuts the younger phyllite-schist unit—Grew's unit De and Peck's units 3 and 4. The analyses indicate a Silurian to possibly earliest Devonian age for the granitic rocks and an older age for the metasedimentary rocks that they cut. Thus, the radiometric analyses currently available would suggest a pre-Silurian age for the Hebron Formation, its correlatives, and overlying phyllite-schist rocks. The lithologic differences between the SD? rocks (Fig. 1) and the underlying pre-Silurian metasedimentary rocks may in part be the result of differences in metamorphic grade rather than age.

The age of the Brimfield sequence of Peper and Pease (this volume) also is not definitely established. Peper and Pease suggest that the rocks in their area may correlate with Silurian-Devonian fossiliferous rocks on the east flank of the Merrimack synclinorium in Maine, but, as mentioned above, a possible pre-Silurian age would be indicated by correlations with the Tatnic Hill and other rocks. Unit D of Peper and Pease (this volume) may correlate with the younger phyllite-schist units in Grew's (1973) and Peck's (this volume) areas, as mentioned above, or, more likely with similar rocks in the Devonian sequence of the Bronson Hill anticlinorium (see Peper and Pease, this volume).

REFERENCES CITED

Aitkin, J. M., 1955, Bedrock geology of the Rockville quadrangle: Connecticut Geol. and Nat. History Survey Quad. Rept. 6, 55 p.

Bell, K. G., and Alvord, D., 1976, Pre-Silurian stratigraphy of northeastern Massachusetts, *in* Page, L. R., ed., Contributions to the stratigraphy of New England: Geol. Soc. America Mem. 148 (this volume).

Billings, M. P., 1929, Structural geology of the eastern part of the Boston Basin: Am. Jour. Sci., 5th ser., v. 18, p. 97-137.

Billings, M. P., and Rahm, D. A., 1966, Geology of the Malden Tunnel, Massachusetts: Boston Soc. Civil Engineers Jour., v. 53, no. 2, p. 116-141.

Dixon, H. R., 1964, The Putnam Group of eastern Connecticut: U.S. Geol. Survey Bull. 1194-C, 12 p.

——1968, Bedrock geology of the Plainfield-Danielson area, Connecticut: U.S. Geol. Survey Open-File Rept., 308 p.

——1974, Bedrock geologic map of the Thompson quadrangle, Windham County, Connecticut, and Providence County, Rhode Island: U.S. Geol. Survey Geol. Quad. Map GQ-1165, scale 1:24,000.

Dixon, H. R., and Lundgren, L. W., Jr., 1968, Structure of eastern Connecticut, *in* Zen, E-an, and others, eds., Studies of Appalachian geology—Northern and maritime: New York, Interscience Pubs., p. 219-229.

Emerson, B. K., 1917, Geology of Massachusetts and Rhode Island: U.S. Geol. Survey Bull. 597, 289 p.

Fairbairn, H. W., Moorbath, S., Ramo, A. O., Pinson, W. H., Jr., and Hurley, P. M., 1967, Rb-Sr age of granitic rocks of southeastern Massachusetts and the age of the Lower Cambrian at Hoppin Hill: Earth and Planetary Sci. Letters, v. 2, no. 4, p. 321-328.

Goldsmith, Richard, 1966, Stratigraphic names in the New London area, Connecticut: U.S. Geol. Survey Bull. 1224-J, 9 p.

——1967, Bedrock geologic map of the Montville quadrangle, New London County, Connecticut: U.S. Geol. Survey Geol. Quad. Map GQ-609, scale 1:24,000.

——1976, Pre-Silurian stratigraphy of the New London area, southeastern Connecticut, *in* Page, L. R., ed., Contributions to the stratigraphy of New England: Geol. Soc. America Mem. 148 (this volume).

Grew, E. S., 1973, Stratigraphy of the Pennsylvanian and pre-Pennsylvanian rocks of the Worcester area, Massachusetts: Am. Jour. Sci., v. 273, p. 113-129.

Lundgren, Lawrence, Jr., 1966a, Muscovite reactions and partial melting in southeastern Connecticut: Jour. Petrology, v. 7, p. 421-453.

——1966b, The bedrock geology of the Hamburg quadrangle: Connecticut Geol. and Nat. History Survey Quad. Rept. 19, 41 p.

Lundgren, Lawrence, Jr., Ashmead, Lawrence, and Snyder, G. L., 1971, The bedrock geology of the Moodus and Colchester quadrangles: Connecticut Geol. and Nat. History Survey Quad. Rept. 27, 24 p.

Peck, J. H., 1976, Silurian and Devonian stratigraphy in the Clinton quadrangle, central Massachusetts, *in* Page, L. R., ed., Contributions to the stratigraphy of New England: Geol. Soc. America Mem. 148 (this volume).

Peper, J. D., and Pease, M. H., Jr., 1976, Summary of stratigraphy in the Brimfield area Connecticut and Massachusetts, *in* Page, L. R., ed., Contributions to the stratigraphy of New England: Geol. Soc. America Mem. 148 (this volume).

Shride, A. F., 1976, Stratigraphy and correlation of the Newbury Volcanic complex, northeastern Massachusetts, *in* Page, L. R., ed., Contributions to the stratigraphy of New England: Geol. Soc. America Mem. 148 (this volume).

Skehan, J. W., S. J., 1969, Tectonic framework of southern New England and eastern New York, *in* Kay, Marshall, ed., North Atlantic—Geology and continental drift: Am. Assoc. Petroleum Geologists Mem. 12, p. 793-814.

Skehan, J. W., S. J., and Abu-moustafa, A. A., 1976, Stratigraphic analysis of rocks exposed in the Wachusett-Marlborough Tunnel, east-central Massachusetts, *in* Page, L. R., ed.,

Contributions to the stratigraphy of New England: Geol. Soc. America Mem. 148 (this volume).

Snyder, G. L., 1970, Bedrock geologic and magnetic maps of the Marlborough quadrangle, east-central Connecticut: U. S. Geol. Survey Geol. Quad. Map GQ-791, scale 1:24,000.

Zartman, R. E., and Marvin, R. F., 1971, Radiometric age (Late Ordovician) of the Quincy, Cape Ann, and Peabody Granites from eastern Massachusetts: Geol. Soc. America Bull., v. 82, p. 937-958.

Zartman, R. E., and Naylor, R. S., 1972, Structural implications of some U-Th-Pb zircon isotopic ages of igneous rocks in eastern Massachusetts: Geol. Soc. America Abs. with Programs v. 4, p. 54-55.

Zartman, R. E., Snyder, G. L., Stern, T. W., Marvin, R. F., and Bucknam, R. C., 1965, Implications of new radiometric ages in eastern Connecticut and Massachusetts: U.S. Geol. Survey Prof. Paper 525-D, p. D1-D10.

MANUSCRIPT RECEIVED BY THE SOCIETY APRIL 18, 1974
REVISED MANUSCRIPT RECEIVED MARCH 27, 1975
MANUSCRIPT ACCEPTED APRIL 11, 1975

Printed in U.S.A.

Geological Society of America
Memoir 148

Pre-Silurian Stratigraphy in South-Central Massachusetts and North-Central Connecticut

Robert W. Schnabel
U.S. Geological Survey
Federal Center
Denver, Colorado 80225

ABSTRACT

A sequence of 13 lithologic units persists through the New Hartford, West Granville, Tariffville, and Southwick 7-1/2′ quadrangles. The sequence of deposition was deduced from the composition of graded beds and cross-beds that locally are well preserved in many of the units. The units from top to bottom are (1) Oh_1; schist, locally graphitic, with thin quartzite, calc-silicate rocks, and coticule rock; minimum 900 m (3,000 ft) thick; (2) Oh_2; coarse-grained kyanite-garnet schist with abundant sillimanite; 1,200 m (4,000 ft) thick; (3) Oh_3; schist with locally abundant calc-silicate rock and amphibolite; 0 to 600 m (0 to 2,000 ft) thick; (4) Os_4; sulfidic and graphitic schist; 600 m (2,000 ft) thick; (5) Os_5; graphitic schist with abundant lenses and pods of zoisite amphibolite; 300 m (1,000 ft) thick; (6) Os_6; sulfidic nongraphitic schist with abundant thin coticule layers; 180 m (600 ft) thick; (7) Oc_7; interlayered felsic and mafic gneiss; minimum 450 m (1,500 ft) thick; (8) O-Єm_8; interlayered fine-grained schist and granular schist with laminated hornblende-epidote amphibolite abundant toward base; minimum 900 m (3,000 ft) thick; (9) Єh_9; schist, commonly containing large muscovite porphyroblasts; 900 m (3,000 ft) thick; (10) Єh_{10}; light-gray gneiss; 1,500 m (5,000 ft) thick; (11) Єh_{11}; schist with feldspar porphyroblasts; 1,500 m (5,000 ft) thick; (12) Єh_{12}; light-gray gneiss with schist layers; 0 to 1,050 m (0 to 3,500 ft) thick; and (13) Єh_{13}; heterogeneous interlayered schist and gneiss; 3,000 m (10,000 ft) thick. Units 1, 2, and 3 are members of the Hartland Formation of Connecticut; units 4, 5, and 6 are equivalent to the Straits Schist of Connecticut; and units 1 through 6 are probably equivalent to the Partridge Formation of New Hampshire. Unit 7 is equivalent to the dome gneiss of Connecticut, the Ammonoosuc Volcanics of New Hampshire, and the Hawley Formation of Massachusetts. Unit 8 is equivalent to the Moretown Formation and Rowe Schist of Vermont and Massachusetts; units 9, 10, 11, 12, and 13 are equivalent to the Waramaug Formation (of local usage) of Connecticut and to the Hoosac Formation of western Massachusetts and Vermont.

INTRODUCTION

In the New Hartford, West Granville, Tariffville, and Southwick 7-1/2′ quadrangles in south-central Massachusetts and north-central Connecticut, 13 major distinctive lithologic units persist for long distances across the area (Fig. 1). Many minor units have been recognized, but they cannot be shown at the scale of Figure 1.

The oldest rocks (p€ on Fig. 1) crop out in the northwest corner of the area. They are light-gray feldspathic gneiss interlayered with amphibolite, and they are lithologically similar to and thus tentatively correlated with Precambrian rocks to the north and west (Harwood, 1971; Norton, 1971b). Interlayered schist and gneiss (units $€h_9$ to $€h_{13}$) about 8,100 m (26,500 ft) thick, overlie the Precambrian(?) rocks and are tentatively correlated with the Hoosac Formation of Early Cambrian age or older. Overlying these are a group of granular schist and amphibolite rocks ($O€m_8$) that are correlated with the Moretown Formation and Rowe Schist of Ordovician and Cambrian ages. This group of rocks is bounded on the east by a major north-northeast-striking fault herein called the Barkhamsted fault for the Barkhamsted Reservoir.

East of the Barkhamsted fault, the oldest rocks exposed in the North Granby dome (Rodgers and others, 1956) are a series of bedded mafic and felsic gneiss, schist, and sparse coticule rocks (Oc_7) that are correlated with the Collinsville Formation of Connecticut (Stanley, 1964) and with part of the Hawley Formation of Massachusetts (Hatch, 1967). Overlying the Collinsville Formation are a series of rusty-weathering graphitic and nongraphitic schists (Os_6, Os_5, Os_4) that are correlated with the Straits Schist of Connecticut (Rodgers and others, 1956) and with part of the Hawley Formation of Massachusetts (Hatch, 1967). Above these rusty-weathering schists is another sequence of rusty-weathering schist, amphibolite, and calc-silicate rocks (Oh_3) correlative with the Rattlesnake Hill Formation and the lower part of the Satan's Kingdom Formation of Connecticut (Stanley, 1964). Overlying these is a thick sequence of non-rusty-weathering kyanite-rich schist (Oh_2) that is correlated with the upper part of the Satan's Kingdom Formation and the Slasher's Ledges Formation of Connecticut (Stanley, 1964). The uppermost unit is a heterogeneous sequence of graphitic and nongraphitic schist, calc-silicate rocks, amphibolite, and coticule rocks (Oh_1) correlated with part of the Rattlesnake Hill and Satan's Kingdom Formations of Connecticut (Stanley, 1964). I believe that these last three units are not present to the north in Massachusetts.

Intrusive rocks are abundant throughout the area; most are small and crosscut granitic and granodioritic rocks that are not shown in Figure 1. The intrusive rocks indicated in Figure 1 include a large diorite body (Dd) in the northwest part of the area, and two smaller bodies to the east. The large body has been dated at 356 ± 11 m.y. B.P., using biotite, and at 373 ± 12 m.y. B.P., using hornblende, by the K-Ar method (Zartman and others, 1970). The other intrusive bodies shown on the map (Ou) are ultramafic rocks of probable Ordovician age that have been altered to serpentinite and talc rock.

Figure 2 is a simplified stratigraphic section for the area. The determination of stratigraphic position of the units has been based on the lack of repetition of units and on observations of primary sedimentary structures that are well preserved in some of the units. In places these structures are near enough to the contacts so that the relative ages of the units throughout the area could be determined with confidence. Thicknesses have been calculated both from measured sections and from cross sections constructed in several places.

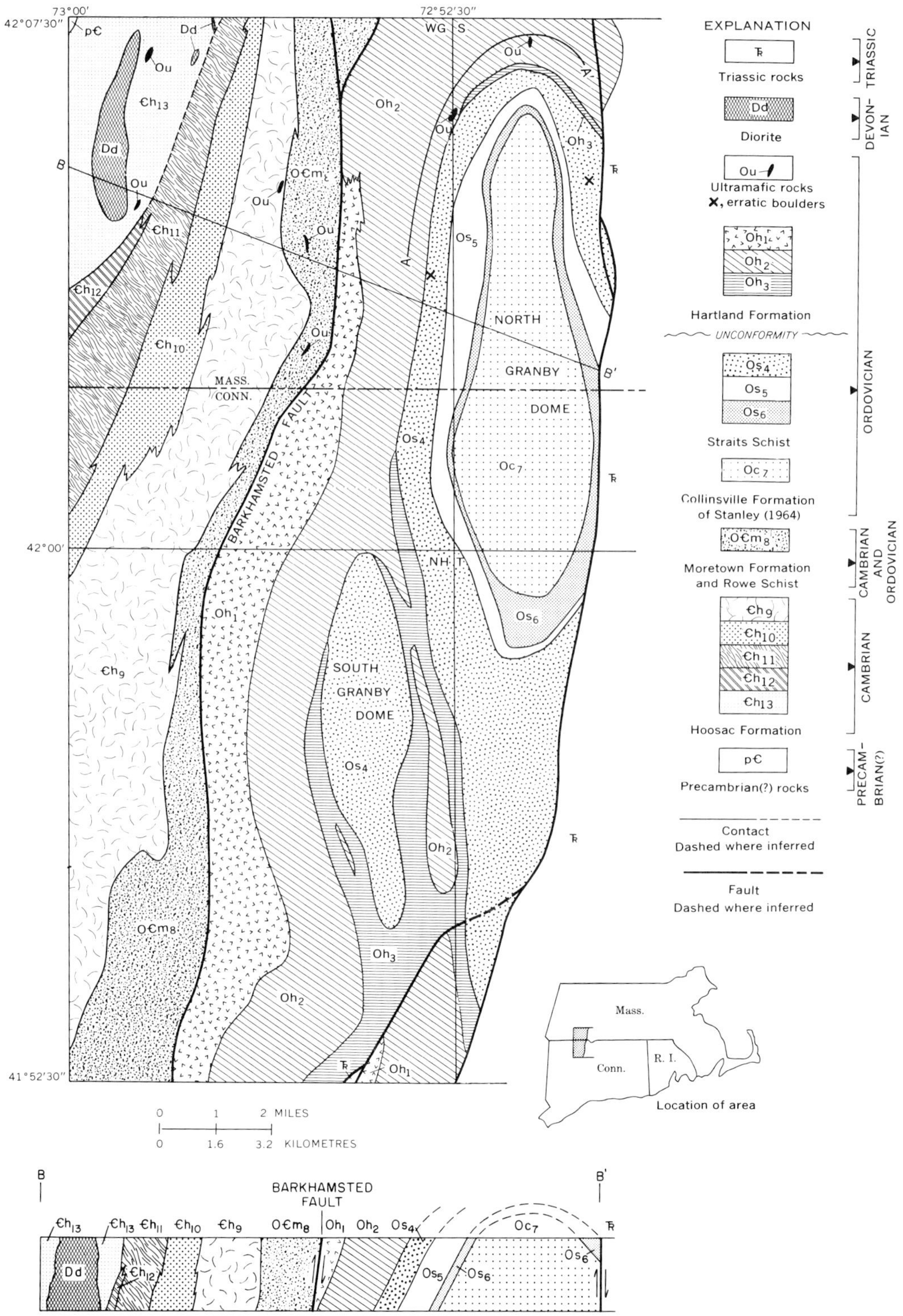

Figure 1. Geologic map of West Granville (WG), New Hartford (NH), Southwick (S), and Tariffville (T) quadrangles, Massachusetts and Connecticut. Section A-A′ is given in Figure 3.

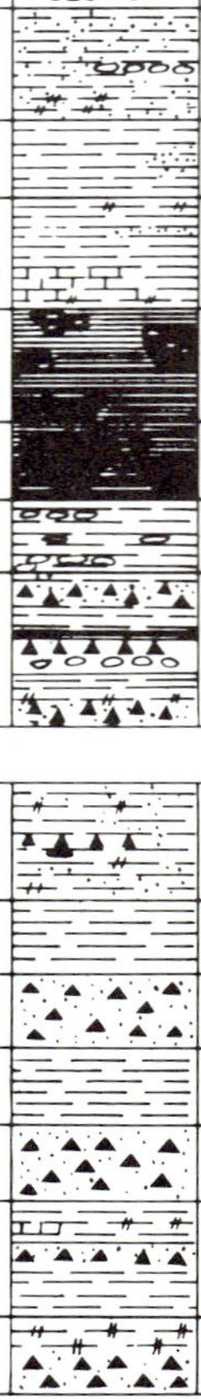

UNIT	COLUMNAR SECTION	THICKNESS (m)	PRESENT LITHOLOGY	ORIGINAL LITHOLOGY	CRITICAL PRIMARY STRUCTURES	ENVIRONMENT OF DEPOSITION
Oh_1		1,000+	Schist, amphibolite, calc-silicate rock, and coticule rock	Sandy and limy shale, mafic volcaniclastic rock, and chert	Beds 2.5 cm to 1.5 m thick; graded beds, channels	Normal marine, locally volcanic
Oh_2		1,330	Kyanite-garnet schist	Massive gray shale	Vague beds 15 cm to 1.5 m thick	Normal marine
Oh_3		0–660	Schist, calc-silicate rock, amphibolite, and quartzite	Gray shale, limy shale, volcaniclastic rock, and sandstone	Beds 2.5 cm to 60 cm thick; graded beds, channels	Normal marine, locally volcanic
Os_4		660	Sulfidic and graphitic schist	Black shale	Beds 1.3 cm to 30 cm thick; graded beds at top	Anerobic marine; turbidite at top
Os_5		330	Graphitic schist and zoisite amphibolite	Black shale with limy concretions or lenses	Beds 2.5 cm to 3 m thick; rare graded beds	Anerobic marine
Os_6		200	Sulfidic nongraphitic schist and coticule rocks	Gray shale with chert layers	Beds 1 mm to 1.5 m thick; rare graded beds	Normal marine
Oc_7		500 +	Amphibolite, light and dark gneiss, schist, and coticule rock	Mafic and felsic volcaniclastic rock, gray shale, black shale, and chert	Beds 1.3 cm to 1.5 m thick	Submarine volcanic, normal marine
FAULT—Unknown thickness of section missing						
O€m_8		1,000+	Schist, granular schist, and amphibolite	Gray shale, siltstone, and mafic volcanic rock	Beds 5 cm to 6 m thick; laminations, crossbeds, graded beds	Normal near shore marine, submarine volcanic
€h_9		1,000	Upper rusty-weathering schist	Massive gray shale	Vague to distinct beds 15 cm to 1.5 m thick	Normal marine
€h_{10}		2,330	Upper light-gray gneiss	Felsic volcanic rock or feldspathic siltstone	Very vague beds	Submarine volcanic
€h_{11}		2,330	Lower rusty and nonrusty schist	Gray shale	Vague to distinct beds, locally graded	Normal marine
€h_{12}		1,160	Lower light-gray gneiss	Felsic volcanic rock or feldspathic siltstone	Very vague beds	Submarine volcanic
€h_{13}		3,330	Schist, gneiss, and sparse amphibolite and calc-silicate rock	Gray shale, limy shale, and volcanic rock	Beds 2.5 cm to 3 m thick; graded beds, channels	Normal marine, submarine volcanic
p€		266 +	Light and dark gneiss	Felsic and mafic volcanic rocks	Beds 2.5 cm to 3 m thick; very few graded beds	Submarine volcanic

Figure 2. Schematic columnar section of pre-Silurian rocks in West Granville, New Hartford, Southwick, and Tariffville quadrangles, Massachusetts and Connecticut. Thicknesses shown are maximum thicknesses observed in area of this report; ranges of thickness are given for units that are absent locally.

STRATIGRAPHY WEST OF THE BARKHAMSTED FAULT

Precambrian(?) Rocks

The oldest rocks are generally well-bedded, interlayered light-gray plagioclase-microcline-quartz-biotite gneiss and hornblende-plagioclase amphibolite exposed in the northwest corner of the area (p€). Only about 250 m (800 ft) of these rocks are known in the area, but they persist to the west and north in the Tolland Center, Otis, and Blandford quadrangles; their total thickness is unknown. This gneiss differs from the gneiss in the overlying Hoosac Formation only by the presence of amphibolite layers in the underlying group. Primary sedimentary structures, including very few graded beds, show tops to both the east and west. These rocks were presumably deposited as submarine volcaniclastic debris.

In the Tolland Center quadrangle to the west, these Precambrian(?) rocks contain several layers of schist similar in composition to schist in the Hoosac Formation. If the presence of schist is taken as the principal criterion for defining the Hoosac Formation, these rocks are a part of that formation and are not Precambrian in age.

Hoosac Formation

A series of schist and gneiss in the western part of the area is here tentatively correlated with the Hoosac Formation of Early Cambrian age or older. The rocks have an aggregate thickness of 8,100 m (26,500 ft) in this area. To the north, however, in the Blandford quadrangle, the entire Hoosac Formation thins rapidly to less than 300 m (1,000 ft) (N. L. Hatch, Jr., 1970, oral commun.).

The distribution of mapped units within lithologies in the Hoosac Formation is generalized in Figure 1; individual units commonly are discontinuous across the map and are too small to be shown separately. In addition, outcrops are uncommon throughout much of the area, but in an area of abundant outcrops in the northwesternmost corner of the map area, the interlayering of the lithologies is clearly more complicated than is shown.

Contacts between schist and gneiss within the mapped units shown in Figure 1 are exposed in a few places. No evidence of faulting, such as gouge or breccia, nor relict textures indicating premetamorphic gouge or breccia, has been found in this area at the contacts of the schist and gneiss; rather, where exposed, they have normal sedimentary contacts. Sedimentary structures (see Fig. 2) indicate an interlayered homoclinal sequence.

The Hoosac Formation includes units 9 through 13 of this report. Units 9, 11, and 13 are dominantly schist; units 10 and 12 are dominantly gneiss. I consider the gneiss to be an accumulation of volcaniclastic material deposited at intervals in a basin that was receiving normal land-derived sediments during the intervolcanic periods. Such an interpretation requires repeated intervals of volcanic activity, from Precambrian(?) into Cambrian time, in a basin that was deepening rapidly to the south.

Schist, Gneiss, Amphibolite, and Calc-silicate Rocks (Unit 13). Overlying the Precambrian(?) gneiss in the northwestern corner of the map area is a sequence of discontinuous bands of schist, gneiss, and sparse amphibolite and calc-silicate rock (€h_{13}) about 3,000 m (10,000 ft) thick. The succession of lithologies from bottom to top is (1) light- to medium-gray, medium-grained plagioclase-muscovite-quartz-biotite schist characterized by plagioclase porphyroblasts averaging about 3 mm (1/8 in.) in diameter; (2) light- to medium-brown, medium-grained plagioclase-muscovite-quartz-biotite-kyanite-garnet schist; (3) light-gray, medium- to fine-grained quartz-plagioclase-microcline-biotite gneiss; (4) light- to medium-brown, medium-grained quartz-plagioclase-muscovite-biotite schist, locally containing muscovite porphyroblasts that average about 3 mm (1/8 in.) in diameter and are at random angles to the foliation, and locally having rust staining that penetrates about 2.5 cm (1 in.) into the rock from weathered surfaces; (5) schist, similar to band 1 above, containing scattered lenses of gneiss similar to band 3 above; (6) light- to medium-brown, medium-grained quartz-plagioclase-muscovite-biotite-garnet schist characterized by subhedral garnet porphyroblasts about 6 mm (1/4 in.) in diameter; (7) schist similar to band 4 above; (8) schist similar to band 1 above; and (9) schist similar to band 4 above. The amphibolite rocks are commonly less than 1.5 m (5 ft) thick and are mostly hornblende-plagioclase rocks that in a few places contain small garnets. The amphibolite is scattered throughout the schist of this unit and appears to be more abundant toward the base. The calc-silicate rocks commonly contain amphibole, epidote, clinozoisite, microcline, and sparse calcite. They are very rare and occur mostly as thin lenses near the base of the unit. The light-gray gneiss is typically poorly exposed.

Lower Light-gray Gneiss (Unit 12). This unit ($€h_{12}$) is a layer of generally poorly exposed light-gray gneiss, with a maximum thickness of about 1,050 m (3,500 ft) on the northwest edge of the area. It pinches out about 3.2 km (2 mi) to the northeast. This gneiss is composed of plagioclase, microcline, quartz, and biotite with very few flakes of muscovite (seen only in thin sections). It is similar in composition and texture to the light-gray gneiss of unit 13. Primary structures are very rare in this unit and consist generally of only very indistinctly defined beds.

Lower Schist (Unit 11). Overlying the lower light-gray gneiss is about 1,500 m (5,000 ft) of rusty-weathering and non-rusty-weathering quartz-plagioclase-muscovite-biotite schist ($€h_{11}$). In the northern part of the area, this unit is mostly non-rusty weathering and is characterized by kyanite or sillimanite knots about 6 mm (1/4 in.) in diameter, and by abundant small garnets. In the southern part, it is rusty weathering and is characterized by muscovite porphyroblasts as large as 3 mm (1/8 in.) in diameter that are oriented at diverse angles to the foliation. Primary sedimentary structures in this schist are rare and consist mainly of indistinct beds that are graded in a very few places.

Upper Light-gray Gneiss (Unit 10). The upper light-gray gneiss ($€h_{10}$) about 1,500 m (5,000 ft) thick, overlies the lower schist. It is lithologically and texturally identical to the light-gray gneiss and, like that gneiss, is very poorly exposed in most places. Near the northern map boundary, it contains three lenses of schist similar in composition to unit 11. Another very small schist lens is in about the middle of the map area. One amphibolite was seen in the gneiss near the northern boundary of the map area.

Upper Schist (Unit 9). Rusty-weathering plagioclase-quartz-muscovite-biotite schist ($€h_9$), about 900 m (3,000 ft) thick, overlies unit 10. This unit consists of two distinctive rock types. One is a rusty-weathering schist characterized by muscovite porphyroblasts that are about 3 mm (1/8 in.) in diameter and oriented at angles to the foliation; the other is characterized by a "rusty rind" that is about 2.6 cm (1 in.) thick on most natural exposures. In the southern part of the area, these two rock types could be mapped separately—about 300 m (1,000 ft) of the rusty-rind rock overlies about 600 m (2,000 ft) of the schist containing muscovite porphyroblasts. In the northern part of the area, the two schists are intimately interlayered, and it is not possible to map them separately.

Rowe Schist and Moretown Formation (Unit 8). Overlying the Hoosac Formation is a series of rocks ($O€m_8$) that are here tentatively correlated with the Rowe Schist and Moretown Formation of Massachusetts (Hatch and others, 1966; Hatch, 1969). This sequence of rocks is at least 900 m (3,000 ft) thick. It is bounded on the east by the Barkhamsted fault.

The lowest rock type in unit 8 is a delicately layered hornblende-epidote amphibolite. Near the southern boundary of the map area, the amphibolite seems to be less than 3 m (10 ft) thick and is represented only by angular fragments in the till. It thickens to about 300 m (1,000 ft) in the north-central part of the West Granville quadrangle and pinches out just south of the northern map boundary. Similar amphibolites are present through a short interval in the overlying rocks in the southern part of the map area. Presumably this rock was deposited as a blanket of mafic volcanic ash in a submarine environment.

In the northern part of the area, a lens, in a few places as much as 30 m (100 ft) thick, of very coarse grained massive hornblende-anthophyllite amphibolite overlies the hornblende-epidote amphibolite. This coarse-grained amphibolite is characterized by 5- to 7.5-cm (2- to 3-in.) crystals in a crisscross network surrounded by a fine-grained matrix. The lens crops out intermittently for about 3 km (2 mi) and apparently pinches out both to the north and south. This rock could have been a flow, a sill, or a magnesium-rich limestone.

A layer of very coarse grained quartz-plagioclase-biotite-muscovite-staurolite-kyanite-sillimanite-garnet schist overlies the amphibolite. It extends for about 5 km (3 mi) north and south and has a maximum thickness of about 300 m (1,000 ft). Bedding is poorly defined. The layer presumably was deposited as an aluminum-rich gray shale in a normal marine environment.

Overlying the coarse-grained schist is a delicately pin-striped quartzose rock that has very thin quartzofeldspathic layers separated by paper-thin partings of biotite-muscovite schist. In the southern part of the area, this rock is at least 1,050 m (3,500 ft) thick. To the north it either pinches out or interfingers with stratigraphically equivalent rocks; exposures are too poor to permit an accurate evaluation. This unit is typical of the pin-stripe rock in the Moretown Formation, described by Cady (1956), in Vermont.

The environment of deposition for this rock is difficult to envision. Presumably, the pin striping represents cyclic deposition of some sort in a massive gray shale. In many outcrops, very delicate sedimentary structures are preserved. These include cross-beds and graded beds, and they suggest quiet-water deposition. In the southern part of the area, such structures are preserved close enough to the contact to indicate the stratigraphic superposition of the pin-stripe rock over the hornblende-epidote amphibolite; there is no repetition of beds, indicating that the sequence is older to the west.

In the northern part of the area, the pin-stripe rock is not recognized, and an interbedded sequence of more thickly bedded quartz-plagioclase-biotite-muscovite schist and granular schist seems to be in its place. This part of unit 8 contains many thin layers and lenses of coarse-grained hornblende-plagioclase-garnet amphibolite; the garnets are euhedral and commonly 6 to 13 mm (1/4 to 1/2 in.) in diameter. These rocks are at least 900 m (3,000 ft) thick at the northern end of the area, where they make up most of the unit.

Cross-bedding, graded beds, and small channels abound in the schist and granular schist. The environment of deposition was probably nearshore marine, with the granular schist representing siltstone and the schist representing shale. The amphibolites in this part of the unit may have been impure calcareous sediment.

STRATIGRAPHY EAST OF THE BARKHAMSTED FAULT

East of the Barkhamsted fault is a series of rocks that are probably younger than any of those described to the west of the fault. These rocks are exposed around a pair of en echelon domes, the Granby domes (Rodgers and other, 1956). The oldest rocks in this sequence (Oc_7), exposed in the central part of the northern dome, are correlated with the Bristol Member of the Collinsville Formation of Stanley (1964), and with part of the Hawley Formation of Massachusetts (Hatch, 1967). In the Collinsville quadrangle, the Collinsville Formation overlies the Taine Mountain Formation, which is considered to be equivalent to the Moretown Formation of Massachusetts (Stanley, 1968). Top-facing directions around the domes, as determined from sedimentary structures, indicate that the units are successively younger outward from the center of the domes; therefore, on the basis of this correlation, all the rocks east of the fault are younger than the rocks west of the fault.

Collinsville Formation (Unit 7)

Interlayered amphibolite, light- and dark-colored gneiss, schist, and coticule rock (Oc_7) are here correlated with similar rocks mapped as the Collinsville Formation by Stanley (1964) in the Collinsville and Bristol domes in the Collinsville quadrangle, Connecticut. These rocks are a minimum of 450 m (1,500 ft) thick in the center of the North Granby dome, but only the upper 150 m (500 ft) is exposed; they are not exposed in the southern dome.

The composition and the thickness of the individual layers vary greatly from layer to layer, and exposures are poor. Amphibolite is probably the most abundant rock; light- and dark-colored (felsic) gneiss is less abundant; schist and coticule rock seem to be most abundant near the top of the formation. Sedimentary structures other than beds are rare. Beds range from 1.3 cm to 1.5 m (1/2 in. to 5 ft) in thickness; some of the schist layers were graded beds. These rocks probably were originally deposited in a marine environment as bedded volcaniclastic rocks interlayered with gray shale and calcareous chert.

Straits Schist

Overlying the Collinsville Formation is a sequence of rusty-weathering mica schist (Os_6, Os_5, Os_4) about 490 m (3,600 ft) thick; this sequence is here correlated with the Straits Schist and the Sweetheart Mountain Member of the Collinsville Formation of Stanley (1964). It consists of three distinctive units around the North Granby dome; only the upper unit is exposed in the South Granby dome.

Sulfidic Schist and Coticule Rocks (Unit 6). The lowest unit in the Straits Schist (Os_6), 180 m (600 ft) thick, is a rusty-weathering, nongraphitic quartz-plagioclase-biotite-garnet schist containing sparse kyanite or sillimanite. Interlayered with the schist are beds of coticule rocks, ranging from 1 mm to 7.5 cm (1/32 to 3 in.) in thickness, which vary in abundance from place to place. In general the schist is well bedded, but other primary structures are rare and consist mainly of graded beds. Originally this rock was probably a gray shale containing thin layers of calcareous chert or sandstone. It was probably deposited in a normal marine environment.

Graphitic Schist and Zoisite Amphibolite (Unit 5). An indistinctly defined unit (Os_5), about 300 m (1,000 ft) thick, of rusty-weathering, graphitic quartz-plagioclase-muscovite-biotite-kyanite-sillimanite-staurolite-garnet schist containing abundant pods and lenses of hornblende-zoisite-microcline-garnet amphibolite overlies unit 6 around the northern dome. This unit is defined by the presence of zoisite amphibolite pods, and the upper contact is well defined only where the pods are abundant and well exposed, as on the south and east sides of the dome. Primary structures are uncommon in this unit; they consist of ill-defined beds 2.5 cm to 3 m (1 in. to 10 ft) thick in the schist. There are no bedding planes in the amphibolite. A few beds in the schist are graded. Originally the rocks were probably massive black shale containing lenses, or perhaps giant concretions, of impure limestone deposited in an anaerobic marine environment.

Sulfidic and Graphitic Schist (Unit 4). Overlying unit 5 is a monotonous sulfidic and graphitic quartz-plagioclase-muscovite-biotite-kyanite-sillimanite-staurolite-garnet schist (Os_4) very similar in composition to the schist in the underlying unit. It is about 600 m (2,000 ft) thick in most places. Sedimentary structures are ill defined in the lower part of this unit, but toward the top, bedding becomes distinct and characterized by graded beds that range from 13 mm to 30 cm (1/2 in. to 1 ft) in thickness. This unit was probably originally a black shale that graded upward into a turbidite; it was probably deposited in an anaerobic marine environment.

Hartland Formation

Units 1, 2, and 3 (Oh_1, Oh_2, Oh_3) are here considered to be part of the Hartland Formation of Connecticut. The maximum thickness of the Hartland Formation in this area is about 2,700 m (9,000 ft), but the top is not exposed. The Hartland Formation is a heterogeneous group of rocks, primarily schist throughout the area.

Schist, Calc-silicate Rock, Amphibolite, and Quartzite (Unit 3). The lowest unit (Oh_3) in the Hartland Formation is a heterogeneous sequence of rocks 0 to 600 m (0 to 2,000 ft) thick. Most of the rocks in this unit are rusty-weathering schist, but it also includes some thin amphibolite layers and is characterized by thin layers and lenses of well-bedded calc-silicate rock containing quartz, microcline, clinozoisite, epidote, sphene, plagioclase, and calcite in varying proportions. In one place, near the southern part of the map area, a bed of nearly pure calcite marble is exposed in a small roadcut. In another place, at the southern boundary, a bed of vitreous quartzite about 60 cm (2 ft) thick is exposed in a small stream valley. Sedimentary structures other than bedding commonly are not well developed in this unit. Originally the unit probably consisted of a sequence of interlayered gray shale, calcareous gray shale, limestone, sandstone or quartz-rich siltstone, and volcaniclastic debris. It was deposited in a marine environment, probably near shore.

Exposures of contacts between unit 3 and the overlying and underlying rocks are very rare. Where they exist, both contacts appear to be conformable, at least at the scale of the exposures. The map pattern, however, suggests an unconformity (Fig. 1). The unconformity probably is at the base of this unit, but another unconformity may occur at the top. In general aspect, the schist in this unit is most like the schist of the underlying unit, except in the northern part of the area.

Kyanite-Garnet Schist (Unit 2). A monotonous coarse-grained, non-rusty-weathering quartz-plagioclase-muscovite-biotite-kyanite-garnet-sillimanite-staurolite schist (Oh_2) that locally contains abundant magnetite overlies unit 3. It is uniformly 1,200 m (4,000 ft) thick from the southern end of the area to near the northern end, where it thickens. Characteristic of this unit is an abundance of aluminosilicate minerals; locally, kyanite crystals as much as 16 cm (6 in.) long make up as much as 30 percent of the rock. In places, sillimanite is pseudomorphous after kyanite.

In the northern part of the area, the homogeneity of this unit disappears, and several additional rock types appear to interfinger with the coarse-grained kyanite-garnet schist. Figure 3 is a restored facies diagram of unit 2 parallel to strike along line A-A′ of Figure 1; this diagram was constructed by drawing cross sections perpendicular to A-A′ and measuring the thicknesses of each rock type along them. These thicknesses were then transferred to A-A′, using the top of unit 3 as a reference datum. This procedure introduced some geometric distortions, but the illustration is an approximate representation of the distribution of the various rock types.

Layers of the coarse-grained kyanite-garnet schist (aluminum-rich schist of Fig. 3) persist to the northern edge of the map area. Interfingering with these layers is a schist similar in texture and composition, except for a great decrease in the amount of aluminosilicate minerals (aluminum-poor schist of Fig. 3); a quartzofeldspathic schist; and a rusty-weathering muscovite-rich schist. These alternations in lithology imply an offlap-onlap relationship between the various lithologies. The aluminous rocks represent clay-rich massive shale, the quartzofeldspathic rocks represent more sandy or silty rocks, and the rusty and nonaluminous schist represents rocks of intermediate composition. A further implication is that the source area for the sediment was to the north or northeast.

Schist, Amphibolite, Calc-silicate Rock, and Coticule Rock (Unit 1). The uppermost unit (Oh_1) in the Hartland Formation is a sequence of extremely variable rocks that is a minimum of 900 m (3,000 ft) thick. The unit contains schist, granular schist, calc-silicate rock, coticule rock, and thin amphibolite. For the most part, it is concealed by a thick blanket of glacial drift; it is very poorly exposed along the east-facing slope of the valley of the East Branch of the Farmington River. None of the individual lithologies exceeds 30 m (100 ft) in thickness, and few can be traced for more than 1.6 km (1 mi). This unit is very well bedded and has sparse graded beds and small channels. The original lithology was probably as varied as the present rock and probably consisted of shale, siltstone, limy shale, mafic volcaniclastic sediment, and chert. These materials were probably deposited as normal marine sediment, and may represent an onlap relation with the underlying rocks.

CORRELATION

Correlation between rocks mapped on the east flank of the Berkshire anticlinorium in Massachusetts and Vermont with rocks mapped in what appears to be the same stratigraphic position in Connecticut has long been a problem. The early, simple version, which correlated the Hoosac Schist of Massachusetts with the Hartland Formation of Connecticut (Rice and Gregory, 1906), has not been substantiated by recent detailed mapping. Geologists working in the area have wrestled with the problem for several years and have come to differing conclusions (Fritts, 1962; Gates and Christensen, 1965; Gates and Martin, 1967, and this volume; Hatch and Stanley, 1970; Hatch and others, 1968; Martin, 1970; Rodgers and others, 1956; Stanley, 1964, 1967, 1968). The correlations presented here are, with minor modifications, those arrived at by Hatch and others (1968).

Figure 4 shows the correlations proposed here with areas nearby to the north and south. The units shown in Figure 4 are drawn approximately to vertical scale, and the thicknesses are taken from the reports cited or scaled from maps or cross sections in those reports. Each column shows the stratigraphic sequence indicated by the author or authors cited.

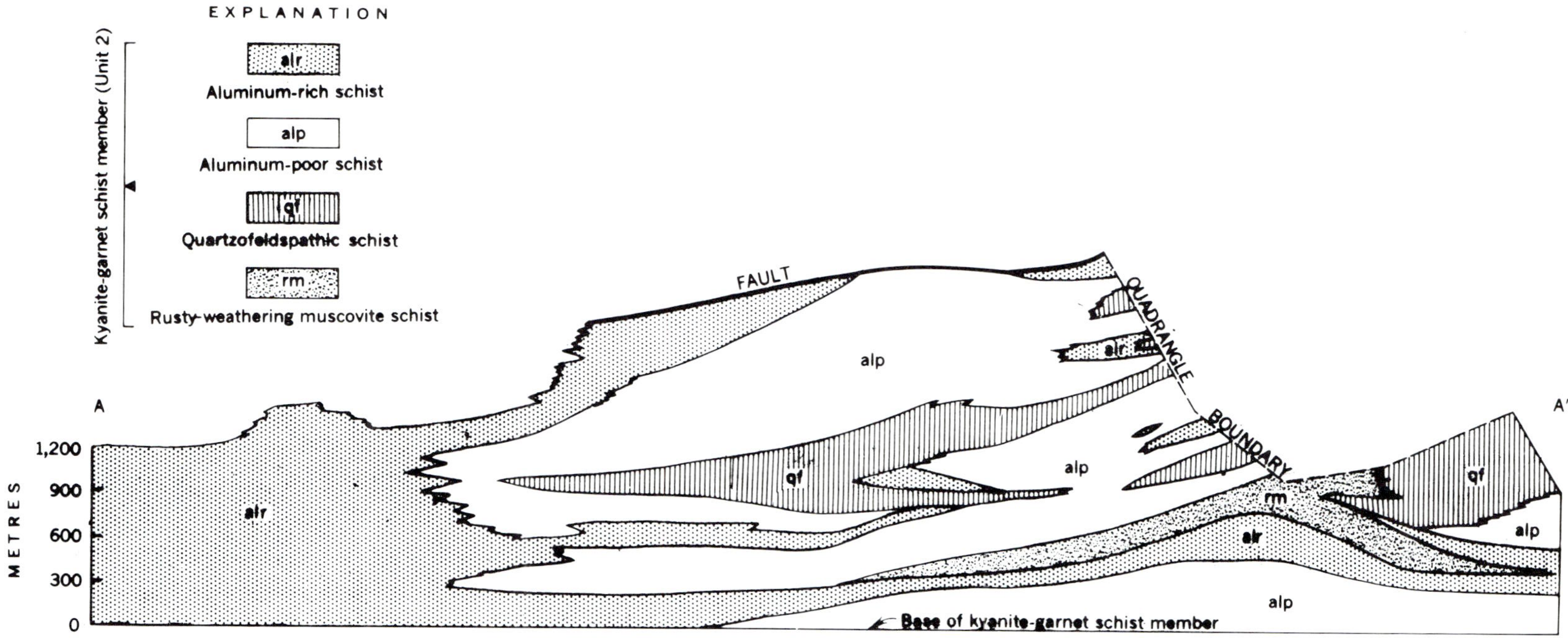

Figure 3. Restored facies diagram of kyanite-garnet schist member (unit 2) of Hartland Formation, along line A–A′ of Figure 1, West Granville and Southwick quadrangles, Massachusetts and Connecticut.

Correlations of units mapped in the area west of the Barkhamsted fault with units mapped to the north are based on continuous mapping. Rocks in the Hoosac Formation are lithologically the same as those described by Norton (1971a, 1971b), although the stratigraphic section in this area is not the same as his. They are also the same as those described by Hatch (1970, oral commun.) in the Blandford quadrangle. The Rowe Schist has been described as pinching out in the southern part of the Blandford quadrangle (Hatch and others, 1968). At the time, however, the coarse-grained kyanite schist and the hornblende-anthophyllite amphibolite of unit 8 (O€m_8) had not been discovered in the area shown in Figure 1. It now seems reasonable to equate the coarse-grained schist, the hornblende-anthophyllite amphibolite, and the underlying laminated hornblende-epidote amphibolite with units mapped as Rowe to the north and to retain the correlation with the Moretown

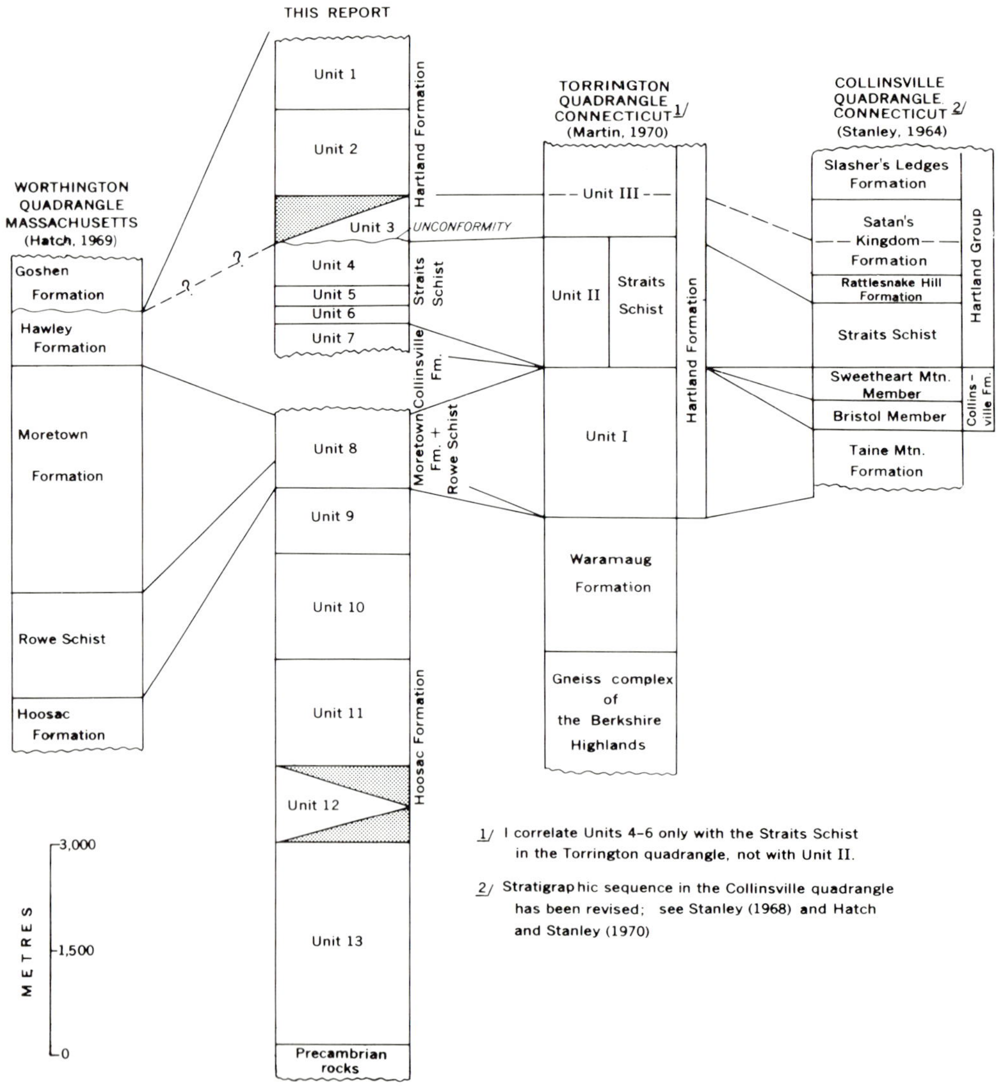

Figure 4. Correlation diagram showing proposed correlations between units mapped in West Granville, Southwick, Tariffville, and New Hartford quadrangles and areas mapped to north and south.

Formation and the overlying rocks in unit 8. The part of unit 8 north of the middle of the West Granville quadrangle is unlike the pin-stripe rocks to the south; it may represent a different, younger formation or it may be a facies equivalent of rocks to the south.

Correlation of rocks mapped west of the Barkhamsted fault with rocks mapped to the south is based on nearly continuous mapping. The Hoosac Formation is equivalent to the Waramaug Formation and part of the gneiss complex of the Berkshire Highlands as mapped by Martin (1970) and by Gates and Christensen (1965). The laminated hornblende-epidote amphibolite thins greatly toward the southern end of the area and was not mapped separately by Martin (1970) nor by Gates and Christensen (1965). It was, however, mapped separately by J. B. Gonthier (1962, written commun.), who called it the Chester amphibolite, in the northern part of the Torrington quadrangle. The Moretown Formation correlates directly with rocks mapped as unit 1 (Oh_1) of the Hartland Formation by Martin (1970) and by Gates and Christensen (1965). It also correlates directly with the part of the Rattlesnake Hill Formation designated as "Rocks on Jones and Yellow Mountain" by Stanley (1964). Gonthier (1962, written commun.) referred to the unit as the Savoy Formation and correlated it with the Moretown, Stowe, and Ottaqueche Formations of Vermont.

Correlations of units mapped east of the Barkhamsted fault with formations mapped in the Collinsville quadrangle by Stanley (1964) are based on continuous mapping. The differences in names and relative stratigraphic positions of the various units are largely differences in structural interpretations.

The Collinsville Formation of this report (unit 7—Oc_7) is correlated with the Bristol Member of the Collinsville Formation of Stanley (1964). The Straits Schist (units 4 to 6—$Os_{4\text{-}6}$) is correlated with the Sweetheart Mountain Member of Stanley's Collinsville Formation and with the Straits Schist of Stanley (1964) and Martin (1970). The Hartland Formation (units 1 to 3—$Oh_{1\text{-}3}$) is correlated with the Rattlesnake Hill Formation, the Satan's Kingdom Formation, and the Slasher's Ledges Formation.

Correlation of the units mapped east of the Barkhamsted fault with formations mapped to the north is based on the correlation of the Taine Mountain Formation of Connecticut with the Moretown Formation of Massachusetts (Stanley, 1968). If this correlation is correct, and if the correlations of younger units are correct, then the Moretown Formation is present at depth in the Granby domes. With the knowledge that the rocks are successively younger away from the core of the dome, as deduced from primary structures, a reasonable correlation can be made between the Straits Schist and Collinsville Formation of Connecticut and the Hawley Formation of Massachusetts, because this part of the section is marked in both places by the presence of metamorphosed mafic volcaniclastic rocks and by graphitic muscovite schist.

In most of central Massachusetts, the Hawley Formation is unconformably overlain by the Goshen Formation (Hatch, 1967). In the southern part of the Blandford and Woronoco quadrangles, however, the Russell Mountain Formation (Hatch and others, 1970) appears between the two. Unit 3 (Oh_3) of the Hartland Formation is similar to the Russell Mountain Formation, and it too seems to mark an unconformity. If unit 3 and the Russell Mountain Formation are correlated, then the rocks above unit 3 should be equivalent to the Goshen Formation. The rocks above unit 3, however, do not resemble the Goshen Formation. This problem was resolved by Hatch and others (1968) by considering the Hartland Formation to postdate emplacement of the Hawley Formation and to predate emplacement of the Shaw Mountain Formation.

REFERENCES CITED

Cady, W. M., 1956, Bedrock geology of the Montpelier quadrangle, Vermont: U.S. Geol. Survey Geol. Quad. Map GQ-79, scale 1:62,500.

Fritts, C. E., 1962, Age and sequence of metasedimentary and metavolcanic formations northwest of New Haven, Connecticut: U.S. Geol. Survey Prof. Paper 450-D, p. D32-D36.

Gates, R. M., and Christensen, N. I., 1965, The bedrock geology of the West Torrington quadrangle: Connecticut Geol. and Nat. History Survey Quad. Rept. 17, 38 p.

Gates, R. M., and Martin, C. W., 1967, The bedrock geology of the Waterbury quadrangle: Connecticut Geol. and Nat. History Survey Quad. Rept. 22, 36 p.

——1976, Pre-Devonian stratigraphy of the central section of the western Connecticut highlands, *in* Page, L. R., ed., Contributions to the stratigraphy of New England: Geol. Soc. America Mem. 148 (this volume).

Harwood, D. S., 1971, Recumbent folding and southwestward thrusting of Precambrian and lower Paleozoic rocks in the South Sandisfield quadrangle, Massachusetts and Connecticut: Geol. Soc. America Abs. with Programs, v. 3, p. 36-37.

Hatch, N. L., Jr., 1967, Redefinition of the Hawley and Goshen Schists in western Massachusetts: U.S. Geol. Survey Bull. 1254-D, p. D1-D16.

——1969, Geologic map of the Worthington quadrangle, Hampshire and Berkshire Counties, Massachusetts: U.S. Geol. Survey Geol. Quad. Map GQ-857, scale 1:24,000.

Hatch, N. L., Jr., and Stanley, R. S., 1970, Stratigraphic continuity and facies changes in formations of early Paleozoic age in western Massachusetts and tentative correlations with Connecticut: Geol. Soc. America Abs. with Programs, v. 2, p. 23-24.

Hatch, N. L., Jr., Chidester, A. H., Osberg, P. H., and Norton, S. A., 1966, Redefinition of the Rowe Schist in northwestern Massachusetts, *in* Cohee, G. V., and West, W. S., eds., Changes in stratigraphic nomenclature by the U.S. Geological Survey, 1965: U.S. Geol. Survey Bull. 1244-A, p. A33-A35.

Hatch, N. L., Jr., Schnabel, R. S., and Norton, S. A., 1968, Stratigraphy and correlation of the rocks on the east limb of the Berkshire anticlinorium in western Massachusetts and north-central Connecticut, *in* Zen, E-an, White, W. S., Hadley, J. B., and Thompson, J. B., Jr., eds., Studies of Appalachian geology—Northern and maritime: New York, Interscience Pub., p. 177-184.

Hatch, N. L., Jr., Stanley, R. S., and Clark, S. F., Jr., 1970, The Russell Mountain Formation, a new stratigraphic unit in western Massachusetts: U.S. Geol. Survey Bull. 1324-B, p. B1-B10.

Martin, C. W., 1970, The bedrock geology of the Torrington quadrangle: Connecticut Geol. and Nat. History Survey Quad. Rept. 25, 53 p.

Norton, S. A., 1971a, The Hoosac Formation (Lower Cambrian or older) on the east limb of the Berkshire anticlinorium, western Massachusetts: Geol. Soc. America Abs. with Programs, v. 3, p. 45-46.

——1971b, Possible thrust faults between Lower Cambrian and Precambrian rocks, east edge of the Berkshire Highlands, western Massachusetts: Geol. Soc. America Abs. with Programs, v. 3, p. 46.

Rice, W. N., and Gregory, H. E., 1906, Manual of the geology of Connecticut: Connecticut Geol. and Nat. History Survey Bull. 6, 273 p.

Rodgers, John, Gates, R. M., Cameron, E. N., and Ross, R. J., Jr., 1956, Preliminary geological map of Connecticut: Connecticut Geol. and Nat. History Survey, scale 1:253,440.

Stanley, R. S., 1964, The bedrock geology of the Collinsville quadrangle: Connecticut Geol. and Nat. History Survey Quad. Rept. 16, 99 p.

——1967, Geometry and age relations of some minor folds and their relation to the Woronoco nappe, Blandford and Woronoco quadrangles, Massachusetts, *in* New England Intercollegiate Geol. Conf., 59th Ann. Mtg., 1967, Guidebook for field trips in the Connecticut Valley of Massachusetts: p. 48-60.

——1968, Metamorphic geology of the Collinsville area, *in* New England Intercollegiate Geol. Conf., 60th Ann. Mtg., 1968, Guidebook for field trips in Connecticut: Connecticut Geol. and Nat. History Survey Guidebook 2, 17 p.

Zartman, R. E., Hurley, P. M., Krueger, H. W., and Giletti, B. J., 1970, A Permian disturbance of K-Ar radiometric ages in New England: Its occurrence and cause: Geol. Soc. America Bull., v. 81, p. 3359-3374.

MANUSCRIPT RECEIVED BY THE SOCIETY APRIL 18, 1974
REVISED MANUSCRIPT RECEIVED MARCH 3, 1975
MANUSCRIPT ACCEPTED MARCH 18, 1975

Printed in U.S.A.

Geological Society of America
Memoir 148

Pre-Devonian Stratigraphy of the Central Section of the Western Connecticut Highlands

ROBERT M. GATES
University of Wisconsin
Madison, Wisconsin 53706

AND

CHARLES W. MARTIN
Earlham College
Richmond, Indiana 47374

ABSTRACT

The rocks of western Connecticut can be divided into three general categories: (1) the Precambrian complexes (Housatonic and Berkshire Highlands Gneiss complexes), (2) the metasedimentary rocks west of a major structural break known as Cameron's line, and (3) the lower Paleozoic rocks east of Cameron's line. The core of the Waterbury dome (Waterbury Formation), though east of Cameron's line, is older than the mantling lower Paleozoic rocks.

The Precambrian rocks are interlayered granitic to dioritic gneisses and amphibolites and mafic gneisses with subordinate terrigenous metasediments and are separated readily from the lower Paleozoic sequence.

The Waramaug Formation, probably the oldest part of the Paleozoic sequence, is a belt of interlayered and intergradational mica-plagioclase-quartz gneiss and sillimanite-garnet-mica-plagioclase-quartz gneiss, representing original graywackes and shales. This belt wraps around the Precambrian highlands and lies to the west of Cameron's line between the Precambrian rocks and the eastern Paleozoic rocks.

The Waterbury Formation, exposed in the core of the Waterbury dome, is an assemblage of metasedimentary and metavolcanic rocks of unknown thickness representing sedimentary mixtures of quartz and plagioclase sands, clay minerals, volcanogenic material, coarser clastic material, and possibly flows. These rocks are most similar in lithology to the belt of miogeosynclinal metasedimentary rocks west of Cameron's line and may be their eugeosynclinal equivalent. The Rb-Sr whole-rock data on these rocks give an isochron age of 465 ± 50 m.y. (Clark and Kulp, 1968).

The rocks in the belt east of Cameron's line constitute a continuous sequence of predominantly eugeosynclinal sediments that is divided into five mappable units. The lowest unit (Hartland Unit I) is predominantly fine-grained gray mica-plagioclase-quartz granulite and granulitic gneiss with subordinate schist layers. The second unit (Hitchcock Lake) is a banded and streaked rock composed of alternating thick and thin layers of a fine-grained plagioclase-quartz granulite and coarsely streaked micaceous plagioclase-quartz gneiss. The third unit (Hartland Unit II and The Straits Schist) is a lustrous muscovite-plagioclase-quartz schist with large porphyroblasts of garnet and kyanite or staurolite. The fourth unit (Hartland Unit III) is a thinly interlayered, heterogeneous assemblage of fine-grained muscovite-biotite-plagioclase-quartz granulite and muscovite-plagioclase-biotite-quartz schist; many of these rocks are unusually graphitic. The fifth unit (Hartland Unit IV) is also as assemblage of thick and thin layers of granulites and schists with minor garnet-staurolite-bearing units.

INTRODUCTION

A major remapping project of the bedrock geology of Connecticut was begun after World War II and is currently in its final stages. We have been involved directly or indirectly in the mapping of eleven 7½′ quadrangles in the Highlands of Western Connecticut. During this time there have been many changes in nomenclature, and some formations have been subdivided. The purpose of this paper is to summarize the stratigraphy of this area with the ultimate objective of correlating this section with other areas in Connecticut and in Massachusetts. The stratigraphic units and the nomenclature are those currently used on recent maps of the area. The environments of the original sedimentary rocks have been inferred from the compositions and structure of the metamorphic rocks of the various formations.

Background

Three geologic maps of Connecticut have been published in the past 130 yr. The first of these was the classic work of Percival (1842), which still serves as a reference for current mapping. In 1906, Rice and Gregory published a report on the geology of Connecticut, and in 1907 (Gregory and Robinson), the second "preliminary" geologic map appeared. In 1956, Rodgers and others prepared another "preliminary" geologic map of Connecticut that was essentially a status report reflecting some of the changes based on the mapping since 1907. Since World War II, the Connecticut Geological and Natural History Survey and the U.S. Geological Survey have nearly completed remapping the bedrock geology of Connecticut on 7½′ quadrangle sheets. We have been involved in the Connecticut Geological Survey program since 1948 and 1957, respectively, mapping and studying the central and northern part of the Western Connecticut Highlands. The area covered extends from near Danbury and Naugatuck on the south to Winsted and Falls Village on the north.

Martin (1962) made a special study of the Hartland stratigraphy in the area between Roxbury and Litchfield and traced and confirmed the subdivisions of the Hartland as established in the Roxbury quadrangle (Gates, 1959).

Cassie (1965) mapped the Thomaston quadrangle and established the stratigraphy of the Hartland Formation associated with the line of gneiss domes along the eastern border of the crystalline belt. His work was critical in setting up the stratigraphy that we extended into the contiguous quadrangles north and south of Thomaston.

Much critical information has been added to the published maps by C. W. Martin (1959), Thede (1958), D. Martin (1960), Smith (1960), Christensen (1961, 1963), Vogel (1961), Alsobrook (1965), Dewees (1966), Howe (1966), Brauner (1969), and Schoenwald (1971).

Method of Study

All geologic field mapping was done on a scale of 1:24,000, and essentially all outcrops were located and examined. Most maps show the outcrop areas so that inferred contacts and areas of geologic uncertainty can be readily determined. The present summary is based on a review of notes taken at more than 13,000 localities and petrographic studies of more than 2,000 thin sections.

Modal analyses were made using the Chayes method of point counting with 1,000 points per standard thin section. Plagioclase compositions were determined by measuring extinction angles in sections oriented normal to "$\underset{\rightarrow}{a}$" using the five-axis universal stage in most cases. Some analyses of muscovite, biotite, garnet, and plagioclase from the Waterbury Formation were obtained by partial electron microprobe analyses (Schoenwald, 1971). The calculations of chemical compositions were made from the modal analyses using standard compositions of minerals with relatively fixed compositions and representative compositions for the micas using analyses (Deer and others, 1966) of appropriate rock types. Such approximations are considered adequate for this purpose in view of the wide range of compositions in the modal analyses, the variability of the textures in the rocks analyzed, and the irregularity in spacing of samples resulting from the distribution of outcrops. Minor accessory minerals are not included in summaries of modal analyses or in calculated chemical compositions.

GENERAL GEOLOGY

The general geologic setting of the metasedimentary and volcanic rocks of the Western Connecticut Highlands can be given in terms of a northwestern, a central, and a southeastern province or belt (Figs. 1A, 1B). The northwestern belt is a sequence of sandstone, carbonate, and shale that is part of the Cambrian and Ordovician zone extending from Newfoundland to Alabama (Rodgers, 1968). Here it includes the Dalton, Stockbridge, Woodville, and Walloomsac Formations. In the central belt, pelitic rocks structurally overlie the Precambrian rocks and the Cambrian and Ordovician carbonate rocks. The central belt includes the Berkshire and Housatonic Highlands complexes and the Waramaug and Canaan Mountain Formations. The southeastern belt is a sequence of thick graywacke, shale, and volcanic rock of Paleozoic age containing on the east a line of mantled gneiss domes (Rodgers and others, 1956). It is bounded by a Triassic basin on the east and, on the west, by a major tectonic lineament (Cameron's line) that, for a considerable distance, separates rocks at kyanite grade from rocks at sillimanite grade to the west.

The Precambrian Housatonic and Berkshire Highlands Gneiss complexes form a typically well-layered sequence of metamorphosed acid volcanic materials with subordinate interbedded mafic volcanic and terrigenous clastic sediments. The Precambrian gneiss complex of the northwestern belt is overlain unconformably, at least in part, by the metamorphic equivalents of the sequence of sandstone,

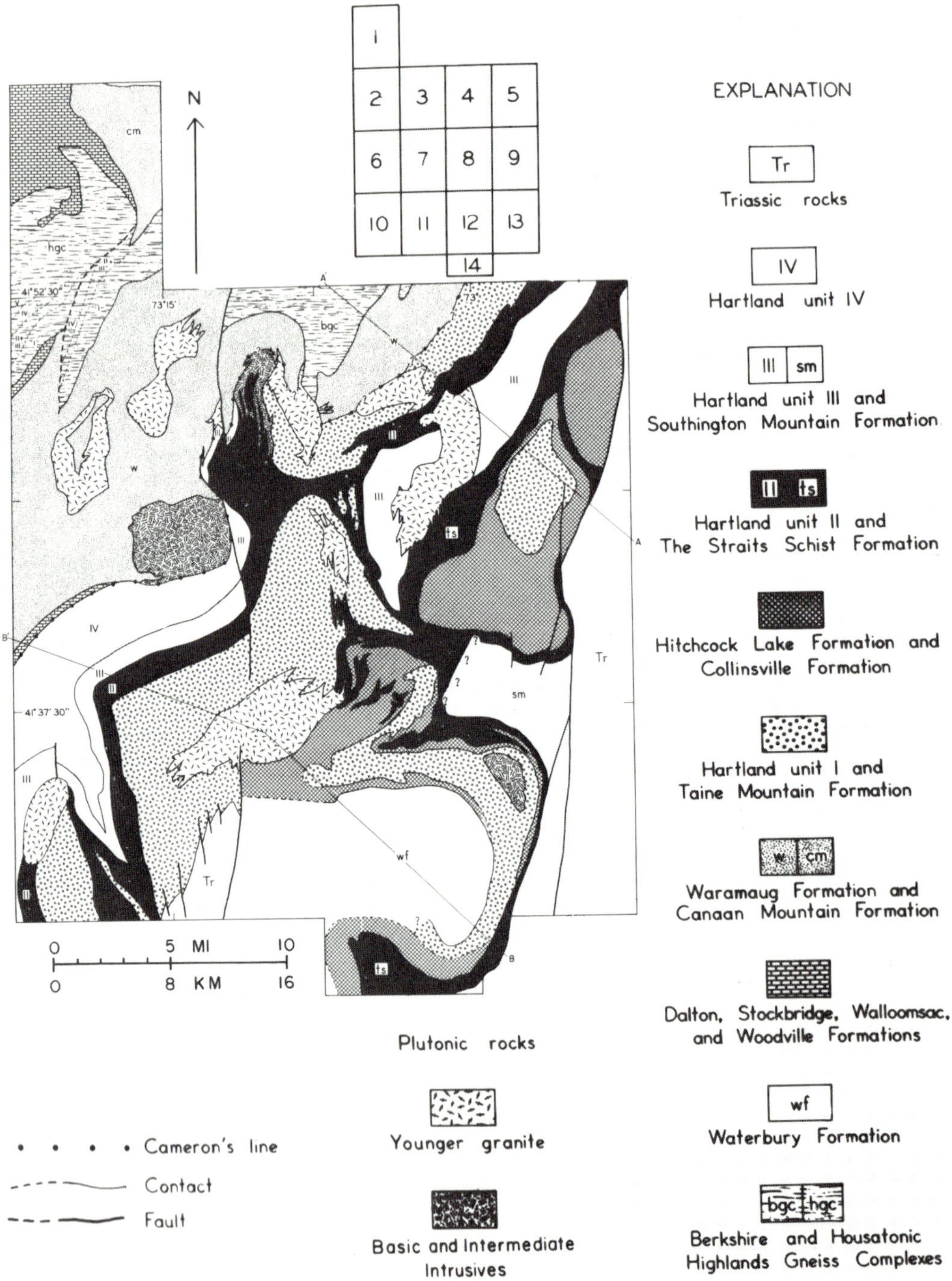

Figure 1A. Geologic map of central-western Connecticut (R. M. Gates and C. W. Martin). Sources of data (see index to 7-1/2′ quadrangles on Fig. 1A): 1. South Canaan, Gates (1975). 2. Cornwall, Gates (1961). 3. West Torrington, Gates and Christensen (1965). 4. Torrington, Martin (1970). 5. Collinsville, Stanley (1964). 6. New Preston, Gates and Bradley (1952); Martin (1962). 7. Litchfield, Gates (1951); Martin (1962). 8. Thomaston, Cassie (1965). 9. Bristol, Rodgers and others (1959). 10. Roxbury, Gates (1959). 11. Woodbury, Gates (1954). 12. Waterbury, Gates and Martin (1967). 13. Southington, Fritts (1963). 14. Naugatuck, Carr (1960).

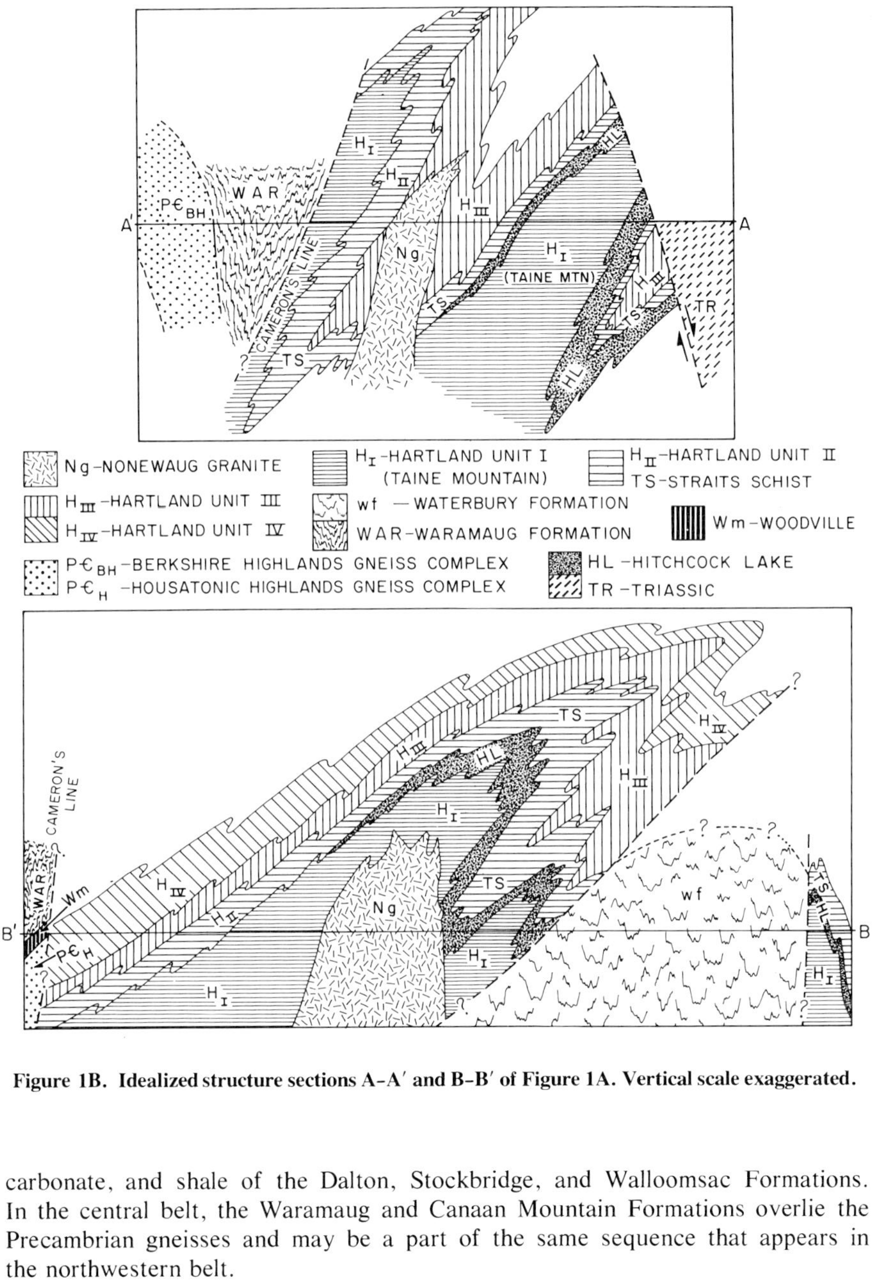

Figure 1B. Idealized structure sections A–A′ and B–B′ of Figure 1A. Vertical scale exaggerated.

carbonate, and shale of the Dalton, Stockbridge, and Walloomsac Formations. In the central belt, the Waramaug and Canaan Mountain Formations overlie the Precambrian gneisses and may be a part of the same sequence that appears in the northwestern belt.

The unquestioned Cambrian and Ordovician rocks of the Dalton, Stockbridge, Walloomsac group (Zen, 1967) are restricted largely to the northwestern belt and are probably autochthonous (Rodgers, 1968). A thin wedge of marble (Woodville), possibly Stockbridge, is exposed on the east side of the central belt adjacent to Cameron's line.

The lower Paleozoic rocks in the central belt, which structurally overlie the Precambrian rocks, are the Waramaug Formation and, to the west and separated from the Waramaug by a fault, the Canaan Mountain Formation. A general lack of bedding features, especially in the Waramaug Formation, is characteristic of these rocks.

The lower Paleozoic rocks of the eastern belt can be divided into two groups: (1) the rocks of the central part of the Waterbury dome and (2) the mantling rocks. The oldest unit is the Waterbury Formation, an interlayered assemblage of metamorphosed shales, graywackes, and volcaniclastic sediments that has an isochron age of 465 ± 50 m.y. (Clark and Kulp, 1968). These paragneisses and migmatites have a structural style very different from that of the overlying rocks. The oldest rock in the Bristol dome is the Waterbury Formation (Cassie, 1965, oral commun.), but most of the dome is underlain by the two lower units mantling the Waterbury Formation in the Waterbury dome. In the Collinsville dome to the north, the second oldest unit (Taine Mountain Formation) forms the center of the structure.

Around the domes and throughout most of the belt east of Cameron's line is a continuous sequence of rocks divided into five mappable units; from oldest to youngest they are: (1) Hartland Unit I (Taine Mountain Formation), (2) Hitchcock Lake Formation (Collinsville Formation), (3) Hartland Unit II—The Straits Schist, (4) Hartland Unit III (Southington Mountain Formation), and (5) Hartland Unit IV. The parental rocks of this sequence were interlayered shales, graywackes, and volcaniclastic rocks, types common to eugeosynclines. The rocks of the eastern belt are intruded by several discordant granites one of which, the Nonewaug granite, has an isochron age of 382 ± 64 m.y. (Besancon, 1970). Structurally, this is one of the youngest plutons in the Western Connecticut Highlands and should date the last major tectonic event.

PRECAMBRIAN GNEISS COMPLEXES

The rocks of Precambrian age exposed in the Housatonic and Berkshire Highlands of west-central Connecticut are essentially a southern extension of those in the Green Mountains of Vermont. They are largely metavolcanic granitic to granodioritic gneisses, two-mica-quartz-plagioclase gneisses, mafic gneisses and amphibolites, and subordinate calc-silicate rocks. Normal terrigenous sediments are represented by sillimanite- and kyanite-bearing two-mica-quartz-feldspar gneisses. The granitic gneisses, the most abundant rock, are typically fine-grained and layered. The amphibolites and mafic (biotite-hornblende-rich) gneisses are ubiquitous, but relatively subordinate.

Structure and Internal Characteristics

In both the Berkshire and Housatonic Highlands, a semblance of a stratigraphic sequence can be worked out, but although the lithic types are similar, the two sequences cannot be correlated readily. In both areas, the layering appears to parallel the contact with the overlying Waramaug Formation (see Gates, 1961; Gates and Christensen, 1965; Martin, 1970). In the Housatonic Highlands, a sequence of five units has been mapped for a distance of about 8 km; then structural complexities and the absence of critical marker horizons preclude separation of these units. Small isoclinal folds can be seen and one large fold has been mapped in the Berkshire Highlands. In the Housatonic Highlands, although intralayer isoclinal folds have been observed, there is apparently no duplication of mappable units (see Fig. 1A, units I-V). Here the bed dips 60° to 85°SE and the outcrop width is greater than 7,000 m; a minimum thickness of 5,000 m is reasonable.

Relation to Other Formations

The Housatonic and Berkshire Gneiss complexes are structurally overlain by the Waramaug Formation. The nature of the contact with the Waramaug is in doubt because the contact relations are reasonably well-exposed at only a few places. In the Cornwall quadrangle (Gates, 1961), the layering of the Housatonic Highlands Gneiss complex is parallel to the contact with the Waramaug Formation, and in two places there appears to be an interlayering of gray quartz-feldspar gneiss (a rock type common in the Housatonic Gneiss complex) and rusty-weathering biotite-quartz-feldspar gneiss characteristic of the Waramaug Formation. In the Torrington and West Torrington quadrangles (Martin, 1970; Gates and Christensen, 1965), the same parallelism of the stratigraphy in the gneiss complex with the Waramaug Formation contact persists, and in two places the same interlayering of gray quartz-feldspar gneiss and rusty biotite-quartz-feldspar gneiss has been observed. Thus, it would appear that this is a normal sedimentary contact from these limited exposures. However, no mappable stratigraphic units in the Waramaug Formation have been identified. Imbricate faulting could produce the same geometric relation.

Elsewhere in the Cornwall and South Canaan quadrangles, the Housatonic Gneiss complex is overlain by the Dalton, Stockbridge, and Walloomsac Formations, but the contact relations are equivocal. In the South Canaan quadrangle, the Housatonic Gneiss complex is unambiguously unconformably overlain by the Dalton Formation. The gneiss complex has been thrust into its present position in much of the South Canaan quadrangle.

Petrography

In a general way, the Precambrian rocks can be divided into three mineralogical categories: granitic gneisses, quartz-plagioclase-biotite gneisses, and mafic to amphibolitic gneisses. Figure 2 shows the averages and ranges of composition of the three types. The ranges of composition within the groups are about as large as the differences between the groups; however, the lithologic basis for the classification in the field is confirmed by microscopic study. The granitic gneisses are typically finer-grained, less micaceous, and weather gray as compared with the quartz-plagioclase-biotite gneisses that are rusty on weathered surfaces. The plagioclase in the granitic gneisses averages An_{20} compared with An_{25} in the quartz-plagioclase-biotite gneisses. The biotite-muscovite ratio of approximately 3:1 is nearly the same in both gneisses. The mafic gneisses and amphibolites are characterized by the presence of hornblende and sphene. Plagioclase averages An_{43}, and biotite is abundant where hornblende is minor. The composition of these gneisses ranges continuously between these three modal types.

Interlayered subordinate rock types are the diopside-quartz-graphite granulite and other calc-silicate rocks and kyanite-sillimanite-mica-quartz-plagioclase gneisses. The kyanitic or sillimanitic gneisses are very similar lithically to the aluminous rock types of the overlying Waramaug Formation. Neither type is quantitatively important, but they are good marker horizons.

The rocks of Precambrian age in this area are predominantly fine grained to fine-medium grained, but the textures vary as widely as the mineralogy. They are well layered on a scale ranging from a few centimetres to a few metres. This layering, primarily the result of differences in mica content, is particularly characteristic of the granitic gneisses and is in marked contrast to the lack of layering in the Waramaug Formation. The layered granitic gneisses are some of the best

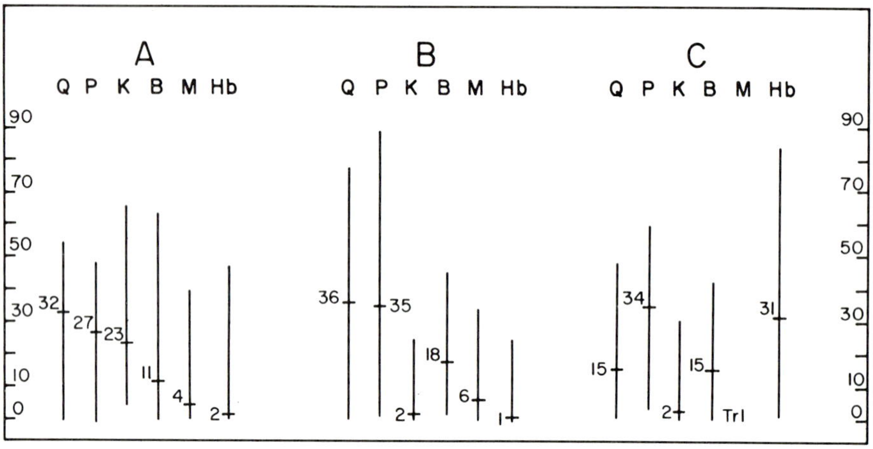

Figure 2. Modal analyses, Housatonic Gneiss complex. Range of composition indicated by bar; average composition by tick mark. A. Granitic or potash feldspar-rich gneiss (51 samples). Plagioclase composition An_{19} to An_{29}; average An_{20} (66 samples). B. Biotite-feldspar-quartz gneiss (62 samples). Plagioclase composition An_7 to An_{58}; average An_{25} (53 samples). C. Amphibolitic and mafic gneiss (43 samples). Plagioclase composition An_{13} to An_{75}; average An_{43} (41 samples). The following abbreviations are used in all figures. Q-quartz, P-plagioclase, K-potash feldspar, B-biotite, M-muscovite, Hb-hornblende, Ky-kyanite, G-garnet, Sil-sillimanite, Mag-magnetite, St-staurolite, Ch-chlorite, Gr-graphite.

horizon markers in the Precambrian assemblage. The quartz-plagioclase-biotite rocks are strongly gneissic, but are not layered; they have abundant micas and a variety of textures, none sufficiently distinctive to make mappable units. The micas occur (1) as continuous folia or thin layers alternating with the granular quartz-plagioclase, (2) as discontinuous streaks and wisps in the quartz-plagioclase host, and (3) as knots or clots in an obscurely gneissic rock. These variations may reflect the original composition or subsequent recrystallization and deformation.

Origin

The Housatonic and Berkshire Highlands Gneiss complexes in Connecticut are not identical, but have many lithic types in common. The distinctive layered granitic gneisses ranging from 100 to 235 m in thickness represent reworked felsic volcanic tuffs. The amphibolites and mafic gneisses represent mafic volcaniclastic material deposited along with the felsic volcanic tuffs. The aluminous schists and gneisses represent sediments derived from a chemically weathered land surface. The well-layered character of the assemblage combines with the compositions to leave little doubt that in this area the Precambrian section is represented by a volcaniclastic mixture with subordinate terrigenous clastics.

WATERBURY FORMATION

The oldest unit in the core of the Waterbury dome is the Waterbury Formation, a heterogeneous assemblage of thin-to-thick interlayered biotite-streaked granulitic gneisses, kyanitic-garnetiferous-biotite-plagioclase-quartz gneisses and schists, muscovite-rich gneisses, and plagioclase-quartz granulites. Commonly, the para-gneisses have a migmatitic texture that reflects the original layering and are believed to result from simple recrystallization and segregation (Schoenwald, 1971). Trondhjemitic to granitic rocks intrude the Waterbury Formation partly as small, discrete, mappable plutons and are unrelated to the migmatitic textures in the Waterbury Formation. The trondhjemites intrude only the Waterbury Formation and are not known to intrude the overlying rocks.

The Rb-Sr whole-rock data on this formation give an isochron age of 465 ± 50 m.y. (Clark and Kulp, 1968).

Structure and Internal Characteristics

In sharp contrast with the overlying rocks, the layering of the gneisses is not consistent over any appreciable distance. The axes of small folds, a few centimetres to a few metres in amplitude, are the most obvious structural feature. In the migmatites, the linear features are obscure, and a turbulent flow pattern best describes the appearance of these rocks.

The thickness of the Waterbury Formation is unknown because no underlying older rocks are exposed. The various rock types are in layers ranging from less than 3 cm to as much as 3 m thick, but mappable units have not been recognized. Because evidence of repetition of beds and mappable folds is lacking, the formation is considered several hundreds of metres thick.

No contact between the Waterbury Formation and the overlying Hartland Formation has been observed because a rather broad belt that is devoid of outcrops lies between them. Gates and Martin (1967), however, concluded, from stratigraphic interpretations of the mantling rocks and apparent structural discontinuities, that a fault marked the contact.

Petrography

The Waterbury Formation is primarily a heterogeneous paragneiss composed of individual layers that range greatly in thickness, are commonly intricately folded, and are not mappable lithic units. Figures 3 through 6 show the various stages of folding and migmatization in the thinly layered gneisses; Figures 7 and 8 show more thickly layered granulites and migmatites. Modal analyses of 46 samples indicate the mineralogic variation from layer to layer and the probable parental material. The mineralogy of the various layers as well as the composition of the plagioclase is given in Figure 9.

The granulites, granulitic gneisses, and layered gneisses are essentially fine- to medium-grained plagioclase-quartz rocks with variable amounts of mica, principally biotite, and with subordinate kyanite and garnet in the biotite-rich layers. The granulites contain more than 10 percent micas, and the flakes are more or less disseminated and unoriented. The most common paragneiss averages 20 percent biotite; a typical hand sample contains both biotite-rich streaks or layers and granulitic layers. The biotite-rich layers are normally less than 5 to 8 cm in thickness and are characterized by felted masses of kyanite interleaved with the biotite. Quartz predominates over plagioclase in 70 percent of the samples.

The chemical compositions of the various rock types were calculated from the modal analyses in Figure 9 and are tabulated in Table 1. The analysis of the biotite is based on partial electron microprobe of 37 samples for Ti, Fe, and Mg. The muscovite analysis is based on microprobe of 27 samples for K, Na, Al, Mg, Fe, and Ti (Schoenwald, 1971).

The textures of these rocks range from fine- to medium-grained, equigranular to inequigranular, and are generally gneissic. The quartz and plagioclase are anhedral;

Figure 3. Undeformed thinly interlayered biotite-plagioclase-quartz granulite (light layers) and garnet-kyanite-biotite-plagioclase-quartz schist (dark layers), Waterbury Formation.

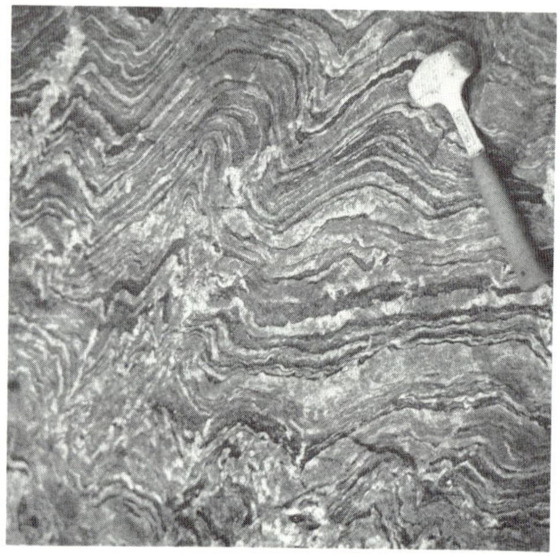

Figure 4. Moderately folded, thinly interlayered granulite and schist as in Figure 3, Waterbury Formation.

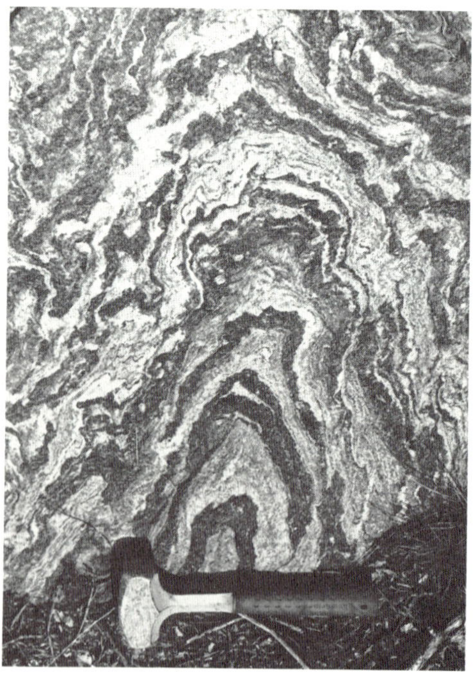

Figure 5. Migmatitic texture in intensely folded and recrystallized thinly interlayered granulite and schist as in Figures 3 and 4, Waterbury Formation.

Figure 6. A common migmatitic texture derived from the crumpling and recrystallization of thinly interlayered granulite and schist, Waterbury Formation.

← Figure 7. Interlayered granulite and biotite-rich schist, Waterbury Formation.

Figure 8. Migmatite formed from relatively thick layers of granulite and thin layers of biotite-rich schist, Waterbury Formation. →

they form a granular mosaic in the granulitic layers and are interspersed grains or stringers in the micaceous layers. The micas range from well oriented to unoriented. The mica folia in the micaceous layers are highly contorted, but still show a poor orientation. Abundant, irregularly shaped, ragged biotite flakes are transverse to the layering and the schistosity. Fine-grained, granular kyanite crystals are in streaks, irregular grains, or convoluted clusters in the biotitic layers. Garnet, in small, rounded to angular grains, is intimately associated with the kyanite. Calc-silicate pods, 2 to 15 cm thick, are sparsely scattered throughout the paragneiss.

All rocks of the Waterbury Formation are hard, tough, and coherent and tend to weather with a rough surface because of the alternation of felsic and biotitic layers. The biotitic layers are commonly prickly to the touch owing to the small kyanite crystals. The rocks break across the layering as easily as along it. Generally, the outcrops are shaped rather than plucked by glacial abrasion, and consequently most outcrops are irregularly rounded. The Waramaug Formation shares this characteristic, but the Hartland Formation does not.

The intrusives in the Waterbury Formation aid in separating the formation from the overlying rocks. The trondhjemite, particularly, forms mappable areas in the central part of the Waterbury dome and is not found in the formations overlying the Waterbury Formation. The granitic rocks in the Waterbury Formation are texturally distinguishable from those in the overlying formations but are not demonstrably different in age.

Origin

The Waterbury Formation is considered to be an assemblage of metasedimentary and metavolcanic rocks representing sedimentary mixtures of quartz and plagioclase

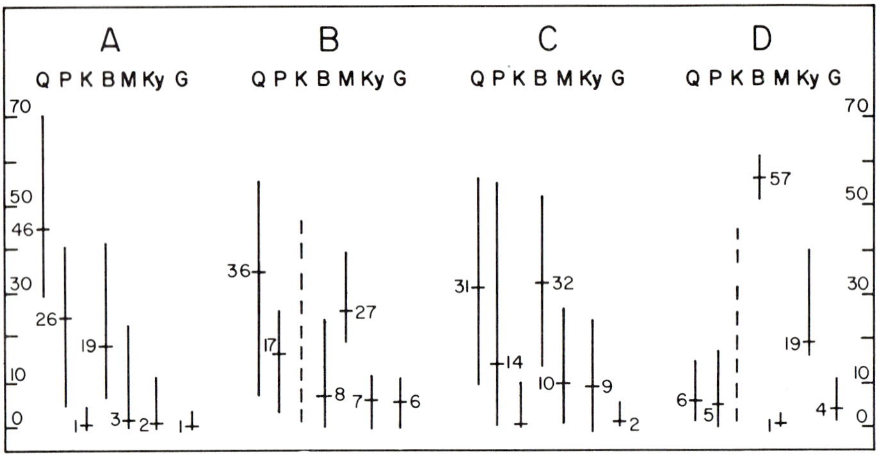

Figure 9. Modal analyses, Waterbury Formation. Range of composition indicated by bar; average composition by tick mark; dashed line indicates mineral not present. A. Granulites and granulitic gneisses (17 samples). Plagioclase composition An_{15} to An_{32}; average, An_{23} (74 determinations in 13 samples). Electron microprobe analyses of 5 grains in one sample show a range in anorthite content of 16 to 22 percent; average, 18.8 (Schoenwald, 1971). B. Muscovite-rich granulite gneiss (7 samples). Plagioclase composition An_0 to An_{22}; average, An_{12} (55 determinations in 5 samples). C. Thinly interlayered paragneiss as shown in Figure 3 (22 samples). Plagioclase composition An_{10} to An_{24}; average, An_{16} (108 determinations in 13 samples). Electron microprobe analyses of 6 grains in 2 samples give average anorthite content of 17 (Schoenwald, 1971). D. Kyanite-biotite-rich layers in thinly interlayered paragneiss as shown in Figure 3.

TABLE 1. CHEMICAL COMPOSITIONS CALCULATED FROM MODAL ANALYSES (FIG. 9), WATERBURY FORMATION

	A	B	C	D
SiO_2	72.8	64.4	59.3	40.5
Al_2O_3	11.9	20.8	20.4	29.2
$FeO + Fe_2O_3$	5.5	5.5	8.3	14.3
MgO	2.0	1.6	3.2	5.6
CaO	0.5	0.4	0.4	0.2
Na_2O	3.4	1.8	1.6	0.8
K_2O	2.5	3.9	4.0	5.2
TiO_2	0.8	0.6	1.2	2.1
H_2O	0.9	1.4	1.6	2.0

A. Granulite and granulitic gneisses.
B. Muscovite-rich granulitic gneiss.
C. Thinly interlayered paragneiss.
D. Kyanite-biotite-rich layers in thinly interlayered paragneiss.

sands, clay minerals, volcanic ash, coarser volcaniclastic debris, and possibly flows. The flows may be represented by some of the layered (sill-like) and gneissic trondhjemitic rocks and by some amphibolites. Both amphibolites and gneissic trondhjemite occur as lenses or pods in the Waterbury Formation, but are not quantitatively abundant. The migmatites of the Waterbury Formation are considered to have formed by simple recrystallization of the paragneiss with slight metamorphic differentiation (Schoenwald, 1971). The development of the migmatitic textures and the apparent changes in the rheologic state during deformation and metamorphism is shown in Figures 3 through 8.

The assemblage of lithic types and the nature and scale of their interlayering are characteristic of eugeosynclinal rocks. The calculated compositions have been compared with unmetamorphosed sedimentary rocks in an attempt to determine the parental sediments. The average granulite and granulitic gneiss (Rock type A, Fig. 9; Table 1) have a K_2O:Na_2O ratio of less than one and a relatively high alkali/alumina ratio typical of eugeosynclinal graywackes (Middleton, 1960). The muscovite-rich granulitic gneisses (Rock type B, Fig. 9; Table 1) have the composition of aluminous shales. The thinly layered granulites and schists (Rock C, Fig. 9; Table 1) represent ferruginous shales and sands. The biotite-kyanite-rich layers (Rock type D, Fig. 9; Table 1) represent iron- and alumina-rich shales with a high potash/soda ratio. The amphibolites are derived from volcanic material of basic or intermediate composition, and the gneissic trondhjemites probably represent a more felsic volcanic contribution to the pile. The scattered, small bodies of serpentinite around the dome (not previously mentioned nor quantitatively abundant) complete the assemblage.

WARAMAUG FORMATION

The Waramaug Formation is principally in the central belt west of Cameron's line and east of the western belt where carbonate rock predominates. It structurally overlies the Stockbridge Formation along its western edge and the Woodville marble (probably Stockbridge equivalent) along the southeastern edge. The Waramaug Formation appears to overlie the Precambrian rocks in the area between the Housatonic and Berkshire Highland Complexes and east of the Berkshire Complex.

The Waramaug Formation has been characterized by several workers as lacking good bedding features (Gates, 1952, 1961; Gates and Christensen, 1965; Hall, 1968; Martin, 1970). Certainly, the layering in the Waramaug Formation is obscure compared to that in the Precambrian gneisses, the Waterbury Formation, and the younger Hartland Formation. There are two "end member" rock types: (1) a muscovite-biotite-plagioclase-quartz gneiss with minor kyanite, sillimanite, and garnet and (2) a biotite-plagioclase quartz gneiss. Numerous intermediate varieties occur, but none is conspicuously interlayered. In general, the rock is rich in quartz and plagioclase with relatively subordinate micas and with inconspicuous garnet, or it is characterized by garnet and sillimanite and (or) kyanite with relatively abundant micas and weathers with a rough nubbly or ribbed (corrugated) surface.

The amphibolites in this unit are generally discrete bodies of uniformly basaltic composition characteristic of orthoamphibolites. These may represent volcaniclastic material, but are, at least in part, intrusive bodies related to the post-depositional tectonic history (Gates, 1967). They are relatively small bodies and, though abundant locally (Gates, 1961), do not constitute a mappable stratigraphic horizon.

The granite bodies in this unit are not significantly different from the youngest granites in the region and consequently have little stratigraphic relevance.

Structure and Internal Characteristics

The various lithic types in this formation do not provide much information on the internal structure or stratigraphy. In general bedding is absent or at least can be traced for only a few metres even in a large outcrop. Possible correlatives of this unit farther northwest (Canaan Mountain Formation) and southwest (Manhattan C) have more definitive bedding characteristics. Because bedding and other indications of internal structure are lacking, we can estimate the thickness of this unit only in terms of the width of the exposure. The width ranges from about 2,700 to 4,700 m. The true thickness is less but cannot be determined. In general, the structural style of this unit is similar to that of the Waterbury, Manhattan C, and Canaan Mountain Formations (Gates, 1952; Gates and Christensen, 1965).

Relation to Other Formations

For several kilometres along the southern part of the contact, the Waramaug Formation is underlain along the western edge by the Stockbridge Formation and along the eastern edge by the Woodville marble (probably Inwood-Stockbridge equivalent). The general structure is a syncline with the Waramaug Formation in the keel.

Farther northeast, the western edge of the Waramaug Formation overlies the Housatonic Highlands Gneiss complex directly with what appears to be a sedimentary contact. The contact relations, however, are exposed in only three areas of limited outcrop. The Waramaug Formation has similar contact relations on the east side of the Berkshire Highlands Gneiss complex.

Part of the eastern edge of the Waramaug Formation abuts Cameron's line and its relations to the Hartland Formation are obscured by this tectonic break. Farther north, the eastern edge of the Waramaug Formation lies adjacent to Hartland Unit I (Taine Mountain Formation), but the contact is not exposed.

Petrography

The principal minerals of the Waramaug rocks are muscovite, biotite, plagioclase, and quartz typically with subordinate sillimanite, kyanite, and garnet. The accessories are chlorite, zircon, staurolite, apatite, tourmaline, and sphene. The amount of magnetite ranges widely, but it is significant in the sillimanite-garnet-bearing rocks. Although muscovite is ubiquitous, it is clearly subordinate to biotite. Microcline is rare and probably genetically related to contiguous granite intrusions. Figure 10 gives the modal composition for the two predominant rock types.

The two end members of this lithic series are readily distinguished in the field by the nature of the weathered surface. The rocks lacking garnet and sillimanite tend to have a smooth, glacially abraded surface and are somewhat sandy and friable. The sillimanite-garnet-bearing gneiss has a rough, nubbly, or corrugated weathered surface with the positively weathering nubs or ribs related to the sillimanite. The sillimanite-bearing rock is a tough, coherent rock that breaks into rounded hand samples where good foliation is lacking. The kyanite-bearing rocks of the Waterbury Formation break in a similar manner. Gradational rock types show weathered surfaces intermediate between the end members. Interlayering of the two types is not common, but is found locally.

The questionable position of the Waramaug Formation in the stratigraphic section makes it advisable to include a more detailed description of the typical rock types.

Muscovite-biotite-plagioclase-quartz gneiss (Rock type A, Fig. 10; Table 2) typically weathers gray to light rust. The foliation ranges from poor to good depending on amount of thin biotite folia and diffuse quartz-plagioclase streaks. The texture is lepidoblastic without a marked segregation of felsic and mica minerals.

Quartz and plagioclase are typically in equant to slightly elongate grains. Some large grains of plagioclase poikilitically include quartz, mica, and magnetite. Plagioclase ranges in composition from calcic oligoclase to intermediate andesine; the average is sodic andesine (An_{33}).

Biotite exhibits pale yellow to dark brown pleochroism and occurs in irregular,

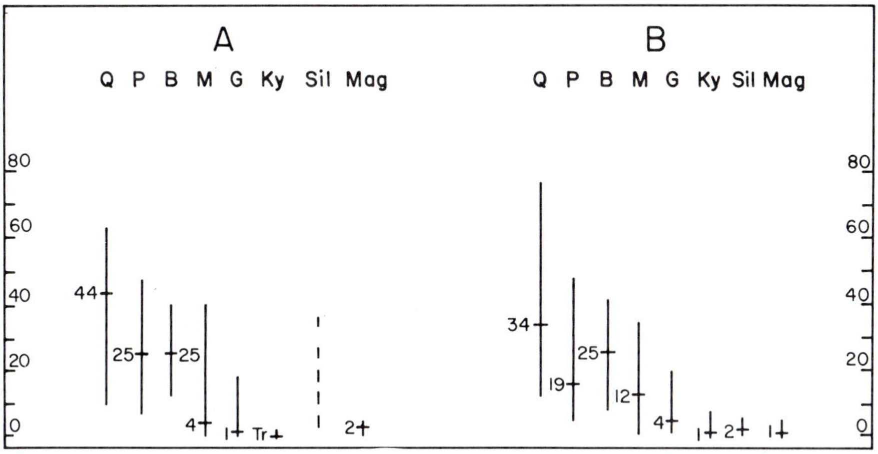

Figure 10. Modal analyses, Waramaug Formation. Range of composition indicated by bar; average composition by tick mark; dashed line indicates mineral not present. A. Biotite-plagioclase-quartz gneiss (44 samples). Plagioclase composition An_{22} to An_{42}; average, An_{33} (44 samples). B. Sillimanite or kyanite-garnet-plagioclase-quartz gneiss (61 samples). Plagioclase composition An_{20} to An_{50}; average, An_{33} (61 samples).

TABLE 2. CHEMICAL COMPOSITIONS CALCULATED FROM MODAL ANALYSES (FIG. 10), WARAMAUG FORMATION

	A	B
SiO_2	67.6	61.2
Al_2O_3	12.9	16.7
$FeO + Fe_2O_3$	9.0	10.4
MgO	2.5	2.8
CaO	1.6	1.2
Na_2O	2.1	1.6
K_2O	2.6	3.4
TiO_2	1.0	1.0
H_2O	1.1	1.4

A. Biotite-plagioclase-quartz gneiss.
B. Sillimanite or kyanite-garnet-plagioclase-quartz gneiss.

discontinuous folia as in the Waterbury Formation and in contrast to the micas in the Hartland Formation. Muscovite is commonly in large flakes oriented transverse to the foliation.

Sillimanite-kyanite-garnet-plagioclase-quartz gneiss (Rock type B, Fig. 10; Table 2) characteristically shows a rusty, rough-weathered surface. Mineralogically, this rock differs from the biotite-plagioclase-quartz gneiss in having more biotite, muscovite, sillimanite or kyanite, and garnet. Texturally, this is reflected in knots, augen, and crenulations around garnet, biotite, and sillimanite or kyanite.

Quartz and plagioclase are irregularly equant to slightly elongate parallel to foliation. The plagioclase composition averages An_{33} and ranges from An_{20} to An_{50}.

Biotite and muscovite constitute a major part of the rock and occur in streaks, folia, and irregularly convoluted folia. Biotite predominates and is pleochroic in pale yellow to deep reddish-brown. Most muscovite occurs in blocky, unoriented flakes.

Garnet crystals are fine grained to coarse grained, generally anhedral, and poikilitically include fine magnetite grains and to a lesser extent all other minerals in the rock.

Sillimanite is both fibrolitic and prismatic and occurs largely in quartz and biotite. Some biotite containing fibrolitic sillimanite is "bleached" or only slightly pleochroic, whereas other biotite is normal. Where kyanite occurs with sillimanite, there is no indication of any reaction relationship.

Origin

The chemical compositions of the rocks of the Waramaug Formation as calculated from the modal analyses (Table 2) show that the biotite-plagioclase-quartz gneiss compares favorably with an argillaceous graywacke and the sillimanitic gneiss with a common shale. More significant than the composition is the general lack of good bedding features that are exhibited in most other formations. The lack of compositional layering could be interpreted as a wildflysch or a tectonic melange of disrupted folds. In most outcrops, "sandy" layers or streaks can be traced for a few metres in the sillimanitic gneiss and vice versa, but well-defined bedding is not common. The lack of coarse clastics suggests that the Waramaug Formation accumulated in quiet water remote from a land mass, but the high alumina and iron oxides suggest that the fine sediment was land-derived.

HARTLAND UNIT I
TAINE MOUNTAIN FORMATION

The eastern belt of rocks, between Cameron's line and the Connecticut Valley Triassic boundary fault, consists of the Waterbury Formation and a sequence of five apparently conformable lithologic units. Hartland Unit I (Taine Mountain Formation) is the lowest and probably the thickest unit of the series. It is predominantly a fine-grained, gray two-mica-oligoclase-andesine-quartz granulite or granulitic gneiss with subordinate schist. It is designated granulite where micas are evenly disseminated and granulitic gneiss where they are concentrated in thin discontinuous or pin-stripe layers. Schist and amphibolite layers and lenses, from less than 2 cm to several metres in thickness, are ubiquitous but subordinate. Kyanite-bearing gneisses are found in several parts of Hartland Unit I.

Structure and Internal Characteristics

Hartland Unit I characteristically contains well-oriented micas producing an excellent foliation. Most of the rocks are friable and form smooth-weathering rounded to slabby outcrops. The rocks with abundant kyanite are distinctly more coherent and have prickly weathered surfaces owing to protruding resistant kyanite blades. These kyanite-bearing layers differ from those in the Waterbury Formation in having more quartz and plagioclase, less biotite, and megascopic kyanite.

Schist layers are more abundant near the top of the unit, range from less than 2 cm to several metres, and are everywhere subordinate to the granulitic types. This interlayering of granulite and schist is interpreted as relict bedding. Textural variations within the granulitic rocks result mainly from the arrangement of mica flakes, and all gradations are found from rocks in which micas are almost perfectly disseminated to those in which they are in thin pin-stripe layers or folia. The former are massive, homogeneous rocks that form rounded outcrops; the latter are thinly layered and commonly have slabby outcrops. The orientation of the individual mica flakes, pin-stripe layering, and relict bedding are all parallel except at the crests of small folds where layering and schistosity cross at angles of as much as 90°.

Minor folds are common in Hartland Unit I and range in size from microscopic to those with wave lengths and amplitudes of more than 15 m; some have been refolded. The rocks are distinctly spindled where minor folds are abundant.

The best calculation of the thickness of Hartland Unit I can be made in the Torrington quadrangle where both the top and bottom can be fairly accurately located; the unit is about 2,200 m thick. East of Torrington, there is a discontinuous kyanitic layer with a maximum thickness of 200 to 220 m; thinner layers are found in several areas.

Relation to Other Formations

The relative age of Hartland Unit I is based on (1) its location in the cores of domes and anticlines with respect to both foliation and layering, (2) its occurrence at the base of a consistently westward-dipping sequence, and (3) its position above the Waterbury Formation in the Waterbury and Bristol domes. A large part of Hartland Unit I abuts Cameron's line and (or) the Waramaug Formation along its west side. The truncation of some units shows the contact to be a fault (Gates and Christensen, 1965); farther to the northeast the contact is not exposed, but on petrographic evidence, Martin (1970) interpreted the contact to be either a fault or an unconformity. The upper contact is arbitrarily placed where the gray granulitic rocks cease to be the dominant rock type. The structural styles of Hartland Unit I and the Waterbury and Waramaug Formations are markedly different; for example, axial plane folding is apparent in Hartland Unit I and foliation is not consistent in the other two formations. Other petrographic comparisons of Hartland Unit I and the Waramaug and Waterbury Formations are given in Gates and Christensen (1965), Gates and Martin (1967), and Martin (1970). We are virtually certain that Hartland Unit I is younger than the other two formations.

Petrography

The rock in Hartland Unit I is both mineralogically and texturally simple. The essential minerals are quartz, plagioclase, biotite, and muscovite with garnet, kyanite, and chlorite in some places. Apatite and zircon are ubiquitous accessory minerals, and tourmaline, sphene, rutile, and magnetite are less common. The total mica content is normally less than 20 percent except in thin schist layers. Modal analyses of 61 thin sections of this unit are given in Figure 11. Table 3 shows the calculated chemical analysis. The textures range from lepidoblastic in the more granulitic rocks to gneissic where micas are in streaks or layers. Nearly all grains are less than 1 mm in the longest dimension.

Quartz and plagioclase commonly constitute 80 percent or more of the rock and occur as anhedral grains, either equidimensional or slightly elongate parallel to the foliation. Most quartz shows mild strain shadows; the plagioclase grains appear fresh and most are untwinned. The plagioclase ranges from An_{11} to An_{40}; the average is An_{22}. A frequency distribution diagram of the compositions, however, shows two peaks, one at An_{15} and the other between An_{27} and An_{30}. The reason for this is unknown. The peaks appear to be unrelated to stratigraphic position or geographic locality.

Biotite, commonly partially altered to chlorite, is the dominant mica in most thin sections studied. Both micas are oriented parallel to foliation; only a few

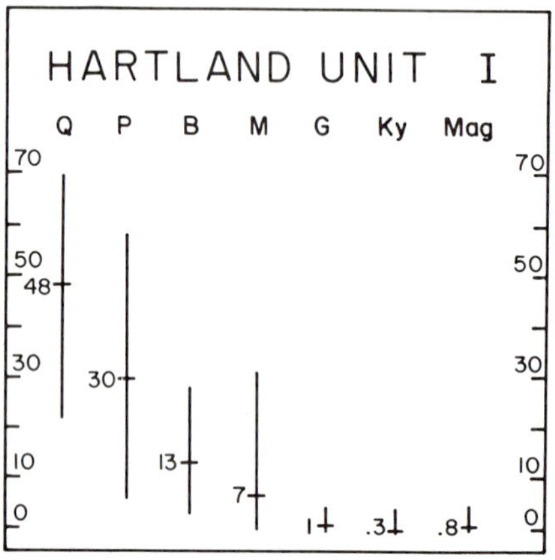

Figure 11. Modal analyses of granulite and granulitic gneiss (61 samples), Hartland Unit I. Range of composition indicated by bar; average composition by tick mark. Plagioclase composition An_{11} to An_{40}; average, An_{22} (199 determinations in 58 samples).

TABLE 3. CHEMICAL COMPOSITION CALCULATED FROM MODAL ANALYSIS (FIG. 11) IN HARTLAND UNIT I

SiO_2	72.85
Al_2O_3	12.69
$FeO + Fe_2O_3$	5.49
MgO	1.66
CaO	1.29
Na_2O	2.66
K_2O	1.92
TiO_2	0.56
H_2O	0.87

flakes are transverse. They are scattered uniformly throughout some rocks and in discontinuous streaks in others.

Garnet occurs as small, irregularly anhedral, poikilitic grains. In some places, it is stratigraphically controlled; in others its distribution appears to be related to pegmatitic patches. Most of the inclusions are opaque minerals or quartz, but plagioclase, biotite, muscovite, zircon, and tourmaline have been observed. In a few grains, trails of opaque minerals indicate that garnet has been rotated. The kyanitic rocks are characterized by clusters or stringers of small kyanite grains and mica as much as 2 cm thick.

Throughout this unit, there are very subordinate thin schist layers that are lithically similar to those that make up the overlying Hartland Unit II.

Origin

The similarity of granulitic rocks of Hartland Unit I and the granulitic layers in the Waterbury Formation is shown by the modal analyses (Rock type A, Fig. 9; Fig. 11). Hartland Unit I is therefore also considered to have been predominantly graywacke with subordinate shale (Table 3). The Waterbury Formation, however, has a much greater proportion of pelitic rocks. The well-bedded aspect of the rocks favors a marine environment. The fine grain size and the high plagioclase content of the granulitic rocks probably require a contribution of volcaniclastic material. The amphibolites may be flows or intrusives.

HITCHCOCK LAKE FORMATION (COLLINSVILLE FORMATION)

Hitchcock Lake Formation is a very heterogeneous assemblage of light and dark, layered and streaked granulites, gneisses, and amphibolites that range considerably in the thickness of the individual layers as well as in the thickness of the assemblage. There are rapid changes in lithology both vertically and laterally, but there is little doubt regarding the unity of the formation as it is clearly delimited by the more uniform underlying and overlying formations. The light and dark layers and streaks in this formation make it readily recognizable.

Structure and Internal Characteristics

The compositional and textural layering of this assemblage is characteristically parallel to the schistosity and in marked contrast to the Waterbury Formation and to a lesser degree the Waramaug Formation. Isoclinal folding with high amplitude to wavelength ratios can be seen in some large exposures on the north side of the Waterbury dome. Map patterns show that the assemblage is tightly folded and that a later open fold developed almost coaxially.

The formation is fairly persistent, though variable in thickness, along the eastern part of its outcrop belt and apparently disappears to the west. In several small areas where outcrops are abundant, such as the north side of the Waterbury dome, the formation ranges from 0 to 500 m in thickness; the average is less than 300 m. In other areas, the outcrop width is much greater because of open folding. The variable thickness is attributed to nondeposition and erosional disconformity.

Relation to Other Formations

Hitchcock Lake Formation has a gradational contact with the conformably underlying Hartland Unit I. The friable gray granulite characteristic of Hartland Unit I occurs as thin interlayers more than 2 m thick in Hitchcock Lake Formation; therefore the contact is placed at the last thick gray granulite and the beginning of the layered and streaked rock. The upper contact is marked by a lustrous muscovite-quartz schist that may disconformably overlie Hitchcock Lake Formation. Immediately underlying this muscovite-quartz schist are erratically distributed amphibolites, marbles, calc-silicate rocks, hornblende gneisses, and garnetiferous gneisses. They form a discontinuous horizon at the top of the formation. The erratic distribution of these uppermost lithologies and the variable thickness of Hitchcock Lake Formation are indicative of nondeposition or a disconformity.

Petrography

The petrographic characteristics of rocks from Hitchcock Lake Formation are most easily described layer by layer. Locally, there has been significant recrystallization and diffusion and probably introduction of material (Cassie, 1965). Consequently, the less altered layers were selected for study.

Hitchcock Lake Formation is predominantly a layered and streaked assemblage of alternating thick and thin layers of hard, fine-grained plagioclase-quartz granulite and coarsely streaked biotite-plagioclase-quartz gneiss. Both lithic types are streaked or gneissic—the granulite because of the quartz and plagioclase, and the gneiss because of the coarse biotite and muscovite (Figs. 12, 13).

Garnetiferous gneisses, calc-silicate rocks, amphibolites, hornblende gneisses, and marbles are erratically distributed laterally, but tend to be more common toward the upper contact. The marbles generally are at the contact with the overlying rock or within a few metres of it. The largest marble layer (Cassie, 1965) is 2 m thick and 300 m long. The amphibolite interlayered with quartz-plagioclase gneiss ranges from a few centimetres to large lenses more than 30 m thick.

The modal analyses of the various layers of the gneisses of Hitchcock Lake

Figure 12. Finely streaked, fine-grained, plagioclase-quartz granulite, Hitchcock Lake Formation. Modal analyses in A of Figure 14.

Figure 13. Coarse biotite-streaked plagioclase-quartz gneiss, Hitchcock Lake Formation. Modal analyses in B of Figure 14.

Formation are given in Figure 14 and the computed chemical analyses in Table 4. The modes do not show some of the local variations in the quartz-feldspar granulites and biotite gneisses where microcline is a major constituent. The microcline-bearing rocks are in areas of granitic intrusives and not directly related to the parental rock.

Finely-streaked plagioclase-quartz granulite is one of the two most abundant rocks in this sequence. It occurs in layers ranging in thickness from a centimetre to over 7 m. It is characterized by fine grain size (average grain is 1 mm), fine streaks, and general hardness and toughness. It breaks smoothly across foliation. The streaking is caused by light-colored, mica-free, slightly coarser-grained, plagioclase-rich streaks in a finer-grained, darker, mica-bearing granulite. The mica-bearing layers average less than 1 mm in grain size, whereas the subordinate felsic streaks range from 1 to 3 mm and average about 1.5 mm (Fig. 12; Rock type A, Fig. 14).

Coarse mica-plagioclase-quartz gneiss is more micaceous and coarser grained (average 1.5 mm) and has a more heterogeneous distribution and streaking of the micas than the plagioclase-quartz granulite. Contrasting mica-rich and mica-poor streaks and layers make the rock a striking black and white gneiss. The mineralogic variation of the micaceous and felsic layers is given in Rock types D and E in Figure 14. Some distinctive layers contain abundant pink garnet crystals, 5 mm in diameter, but the major variations are in the ratios of micas to felsic minerals and in the thickness and distribution of the contrasting layers. The texture is strongly lepidoblastic in the micaceous layers and xenomorphic inequigranular in the felsic layers (Fig. 13; Rock type B, Fig. 14).

Friable, gray plagioclase-quartz granulite differs from the two rock types described above mainly in its massive aspect and its friable nature. The micas are more commonly disseminated and do not make good folia. The mineralogy is essentially the same as the streaked granulite and gneiss (Rock type C, Fig. 14). It is more typical of the rocks of the underlying unit into which it grades.

Amphibolites, calc-silicate rocks, and marbles cannot be considered a separate unit because quantitatively they are subordinate; they are most abundant toward the top of the sequence. Compositionally, the amphibolites are more variable than the other rocks in that garnet, epidote, and biotite may be present in significant amounts. Usually they are simple hornblende-plagioclase-quartz gneiss. The calc-silicate rocks differ from the amphibolites in that diopside is the major mineral and zoisite is common. The calc-silicate rocks are most common in small pods or lenses associated with the calcite marbles. The textures of these rocks range from fine grained to coarse grained and from granulitic to gneissic. Hornblende is typically in prisms; diopside in blocky crystals, and zoisite in granular aggregates.

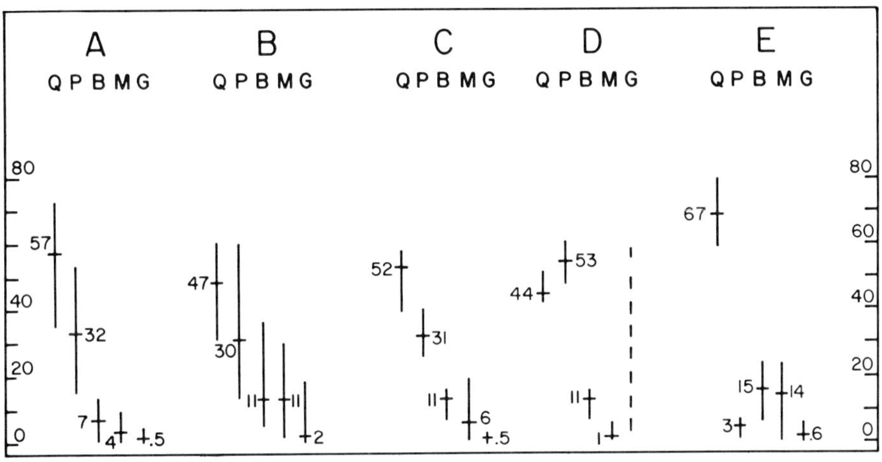

Figure 14. Modal analyses, Hitchcock Lake Formation. Range of composition indicated by bar; average composition by tick mark; dashed line indicates mineral not present. A. Finely streaked plagioclase-quartz granulite (12 samples). Plagioclase composition An_{10} to An_{33}; average, An_{21} (156 determinations in 12 samples). B. Coarse biotite-streaked plagioclase-quartz gneiss (16 samples). Plagioclase composition An_{10} to An_{34}; average, An_{23} (213 determinations in 16 samples). C. Friable mica-quartz-plagioclase granulite (5 samples). Plagioclase average composition is An_{21} (60 determinations in 5 samples). D. Felsic layers in the coarse biotite-streaked plagioclase-quartz gneiss (4 layers from 4 samples). E. Micaceous layers contiguous to felsic layers of D (4 layers).

TABLE 4. CHEMICAL COMPOSITIONS CALCULATED FROM MODAL ANALYSES (FIG. 14), HITCHCOCK LAKE FORMATION

	A	B	C	D	E
SiO_2	81.6	73.5	76.8	78.7	79.8
Al_2O_3	10.3	13.7	11.8	13.2	8.9
$FeO + Fe_2O_3$	1.9	3.7	3.0	0.3	4.1
MgO	0.7	1.3	1.2	0.1	1.4
CaO	1.4	1.4	1.3	2.4	0.1
Na_2O	3.0	2.7	2.9	4.9	0.4
K_2O	1.1	2.2	1.7	0.2	2.9
TiO_2	0.3	0.5	0.6	0.1	0.7
H_2O	0.4	0.9	0.7	0.1	1.2

A Finely streaked plagioclase-quartz granulite.
B. Coarsely streaked biotite-plagioclase-quartz gneiss.
C. Friable mica-quartz-plagioclase granulite.
D. Felsic layers in the coarsely streaked biotite-plagioclase-quartz gneiss.
E. Micaceous layers contiguous to felsic layers of D.

Origin

The most significant difference between the rocks of Hitchcock Lake Formation and the underlying Hartland Unit I is the scale of the interlayering and the highly variable texture. Their origins may differ primarily in the greater importance of volcanic additions to the Hitchcock Lake Formation. Based on chemical considerations, the rocks were predominantly graywackes with subordinate mica-rich layers representing the interbedded shales in this volcanic-marine assemblage.

The variable textures of the quartz-plagioclase granulite and coarse-grained biotite-quartz-plagioclase gneiss are considered to be relicts from the primary sediments. Though it is possible that the parental rocks were alternating fine-grained and coarse-grained graywackes with interlayered shale, it seems equally reasonable to assign the fine-grained granulite to parental rocks composed in large part of acid volcanic ash or other fine volcanic detritus. Many of the Precambrian rocks of the Housatonic and Berkshire Highlands are fine grained and felsic and are a suitable source for fine-grained quartz-feldspar sands as would be an emerging island arc.

The amphibolites, biotite-hornblende gneisses, calc-silicate rocks, and marbles are subordinate and represent basic volcanics, marls, and limestones.

HARTLAND UNIT II AND THE STRAITS SCHIST

The rocks here designated as Hartland Unit II and The Straits Schist form the most distinctive and readily recognized lithic units in the sequence east of Cameron's line. The term Hartland Unit II has been used for the western belt, and The Straits Schist for the eastern belt. Lustrous plagioclase-mica-quartz schist with porphyroblastic garnet, staurolite, kyanite, and plagioclase is the dominant rock with subordinate amounts of granulitic layers and white quartz pods. Figure 15 shows the range in mineralogy, and the average composition of these rocks is given in Table 5. The coarse grains, porphyroblastic texture, and quartz pods make the compositional range less significant than the average composition.

Because of the abundance of kyanite and(or) staurolite and garnet, this unit is resistant and crops out well. Exposures are uniform and rounded; they range from silver-gray to spangly where muscovite flakes are coarse. Weathered surfaces are rough and prickly where kyanite blades and coarse muscovite flakes protrude. In some parts of the New Preston and Roxbury quadrangles, very fine grains of muscovite are abundant, and the rocks are crumbly.

Relation to Other Formations

The correlation of Hartland Unit II with The Straits Schist is controversial, but they do contain very similar rocks and occupy almost the same position in the section. Based on our present knowledge, we consider them correlative and find the implications of other alternatives less attractive. We are more impressed by their similarities than by their differences. Hartland Unit II in the southeastern belt, east of Cameron's line, lies between Hartland Units I and III. To the east, The Straits Schist, equivalent of Hartland Unit II, overlies Hartland Unit I directly or is separated from it by Hitchcock Lake Formation. In the Waterbury and Thomaston areas, the contact is marked by discontinuous lenses of amphibolitic rocks, thin marbles, and crumbly schists of Hitchcock Lake Formation. Where Hartland Unit II rests upon Hartland Unit I, the contact is drawn above the last

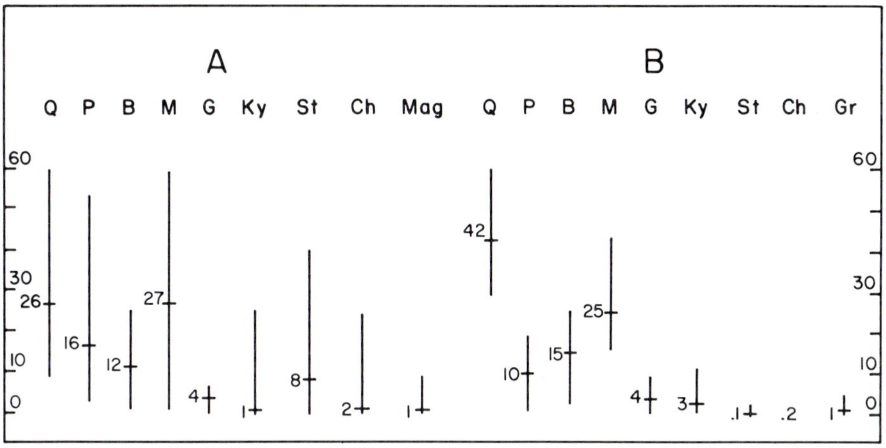

Figure 15. Modal analyses of Hartland Unit II and The Straits Schist. Range of composition indicated by bar; average composition by tick mark. A. Plagioclase-mica-quartz schist (40 samples), Hartland Unit II. Plagioclase composition An_{10} to An_{32}; average, An_{22} (21 determinations). B. Plagioclase-mica-quartz schist (17 samples, Straits Schist). Plagioclase composition An_{21} to An_{30}; average, An_{24} (21 determinations).

TABLE 5. CHEMICAL COMPOSITIONS CALCULATED FROM MODAL ANALYSES (FIG. 15), HARTLAND UNIT II AND THE STRAITS SCHIST

	A	B
SiO_2	56.11	66.36
Al_2O_3	22.97	17.36
$FeO + Fe_2O_3$	9.47	6.15
MgO	2.53	2.10
CaO	0.71	0.48
Na_2O	1.59	1.05
K_2O	3.84	3.97
TiO_2	0.83	0.86
H_2O	1.97	1.62
C		0.6

A. Plagioclase-mica-quartz schist, Hartland Unit II.
B. Plagioclase-mica-quartz schist, The Straits Schist.

granulitic layer. Locally, the contact is marked by several amphibolite layers that range from 2 cm to a maximum of 2 m in thickness near the upper part of Hartland Unit I. This type of contact is particularly well exposed in the Torrington and West Torrington quadrangles near East Litchfield. Also, in the Litchfield quadrangle, amphibolites abound in the upper part of Hartland Unit I and in Unit II a mile east of Litchfield and a mile southeast of Morris. The upper contact is gradational with Hartland Unit III and is placed just below the lowermost quartzose, graphitic, finer-grained, and slabby rocks of this unit. The contact between The Straits Schist and Southington Mountain Formation (Hartland Unit III equivalent) is similar.

Structure and Internal Characteristics

On an outcrop scale, both outcrop belts of this formation appear to be one of the most uniform and homogeneous units in the sequence east of Cameron's line; however, both vertical and lateral textural variations are present. In most areas, relatively few layers of other rock types occur in this quartz-mica schist sequence, but in the New Preston and Roxbury quadrangles, subordinate granulitic layers from a few centimetres to 1 to 2 m thick are more abundant. The layering is considered to be relict bedding. In the Torrington and West Torrington quadrangles, coarse porphyroblasts of garnet, staurolite, and kyanite are relatively abundant in the upper half of Hartland Unit II; otherwise the rocks are the same throughout the section.

On a smaller scale, textural and compositional variations take the form of thin, discontinuous quartz-plagioclase lenses and stringers, generally 1 cm or less thick and as much as 3 m long. These are particularly well developed in areas where discrete granulitic layers are rare, and it is possible that they represent thoroughly disjointed and recrystallized granulitic layers. The regular size and distribution of these streaks and lenses accounts for the overall uniform and homogeneous appearance of this formation. White quartz pods up to 2 to 3 m long are common.

The map configuration of Hartland Unit II and The Straits Schist and their relationship to adjacent units clearly show that they have been involved in fairly large-scale folding. The absence of observable minor folds is attributed to the uniformity of the rocks and the lack of distinctive beds and to the possibility of obliteration of folds by slippage along foliation surfaces.

Figure 1A shows the outcrop width to be fairly uniform where least complicated by large-scale folding. In most areas, the thickness is about 1,000 m; although the range is from 235 to 1,100 m.

Petrography

Hartland Unit II and The Straits Schist are fine- to coarse-grained biotite-muscovite-plagioclase-quartz schists with varying combinations of porphyroblastic garnet, staurolite, plagioclase, and kyanite. Accessory minerals include chlorite, magnetite, graphite, tourmaline, zircon, apatite, rutile, and epidote. Modal analyses are given in Figure 15, and computed chemical compositions are in Table 5.

The texture is lepidoblastic or porphyroblastic. The distinct preferred orientation of the mica flakes combined with quartz-plagioclase stingers and mica folia give the rock a pronounced foliation. Porphyroblasts of garnet, staurolite, and kyanite range from microscopic to several centimetres in length. They are usually prominent on weathered surfaces.

Quartz occurs both as medium-sized grains within the schist and in coarse-grained pods. The smaller grains are commonly elongate parallel to the foliation and lack strain shadows; the larger grains are nearly equidimensional and some are strained.

Plagioclase occurs both as poikilitic porphyroblasts and as small grains usually in a mosaic with quartz. Twinning, nearly absent in the small grains, is abundant in the porphyroblasts. Both types are anhedral. There is no appreciable difference in range, average, or distribution of composition between small grains and porphyroblasts or between samples from different stratigraphic or geographic positions. In some rocks, the shape and location of plagioclase porphyroblasts and the pattern of inclusions indicate that the porphyroblasts have replaced mica.

The micas occur typically as well-oriented flakes; muscovite is more abundant than biotite in most rocks. In The Straits Schist, muscovite is typically coarse-grained and gives the rock a spangled aspect. In Hartland Unit II, fine-grained, almost powdery, muscovite occurs in increasing abundance from the West Torrington quadrangle southwestward to the Roxbury quadrangle. Biotite is more randomly oriented and distributed than muscovite; both micas are in thin folia. Biotite is commonly partly altered to chlorite.

Garnet is in anhedral to euhedral crystals, most of which are poikilitic. In some of the crystals the inclusions are in vague zones; their abundance is not consistent in either the core or the rim of the garnet. Quartz and opaques are the main inclusions; in some garnets they form curved trails indicating rotation of the garnet during growth.

Staurolite occurs as anhedral to subhedral crystals that locally attain lengths of 7 to 10 cm. Many staurolite crystals, in micaceous portions of the rock, appear to have replaced the mica, inherited the texture of the mica, and retained the quartz and opaques as inclusions. Some staurolite has been altered to fine-grained muscovite along its edges.

Kyanite is widely variable in amounts and forms either clusters of small grains or coarse poikilitic porphyroblasts. It is more abundant in The Straits Schist where it forms blades as long as 3 cm. In the Hartland Unit II, it is less common but locally forms spectacular crystals several cm long.

Graphite is ubiquitous in The Straits Schist, but the amount ranges from a mere trace to as much as 10 percent. Some discontinuous lenses more than 30 m thick and 100 to 200 m long are exceptionally graphitic. Graphite is not a characteristic mineral in Hartland Unit II, but in the West Torrington and Torrington quadrangles, graphite-rich zones can be traced for 800 m. Farther southwest, graphite is not apparent.

Origin

Any theory of the origin of Hartland Unit II and The Straits Schist must take into account the following facts: (1) the calculated chemical compositions based on the average modes are those of aluminous shales (Table 5) representing maturely weathered material probably derived from a nearby terrigenous area; (2) if the correlation of Hartland Unit II and The Straits Schist is valid, the unit grades from variably graphitic and quartzose in the east to nongraphitic and ferruginous in the west; (3) the uniquely homogeneous nature of Hartland Unit II and The Straits Schist over a wide area; and (4) The Straits Schist overlies a heterogeneous assemblage of rocks (amphibolites, marble, crumbly schists, calc-silicate rocks, granulites, and gneisses), which suggests it was deposited at an unconformity.

HARTLAND UNIT III (SOUTHINGTON MOUNTAIN FORMATION)

The rocks of Hartland Unit III (Southington Mountain Formation) are a heterogeneous assemblage of mica-plagioclase-quartz granulite and plagioclase-quartz-mica schists. Some textural and mineralogical variations exist, but overall the units are characterized by thin interlayering with some unusually graphitic layers. Some of the more micaceous rocks contain garnet, staurolite, and kyanite.

The layers, relict bedding, differ in the proportion of mica, quartz, and plagioclase components and in the abundance of graphite. The texture and grain size of the rocks show some variation in different areas, but the thickness of layers, the nature of their differences, and their essential mineralogy remain constant. The thin interlayering causes most outcrops to be very slabby. Where graphite is abundant, they also weather rusty and punky.

Relation to Other Formations

Hartland Unit III conformably overlies Hartland Unit II. The upper contact with Hartland Unit IV is not well exposed except in the Roxbury quadrangle. There, for a distance of 5 km, the contact is marked by a coarsely porphyroblastic staurolite-kyanite-garnet-bearing granulitic gneiss. The overlying unit normally also consists of interlayered schist and granulite, and it is difficult to mark the precise location of the contact. The contact is placed at the change from slabby, thinly layered rocks to more thickly layered and massive rocks.

Structure and Internal Characteristics

The major structural feature that sets Hartland Unit III apart from the underlying unit and from the overlying Hartland Unit IV is thin interlayering of different lithologies. Individual layers range from 0.5 cm to 2 m in thickness, but most are between 2 and 8 cm thick. Thin layers are very regular, giving many outcrops a striped aspect.

In most of the rocks, the mica flakes show good parallel orientation, and in some areas, there is nearly perfect parallel orientation. Mica foliation and layering are parallel except in hinge zones. The relatively high total mica content, excellent orientation, and the thin interlayering causes this rock to form slabby outcrops (Fig. 16).

Northeast of the Waterbury dome, the rocks in the part of this unit mapped as Southington Mountain Formation tend to be somewhat coarser grained and less well foliated than the predominantly fine-grained rocks of Hartland Unit III.

Minor folds with wave lengths and amplitudes of 1 to 3 m can be found locally. Most are isoclinal with the axial plane parallel to the foliation. The thickness is best determined in the Litchfield and New Preston quadrangles where the dip is uniformly to the west. In this belt, the thickness shows little variation, being consistently about 900 m. The range in thickness as calculated from the map pattern is 700 to 1,500 m.

Petrography

Hartland Unit III is a heterogeneous sequence of thinly interlayered mica-quartz granulite and granulitic gneiss, and plagioclase-quartz-mica schist. Considerable variation in the proportion of mica, quartz, and plagioclase is found in both the granulitic and schistose rocks; the granulitic and schistose rocks grade into each other. Graphite is abundant in some granulitic rocks and schist, but is absent in others. Garnet, staurolite, and kyanite, in variable combinations and abundance,

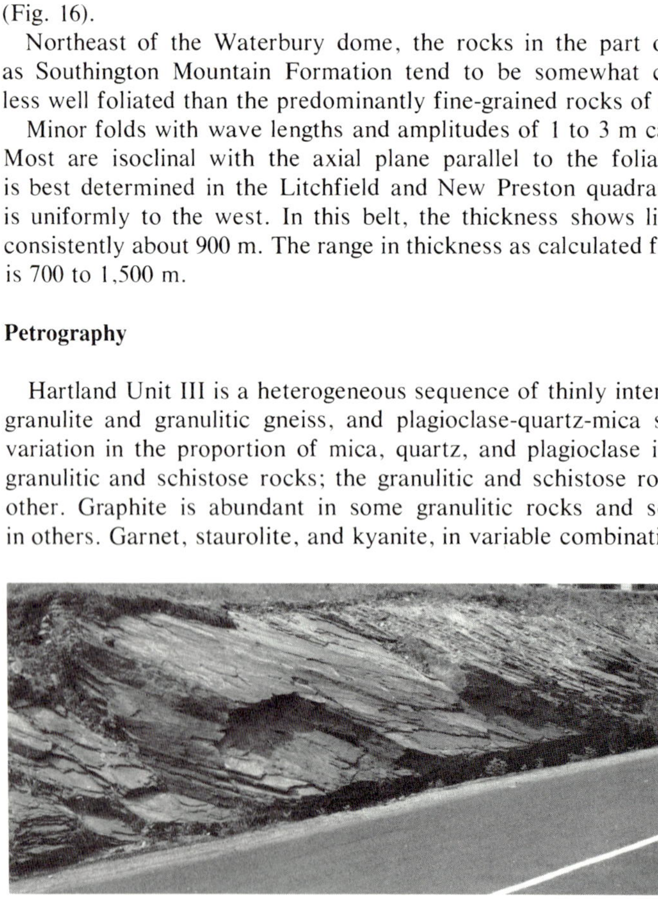

Figure 16. Typical thin layering and slabby character of Hartland Unit III.

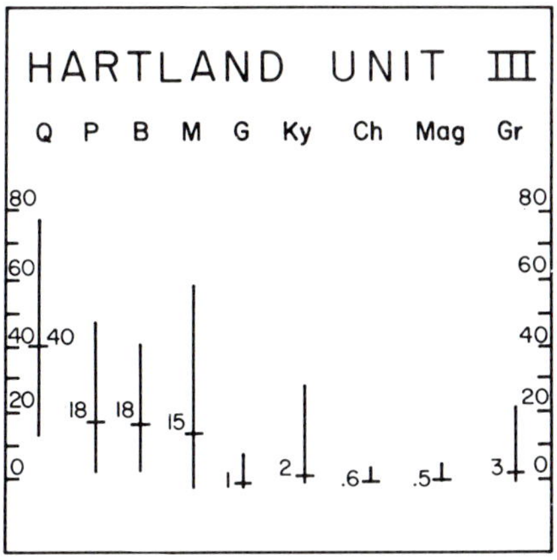

Figure 17. Modal analyses of thinly interlayered granulite and schist (33 samples), Hartland Unit III. Range of composition indicated by bar; average composition by tick mark. Plagioclase composition estimated to average An_{24} based on preliminary studies.

TABLE 6. CHEMICAL COMPOSITIONS CALCULATED FROM MODAL ANALYSES OF THINLY INTERLAYERED GRANULITE AND SCHIST (FIG. 17), HARTLAND UNIT III

SiO_2	65.19
Al_2O_3	15.19
$FeO + Fe_2O_3$	6.47
MgO	2.24
CaO	0.89
Na_2O	1.76
K_2O	3.26
TiO_2	0.86
H_2O	1.39
C	2.78

occur mostly in schist layers. The graphitic rocks are finer grained than other parts of the sequence.

In most rocks of this unit, the texture is strongly lepidoblastic; some staurolite and kyanite rocks are coarsely porphyroblastic. Mica flakes are well disseminated; adjacent layers differ in total mica content rather than in arrangement of micas. Pin-stripe or mica-streaked rocks are not abundant in Hartland Unit III.

Quartz, plagioclase, biotite, and muscovite are the essential minerals. Garnet, staurolite, kyanite, and graphite occur locally. Trace amounts of magnetite, zircon, apatite, sphene, rutile, tourmaline, and chlorite are found. Modal analyses are shown in Figure 17, and Table 6 is a computed composition.

Quartz and plagioclase are typically fine grained to medium grained and elongated parallel to the foliation. Quartz is unstrained and plagioclase is untwinned and unzoned. The micas are well oriented except in a few rocks in which coarse biotite porphyroblasts are unoriented. In some layers, muscovite is most abundant; in others, biotite predominates. There is little tendency for micas to form thin folia.

Garnet, staurolite, and kyanite generally occur as coarse porphyroblasts in schistose layers; most of the porphyroblasts are poikilitic and subhedral. Kyanite is locally very abundant, and in the Torrington quadrangle it occurs in blades as long as 2 cm.

Graphite is a distinctive mineral in Hartland Unit III and is unusually abundant in some layers. It occurs as "clouds" of tiny flakes which are either disseminated or are in bands or streaks parallel to the foliation and layering. Graphite forms inclusions in all the essential minerals; it is most abundant in mica flakes. The presence of even a small amount of graphite causes the rocks to be much darker and generally finer grained than the equivalent nongraphitic rocks.

Origin

The variability of both the granulitic and schistose layers makes it difficult to calculate representative chemical compositions for the original sediments. However, it seems clear that they must have been thinly interlayered shale and siltstone or impure fine-grained sandstone or subgraywacke (Table 6). The very fine grain size of many of the granulitic rocks would suggest siltstone. The usually high graphite content of some layers indicates organic material was abundant, and some layers may have once been black shales. This thinly bedded assemblage contrasts strongly with the rather massive or uniform muscovite schist below it.

HARTLAND UNIT IV

Hartland Unit IV is the most heterogeneous and least distinctive of the units in the sequence southeast of the Berkshire Highlands gneisses. It consists of thinly to thickly interlayered granulites, granulitic gneisses, schist, and subordinate graphitic rocks, calc-silicate rocks, and amphibolites. Variations are found in all of these lithic types, and the unit is best described as a mixture of lithic types that includes layers typical of those in all other units of the Hartland Formation sequence. Hartland Unit IV differs from Hartland Units I and II in that it has about equal proportions of granulitic rocks and schist, and from Hitchcock Lake Formation and Hartland Unit III in that it has thicker layers (as much as 30 m thick).

Structure and Internal Characteristics

Hartland Unit IV is characterized by its lack of uniformity. Individual layers vary in thickness from a few centimetres to 30 m or more. Granulitic layers are seldom over 3 to 5 m thick, whereas schist layers, though fewer in number, are commonly 15 to 30 m thick. None of the layers can be traced with certainty for more than 200 m, and sequences or assemblages of lithic types can be followed only slightly farther.

The rocks in Hartland Unit IV range from fine grained to medium grained; the graphitic rocks have the finest grain. Large porphyroblasts are found in a few rocks near amphibolites that are considered to be syntectonic intrusives (Gates, 1967). In most rocks, foliation depends upon parallelism of mica flakes and typically is well developed. The flakes range from being disseminated throughout the rock to aggregates in thin pin-stripe folia. The foliation and layering are parallel except in hinge zones where mica folia cross the layering. Several folds, 3 to 10 m across, can be seen in Hartland Unit IV; larger folds have not been recognized. The thickness of Hartland Unit IV cannot be accurately determined because the upper contact is a fault. The minimum thickness is 3 km, if the sequence is not thickened by folding.

Relation to Other Units

Hartland Unit IV is exposed in the Roxbury and New Preston quadrangles where it is conformably above Hartland Unit III. The northwest contact between this unit and the Woodville marble and the Waramaug Formation is gradually truncated and at its eastern end is cut off by the fault at Cameron's line. Cameron's line forms the boundary between Hartland Unit IV and the Woodville marble, the Waramaug Formation, and a basic intrusive body at Mount Prospect. Around the Waterbury and other domes to the east, it is missing.

Petrography

Hartland Unit IV is a heterogeneously interlayered assemblage consisting predominantly of granulitic rocks and schist with very subordinate graphitic and calc-sili-

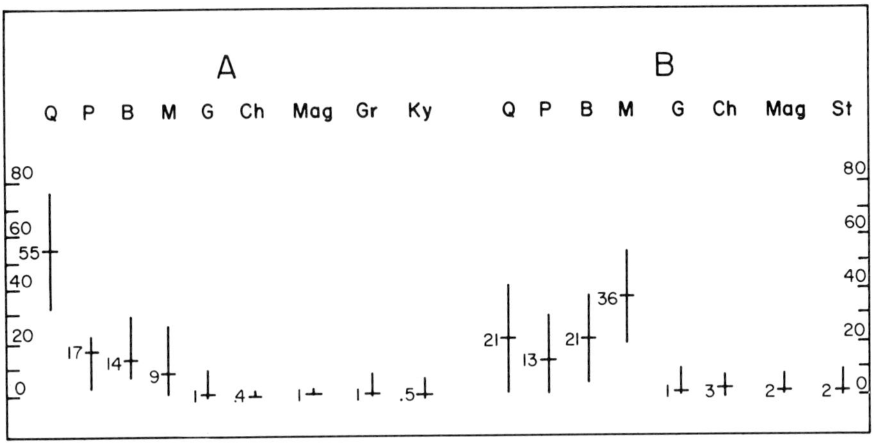

Figure 18. Modal analyses, Hartland unit IV. Range of composition indicated by bar; average composition by tick mark. A. Mica-plagioclase-quartz granulite (11 samples). Plagioclase composition estimated to average An_{23} based on preliminary studies. B. Plagioclase-quartz-mica schist (9 samples). Plagioclase composition estimated to average An_{24} based on preliminary studies.

cate-bearing rocks. Within each of these major types, there are layers of differing mineralogy and texture. None of the individual layers or major lithic types form mappable units, and none of the individual rock types is diagnostic of the unit.

The major minerals are quartz, plagioclase, muscovite, and biotite; garnet, staurolite, graphite, calcite, and hornblende are found in some rocks (Fig. 18).

The granulitic rocks are similar to the granulite and granulitic gneiss found in Hartland Unit I. They are characterized by containing at least 70 percent equidimensional quartz and plagioclase grains. Most of the rest of the rock is mica; either muscovite or biotite may predominate. Variations in both the relative abundance and the arrangement of the mica flakes result in all gradations between granulite and granulitic gneiss and schist. Porphyroblastic garnet and biotite occur in some granulitic rocks. Granulitic rocks make up about half of the total thickness of Hartland Unit IV and form light gray, slabby to massive outcrops.

With an increase in the proportion of mica, the granulitic rocks grade into quartz-mica schist. In most schists, muscovite is the most abundant mica; it is commonly in very fine flakes, which give the rock a soapy feel. X-ray analysis indicates it contains 20 percent paragonite. Biotite is commonly coarse grained and randomly oriented. Fine-grained quartz and plagioclase grains typically are elongated parallel to the foliation. Porphyroblasts of garnet and staurolite are relatively abundant; kyanite occurs locally. The texture is lepidoblastic; the rocks form slabby or crumbly outcrops, except where convolutely folded.

Graphitic rocks form subordinate, but very distinctive, layers in which fine-grained graphite is abundant. These rocks are typically very fine grained and slabby to almost slaty. Fresh surfaces are dark gray; weathered surfaces are rust colored. Pyrite is an abundant accessory mineral in graphitic rocks, but otherwise their mineralogy is similar to that of the more normal granulite or schist of this unit. The graphite occurs as clouds of very fine flakes in layers parallel to the foliation. These layers are widely scattered, as much as 3 m thick, and at most 100 to 200 m long.

The calc-silicate rocks contain calcite, epidote, zoisite, garnet, diopside, tremolite, and sphene mixed with the dominant rock minerals quartz and plagioclase. Micro-

TABLE 7. CHEMICAL COMPOSITIONS CALCULATED FROM MODAL ANALYSES (FIG. 18), HARTLAND UNIT IV

	A	B
SiO_2	73.74	53.24
Al_2O_3	10.84	21.96
$FeO + Fe_2O_3$	6.32	10.98
MgO	1.69	3.31
CaO	0.79	0.59
Na_2O	1.6	1.36
K_2O	2.28	5.55
TiO_2	0.64	1.21
H_2O	0.96	2.65
C	1.12	

A. Mica-plagioclase-quartz granulite.
B. Plagioclase-quartz-mica schist.

cline, muscovite, biotite, and chlorite occur in places. These rocks are white to light gray and often have pink and green patches of garnet and epidote. They are massive and brittle and typically have granoblastic textures.

Calc-silicate-bearing rocks are found sparingly throughout this unit as thin layers or lenses less than 30 cm thick; one layer reaches a thickness of 3 m. None of the layers can be traced laterally for more than 7 m. Identical rock types occur in Hartland Unit I.

Origin

The granulite and schist of Hartland Unit IV are mineralogically the same as the granulite and schist found in Hartland Units I and II. This indicates that interbedded subgraywacke and shale originally formed the bulk of Unit IV (Table 7). The graphitic rocks probably were black shales. The calc-silicate rocks were thin lenses of impure limestone.

CORRELATIONS

The three belts of metamorphic rocks described above pose major problems of correlation that are best discussed in terms of protoliths. Correlations within belts are much less controversial than the correlations between contiguous belts or from the western belt to the eastern belt. The western belt constitutes a sandstone-carbonate-shale sequence unconformably above the Precambrian gneiss complex. The central belt involves graywackes and shales and their uncertain stratigraphic or structural relations to the western carbonate sequence and to the Precambrian rocks. Their correlation with rocks of the eastern sequence is even more in doubt. The eastern sequence of volcaniclastic rocks, graywackes, and shales poses a problem not only of its own internal correlation, but also of its relation to the central and western belts across a major tectonic line. A summary of our views is given in Figure 19.

The Northwestern Belt

In the northwestern belt, a Cambrian and Ordovician sequence of sandstones (Dalton) and carbonate rocks (Stockbridge) unconformably overlies the Precambrian

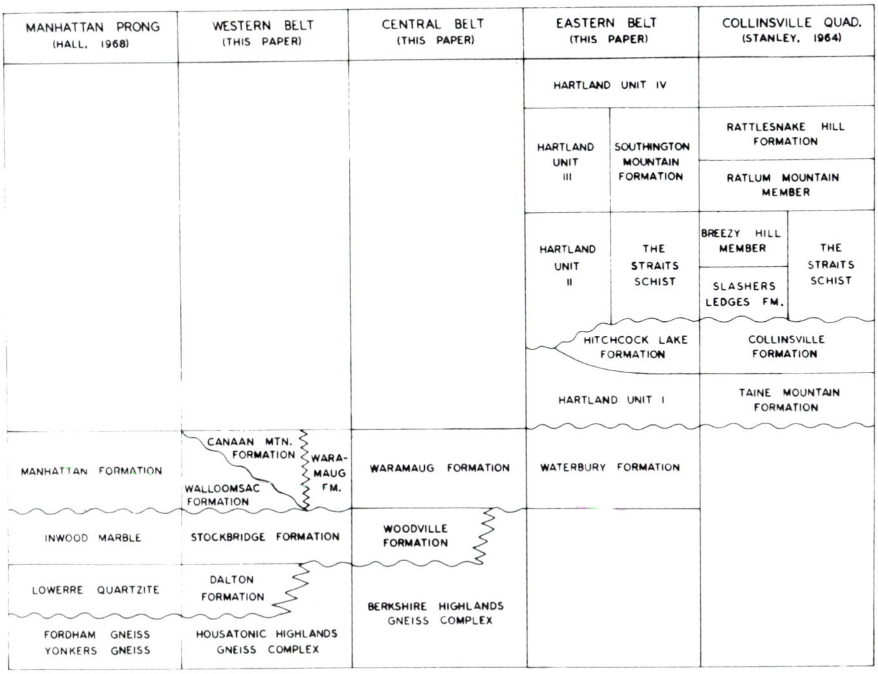

Figure 19. Correlation chart for western Connecticut.

gneiss complex. Truncating the sandstones and carbonate rocks is a heterogeneous shale-carbonate unit (Walloomsac). Overlying this entire assemblage is a shale-graywacke formation variously identified as Manhattan C (Hall, 1968), the Waramaug Formation, and the Canaan Mountain Formation. Although these formations may be correlative, their relations to the underlying units are conjectural. Lithologically they are similar and lack good bedding features, although Gates has mapped members in the Canaan Mountain Formation. In several places, the Waramaug Formation directly overlies both the Stockbridge Formation and the Precambrian rocks and is in fault contact with members of the Hartland Formation. The Canaan Mountain Formation is in direct contact with all rocks from the Precambrian rocks to the Walloomsac Formation. These relations can be reasonably interpreted as an unconformity or a fault.

The Central Belt

The central belt can be considered as transitional between the northwestern carbonate shelf sequence and the southeastern volcanic-graywacke-shale sequence. In this belt, the Waramaug Formation overlies directly the Precambrian gneisses, the Stockbridge Formation, and the Woodville marble and is in fault contact with the southeastern sequence along Cameron's line. Farther to the northeast, the position of Cameron's line is uncertain, and its existence between the Waramaug Formation and the southeastern belt is questionable.

The Waramaug Formation lacks the good bedding features of both the Precambrian gneisses and the Dalton, Stockbridge, and Walloomsac Formations. Foliation and compositional layering are well defined in the latter formations, but are generally lacking or obscure in the Waramaug Formation. The structural style and lack of lithic contrasts gives the Waramaug Formation a homogeneity lacking in the other contiguous formations. This homogeneity apparently represents a different sedimentational environment that must be resolved in any reasonable reconstruction of the depositional, tectonic, and metamorphic history of the region.

The Southeastern Belt

The southeastern belt is composed of five mappable units representing essentially continuous marine volcaniclastic deposition with a slight disconformity below Hartland Unit II (The Straits Schist) horizon. Our correlation of these five units with other mapped units is given in Figure 19.

We have correlated Hartland Unit II and The Straits Schist for the following reasons. Mineralogically and compositionally, they are aluminous shales that are readily recognized (that is, distinctive) in the field and are the most easily traced formations in west-central Connecticut. Their compositional variations are minor and easily explained in terms of expectable facies changes in a shale sequence. Both The Straits Schist and Hartland Unit II are in contact with Hartland Unit I. The lithic types on the other side of both The Straits Schist and Hartland Unit II are very different from Hartland Unit I. Hitchcock Lake Formation, which lies between The Straits Schist and Hartland Unit I, is discontinuous and known to "pinch out," allowing The Straits Schist to come in direct contact with Hartland Unit I. The variable thickness of Hitchcock Lake Formation and the absence of Hitchcock Lake Formation adjacent to Hartland Unit II are considered to represent a disconformity, although they could represent nondeposition in view of its heterogeneous composition—amphibolites, marbles, and felsic gneisses. The change from a volcaniclastic sequence to a terrigenous shale sequence is very marked. In summary, the lithologic sequence of Hartland Units I, II, and III is considered similar to Hartland Unit I, Hitchcock Lake Formation, The Straits Schist, and Southington Mountain Formation.

The correlation of the Waterbury Formation with other formations in western Connecticut is fraught with problems and contradictions; however, for several reasons, we would like to suggest a correlation with the Waramaug and Canaan Mountain Formations. First, and most obvious, is the lithic similarity of these formations. They are all tough, coherent rocks that are rounded by glacial abrasion and break as readily across foliation as parallel to it in contrast with most other rocks in this area. The Waterbury Formation is well layered, and the Waramaug Formation notably lacks good bedding. The Canaan Mountain Formation has mappable subunits and is different from the Waramaug and Waterbury Formations in this respect. The structural style of these three formations is similar, and it contrasts with that of other formations in this region. The migmatitic texture in the Waterbury Formation results largely from metamorphic recrystallization (Schoenwald, 1971), not injection. The three formations are also similar in their relations to other formations, but are, at least in part, in a faulted relation to most other units. The Canaan Mountain, Waramaug, and Waterbury Formations may be shallow water, transitional, and deep water facies, respectively.

Regional Correlations

Several schemes have been proposed for correlating the rocks in western Connecticut (particularly the Waramaug Formation and the units of the southeastern belt) with the established section to the north in Massachusetts and Vermont. Stanley (1964, 1968) and Hatch and Stanley (1970) presented a series of correlations of the rocks in the Collinsville quadrangle with those farther north; each subsequent correlation resulted in reinterpretation and rearrangement of the stratigraphic sequence in the Collinsville quadrangle (Fig. 20). One of the more significant changes in their recent correlation is the placement of The Straits Schist of the Collinsville quadrangle at the top of the stratigraphic section. Because it is now possible to trace units from Collinsville south to the Roxbury quadrangle, future correlations must take into account not only the lithic characteristics of the rocks near Collinsville, but also the lithic characteristics and stratigraphic succession throughout the central section of western Connecticut. The rock units, their characteristics and relations to other formations, and the evidence leading to the establishment of a pre-Devonian stratigraphic sequence in west-central Connecticut have been detailed in this paper. These conclusions are not consistent with the placement of The Straits Schist or its equivalents at the top of the Hartland Formation nor with the proposal (Stanley and Hatch, this volume) that Hartland Units III and IV are equivalent to the Waterbury Formation, Hartland Unit I, and Hitchcock Lake Formation.

Hatch and others (1968) proposed correlations from Massachusetts into northern Connecticut (Columns B and C, Table 8) and included the only recent map showing distribution of stratigraphic units in western Massachusetts and northern Connecticut. The southern edge of their map coincides with the northern boundary of the Torrington quadrangle (Martin, 1970). Their map thus adjoins the northeast part of Figure 1A, and comparison of the two maps shows the following direct correlations based on mapping of units on both sides of the common boundary (Columns A and B, Table 8). Correlation of the units of southern Massachusetts and north-central Connecticut with those of northern Massachusetts is taken from Hatch and others (1968). Both The Straits Schist belt, as we have mapped it in central Connecticut, and Hartland Unit II belt trace into separate bands of unit 7 of Hatch, Schnabel, and Norton, thus substantiating the correlation of The Straits Schist and Hartland Unit II. However, Schnabel (1971; 1971, oral commun.) correlated unit 5 on the Hatch, Schnabel, and Norton map with The Straits Schist and noted that unit

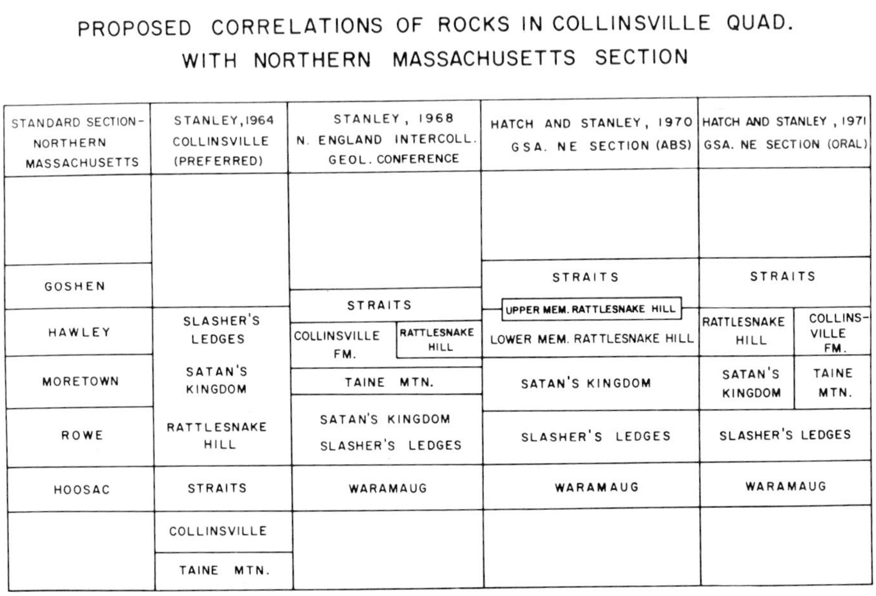

Figure 20. Regional correlation chart. Proposed correlations of rocks in Collinsville quadrangle with northern Massachusetts section.

TABLE 8. CORRELATION OF WEST-CENTRAL CONNECTICUT AND SOUTHERN MASSACHUSETTS AND NORTH-CENTRAL CONNECTICUT WITH NORTHERN MASSACHUSETTS

A	B	C
The Straits Schist	5, 7	Hawley Formation
Hartland Unit III	6	no equivalent
Hartland Unit II	7	no equivalent
Hartland Unit I	3	Moretown Formation
Waramaug Formation	1, 2	Hoosac, Rowe Formations

A. West-central Connecticut (this paper and Martin, 1970).
B. Southern Massachusetts and north-central Connecticut (Hatch and others, 1968).
C. Northern Massachusetts (Hatch and others, 1968).

7 overlies The Straits Schist. Although there are some discrepancies between the stratigraphic section and interpretations presented here and those of Hatch, Schnabel, and Norton for north-central Connecticut, there are several significant areas of agreement, including the following:

1. Both maps show rock units to be progressively younger eastward from the Precambrian gneiss complexes for about half the distance to the Triassic border fault. This includes, in order of decreasing age, the Waramaug Formation, Hartland Unit I, and Hartland Unit II in central Connecticut; and units 1, 2, and 3 in north-central Connecticut. The top of this sequence is interpreted as a fault by Hatch, Schnabel, and Norton, and as a synclinal keel in Hartland Unit III by us.

2. Eastward, near the Triassic border fault, the oldest rocks occur in the cores of domes with stratigraphically higher rocks outward in all directions. Thus, west of the line of domes, rock units become progressively younger to the west. This includes Hartland Unit I (equivalent to Taine Mountain Formation), Hitchcock Lake Formation (equivalent to Collinsville Formation), The Straits Schist, and Hartland Unit III in west-central Connecticut and units 4, 5, 6, 7, and 8 of Hatch, Schnabel, and Norton in north-central Connecticut. In general, stratigraphically higher rocks occur in the middle of the belt, and lower and older rocks are found in domes to the east and against the Precambrian to the west.

3. Both maps indicate that The Straits Schist and its equivalents occur within the stratigraphic section, above the centers of the domes and below the units in the middle of the belt rather than at the top as suggested by Stanley (1968) and Hatch and Stanley (1970). In fact, Hatch and others (1968) reported graded bedding and channeling that indicate that units 4 and 5 (The Straits Schist equivalent) underlie units 6, 7, and 8 of the middle belt.

REFERENCES CITED

Alsobrook, A. F., 1965, The origin of the granitic gneiss in the northwest quarter of the Woodbury quadrangle, Connecticut [M.S. thesis]: Madison, Wisconsin Univ., 135 p.

Besancon, J. R., 1970, A Rb-Sr isochron for the Nonewaug Granite, *in* Contributions to Geochronology in Connecticut: Connecticut Geol. and Nat. History Survey Rept. Inv. 5, p. 1-9.

Brauner, M. B., 1969, A petrological study of the Housatonic Highlands gneiss and the Dalton Formation of the South Canaan quadrangle, Connecticut [M.S. thesis]: Madison, Wisconsin Univ., 54 p.

Carr, M. H., 1960, The bedrock geology of the Naugatuck quadrangle: Connecticut Geol. and Nat. History Survey Quad. Rept. 9, 25 p.

Cassie, R. M., 1965, The evolution of a domal gneiss (granitic) and its relations to the geology of the Thomaston quadrangle, Connecticut [Ph.D. thesis]: Madison, Wisconsin Univ., 109 p.

Christensen, N. I., 1961, Amphibolites and related rocks in the West Torrington quadrangle, Connecticut [M.S. thesis]: Madison, Wisconsin Univ., 61 p.

——1963, The Hodges Mafic Complex—A study of the structural control of basic intrusives [Ph.D. thesis]: Madison, Wisconsin Univ., 98 p.

Clark, G. S., and Kulp, J., 1968, Isotopic age study of metamorphism and intrusion in western Connecticut and southeastern New York: Am. Jour. Sci., v. 266, p. 865-894.

Deer, W. A., Howie, R. A., and Zussman, J., 1966, An introduction to the rock forming minerals: New York, Wiley, 528 p.

Dewees, D. J., 1966, A structural interpretation of some deformed metasediments in northern Waterbury quadrangle, Connecticut [M.S. thesis]: Madison, Wisconsin Univ., 129 p.

Fritts, C. E., 1963, Bedrock geology of the Southington quadrangle, Connecticut: U.S. Geol. Survey Geol. Quad. Map GQ-200.

Gates, R. M., 1951, The bedrock geology of the Litchfield quadrangle: Connecticut Geol. and Nat. History Survey Quad. Rept. 1 (Misc. Ser. 3), 13 p.

——1952, The geology of the New Preston quadrangle, Part I, The bedrock geology: Connecticut Geol. and Nat. History Survey Quad. Rept. 2 (Misc. Ser. 5), p. 5-34.

——1954, The bedrock geology of the Woodbury quadrangle: Connecticut Geol. and Nat. History Survey Quad. Rept. 3, 32 p.

——1959, The bedrock geology of the Roxbury quadrangle, Connecticut: U.S. Geol. Survey Geol. Quad. Map 121.

——1961, The bedrock geology of the Cornwall quadrangle: Connecticut Geol. and Nat. History Survey Quad. Rept. 11, 35 p.

——1967, Amphibolites: syntectonic intrusives?: Am. Jour. Sci., v. 265, p. 118-131.

——1975, The bedrock geology of the South Canaan quadrangle: Connecticut Geol. and Nat. History Survey Quad. Rept. 32, 33 p.

Gates, R. M., and Bradley, W. C., 1952, The geology of the New Preston quadrangle. Pt. I. Bedrock geology: Connecticut Geol. and Nat. History Survey Quad. Rept. 2 (Misc. Ser. 5), 46 p.

Gates, R. M., and Christensen, N. I., 1965. The bedrock geology of the West Torrington quadrangle: Connecticut Geol. and Nat. History Survey Quad. Rept. 17, 38 p.

Gates, R. M., and Martin, C. W., 1967, The bedrock geology of the Waterbury quadrangle, Connecticut Geol. and Nat. History Survey Quad. Rept. 22, 36 p.

Gregory, H. E., and Robinson, H. H., 1907, Preliminary geological map of Connecticut: Connecticut Geol. and Nat. History Survey Bull. 7, 39 p.

Hall, L. B., 1968, Times of origin and deformation of bedrock in the Manhattan Prong, *in* Zen, E-An, White, W. S., Hadley, J. B., and Thompson, J. B., Jr., eds., Studies of Appalachian geology: northern and maritime (Billings volume): New York, Wiley-Interscience, p. 117-127.

Hatch, N. L., and Stanley, R. S., 1970, Stratigraphic continuity and facies change in formations of early Paleozoic age in western Massachusetts and tentative correlations with Connecticut: Geol. Soc. America Abs. with Programs, v. 2, no. 1, p. 23-24.

Hatch, N. L., Schnabel, R. W., and Norton, S. A., 1968, Stratigraphy and correlation of the rocks on the east limb of the Berkshire anticlinorium in western Massachusetts and north-central Connecticut, *in* Zen, E-An, White, W. S., Hadley, J. B., and Thompson, J. B., Jr., eds., Studies of Appalachian geology: northern and maritime (Billings volume): New York, Wiley-Interscience, p. 177-184.

Howe, D. M., 1966, The origin of the Hitchcock Lake banded gneiss, northern Waterbury quadrangle, Connecticut [M.S. thesis]: Madison, Wisconsin Univ., 78 p.

Martin, C. W., 1959, Minor structures and their significance in the Hartland Formation, Connecticut [M.S. thesis]: Madison, Wisconsin Univ., 52 p.

——1962, Petrology, metamorphism, and the structure of the Hartland Formation in the central western Connecticut Highlands [Ph.D. thesis]: Madison, Wisconsin Univ., 99 p.

——1970, The bedrock geology of the Torrington quadrangle, Connecticut: Connecticut Geol. and Nat. History Survey Quad. Rept. 25, 53 p.

Martin, D.C.H., 1960, The age relation of the Mount Tom hornblende gneiss to the dioritic gneisses of the Mount Prospect Complex [M.S. thesis]: Madison, Wisconsin Univ., 61 p.

Middleton, G. V., 1960, Chemical characteristics of sandstones: Geol. Soc. America Bull., v. 71, p. 1011-1026.

Percival, J. G., 1842, Report of the geology of the state of Connecticut: New Haven, Osborn and Baldwin, 495 p.

Rice, W. N., and Gregory, H. C., 1906, Manual of the geology of Connecticut: Connecticut Geol. and Nat. History Survey Bull. 6, 273 p.

Rodgers, J., 1968, The eastern edge of the North American continent during the Cambrian and Early Ordovician, *in* Zen, E-an, White, W S., Hadley, J. B., and Thompson, J. B., Jr., eds., Studies of Appalachian geology: northern and maritime (Billings volume): New York, Wiley-Interscience, p. 141-149.

Rodgers, John, Gates, R. M., Cameron, E. N., and Ross, R. J., Jr., 1956, A preliminary geological map of Connecticut: Connecticut Geol. and Nat. History Survey.

Rodgers, J., Gates, R. M., and Rosenfeld, J. L., 1959, Explanatory text for preliminary geological map of Connecticut, 1956: Connecticut Geol. Nat. History Survey Bull. 84, 64 p.

Schnabel, R. W., 1971, Pre-Silurian stratigraphy in south-central Massachusetts and north-central Connecticut: Geol. Soc. America Abs. with Programs, v. 3, no. 1, p. 54.

Schoenwald, C. P., 1971, The origin of the Waterbury dome migmatite [M.S. thesis]: Madison, Wisconsin Univ., 72 p.

Smith, J. M., 1960, The bedrock geology of the vicinity of Torrington, Connecticut [M.S. thesis]: Madison, Wisconsin Univ., 49 p.

Stanley, R. S., 1964, The bedrock geology of the Collinsville quadrangle: Connecticut Geol. and Nat. History Survey Quad. Rept. 16, 99 p.

——1968, Metamorphic geology of the Collinsville area, Trip D-4, *in* Guidebook for field trips in Connecticut—New England Intercollegiate Geol. Conf., 60th Ann. Mtg.: Connecticut Geol. and Nat. History Survey Guidebook 2, 17 p.

Stanley, R. S., and Hatch, N. L., Jr., 1976, Discussion: Early Paleozoic stratigraphy of western Massachusetts and Connecticut and southeastern New York, *in* Page, L. R., ed., Contributions to the stratigraphy of New England: Geol. Soc. America Mem. 148 (this volume).

Thede, R., 1958, The petrology of the type Thomaston (Reynolds Bridge) granite, Connecticut [M.S. thesis]: Madison, Wisconsin Univ., 63 p.

Vogel, T. A., 1961, The sedimentary origin of the Housatonic Highlands gneiss complex in the Cornwell area, Connecticut [M.S. thesis]: Madison, Wisconsin Univ., 70 p.

Zen, E-an, 1967, Time and space relationships of the Taconic allochthon and autochthon: Geol. Soc. America Spec. Paper 97, 107 p.

MANUSCRIPT RECEIVED BY THE SOCIETY APRIL 18, 1974
REVISED MANUSCRIPT RECEIVED NOVEMBER 11, 1974
MANUSCRIPT ACCEPTED DECEMBER 4, 1974

Printed in U.S.A.

Geological Society of America
Memoir 148

Preliminary Correlation of Rocks in Southwestern Connecticut

LEO M. HALL
Department of Geology
University of Massachusetts
Amherst, Massachusetts 01002

ABSTRACT

In southwestern Connecticut and southeastern New York, a sequence of Precambrian rocks is unconformably overlain by a Cambrian-Ordovician quartzite-carbonate sequence. This, in turn, is unconformably overlain by a dark-gray fissile sillimanite schist that locally has calcite marble and calcite-rich mica schist at its base. The Paleozoic portion of this stratigraphic sequence correlates with the miogeosynclinal sequence of eastern New York and the western parts of Vermont, Massachusetts, and Connecticut.

A eugeosynclinal sequence is in contact with the fissile schist of the miogeosynclinal sequence. This eugeosynclinal sequence consists of the following rock units in order eastward from the fissile schist and apparently in order from oldest to youngest: (1) feldspathic garnet-muscovite-biotite schist commonly with sillimanite nodules and with some interbedded amphibolite; (2) garnet-muscovite-biotite schist with local sillimanite and (or) kyanite, amphibolite, and light-gray or white biotite-muscovite-quartz-feldspar gneiss locally containing garnet; (3) white biotite-muscovite-quartz-feldspar gneiss locally containing garnet; (4) interbedded garnet-muscovite-biotite-quartz-feldspar schist and muscovite-biotite-quartz-feldspar granulite; (5) dark-gray biotite- and (or) hornblende-quartz-feldspar gneiss with feldspar augen common. A suggested correlation of these units with formations in Connecticut and Massachusetts is as follows: (1) correlates with the Waramaug and Hoosac Formations; (2) and (3) with part of the Hartland Formation, Rowe Schist, and the Trap Falls Formation; (4) with parts of the Hartland, Taine Mountain, Moretown, and Fairfield Formations and the banded schist member of the Southington Mountain Formation; and (5) with parts of the Prospect, Collinsville, and Hawley Formations.

If the above correlation is correct, the eugeosynclinal sequence in southwestern Connecticut is the same age as the Cambrian-Ordovician miogeosynclinal sequence, and its structural position above the miogeosynclinal sequence can be explained by a major premetamorphic thrust fault similar to, if not actually, the Taconic thrust fault. The proposed correlation also suggests that Silurian rocks rest unconformably on overturned pre-Silurian rocks in the vicinity of the Westport, Bridgeport, and Long Hill quadrangles in southern Connecticut.

INTRODUCTION

The sillimanite-grade Precambrian and Paleozoic rocks that have been mapped in detail (Hall, 1968a, 1968b) are in southeastern New York and southwestern Connecticut, south of the Precambrian Hudson highlands and east of the Mesozoic Newark basin. The rocks of the White Plains-Glenville area, named after the two $7^1/_2$-minute quadrangles that cover it (Fig. 1), are interpreted as a Precambrian basement gneiss complex that is unconformably overlain by a Cambrian-Ordovician miogeosynclinal sequence, which is in turn tectonically overlain by a Cambrian-Ordovician eugeosynclinal sequence. The intent of this paper is to describe briefly the stratigraphic column in the White Plains-Glenville area and to suggest preliminary correlations with rocks elsewhere in New York, Connecticut, and Massachusetts.

STRATIGRAPHY

It is possible to describe a stratigraphic column for the White Plains-Glenville area (Fig. 2) by assuming that the rocks are progressively younger with increasing distance across strike and upward from the basement gneiss complex (Fig. 3). Although this at first seems to be a straightforward approach to the stratigraphy in this area, regional considerations point to the presence of at least one, and probably two, major faults in the section. One of these faults involves thrusting of the Cambrian-Ordovician eugeosynclinal sequence onto the miogeosynclinal sequence. This allochthonous eugeosynclinal sequence also appears to decrease in age, across the strike, away from the gneiss complex. The other fault is within the eugeosynclinal sequence and is of uncertain nature. For purposes of description, the units that make up the stratigraphic column are presented in order of increasing distance from the gneiss complex. Thicknesses have been calculated directly from map data, and no factor to account for tectonic effects on thickness was considered. Actual relation of the stated thicknesses to the original thicknesses is not known.

Fordham and Yonkers Gneisses

The rock units that make up the Precambrian Yonkers Gneiss and Fordham Gneiss (Fig. 2) form a basement gneiss complex (Hall, 1966). Radiometric dating (Grauert and Hall, 1973) indicates that there is a significant age difference between the Yonkers and Fordham, but both are truncated by the unconformity at the base of the Paleozoic section and are therefore older than the Lowerre Quartzite.

The Yonkers Gneiss is a relatively homogeneous, commonly pinkish, biotite-hornblende-quartz-feldspar gneiss. The hornblende typically has a $2V$ that is less than 15° and has been optically identified as ferrohastingsite.

The Fordham Gneiss has been subdivided into five rock units in the White Plains quadrangle (Hall, 1966) and into five other rock units in the Glenville quadrangle (Hall, 1968b). It has also been subdivided into three units south of Peekskill, New York (Ratcliffe, 1968).

Subdivisions of the Fordham Gneiss that have been mapped in the White Plains and Glenville quadrangles are briefly described in the following lists. These subdivisions are not indicated separately on the geologic map (Fig. 2) because of restrictions of scale. The spatial relations of the subdivisions allow a sequence to be established, but the relative age of the units is unknown. The sequential relations of subdivisions in the White Plains quadrangle with respect to those in the Glenville quadrangle are unknown, but it is hoped that work in progress will determine these relations.

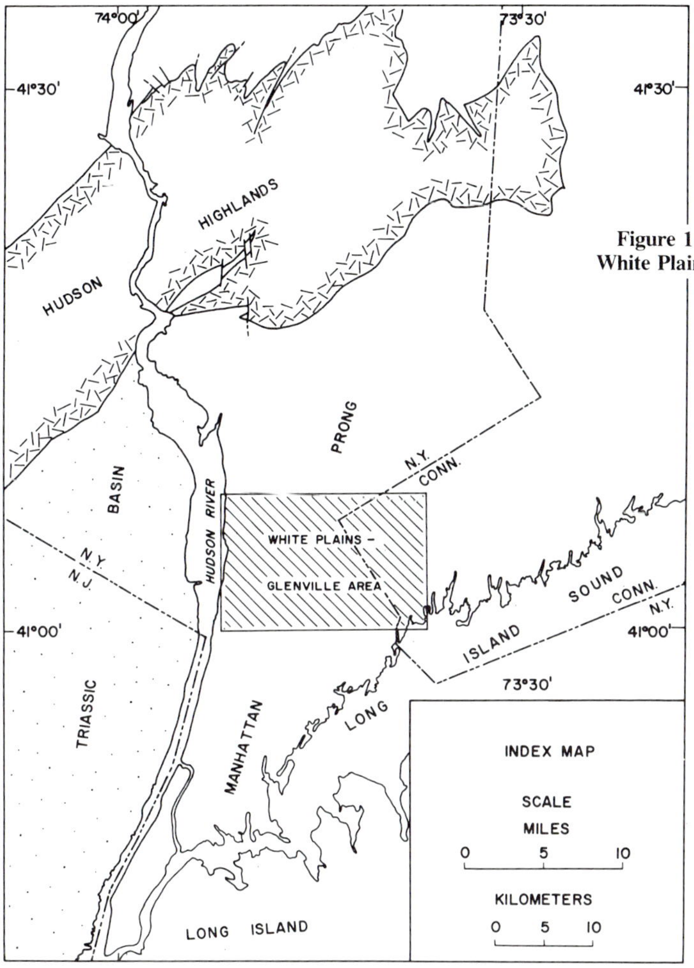

Figure 1. Index map of the White Plains-Glenville area.

Five subdivisions of the Fordham Gneiss in the White Plains quadrangle have been designated by the letters A, B, C, D, and E (Hall, 1968a) and are briefly described below.

1. Fordham Gneiss, member A. This unit consists predominantly of brown-weathering, garnet-biotite-quartz-feldspar gneiss, amphibolite, and gray biotite-hornblende-quartz-feldspar gneiss.
2. Fordham Gneiss, member B. Gray garnet-biotite-quartz-feldspar gneiss, locally containing sillimanite, and amphibolite are the predominant rocks.
3. Fordham Gneiss, member C. Gray biotite-hornblende-quartz-feldspar gneiss with pinkish or white quartz-feldspar layers and lenses constitutes most of this unit. Also commonly present are amphibolite layers and pinkish biotite- and (or) hornblende-quartz-feldspar gneiss that is very similar to the Yonkers Gneiss and is locally extensive enough to be mapped separately.
4. Fordham Gneiss, member D. Rusty-weathering, sulfidic sillimanite-garnet-biotite-quartz-feldspar schist or schistose gneiss with local interbeds of siliceous biotite gneiss or quartzite characterize member D. Calc-silicate and calcite-bearing granulite layers also are locally present.
5. Fordham Gneiss, member E. This is an interbedded assemblage of gray garnet-biotite-quartz-feldspar gneiss, brown- to rusty-weathering biotite-quartz-feldspar gneiss and amphibolite. Gray biotite-quartz-feldspar augen gneiss at the base of member E is locally abundant enough to be mapped separately.

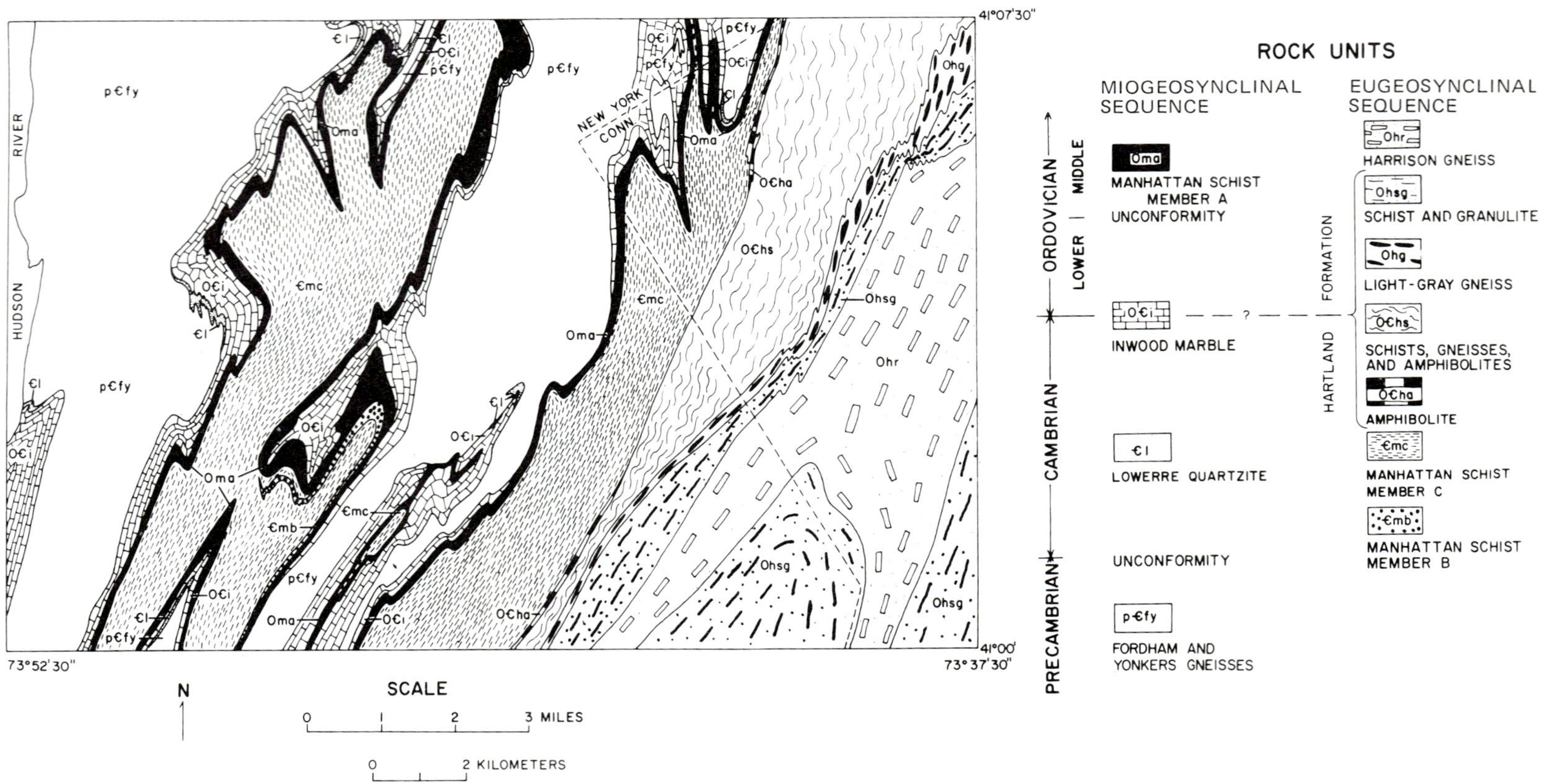

Figure 2. Geologic map of the White Plains–Glenville area.

Five other subdivisions of the Fordham Gneiss have been defined and mapped in the Glenville quadrangle (Hall, 1968b) but have not yet been designated by any nomenclature. They are briefly described, in spatial sequence, as follows:

1. Pinkish biotite and (or) hornblende-quartz-feldspar gneiss lithically similar to the Yonkers Gneiss is the most commonly exposed rock type in the first subdivision of the Fordham Gneiss in the Glenville quadrangle. Dark-gray biotite-hornblende-quartz-feldspar gneiss commonly containing magnetite and (or) ilmenite is locally present with the pinkish gneiss in the Glenville quadrangle.
2. Greenish-black, medium- to coarse-grained biotite amphibolite constitutes nearly all of the second unit.
3. The third subdivision is characterized by light-gray calc-silicate rocks that typically contain quartz, plagioclase, microcline, and diopside. The abundance of calcite in these rocks is varied; some contain more than 50 percent calcite. Amphibolite is locally interbedded with the calc-silicate rocks.
4. Predominantly dark-greenish-black amphibolite with some gray biotite-quartz-feldspar gneiss composes the fourth unit.
5. Interbedded gray garnet-biotite-quartz-feldspar gneiss and amphibolite whose outcrops commonly display a prominent layering are the predominant rocks in this unit.

The Fordham and Yonkers Gneisses are interpreted to represent a metamorphosed Precambrian eugeosynclinal sequence that contains some intrusive rocks. The thickness of this sequence is not yet determined.

Lowerre Quartzite

The Fordham and Yonkers Gneisses are unconformably overlain by the Lowerre Quartzite (Hall, 1966), which is a discontinuous rock unit (Fig. 2) that consists of buff or tan-weathering quartzite, feldspathic quartzite, and micaceous quartzite. Micaceous granulite layers and schistose gneiss layers that contain sillimanite nodules are present at the base of the Lowerre in some places. It is typically a well-bedded rock unit; bedding thickness ranges from less than 2.5 cm to as much as 2 m but is commonly from 10 to 60 cm.

The Lowerre Quartzite represents the beginning of deposition in this part of the Appalachian geosyncline and is interpreted to be Cambrian. The maximum thickness of the Lowerre is 210 m, but it is more commonly about 30 m thick; it is absent in many places (Fig. 2).

Inwood Marble

The Inwood Marble overlies the Lowerre Quartzite and is predominantly well-bedded dolomite marble with some calcite marble near the top. It was previously subdivided into five members (Hall, 1968a), but more recent work indicates that it is more consistent to subdivide it into four members. The subdivision of the Inwood is based largely on composition, presence of other rock types interbedded with the marble units, and on bedding characteristics (Hall, 1968a).

Manhattan Schist

Rocks traditionally mapped as Manhattan Schist have been subdivided into three members (Hall, 1968a). In terms of position with reference to the Fordham and Yonkers Gneisses, member A is physically the lowest member of the Manhattan. Member A is interpreted to have been deposited unconformably upon all of the older rocks mentioned above and is believed to be Middle Ordovician (Hall, 1968a). Members B and C are interpreted to be older than, and thrust onto, member A.

Member A. Member A consists predominantly of dark-gray to black, locally rusty-weathering, fissile schist. Tan-weathering, white or blue-gray calcite marble and phlogopitic calcite marble and dark-gray calcite-bearing schist are interbedded with pelitic schist near the base of member A. The calcite marble at the base of member A is locally extensive enough to be mapped separately at a scale of 1:24,000.

The marble and calcite-bearing schist layers at the base of member A have been interpreted by Prucha (1956) and more recently by Freedman and Kauffman (1971) to represent a gradational contact between the Inwood Marble and the Manhattan Schist. The marble layers of member A do not seem to bear such a relationship to the underlying Inwood because they are found in contact with all of the underlying rock units including the gneiss complex. The distribution of the rock units beneath the Manhattan is highly irregular, and the basal Manhattan rocks, including the marble layers, do not bevel across a sequence of rocks that is regularly and progressively younger as would be expected in a transgressive facies relationship as suggested by Freedman and Kauffman (1971). Hence, the interpretation that these marble and calcite-bearing schist layers are part of member A of the Manhattan and unconformably overlie the Inwood and older rocks in the region (Hall, 1968a; Ratcliffe, 1968) is held to be valid. It should be noted that this unconformity is regionally widespread in New England and New York (Cady, 1945; Thompson, 1959; Zen, 1967) and is not simply a local feature in the Manhattan prong. The average thickness of member A of the Manhattan is 30 m.

Members B and C. A discontinuous amphibolite unit that occurs at and above the base of member C of the Manhattan Schist (Fig. 2) has been mapped separately and designated as member B of the Manhattan. Member B is interpreted as a metamorphosed mafic volcanic rock that probably originated as a lava flow. However, this interpretation was questioned by Blank (1972), who concluded that apparently similar amphibolite units in the Bronx, New York, are metamorphosed calcareous-ferruginous beds. Member B is as much as 60 m thick.

Member C of the Manhattan Schist is typified by feldspathic schist and schistose gneiss layers that weather brown, gray, or rusty. Sillimanite is commonly present in nodules in many of these rocks, and it also occurs as disseminated grains. Kyanite is also present and commonly occurs with sillimanite. The rocks are well foliated, and bedding is difficult or impossible to identify except where thin amphibolite, siliceous schist, or quartzite layers are present. Member C is 120 m thick in the northeast part of the map area (Fig. 2) and appears to be as much as 1,600 m thick farther southwest, but the rocks are still being studied here. Member C is a clastic deposit containing some volcanic material represented by the thin amphibolite layers. Thus, members B and C are part of the eugeosynclinal sequence in the White Plains–Glenville area.

Hartland Formation

The term "Hartland Formation" (Rodgers and others, 1959) has been applied to the following four rock units that lie east of member C of the Manhattan Schist and successively farther from the gneiss complex (Fig. 2). Regional relationships farther north in Connecticut indicate that these rocks are separated from member C of the Manhattan Schist by a major fault known as Cameron's line (Rodgers,

1970). The apparent thinning of member C of the Manhattan Schist toward the northeast in the White Plains-Glenville area (Fig. 2) is explained by the presence of this fault.

Amphibolite Member. Interbedded amphibolite and a minor amount of garnet-muscovite-biotite-schist constitute the amphibolite member, which occurs between member C of the Manhattan Schist and the rocks to the east (Fig. 2). The amphibolite member is as much as 30 m thick and is either absent or obscured by cover in some places.

Schist-gneiss-amphibolite Member. The schist-gneiss-amphibolite member (Fig. 2) is named after the prominent rock types that compose it. It is an interlayered assemblage of brown- and rusty-weathering garnet-muscovite-biotite schist units that locally contain sillimanite and (or) kyanite, light-gray or white garnet-muscovite-biotite-quartz-feldspar gneiss, amphibolite, and gray garnet-muscovite-biotite schist and siliceous granulite. Bedding is well displayed, and individual beds are 10 cm to 2 m thick. The light-gray gneiss layers are interpreted as either felsic volcanic beds or granitic sills because of their composition as well as their apparent parallelism with bedding. The light-gray gneiss is relatively more resistant to erosion than the other rocks in this unit; consequently, approximately 80 percent of the natural exposures of this member are gneiss or pegmatite. However, artificial exposures reveal the true relative abundance of the other rock types and show that schist is more abundant. The thickness of this unit ranges from 210 to 1,500 m.

Light-Gray Gneiss Member. The light-gray gneiss member consists of light-gray to white, moderately well foliated garnet-muscovite-biotite gneiss that is similar to the felsic gneiss layers described from the underlying rock unit. It is essentially a homogeneous felsic gneiss at the top of the schist-gneiss-amphibolite member and, for the same reasons stated for felsic gneiss layers in that unit, is interpreted to be a metamorphosed felsic volcanic horizon or a granitic sill. The light-gray gneiss member ranges from 105 to 240 m in thickness.

Schist and Granulite Member. Alternating layers of gray feldspar-muscovite-biotite schist or schistose gneiss and brownish-tan weathering, gray muscovite-biotite-quartz-feldspar granulite are prominent rocks in the schist and granulite member. Schistosity surfaces in these rocks present a characteristic sparkling appearance because of the reflection of light by coarse muscovite plates that are oriented parallel to the surfaces. The rocks are typically rhythmically layered in beds 15 to 20 cm thick and locally appear to be graded. The schist and granulite member ranges from 300 to 880 m in thickness.

Harrison Gneiss

The Harrison Gneiss consists mainly of dark-gray biotite-hornblende-quartz-feldspar gneiss with sphene as a common accessory. It is commonly rich in biotite and hence very well foliated near the base but is richer in hornblende and feldspar above. Compositional layering is locally well developed in the Harrison and consists of feldspathic schist layers or lenses 5 to 15 cm thick and light- and dark-gray gneiss units as much as 60 cm thick that owe their color to the relative abundance of biotite and hornblende. In other places, the Harrison is more massive and locally contains feldspar megacrysts as much as 5 cm long. Clasts of mafic and felsic rocks from 5 to 120 cm across are locally present and are typically elongated and flattened in the foliation. The Harrison is interpreted as a volcanic assemblage; its thickness is unknown because the top is not exposed in the White Plains-Glenville area.

SUMMARY AND CORRELATION OF THE STRATIGRAPHIC COLUMN

If the stratigraphic sequence above the Precambrian Fordham and Yonkers Gneisses is assumed to be continuously younger upward from that gneiss complex, it is approximately 5,800 m thick (Fig. 3) and consists of a clean clastic-carbonate sequence at the base overlain by a volcanic and clastic sequence that contains some granitic intrusive rocks. However, interpreted in the light of the regional geology, the stratigraphic column described above seems to be best interpreted as a Cambrian-Ordovician miogeosynclinal sequence with a eugeosynclinal sequence thrust on top of it (Hall, 1968a).

Miogeosynclinal Sequence

The section from the Lowerre Quartzite through member A of the Manhattan Schist makes up the miogeosynclinal part of the stratigraphic column that correlates with similar rocks to the north in eastern New York and the western parts of Connecticut, Massachusetts, and Vermont (Hall, 1968a). The miogeosynclinal part of the stratigraphic column in the White Plains-Glenville area is also interpreted to be equivalent in age to most of the eugeosynclinal sequence in this area (Fig. 4), which is described below. In the Lowerre Quartzite, micaceous granulite and schistose gneiss that locally contain sillimanite nodules are very similar to, and in places indistinguishable from, the feldspathic schistose gneiss with sillimanite nodules that is common in member C of the Manhattan Schist. The fact that these rock types are common to both rock units leads to the interpretation that the Lowerre Quartzite is a western facies of members B and C of the Manhattan Schist. In turn, these members of the Manhattan Schist were deposited farther east with the eugeosynclinal sequence and subsequently thrust westward onto the miogeosynclinal sequence. Exposures of the Lowerre Quartzite in the eastern part of the autochthonous miogeosynclinal sequence in the White Plains-Glenville area contain a greater abundance of these rock types than those toward the west. This also indicates the likelihood of the proposed west-to-east facies transition between the base of the miogeosynclinal sequence and the Manhattan Schist part of the eugeosynclinal sequence.

Eugeosynclinal Sequence

Consideration of the eugeosynclinal part of the stratigraphic column—members B and C of the Manhattan Schist and the amphibolite member of the Hartland Formation through the Harrison Gneiss (Fig. 3)—suggests correlation with rocks to the north in western Connecticut and Massachusetts (Hall, 1968b) and with rocks to the east in the vicinity of Westport and Bridgeport, Connecticut (Figs. 5, 6). These proposed preliminary correlations are based in part on tracing out units on the map, on lithic similarity, and on relative position with reference to other rock units. In addition to this, some of the stratigraphic sections in Connecticut are reinterpreted to be in accord with ideas presented here (Figs. 5 and 6).

Correlations with rocks in western Connecticut and Massachusetts are presented in Figure 5. These correlations are based mainly on the map continuity and lithic similarity of member C of the Manhattan Schist with the Waramaug and Hoosac Formations. The schist and granulite member of the Hartland Formation is lithically similar to parts of Hartland I, the Taine Mountain Formation, and the Moretown Formation. The positions of Hartland III and IV (Gates and Christensen, 1965; Martin, 1970) are not defined here, and correlation of these rock units with rocks

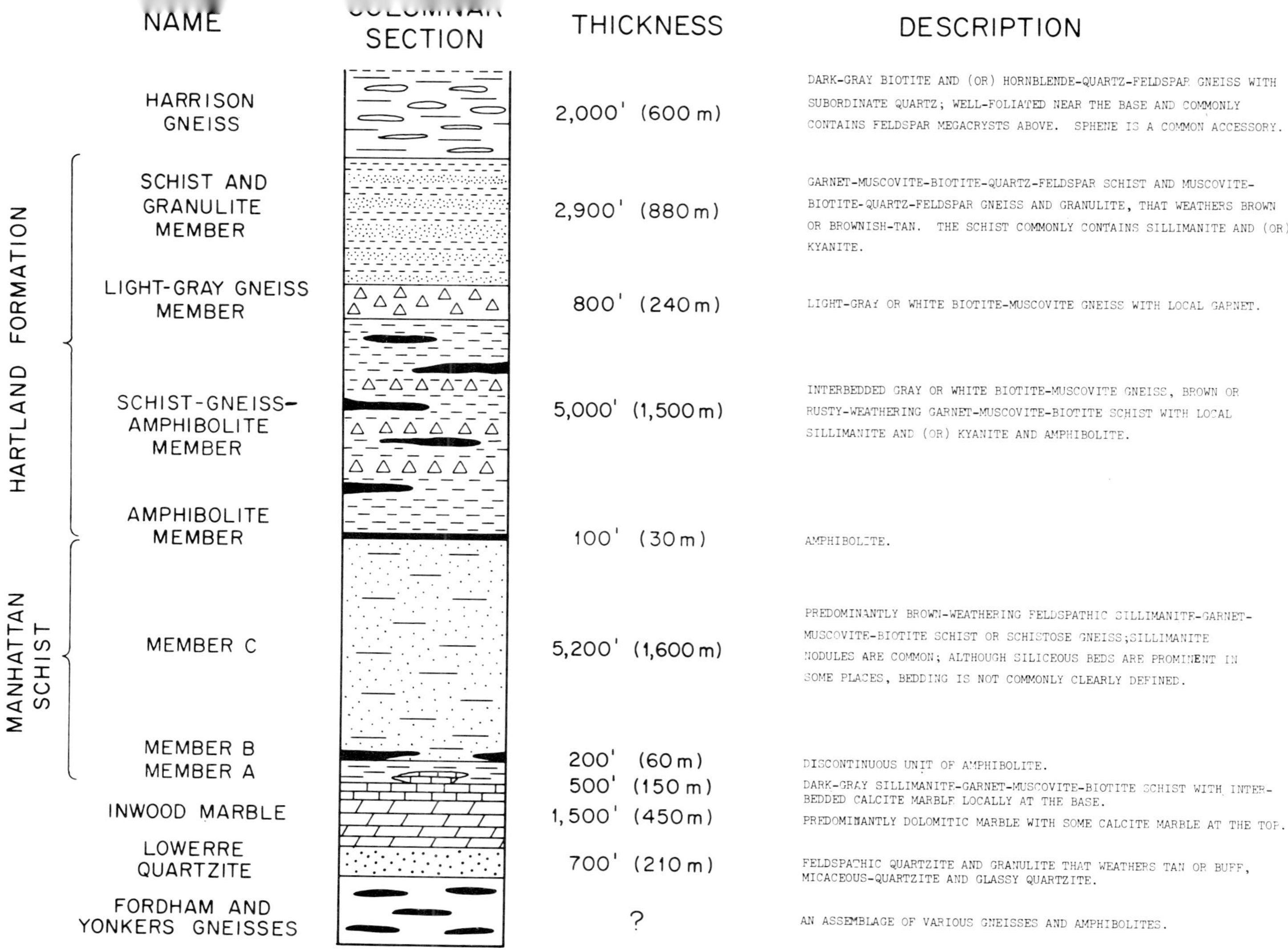

Figure 3. Columnar section for the White Plains–Glenville area.

AGE	MIOGEOSYNCLINAL SEQUENCE		EUGEOSYNCLINAL SEQUENCE	
ORDOVICIAN	MANHATTAN SCHIST MEMBER A		HARRISON GNEISS	
			HARTLAND FORMATION	SCHIST AND GRANULITE MEMBER
		—?—?—?—		LIGHT-GRAY GNEISS MEMBER
	INWOOD MARBLE	— — — —		SCHIST-GNEISS-AMPHIBOLITE MEMBER
CAMBRIAN				AMPHIBOLITE MEMBER
			MANHATTAN SCHIST MEMBERS B AND C	
	LOWERRE QUARTZITE		↓ ↓ ↓	
		— — — —	—?— — —?— — —?— —	
PRECAMBRIAN	FORDHAM GNEISS			

Figure 4. Suggested correlation of the miogeosynclinal and eugeosynclinal sequences in the White Plains-Glenville area.

in the White Plains-Glenville area is not attempted. The remainder of the correlations in Figure 5 are based on position relative to member C of the Manhattan and the schist and granulite member of the Hartland Formation and their correlatives.

Eastward, toward New Haven, the proposed correlation of the White Plains-Glenville column is indicated on Figure 6. This correlation is based on the partial tracing of the Harrison Gneiss into the Prospect Formation and the lithic similarity of the two units. Correlation of the schist and granulite member of the Hartland Formation is based on lithic similarity and position with reference to the Harrison and Prospect. The remainder of the correlations are based on position, although

AGE		LOCATION: WHITE PLAINS - GLENVILLE AREA EUGEOSYNCLINAL SECTION		WESTERN CONNECTICUT GATES AND CHRISTENSEN (1965) AND MARTIN (1970)	COLLINSVILLE QUADRANGLE STANLEY (1964), HATCH AND STANLEY (1970)	WESTERN MASSACHUSETTS HATCH AND OTHERS (1968)
ORDOVICIAN	MIDDLE	HARRISON GNEISS		HITCHCOCK LAKE	COLLINSVILLE FORMATION	HAWLEY FORMATION
		HARTLAND FORMATION	SCHIST AND GRANULITE MEMBER	HARTLAND I	TAINE MOUNTAIN FORMATION	MORETOWN FORMATION
	LOWER		LIGHT GRAY GNEISS MEMBER	HARTLAND II	SATAN'S KINGDOM AND SLASHER'S LEDGES FORMATIONS	ROWE SCHIST
—?—?—?—	?—?—		SCHIST - GNEISS - AMPHIBOLITE MEMBER			
CAMBRIAN			AMPHIBOLITE MEMBER		- - - - - - - -	
		MANHATTAN SCHIST MEMBERS B AND C		WARAMAUG FORMATION		HOOSAC FORMATION
PRECAMBRIAN		- - - - - - - -		BERKSHIRE HIGHLANDS GNEISS COMPLEX		PRECAMBRIAN GNEISS

Figure 5. Suggested preliminary correlation of the eugeosynclinal sequence in the White Plains-Glenville area with rocks in western Connecticut.

AGE \ LOCATION		WHITE PLAINS - GLENVILLE AREA EUGEOSYNCLINAL SECTION	LONG HILL - BRIDGEPORT AREA CROWLEY (1968)	WESTPORT AREA DIETRICH (1968)
SILURIAN-DEVONIAN			THE STRAITS SCHIST	THE STRAITS SCHIST
ORDOVICIAN	MIDDLE	HARRISON GNEISS	PROSPECT FORMATION, ANSONIA GNEISS, AND COLLINSVILLE FORMATION	PROSPECT FORMATION AND MALTBY LAKE VOLCANICS
ORDOVICIAN	MIDDLE	HARTLAND FORMATION: SCHIST AND GRANULITE MEMBER	ORANGE FORMATION, BANDED SCHIST AND BIOTITE GNEISS MEMBERS OF THE SOUTHINGTON MOUNTAIN FORMATIONS	DERBY HILL AND FAIRFIELD FORMATIONS
ORDOVICIAN	LOWER	HARTLAND FORMATION: LIGHT-GRAY GNEISS MEMBER	TRAP FALLS FORMATION	
?		HARTLAND FORMATION: SCHIST-GNEISS-AMPHIBOLITE MEMBER		
CAMBRIAN		HARTLAND FORMATION: AMPHIBOLITE MEMBER		
CAMBRIAN		MANHATTAN SCHIST MEMBERS B AND C		

Figure 6. Suggested preliminary correlation of the eugeosynclinal sequence in the White Plains-Glenville area with rocks in the Long Hill-Bridgeport and Westport areas of southern Connecticut.

the Trap Falls Formation is lithically similar to some of the rocks beneath the schist and granulite member in the White Plains-Glenville area.

Conclusions Based on the Suggested Correlations

If the suggested correlations are true, the following may be concluded:

1. The contact between members B and C of the Manhattan Schist and member A of the Manhattan (Fig. 2) is a thrust fault and is related to Taconic thrusting. The rocks were complexly folded after thrusting, and the folding and metamorphism also seem to have occurred during the Taconic orogeny (Ratcliffe, 1968). However, some deformation and metamorphism almost certainly occurred during the Acadian orogeny (Hall, 1968a), and the Allegheny orogeny may have had a direct effect on the rocks in the area. Another fault, Cameron's line (Rodgers, 1970), disrupts the section along the contact between member C of the Manhattan and the Hartland Formation (Fig. 2). The nature of this fault remains uncertain, but it appears to be of major regional significance.

2. Part of the Straits Schist (Fig. 6) is interpreted to be Silurian-Devonian and to rest unconformably on pre-Silurian rocks in the Bridgeport and Westport areas west of New Haven. The base of the Straits Schist in this region is in contact with the Fairfield Formation and its equivalent, the banded schist member of the Southington Mountain Formation. These rocks, in turn, are apparently in normal

contact with the Prospect Gneiss, which, according to the scheme presented here, is interpreted to be younger than the Fairfield and its equivalents. Thus the Silurian-Devonian rocks appear to rest on an inverted pre-Silurian section west of New Haven. Alternative explanations of this relationship such as a facies change between the Prospect and Fairfield or a fault between them are also possible.

3. The correlation results in a regional map pattern that indicates a major fold system with a trend that is east-northeast between the White Plains-Glenville area and the Westport, Bridgeport, and Long Hill quadrangles in Connecticut. Work in progress indicates that this more east-northeast trend is an early structural system that was deformed by a later northeast-trending system.

ACKNOWLEDGMENTS

I am thankful to James F. Davis of the New York State Museum and Science Service, Geological Survey, and Joe Webb Peoples of the Connecticut Geological and Natural History Survey for continued support of my field work. Peter Robinson read an earlier version of the manuscript, and I am pleased to acknowledge his helpful criticisms. The final manuscript was improved as a result of careful reviews by Avery A. Drake, Robert M. Gates, Charles W. Martin, and Lincoln R. Page, all of whom I gratefully acknowledge.

REFERENCES CITED

Blank, H. R., 1972, Hornblende schists in the Manhattan Formation in the Bronx, New York: Geol. Soc. America Bull., v. 83, p. 1397-1412.

Cady, W. M., 1945, Stratigraphy and structure of west-central Vermont: Geol. Soc. America Bull., v. 56, p. 515-587.

Crowley, W. P., 1968, The bedrock geology of the Long Hill and Bridgeport quadrangles: Connecticut Geol. and Nat. History Survey Quad. Rept., no. 24, p. 1-81.

Dietrich, J. H., 1968, Multiple folding in western Connecticut: A reinterpretation of structure in the New Haven-Naugatuck-Westport area: Connecticut Geol. and Nat. History Survey Guidebook, no. 2, p. D-2, 1-D-2, 11.

Freedman, Jacob, and Kauffman, M. E., 1971, Interfingering and transgressive clastic and carbonate facies along the continental margin of Maryland, Pennsylvania and New York during Cambro-Ordovician time: Geol. Soc. America Abs. with Programs, v. 3, no. 1, p. 31.

Gates, R. M., and Christensen, N. I., 1965, The bedrock geology of the West Torrington quadrangle: Connecticut Geol. and Nat. History Survey Quad. Rept., no. 17, p. 1-38.

Grauert, Borwin, and Hall, L. M., 1973, Age and origin of zircons from metamorphic rocks in the Manhattan Prong, White Plains area, southeastern New York: Ann. Rept. of the Director, Dept. Terrestrial Magnetism 1972-1973, Carnegie Inst. Washington Year Book 72, p. 293-297.

Hall, L. M., 1966, Some stratigraphic relationships within the New York City Group in Westchester County, New York [abs.]: Geol. Soc. America Spec. Paper 87, p. 70.

——1968a, Times of origin and deformation of bedrock in the Manhattan prong, *in* Zen, E-an, White, W. S., Hadley, J. B., and Thompson, J. B., eds., Studies of Appalachian geology: Northern and maritime: New York, Interscience Pubs., Inc., p. 117-127.

——1968b, Geology in the Glenville area, southwesternmost Connecticut and southeastern New York: Connecticut Geol. and Nat. History Survey Guidebook, no. 2, p. D-6, 1-D-6, 12.

Hatch, N. L., Jr., and Stanley, R. S., 1970, Stratigraphic continuity and facies changes in formations of early Paleozoic age in western Massachusetts and tentative correlations

with Connecticut: Geol. Soc. America Abs. with Programs, v. 2, p. 23.
Hatch, N. L., Jr., Schnabel, R. W., and Norton, S. A., 1968, Stratigraphy and correlation of the rocks on the east limb of the Berkshire anticlinorium in western Massachusetts and north-central Connecticut, *in* Zen, E-an, White, W. S., Hadley, J. B., and Thompson, J. B., eds., Studies of Appalachian geology: Northern and maritime: New York, Interscience Pubs., Inc., p. 177–184.
Martin, C. W., 1970, The bedrock geology of the Torrington quadrangle: Connecticut Geol. and Nat. History Survey Quad. Rept., no. 25, p. 1–53.
Prucha, J. J., 1956, Stratigraphic relationships of the metamorphic rocks in southeastern New York: Am. Jour. Sci., v. 254, p. 672–684.
Ratcliffe, N. M., 1968, Stratigraphic and structural relations along the western border of the Cortlandt Intrusives, *in* New York State Geol. Assoc. to Field Excursions, 40th Ann. Mtg., Flushing, New York, 1968: 253 p.
Rodgers, John, 1970, The tectonics of the Appalachians: New York, Interscience Pubs., Inc., 271 p.
Rodgers, John, Gates, R. M., and Rosenfeld, J. L., 1959, Explanatory text for preliminary map of Connecticut, 1956: Connecticut Geol. and Nat. History Survey Bull. 84, 64 p.
Stanley, R. S., 1964, The bedrock geology of the Collinsville quadrangle: Connecticut Geol. and Nat. History Survey Quad. Rept., no. 16, p. 1–99.
Thompson, J. B., 1959, Stratigraphy and structure in the Vermont valley and the eastern Taconics between Clarendon and Dorset, *in* New England Intercollegiate Geol. Conf. Guidebook 51st Ann. Mtg., Rutland, Vermont, 1959: p. 71–85.
Zen, E-an, 1967, Time and space relationships of the Taconic allochthon and autochthon: Geol. Soc. America Spec. Paper 97, 107 p.

Manuscript Received by the Society April 18, 1974
Revised Manuscript Received February 28, 1975
Manuscript Accepted March 14, 1975

Printed in U.S.A.

Geological Society of America
Memoir 148

Discussion: Early Paleozoic Stratigraphy of Western Massachusetts and Connecticut and Southeastern New York

ROLFE S. STANLEY
Department of Geology
University of Vermont
Burlington, Vermont 05401

AND

NORMAN L. HATCH, JR.
U.S. Geological Survey
National Center 926-A
Reston, Virginia 22092

The eugeosynclinal rocks of western Massachusetts and western Connecticut are devoid of fossils and, except for the Goshen Formation and parts of the Straits Schist, contain relatively few reliable primary tops data. Local stratigraphic sequences have been developed, but tracing these sequences over large areas has been complicated by stratigraphic facies changes, complex structures, and conflicting geologic interpretations. Inasmuch as the relative age of any one unit within a sequence rarely can be determined reliably by internal information, we have chosen to rely heavily on lithic correlations of the stratigraphic sequences in western Connecticut and southernmost Massachusetts with the better established sequences of northern Massachusetts and southeastern Vermont.

As discussants of the papers in this volume by Schnabel, Gates and Martin, and L. M. Hall on western Massachusetts and Connecticut, we wish to briefly outline a few points on which we either strongly disagree or believe that the evidence is such that other interpretations should be considered. For purposes of this discussion, we have outlined on Figure 1 the study area in Massachusetts, as well as Schnabel's, Gates and Martin's, and L. M. Hall's study areas, and the Collinsville area.

It should also be noted that the interpretations presented here, which appear in considerably greater detail elsewhere (Hatch and Stanley, 1974), differ markedly from those given earlier by Hatch and others (1968). This change in interpretation results from completion of detailed mapping in all of the pre-Silurian rocks and most of the Silurian and Devonian rocks in the western Massachusetts area (outlined on Fig. 1) by us and by our colleagues, and from extensive reconnaissance studies that we made between the south edge of that area and Long Island Sound.

Schnabel (this volume) has described rocks from northernmost Connecticut and southernmost Massachusetts, the stratigraphic sequence of which he says is supported by sedimentary structures. We have looked, with Schnabel, at many

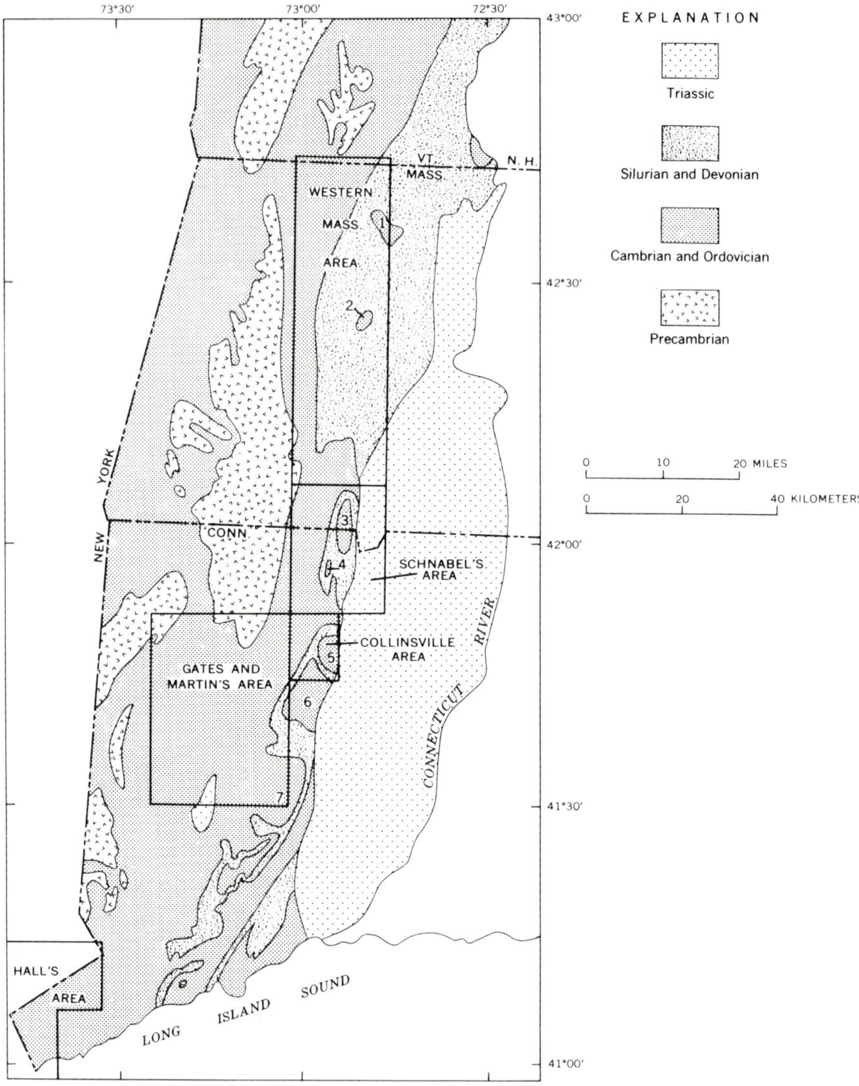

Figure 1. Geologic sketch map of southwestern New England showing our study area in Massachusetts, the Collinsville area, and the study areas of Schnabel, Gates and Martin, and L. M. Hall: (1) Shelburne Falls dome, (2) Goshen dome, (3) North Granby dome, (4) South Granby dome, (5) Collinsville dome, (6) Bristol dome, and (7) Waterbury dome.

of his exposures of these sedimentary structures, and to a considerable extent we disagree with his interpretation. The one group of rocks in which we saw many exposures with graded bedding includes units Os_4 and Os_5 (see Schnabel, this volume, Fig. 1). We have carefully studied these rocks for what we consider reliable data on which tops determinations can be made at their contacts with adjacent units. Although we found many examples of moderately well preserved graded beds across the outcrop belt of these units, which suggests tops in both directions, we did not find any along the contacts with the bounding units that we considered sufficiently reliable to determine the relative age. In addition, we saw no internal structures at any other contacts of Schnabel's units that would refute the alternate stratigraphic interpretation that we suggested here in Figure 2.

Therefore, having no fossils and no definitive primary tops criteria, we have relied heavily on lithic comparisons with rocks of the established sequence in Massachusetts in formulating the correlations in Figure 2.

In brief, Schnabel's units Os_4 and Os_5 (which, along with his unit Os_6, he correlates with the Straits Schist of Connecticut) appear to us to be lithologically identical to the Goshen Formation of Silurian and Devonian age in Massachusetts at the point where it passes under the Triassic cover only a few miles north of Schnabel's area (Fig. 1).The coticule-bearing, generally nongraphitic rocks of unit Os_6 are characteristic of the Hawley Formation and correlative rocks in Massachusetts and are separated from units Os_4 and Os_5 by a sharp contact that we consider to be the Taconic unconformity. Furthermore, as pointed out by Schnabel (this volume), much of his unit Oh_3 (except that south of the South Granby dome) is strikingly similar to our Russell Mountain Formation (Hatch and others, 1970) of western Massachusetts. We propose these correlations (Fig. 2).

The consequences and corollaries of these correlations are as follows. If Schnabel's units Os_4 and Os_5 are Goshen, they are also the youngest rocks in his sequence.

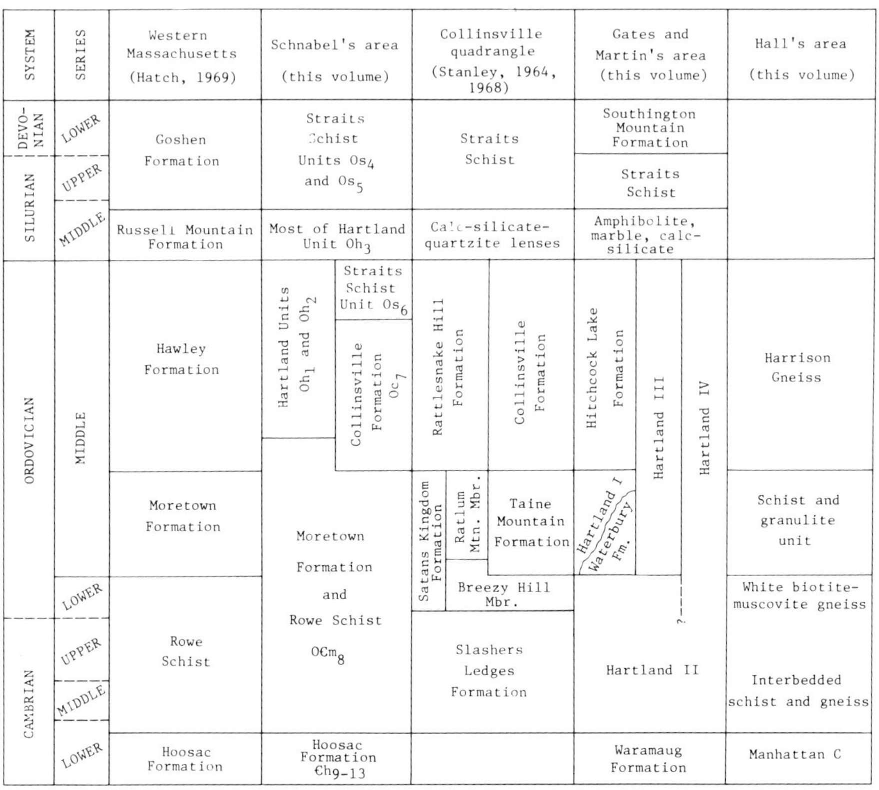

Figure 2. Proposed correlation of stratified units in western Massachusetts with units in the Collinsville quadrangle and units in the stratigraphic sections proposed in this volume by Schnabel, Gates and Martin, and L. M. Hall. The columns of Schnabel and of Gates and Martin have been rearranged according to our interpreted correlation of their units with the western Massachusetts and Collinsville units.

Furthermore, the rocks below the Goshen inside the Granby domes must be, in part, more volcanic facies equivalents of the rocks on which the Goshen rests outside the domes. Specifically, Schnabel's Collinsville Formation and unit Os_6 inside the domes are, we believe, more volcanic facies correlatives of most of his units Oh_1 and Oh_2 of the Hartland outside the domes (Fig. 2). Oh_3 is thus missing along the contact between Os_5 and Os_6, but this is reasonable in light of the highly discontinuous nature of the correlative Russell Mountain Formation to the north.

Finally, rather than viewing them as two stratigraphic sequences both becoming younger toward the Barkhamsted fault eastward from the Precambrian rocks of the Berkshire massif and westward from the Granby domes, we propose instead that the rocks generally become progressively younger eastward from the Berkshires all the way to Os_4 and Os_5 (the Goshen Formation) and thence become older eastward into the cores of the Granby domes. Stratigraphic reversals and repetitions have been mapped in the Collinsville area (Fig. 1; Stanley, 1964, Pl. 1; 1968, p. D4, D8) and doubtless are also present in Schnabel's area, but in no way do these refute either the overall eastward younging of the sequence or the reasonableness of Os_4 and Os_5 being at the top rather than in the middle of Schnabel's sequence.

Extrapolating this interpretation southward, we further propose that most of the rest of the rocks mapped as Straits Schist in western Connecticut are correlative with the Goshen Formation of Massachusetts and are thus Silurian and Devonian in age. These include not only most of the Straits described by Gates and Martin (this volume; see also Gates and Martin, 1967; Martin, 1970), but also most of the Straits mapped by Stanley (1964) and many other workers throughout the rest of the eastern part of western Connecticut. Figure 1 shows, as undifferentiated Silurian and Devonian rocks, the approximate distribution of these Straits and Straits-like rocks as well as the Wepawaug and similar rocks which, although lithically distinct from the Straits, are probably approximately correlative with the Waits River Formation to the north and are, therefore, also interpreted as Silurian and Devonian in age. From a structural point of view, the Straits in western Connecticut thus forms a series of tight synclines that were redeformed at the time of emplacement of the domes.

We have previously proposed (Hatch and Stanley, 1970) a correlation of the stratified rocks of western Massachusetts with those of the Collinsville area. We still hold to this correlation, which is included here in Figure 2. It is based on a combination of lithic similarity of correlated units and their stratigraphic sequence. One of the consequences of this correlation scheme is that, as was the case in Schnabel's area, the generally more feldspathic and gneissic (more volcanic) rocks within the domes of western Connecticut are assumed to be the facies equivalents of some of the more micaceous, quartzose, and schistose (less volcanic) rocks outside the domes. Thus, the metavolcanic rocks of the Bristol Member of the Collinsville Formation (of Stanley, 1964) are inferred to pass westward under the syncline of Straits (Goshen) into the "upper" member of the Rattlesnake Hill Formation (of Stanley, 1964), whereas the Taine Mountain Formation passes westward into part of the Satan's Kingdom Formation.

The stratigraphic sequence that Gates and Martin (this volume) have proposed for the eugeosynclinal rocks east of Cameron's line is based on the assumption, with which we disagree, that the Waterbury Formation in the core of the Waterbury dome is the oldest formation in the sequence, with Hartland Unit I, Hitchcock Lake, Straits Schist, and Southington Mountain Formation forming progressively younger units outward from the core of the dome. Units II and III of the Hartland

Formation are then correlated with the Straits Schist and the Southington Mountain Formation. Hartland Unit IV is considered to be the youngest unit. Although we agree with the sequence of Hartland Unit I, Hitchcock Lake, Straits, Southington Mountain for the northern part of the Waterbury dome, we do not agree that the Waterbury is older than all the rest of the rocks in the central part of western Connecticut. Rather, on the basis of similarities to rocks near Cobble Mountain Reservoir in the southernmost part of our study area in Massachusetts (Fig. 1), we believe that the Waterbury is more probably approximately facies equivalent to part or all of Hartland Unit I, as indicated in Figure 2.

We agree with Gates and Martin in their correlation of Hartland Unit I with the Taine Mountain Formation in the Collinsville area (Stanley, 1964). Equating the Ratlum Mountain Member with most of the Taine Mountain Formation has a profound effect on the relationship among Units I, II, III, and IV of the Hartland Formation. Gates and Martin base the stratigraphic order of the units within this sequence on the position of each unit within the major structures in the central part of western Connecticut (see also Gates and Martin, 1967, p. 24-25; Martin, 1970, p. 26-27). The relationship of Units I, II, and III to the rocks in the Collinsville area can best be seen by comparing the bedrock maps of the Torrington (Martin, 1970) and Collinsville (Stanley, 1964) quadrangles. It can be seen that (1) Hartland Unit II has been mapped into the Slasher's Ledges Formation and the Breezy Hill Member of the Satan's Kingdom Formation, and (2) Hartland Unit III has been mapped into the Ratlum Mountain Member of the Satan's Kingdom Formation and the upper member (carbonaceous schist and quartzite) of the Rattlesnake Hill Formation. These relations are summarized in columns 3 and 4 of Figure 2. As a result of the correlations we have proposed between the Collinsville quadrangle and western Massachusetts, Hartland Unit II as mapped in the Torrington quadrangle is older than Hartland Unit I, and Hartland Unit III is in part equivalent to Hartland Unit I. Unit IV of the Hartland is not present in the Torrington quadrangle, and its relationship to the units in Figure 2 cannot be determined confidently, although we suspect, on the basis of rocks that we have seen in the Roxbury and New Preston quadrangles just east of Cameron's line, that Unit IV may be roughly equivalent to all of the Ordovician section of western Massachusetts.

It is obvious from Figure 2 that we strongly disagree with the correlations of the Straits Schist with Hartland Unit II and of the Southington Mountain with Hartland Unit III proposed by Gates and Martin (this volume). Our disagreement is based on the marked lithologic dissimilarity between these correlated units and on their apparently markedly different stratigraphic positions. Although, as Gates and Martin point out, both the Straits and Hartland Unit II are characterized by aluminous schists, the Straits is characterized by an abundance of graphite whereas Hartland Unit II is characteristically nongraphitic. We have studied the southeast-trending belt, mapped by Cassie (1965) in the Thomaston quadrangle, that supposedly connects Hartland Unit II with the Straits Schist. Graphitic rocks are present in this belt, but they are highly rusty, splintery, sulfate-coated, slabby schist associated with hard, fine-grained quartzite, coticule rocks, and nonrusty gneiss and schist all typical of the Hawley and Moretown Formations (Middle Ordovician) of Massachusetts but unknown in the Goshen Formation. These rocks are the same as the rocks mapped as Hartland Unit III by Martin (1970) a few miles north along strike in the Torrington quadrangle. We believe that the rocks in this critical southeast-trending belt are not the same as the rocks in the Straits Schist in the Collinsville quadrangle or even the Waterbury quadrangle, but rather are stratigraphically lower, Middle Ordovician rocks distinctly different from the Straits Schist.

We have shown Hall's (this volume) columnar section for southwesternmost Connecticut in column 5 of Fig. 2. These units are arranged in the same order as proposed by Hall. We essentially agree with his suggested correlations but plead for some caution as to its general acceptance inasmuch as approximately 40 miles (65 km) separate Hall's area from the Collinsville area, and much more needs to be known about the geology of the intervening area before these correlations can be verified. At present, these correlations are based only on fairly close lithic similarities and approximately comparable stratigraphic sequences; mappable continuity has yet to be proven.

REFERENCES CITED

Cassie, R. M., 1965, The evolution of a domal granitic gneiss and its relation to the geology of the Thomaston quadrangle, Connecticut [Ph.D. thesis]: Madison: Univ. Wisconsin, 109 p.

Gates, R. M., and Martin, C. W., 1967, The bedrock geology of the Waterbury quadrangle: Connecticut Geol. and Nat. History Survey Quad. Rept. 22, 36 p.

——1976, Pre-Devonian stratigraphy of the central section of the western Connecticut highlands, *in* Page, L. R., ed., Contributions to the stratigraphy of New England: Geol. Soc. America Mem. 148 (this volume).

Hall, L. M., 1976, Preliminary correlation of rocks in southwestern Connecticut, *in* Page, L. R., ed., Contributions to the stratigraphy of New England: Geol. Soc. America Mem. 148 (this volume).

Hatch, N. L., Jr., 1969, Geologic map of the Worthington quadrangle, Hampshire and Berkshire Counties, Massachusetts: U.S. Geol. Survey Geol. Quad. Map GQ-857.

Hatch, N. L., Jr., and Stanley, R. S., 1970, Stratigraphic continuity and facies changes in formations of Early Paleozoic age in western Massachusetts and tentative correlations with Connecticut: Geol. Soc. America Abs. with Programs, 1970, v. 2, no. 1, p. 23-24.

——1974, Some suggested stratigraphic relations in part of southwestern New England: U.S. Geol. Survey Bull. 1380, 83 p.

Hatch, N. L., Jr., Schnabel, R. W., and Norton, S. A., 1968, Stratigraphy and correlation of the rocks on the east limb of the Berkshire anticlinorium in western Massachusetts and north-central Connecticut, *in* Zen, E-an, White, W. S., Hadley, J. B., and Thompson, J. B., Jr., eds., Studies of Appalachian geology—Northern and maritime: New York, Interscience Pubs., p. 177-184.

Hatch, N. L., Jr., Stanley, R. S., and Clark, S. F., Jr., 1970, The Russell Mountain Formation, a new stratigraphic unit in western Massachusetts: U.S. Geol. Survey Bull. 1324-B, p. B1-B10.

Martin, C. W., 1970, The bedrock geology of the Torrington quadrangle: Connecticut Geol. and Nat. History Survey Quad. Rept. 25, 53 p.

Schnabel, R. W., 1976, Pre-Silurian stratigraphy in south-central Massachusetts and north-central Connecticut, *in* Page, L. R. ed., Contributions to the stratigraphy of New England: Geol. Soc. America Mem. 148 (this volume).

Stanley, R. S., 1964, The bedrock geology of the Collinsville quadrangle: Connecticut Geol. and Nat. History Survey Quad. Rept. 16, 99 p.

——1968, Metamorphic geology of the Collinsville area, *in* Guidebook for field trips in Connecticut—New England Intercoll. Geol. Conf. 60th Ann. Mtg., New Haven, 1968: Connecticut Geol. and Nat. History Survey Guidebook 2, 17 p.

MANUSCRIPT RECEIVED BY THE SOCIETY APRIL 18, 1974
REVISED MANUSCRIPT RECEIVED NOVEMBER 12, 1974
MANUSCRIPT ACCEPTED DECEMBER 4, 1974

Printed in U.S.A.

Geological Society of America
Memoir 148

Hoosac Formation (Early Cambrian or Older) on the East Limb of the Berkshire Massif, Western Massachusetts

STEPHEN A. NORTON
U.S. Geological Survey
Orono, Maine 04473
and
Department of Geological Sciences
University of Maine
Orono, Maine 04473

ABSTRACT

The Hoosac Formation has been divided into 10 units. These are, from oldest to youngest, (1) quartz-pebble and polymictic-pebble meta conglomerate with a schist matrix, 0 to 160 m thick; (2) quartz-plagioclase-muscovite-biotite-(garnet) schist, 0 to 35 m thick; (3) quartz-muscovite-garnet schist with abundant quartz lenses, 0 to 65 m thick; (4) quartz-plagioclase-muscovite-biotite-(garnet) schist, 0 to 35 m thick; (5) quartz-muscovite-garnet schist, commonly with discontinuous layers composed nearly entirely of Fe-Mg-A1 silicate minerals, 0 to 65 m thick; (6) quartz-plagioclase-muscovite-biotite schist, 350 to 3,500 m thick; (7) slightly calcareous quartz-muscovite schist, interbedded with schist of unit 6 in a zone ranging from 0 to 65 m thick; (8) schist of unit 6, 350 to 3,500 m thick; (9) quartz-plagioclase-garnet-muscovite schist, 0 to 65 m thick; and (10) schist of unit 6 with minor impure quartzite, locally graphitic. The Hoosac Formation is about 6 km thick at the Vermont border, thins to less than 1 km near Becket, Massachusetts, and thickens to about 3 km at the Connecticut border. West of the Precambrian terrane, units 1, 3, and 4 locally underlie, grade into, and are interbedded with the Dalton Formation of Early Cambrian age.

INTRODUCTION

In 1961, the United States Geological Survey, in cooperation with the Massachusetts Department of Public Works, began a mapping program in western Massachusetts to trace the well-defined stratigraphy of southeastern Vermont (Doll and others, 1961) as far as Connecticut along the east side of the Berkshire massif. Chidester and others (1967) initiated this study, and it was continued by Osberg and others (1971), Hatch (1969a), Hatch and Hartshorn (1968), Hatch and others (1971), Norton (1967), and S. F. Clark, Jr. (unpub. data). I have been primarily concerned with the areas where Precambrian and Cambrian rocks were exposed on the east flank of the Berkshire massif. This report deals with (1) detailed stratigraphy of the Hoosac Formation (Early Cambrian or older), the lowest stratigraphic unit in the Paleozoic eugeosynclinal sequence of western Massachusetts (Table 1); (2) the nature of the bounding rocks above and below the Hoosac and the types of contacts between these rocks; (3) correlation of the Hoosac with units to the north, west (across the Precambrian core of the Berkshire massif), and south; and (4) facies relationships within the Hoosac.

GEOLOGIC SETTING

The Hoosac Formation, most of which lies on the east side of the Berkshire massif (Fig. 1), is part of a eugeosynclinal sequence of rocks that ranges in age from Late Precambrian(?) to Early or Middle Devonian (Hatch, 1967). This rock sequence consists predominantly of schistose units with lesser amounts of feldspathic metasandstone, quartzite, metavolcanic rocks, and very subordinate calcareous schist and quartzite. These rocks are intruded by (1) ultramafic intrusive(?) rocks in various states of hydration (Chidester and others, 1967; Norton, 1967; Osberg and others, 1971; Hatch and others, 1971; Chute, 1945; Hatch and Norton, 1969; Hatch and Stanley, unpub. data) that are restricted in distribution to areas of Middle Ordovician age or older; (2) premetamorphic mafic dikes and sills, now amphibolites, sparsely distributed throughout areas of Cambrian and Ordovician age; and (3) premetamorphic felsic rocks, including the Hallockville Gneiss of Devonian or Ordovician age, located in the Plainfield quadrangle (Fig. 1; Osberg and others, 1971); the Middlefield Granite of Middle Devonian age, located in

TABLE 1. STRATIGRAPHIC COLUMN ON THE EAST SIDE OF THE BERKSHIRE MASSIF, WESTERN MASSACHUSETTS

Unit	Age	Reference
Waits River Formation	Late Silurian to Early Devonian	Hatch and Hartshorn (1968)
Goshen Formation	Middle Silurian to Early Devonian	Hatch (1967), Hatch and Hartshorn (1968)
Erosional unconformity(?)		
Russell Mountain Formation	Middle Silurian Devonian	Hatch and others (1970)
Erosional unconformity		
Hawley Formation	Middle Ordovician	Hatch (1967)
Moretown Formation	Middle Ordovician	Chidester and others (1967)
Rowe Schist	Early Cambrian to Early Ordovician	Hatch and others (1966)
Hoosac Formation	Early Cambrian or older	This paper
Granitic gneiss	Late Precambrian(?)	Norton (1967)
Erosional unconformity(?)		
Precambrian massif	Grenville or older	Norton (1967), Ratcliffe (this volume)

the Chester quadrangle (Fig. 1; Hatch and others, 1971); and numerous synmetamorphic to postmetamorphic pegmatite and granite dikes and sills of probable Middle Devonian age, which are especially abundant in the Silurian and Devonian rocks south of the kyanite isograd (Fig. 1).

The Cambrian and Ordovician rocks east of the Berkshire massif are essentially homoclinal from the Massachusetts-Vermont border to the Massachusetts-Connecti-

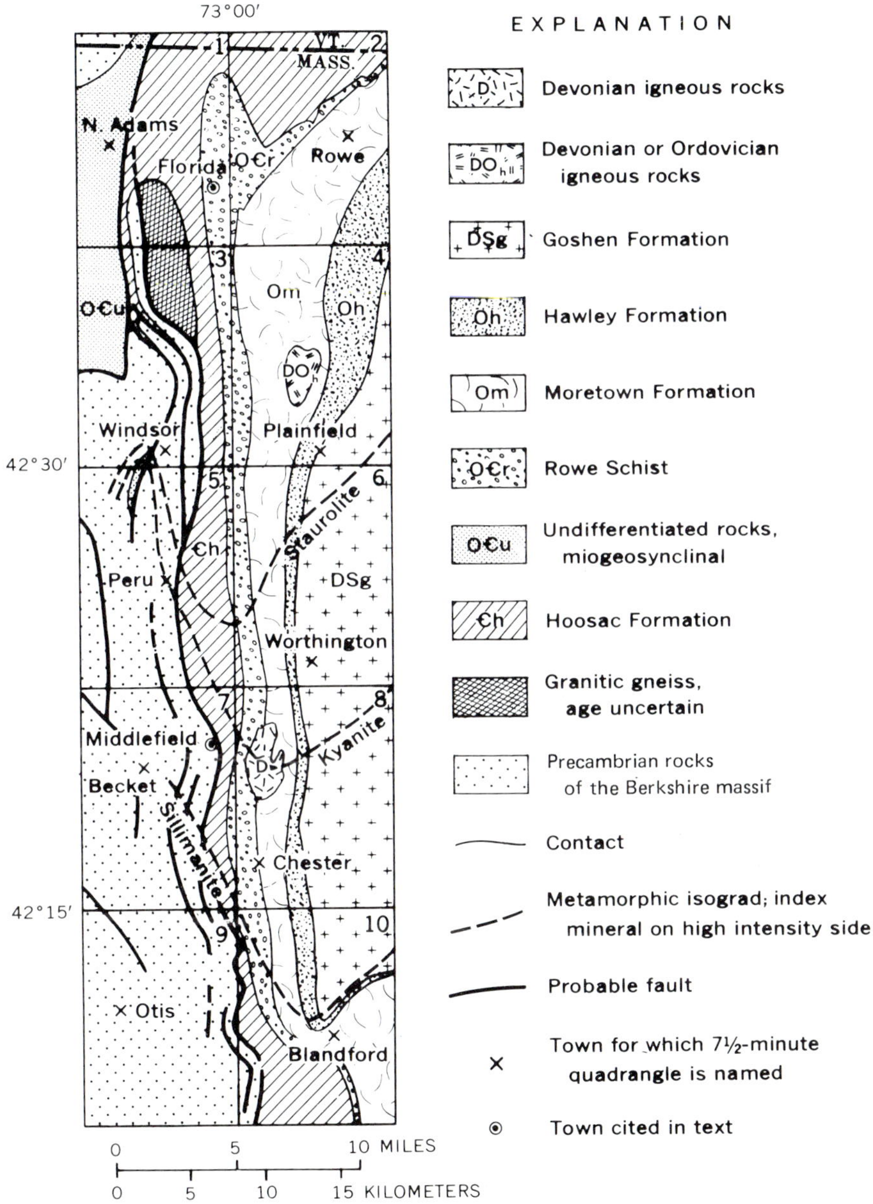

Figure 1. Generalized geologic map of the east side of the Berkshire massif, western Massachusetts. From the work of A. H. Chidester, R. G. Clark, Jr., S. F. Clark, Jr., N. L. Hatch, Jr., Norman Herz, S. Norton, P. Osberg and R. Stanley.

cut border, but in places these rocks are recumbently and isoclinally folded (Stanley, 1967; Norton, 1967, 1975). The Paleozoic-Precambrian contact south of Windsor (Fig. 1) on the east side of the Berkshire massif is not repeated to the east by folding, however. The Silurian and Devonian rocks, although not isoclinally infolded with the Cambrian and Ordovician rocks (Osberg and others, 1971; Hatch, 1969a, 1969b; Hatch and others, 1971; Chidester and others, 1967), are themselves folded into recumbent isoclinal folds (Osberg, 1975) and then refolded into upright, horizontal, isoclinal folds (Hatch, 1969b). These relationships suggest a décollement at the Ordovician-Silurian boundary, developed during Acadian deformation.

At least three and probably four phases of regional deformation can be demonstrated in the Cambrian and Ordovician rocks shown in Figure 1 (see Norton, 1975). The earliest deformation (phase 1) is displayed as folds on a mappable scale in the Blandford-Woronoco area (Stanley, 1967). Possible premetamorphic or synmetamorphic faulting, equivalent in age to phase 1 folding, is present on a regional scale in the inverted limbs of recumbent folds north and west of the Precambrian terrane in the Windsor quadrangle (Norton, 1967, 1969), and thrust faults are present in a zone at the base of the Hoosac Formation (Norton, 1971, 1975) that extends from Windsor, Massachusetts, to at least the Connecticut state line. In this region, schistosity and small-scale isoclinal folds of phase 1 age are common in the Cambrian and Ordovician metasedimentary rocks but are obliterated by later synmetamorphic folding (phase 2) that caused pervasive regional schistosity. The folds are tight to isoclinal with steeply plunging axes. More open, later folds (phase 3) associated with axial-surface cleavage or schistosity deform the earlier schistosity. Lastly, north-trending, very open warps and crinkles (phase 4), generally with no associated foliation, fold all earlier structures. Concentrically arranged folds related to diapiric emplacement (age unknown) of the Hallockville Gneiss (DO_h of Fig. 1, Osberg and others, 1971) fold all structures older than the phase 4 open warps but are of only local importance.

Silurian and Devonian rocks display three phases of deformation, the earliest of which (Hatch, 1969b) produced the pervasive regional schistosity (phase 2) in both these and the Cambrian and Ordovician rocks. The two later fold systems

TABLE 2. STRATIGRAPHIC UNITS IN THE HOOSAC FORMATION, EAST SIDE OF THE BERKSHIRE MASSIF, WESTERN MASSACHUSETTS

Unit*	Mineralogy	Thickness (m)
10	Quartz-plagioclase-muscovite-(chlorite-biotite-garnet-kyanite-sillimanite) schist	65 to 100
9	Quartz-plagioclase-garnet-muscovite-(biotite-staurolite-kyanite-sillimanite) schist	0 to 65
8	Quartz-plagioclase-muscovite-(chlorite-biotite garnet-kyanite-sillimanite) schist	350 to 3,500
7	Quartz-muscovite-calcite schist interbedded with unit 6	0 to 65
6	Quartz-plagioclase-muscovite-biotite-(chlorite-garnet-kyanite-sillimanite) schist	350 to 3,500
5	Quartz-muscovite-garnet-staurolite-(kyanite) schist	0 to 65
4	Quartz-plagioclase-muscovite-biotite-(garnet) schist	0 to 35
3	Quartz-muscovite-garnet-(chlorite-chloritoid-biotite-staurolite-kyanite-sillimanite) schist	0 to 65
2	Quartz-plagioclase-muscovite-biotite-(garnet) schist	0 to 35
1	Quartz-pebble and polymictic-pebble metaconglomerate with schist of type 4	0 to 200

*All units are not present everywhere (see Fig. 2).

include folds with amplitudes generally less than 1 m.

The regional Barrovian-type Paleozoic metamorphism (Fig. 1; Thompson and Norton, 1968) ranges from garnet zone in the north to sillimanite zone in the south. This regional metamorphism reflects at least two events, one of Taconic age (Harwood, 1972; Ratcliffe, 1972) and one (or two) of Acadian age (Zartman and others, 1970). The first metamorphism was nearly synchronous with the most intense deformation (phase 1) of the region on the west side of the Berkshire massif. The second metamorphism was essentially synchronous with phase 2 deformation. Additional higher grade metamorphism locally preceded, was synchronous with, and followed phase 3 deformation (see Norton, 1975, for details).

STRATIGRAPHY

Introduction

The Hoosac Formation in Massachusetts crops out in a continuous band from the Vermont to the Connecticut state line. The breadth of the outcrop ranges from about 6,500 m to less than 800 m; the minimum breadth is near the town of Becket (Fig. 1). Northwest of Florida, Massachusetts, the thickness of the Hoosac is probably less than one-half of the outcrop width because of phase 1 recumbent folding (Herz, 1961).

The Hoosac Formation has been traced around the north end of the Stamford Granite Gneiss body (hereafter called granitic gneiss) in the south end of the North Adams quadrangle (Herz, 1961) and then southerly along the west side of the thin septum of Precambrian gneiss mapped by Herz. The Hoosac rocks that lie to the west of the Precambrian rocks and have been mapped continuously around to the east may in fact be separated from the eastern Hoosac rocks by thrust faults (Fig. 1). West of these possible faults, the Hoosac is probably overturned in most areas, overlain structurally by Precambrian rocks, and underlain (but stratigraphically overlain) by the miogeosynclinal rocks, the oldest of which is the Dalton Formation of Early Cambrian age (Norton, 1969). Southward, west of the Precambrian rocks, the Hoosac grades into and becomes interbedded with the Dalton Formation, with which it is correlated.

The Hoosac Formation consists predominantly of a monotonous quartz-plagioclase-muscovite-biotite-(chlorite) schist[1] and minor, more feldspathic gneiss with lenses and members of other rock types. Ten different map units have been delineated (Table 2) and are here described in detail from oldest to youngest.

Unit 1

In the Windsor quadrangle, the base of the Hoosac Formation both east and west of the Precambrian terrane is marked in a few places by quartz- and gneiss-pebble metaconglomerate with a matrix of quartz-plagioclase-muscovite-chlorite-(microcline) schist. It is thickest at the northeast side of the granitic gneiss body in the southern part of the North Adams quadrangle (Herz, 1961; Pumpelly and others, 1894). There Herz shows a section about 100 m thick; however, the maximum thickness of 200 m is slightly to the north in the Hoosac Tunnel (Pumpelly and

[1]Minerals are listed in order of decreasing abundance. Minerals in parentheses are locally variable depending on bulk rock composition and the metamorphic grade but normally make up less than 10 percent of the total composition.

others, 1894). South of the Windsor quadrangle, conglomerate is absent along the east side of the Berkshire massif both in Connecticut and Massachusetts. In the Windsor quadrangle, the stratigraphic position of the conglomerate is occupied by quartzite, quartz-plagioclase-microcline-biotite gneiss, and microcline-plagioclase-quartz-biotite gneiss. The outcrop width of these rocks ranges from 150 m to 550 m. Muscovite is rare to absent in most of these rocks. Common accessory minerals include magnetite, epidote, and garnet. The quartzite and gneisses are interpreted to be of Precambrian age and to have been faulted into their present stratigraphic position. An alternative interpretation is that they are stratigraphically equivalent to the lower Hoosac Formation.

The typical conglomerate is poorly bedded to non-bedded and non-rusty-weathering. Where abundant, it is interbedded with and grades laterally and vertically into quartz-plagioclase-muscovite-biotite schist similar to most of the Hoosac Formation. Where the conglomerate rests on Precambrian rocks, the pebbles are commonly composed of quartz; but where the conglomerate rests on the granitic gneiss, the clasts are commonly polymictic and as large as 0.3 m in diameter. Most are flattened in the plane of regional foliation. The matrix of the conglomerate is typically gray to green quartz-albite-muscovite-biotite (or chlorite) schist. Microcline is locally abundant in the matrix and as a constituent of some of the polymictic pebbles. Sparse magnetite and garnet are locally present.

Alternatively, the conglomerate on top of the granitic gneiss may be a tectonic synmetamorphic breccia.

Unit 2

From the southern edge of the Windsor quadrangle to the Connecticut state line, the lowermost unit exposed to the east of the early (phase 1 equivalent) faults in the Hoosac Formation consists of gray, non-rusty-weathering to slightly rusty-weathering, quartz-plagioclase-muscovite-biotite-(garnet) schist and gneiss. The unit ranges in thickness from zero east of Windsor to about 30 m east of Peru and southward. Conspicuous porphyroblasts of plagioclase as large as 1 cm across are set in a matrix of 1-mm or smaller grains of quartz, plagioclase, muscovite, and biotite. Bedding is indistinct from Windsor to Becket, but southward, bedding becomes progressively more distinct, with 1- to 5-cm beds, locally flaggy and gneissic, similar to the Dalton Formation. Quartz-rich lenses are rare. A few of the more distinctly bedded sections show graded bedding. North of Becket, bedding may be indicated by quartz-rich lenses as much as 1 m long and 2 cm wide, which commonly weather to positive relief, giving the rock a nubbly to slightly ribbed appearance. An alternative hypothesis is that the quartz-rich lenses were formed during an earlier period of strain. The development of axial-plane cleavage was accompanied by migration of silica into the slip planes; this was followed by refolding and the development of axial-plane schistosity. These features are well shown at the north end of the Berkshire massif.

Unit 3

Coarse-grained, rusty-weathering, quartz-muscovite-garnet-(biotite-chlorite-staurolite-kyanite-sillimanite) schist has been traced nearly continuously along the entire eastern side of the Berkshire massif. Pumpelly and others (1894) traced this unit around the north end of the granitic gneiss in the North Adams quadrangle, where a similar unit was mapped southward (on the west side of the Precambrian terrane) into the Windsor quadrangle by Norton (1969). There it forms the base of the Hoosac, where it contains only minor conglomerate lenses. Furthermore, although its alumina content varies from place to place, the unit is more aluminous than those above and below it. The thickness ranges from 50 m in the Windsor quadrangle west of the Precambrian rocks to nearly zero at the northeast corner of the Windsor quadrangle. South of this point it thickens to 65 m.

Outcrops are extensive and typically have a highly corrugated surface caused by the weathering in relief of numerous granular quartz lenses as much as 3 cm thick and 25 cm long that outline the contorted schistosity. Garnet porphyroblasts as much as 2.5 cm in diameter locally weather out preferentially where they have been crushed by post-metamorphic strain (phase 3); rarely they weather in relief to yield a "hobnail" schist.

Bedding is obscure, indicated(?) only locally by extensive quartz lenses and alternations of more aluminous and less aluminous layers. Bedding is more prominent near contacts with flanking units, where both rocks are apt to be interbedded.

Chloritoid and paragonite are locally abundant, the former only north of the staurolite isograd and the latter only north of the kyanite isograd; paragonite in individual hand specimens may be as abundant as muscovite. Nonporphyroblastic plagioclase is a common accessory mineral, especially at higher metamorphic grades. Graphite inclusions in garnet are common.

This unit is less aluminous to the north and west in Massachusetts but is very aluminous where it reappears in the Sherman dome (Chidester and others, 1967) at the Vermont border. West of the Precambrian terrane in the Windsor quadrangle it typically is a dark brown- to black-weathering, graphitic, quartz-albite-muscovite-garnet-biotite schist, also with numerous quartz lenses. The transition from graphitic to nongraphitic schist within this unit occurs at the north end of the granitic gneiss (Herz, 1961). At the latitude of Becket and at the Blandford-West Granville quadrangle boundary (Table 3), rare, thinly bedded, fine-grained plagioclase-quartz-diopside-(garnet) granulites are found in this unit.

Unit 4

From the latitude of Peru southward to at least Chester, between units 3 and 5, is a non-rusty- to slightly rusty-weathering, poorly bedded, medium-grained, quartz-plagioclase-muscovite-biotite-(garnet) schist, 0 to 35 m thick. Bedding is lacking or obscure, in which case it may be indicated by a few discontinuous quartz ribbons and lenses and by differences in the proportions of the minerals. Unit 4 is transitional between units 3 and 5; gradation and interbedding take place over a vertical distance of several metres. Plagioclase occurs as 2- to 4-mm porphyroblasts in a 0.1- to 1.0-mm groundmass.

Unit 5

Unit 5, a coarse-grained quartz-muscovite-garnet-biotite-kyanite-staurolite schist, 0 to 65 m thick, overlies unit 4. Quartz lenses and ribbons are plentiful and separate layers that are alternately rusty- and non-rusty-weathering, gray, and graphitic. The rock has distinctive orange and dark-gray stripes. The gray bands are composed predominantly of quartz and muscovite with only accessory amounts of garnet, staurolite, and kyanite. The orange bands are composed predominantly of garnet, staurolite, and kyanite. The bands range from 0.5 to 2 cm thick, are laterally discontinuous, and are separated by the quartz-muscovite layers and numerous quartz ribbons and lenses that weather in relief, giving the rock a ribbed and corrugated appearance.

Unit 6

Most of the Hoosac Formation is composed of a rather monotonous quartz-plagioclase-muscovite, fine- to medium-grained schist and minor gneiss, with a combination of accessory minerals that may include chlorite, biotite, garnet, staurolite, kyanite, and sillimanite, depending on local bulk-composition variation of the rock and on the metamorphic grade.

West of the Precambrian terrane in the Windsor quadrangle, the garnet schist (unit 3) is overlain by about 50 m of this rock. East of the Precambrian, along most of the east side of the Berkshire massif, this rock everywhere overlies units 3 or 5 and ranges in thickness from 350 to 3,500 m.

Fresh rock ranges in color from light green to gray to brown, depending on the mineral content and the abundance of graphite, which commonly occurs as inclusions in minerals and as concentrations along foliation surfaces. The amount of graphite increases from Becket southward and from Windsor northward. At garnet grade, the chlorite-bearing varieties tend to weather rusty, whereas biotite-bearing schists commonly are only slightly rusty- or non-rusty-weathering. At higher metamorphic grade, much of the Hoosac weathers to a surface characterized by a rusty rind.

Part of this unit is continuously bedded; beds range from flaggy and poorly graded, 1 to 10 cm thick, to massive and 15 to 30 cm thick. The latter beds are distinguished by slight differences in the amounts of minerals and weathering properties. Much of the schist, however, has an indistinct bedding marked by discontinuous thin quartz lenses and ribbons that range from 0.5 to 2 cm in width and are as much as 3 m in length. These are parallel to the schistosity except in the noses of tight phase 2 folds. Weathered surfaces generally have a distinctly spotty appearance caused by the weathering in relief of the very abundant porphyroblasts of plagioclase, which are usually gray or black at lower grades of metamorphism. Graphite inclusions cause the gray color. These porphyroblasts commonly are 5 to 10 times the size of the matrix minerals. At higher metamorphic grade, the grain size of the matrix gradually coarsens, and at sillimanite grade, the characteristic porphyroblastic texture is lost. The amount of chlorite decreases southward from the Windsor area with a reciprocal increase in garnet because of increasing metamorphic grade.

Unit 7

Interbedded quartz-muscovite-calcite schist and rocks of unit 6 form a zone with a maximum thickness of 65 m in the middle of the Hoosac Formation in the Windsor quadrangle, but this unit cannot be traced to the north or south. Similar lithologies to the south are very thin and more discontinuous.

The calcareous rocks of this zone weather to a punky surface with numerous brown spots. Fresh rocks have a pearly luster on foliation surfaces because of abundant muscovite. Chloritoid is a rare accessory mineral.

Unit 8

Unit 8, composed of rocks similar to unit 6, directly overlies unit 7 where it is present. Where unit 7 is absent, no contact has been mapped between units 6 and 8. The thickness of unit 8 ranges from 350 to 3,500 m, with the maximum in the Windsor area. The minimum thickness occurs between Becket and Blandford (Fig. 1).

Unit 9

Unit 9 is a very coarse grained, massively bedded, quartz-plagioclase-garnet-muscovite-biotite schist with accessory but nearly ubiquitous staurolite and kyanite (or sillimanite). Garnet porphyroblasts as large as 1 cm in diameter are common. Kyanite and staurolite are as large as 5 mm in diameter.

The unit extends from the latitude of Becket southward at least to Chester. South of Chester it is absent, either because of facies changes within the Hoosac or because of thrust faults (Norton, 1971, 1975). The unit ranges from 0 to 65 m in thickness and is commonly interbedded with schist similar to unit 8. It is distinguishable from the coarse-grained garnetiferous schists of Units 3 and 5 by the non-rusty-weathering surfaces, the lack of the abundant quartz lenses and ribbons, the low abundance of muscovite, and the general abundance of plagioclase.

Unit 10

A slightly graphitic, rusty-weathering quartz-plagioclase-muscovite-biotite-garnet schist overlies unit 9 and persists southward beyond the extent of unit 9. Texturally the unit is very similar to units 6 and 8 and to rocks in the Blandford quadrangle. Where units 6 and 8 become more graphitic and units 7 and 9 are absent, units 6, 8, and 10 become indistinguishable.

Gray Gneiss

From approximately the north edge of the Peru quadrangle southward, units 2 to 10 become intercalated with layers of non-rusty-weathering, massive to indistinctly layered gneiss. It is composed predominantly of microcline-quartz-plagioclase-biotite rocks with rare muscovite. The Rb-Sr age for this gneiss is 1.1 b.y. (Brookins and Norton, 1975). Quartzite and metapelite are rare. Intrusive gneissic granite in which the foliation is parallel to regional foliation is common within some of the bands. Preliminary Rb-Sr whole-rock dates on this granite from localities 3 km southwest of Chester (Fig. 1) yield an age of 605 m.y. (Brookins and Norton, 1975). Zircon dates (Zartman, 1973, oral commun.) yield an age of approximately 750 m.y. for the same rocks. These gneissic granite do not have intrusive relationships with the Hoosac schist.

East of the Precambrian-Cambrian contact as mapped by earlier workers, the gneiss layers (hereafter called gray gneiss) make up a small proportion of the total rocks exposed, whereas west of the contact, Hoosac rocks are found only in very thin septa associated with cataclased Precambrian(?) gneiss.

To the south, within the Hoosac Formation, the layers of gray gneiss increase in number and in thickness, and they appear in successively higher units of the Hoosac (Table 3). At the Massachusetts-Connecticut border the upper third of the Hoosac contains layers of gray gneiss as much as 1,500 m thick. Schnabel (this volume) and Martin (1970) have interpreted these relationships as facies changes within the Hoosac Formation and as folds, respectively.

On the basis of reasons given by Norton (1971, 1975) and Brookins and Norton (1975), I interpret these layers of gray gneiss in the Hoosac section to be tectonic slivers of Precambrian rocks intercalated with the Paleozoic section during Taconic time.

The relationships between the rock sequences at different localities have been somewhat simplified in Table 3; only the thicker of the gray gneiss layers are shown.

TABLE 3. CORRELATION OF STRATIGRAPHIC SEQUENCES COMPILED AT QUADRANGLE BOUNDARIES

Wilmington-Rowe	North Adams-Windsor	Windsor-Peru	Peru-Becket	Becket-Otis	Blandford-West Granville
Rowe Schist	Rowe Schist	Rowe Schist	Rowe Schist	Rowe Schist	Rowe Schist or Moretown Formation
Hoosac Formation	10	10	10	10	10
					X
				9	9(?)
				8	8
					X
	7				
	6				
Heartwellville Formation, upper			5	5	5
				X	
			4		4
					X
lower	3	3	3	3	3
			X	X	
			3		
Readsboro Schist					
Sherman Marble					
Readsboro Schist	2	2	2	2	
		X	X		
			2		
Searsburg Conglomerate	1 Granitic gneiss*	1 Granitic gneiss* X Grantitic gneiss*			
Precambrian massif rocks --					

Note: Each column is separated from another by 8.5 mi. X's indicate bands of gray gneiss, mappable on a scale of 1:24,000. See text for explanation of numbered units.

*Equals Stamford Granite Gneiss (Precambrian) of Herz (1961).

Igneous Rocks

Premetamorphic igneous rocks in the Hoosac Formation include numerous beds (or sills or dikes), typically 0.5 to 3 m thick, of fine- to coarse-grained amphibole-plagioclase-epidote-garnet gneiss scattered through the section. None are traceable for any great distance.

Ultramafic rocks of any type are rare; they are located with few exceptions along the Rowe-Moretown contact in western Massachusetts (Fig. 1, Table 1). To the north, in southern Vermont, ultramafic rocks are found in units as young as Middle Ordovician. Chidester (1968) implicitly gives a Late Ordovician age for these presumably intrusive rocks. Thus their absence from the Hoosac Formation is peculiar indeed.

NATURE OF THE BOUNDING FORMATIONAL CONTACTS

Northwest of the Precambrian terrane (Fig. 1) the Hoosac Formation is in part stratigraphically overlain by and in part laterally equivalent to the Dalton Formation. The vertical transition is abrupt, but the lateral transition is made through interbedding and mineralogical gradation (Norton, 1967). The contact with the Precambrian rocks is generally structurally concordant because of isoclinal refolding and shearing of Precambrian structures but locally is an angular unconformity (Norton, 1969) marked by a few metres of quartz-pebble conglomerate. The conglomerate grades abruptly vertically and laterally into unit 3.

In the southeastern part of the North Adams quadrangle (Herz, 1961), the basal conglomerate of the Hoosac rests concordantly on an augen granitic gneiss (Fig. 1 and Table 3). The interpretation of Pumpelly and others (1894) of a Precambrian age (Grenville) for the underlying rocks is in agreement with Herz. I believe that the granitic gneiss may be late Precambrian or Early Cambrian in age and that here the contact is stratigraphic. The two rock units are apparently interbedded and vertically and horizontally gradational. However, southward the nature of the contact everywhere suggests that it is a tectonic break rather than a depositional contact (Norton, 1971). Here evidence suggests that the contact is a zone of premetamorphic low-angle thrust faulting, with intercalation of the cataclased Precambrian-like gneiss (described above) and typical Hoosac lithologies. Where the schist of the Hoosac Formation is in contact with gneiss, the two rock types have been drawn into concordance through intense folding, shearing, and cataclasis. However, regional truncation of metasedimentary units in these gray gneiss layers by Hoosac units has been demonstrated at several localities. This zone of synmetamorphic or premetamorphic imbrication (the Middlefield thrust zone) expands southward from Windsor to the Connecticut-Massachusetts border, where the fault zone may be as much as 5 km wide.

The contact with the overlying Rowe Schist (Hatch and others, 1966) is indistinct in most places because of mineralogical and textural gradation and also because of interbedding, particularly north of Windsor. The gradation is accomplished by a gradual decrease in the amount of plagioclase in the Hoosac Formation, a decrease in grain size, and the loss of any distinct bedding. Southward the contact is marked by graphitic, black to light-gray phyllite, assigned to the Rowe Schist. South of the Becket area, the contact becomes indistinct again because the Hoosac becomes graphitic. The thinning of the Rowe Schist in the Blandford area is probably attributable in part to a facies change in the Rowe unit that resulted in rocks similar to typical Hoosac rocks.

CORRELATION

In east-central Vermont in the vicinity of the Chester dome, the Cavendish Formation plus the Hoosac Formation of Doll and others (1961) are correlated directly with the Hoosac lithologies described in this report (Table 4). The Bull Hill Gneiss member (equivalent to the granitic gneiss of this report) of the Cavendish Formation appears to be a lateral equivalent of the Gassetts member (equivalent to Unit 3 of this report).

There are lithologic correlatives of all my units (1 through 10) of the Hoosac in southern Vermont (Skehan, 1961). However, as can be seen in Table 4, the stratigraphy varies radically from place to place. The two columns shown are from areas separated by only a few miles. These relationships suggest either rapid lateral facies changes or tectonic complications.

In the western part of the Windsor quadrangle (Fig. 1) the Hoosac (units 3 and 4 or 6) in part underlies and in part is laterally equivalent to the Dalton Formation (Norton, 1967).

In western Connecticut the Hoosac has been equated with the lower part of the Hartland Formation (Gates and Martin, this volume), in part with units 9 through 13 of Schnabel (this volume), and with Waramaug Formation rocks exposed east and west of Cameron's line and west of the Woodville Marble (Stanley, 1968). Harwood (1971, personal commun.) has described units in the Waramaug, on the west side of the Berkshire massif in the Norfolk quadrangle, Connecticut, that are similar to the Hoosac units shown in Figure 1. In southern Connecticut, along strike, Hall (this volume) correlates units B and C at the top of the Manhattan Schist with the Waramaug and Hoosac rocks.

FACIES RELATIONSHIPS WITHIN THE HOOSAC FORMATION

From southern Vermont to Connecticut the Hoosac Formation consists predominantly of quartz-plagioclase-muscovite schist with different amounts of biotite, chlorite, garnet, staurolite, kyanite, and sillimanite depending on bulk rock composition and metamorphic grade. The lower part of the Hoosac (units 1 to 5) has two very aluminous units (units 3 and 5) from at least the latitude of Wilmington, Vermont to Chester, Massachusetts (Table 3). Southward, the relationships are less clear because of the intercalation of Precambrian(?) and Hoosac lithologies, but reconnaissance work suggests that these aluminous lithologies continue at least discontinuously to the southern end of the massif and are truncated at that point by faults.

The middle part of the Hoosac Formation (units 6 to 8) is rather monotonous, and with the exception of a general increase in rustiness, graphite content, and effects of increasing metamorphic grade southward, the rock is essentially the same throughout the map area of Figure 1.

The upper two units of the Hoosac Formation reflect significant facies changes. From the Vermont border to the north end of the Peru quadrangle (Fig. 1) the

TABLE 4. CORRELATION OF HOOSAC AND OLDER ROCKS

This paper		Southeastern Vermont* (Skehan, 1961) 1	2	Central eastern Vermont (Doll and others, 1961)	
Rowe Schist		Pinney Hollow Formation	Pinney Hollow Formation	Pinney Hollow Formation	
	10				
	9				
	8	Hoosac Formation	Hoosac Formation	Hoosac	
	7				
Hoosac Fromation	6				
	5	Hartwellville Schist			
	4				
	3				
	2	Readsboro Schist Sherman Marble			
	1		Tyson Formation	Tyson	
Granitic gneiss			Wilmington Gneiss	Cavendish	Gassetts or Bull Hill Gneiss Buff dolomite member
Precambrian gneiss		Mount Holly complex	Mount Holly complex	Mount Holly complex	

*Columns 1 and 2 for southeastern Vermont are separated by 5 mi.

rocks are predominantly quartz-plagioclase-muscovite schist and gneiss with different amounts of biotite, chlorite, and garnet depending on bulk composition and metamorphic grade. Southward, however, the section becomes more aluminous, more graphitic, less sodic, more distinctly bedded, and rustier. Locally, schistose quartzite is common, especially near Chester.

The persistence of the different units of the Hoosac for great distances (for example, unit 3, which may be traced nearly continuously for more than 75 mi) suggests the following: (1) Except for units 1, 9, and 10, the original facies boundaries must have trended roughly north-south prior to deformation. (2) Postulated tectonic breaks at the base of and within the Hoosac Formation and possibly in the Precambrian terrane must be essentially parallel to Hoosac bedding. (3) During deposition of the Hoosac Formation the region was relatively tectonically stable, with the study area steadily subsiding relative to a more slowly subsiding platform to the west.

The present position of the Hoosac-Precambrian boundary along the east side of the Berkshire massif and along the east side of the Green Mountain anticlinorium in Vermont (Doll and others, 1961) is nearly coincident with a major facies boundary that demarks (for lowermost Cambrian units at least) the boundary between marine, nearshore turbulent environments (Tyson Formation, Dalton Formation, Hoosac unit 1) on the west and deeper water environments represented by most of the Hoosac to the east. The original position of this boundary is uncertain because of folding and faulting.

Parts of the Hoosac Formation (units 3, 5, and 9) were probably derived from a highly weathered terrane that supplied clay-rich detritus. Sedimentary structures suggest deposition in quiet, perhaps deep, water. Most of the formation (units 2, 4, 6, 8, and 10) was probably deposited in a similar environment, but the mineralogy indicates that the composition of the sediment was very different because of its derivation from a less mature terrane. Rare cyclical packets of graded beds 2 to 5 cm thick with low original sand/shale ratios suggest deposition in relatively deep water, possibly by turbidity currents. The numerous discontinuous quartz ribbons and lenses, interpreted possibly to be in large part original quartz-rich lenses in the sediments, indicate winnowing either by wave action or bottom currents.

Ubiquitous graphite and very minor amounts of sulfide minerals in some of the rocks suggest the presence of organic material in the sediment, but the amounts would suggest that the Hoosac units were not deposited in restricted basins.

Volcanism was not a dominant factor in this region while Hoosac sediment was being deposited. During the deposition of the sediment of the overlying Rowe Schist, however, the period of stability during which the Hoosac was deposited ended. The rocks of the Rowe Schist are far more siliceous, suggesting a less mature source area, and they include a considerable proportion of volcanic material.

ACKNOWLEDGMENTS

I acknowledge both the fine work of and helpful discussions with A. H. Chidester, D. S. Harwood, N. L. Hatch, Jr., P. H. Osberg, L. R. Page, N. M. Ratcliffe, R. W. Schnabel, and E-an Zen. The manuscript was reviewed thoughtfully by Allison R. (Pete) Palmer.

REFERENCES CITED

Brookins, D. G. and Norton, S. A., 1975, Rb/Sr whole rock ages along the Precambrian-Cambrian contact, east side of the Berkshire massif, Massachusetts: Geol. Soc. America Abs. with Programs, v. 7, p. 30.

Chidester, A. H., 1968, Evolution of the ultramafic complexes of northwestern New England, *in* Zen, E-an, White, W. S., Hadley, J. B., and Thompson, J. B., Jr., Studies of Appalachian geology: Northern and maritime: New York, Wiley-Interscience, p. 343-354.

Chidester, A. H., Hatch,, N. L., Jr., Osberg, P. H., Norton, S. A., and Hartshorn, J. H., 1967, Geologic map of the Rowe quadrangle, Franklin and Berkshire Counties, Massachusetts, and Bennington and Windham Counties, Vermont: U.S. Geol. Survey Geol. Quad. Map GQ-642.

Chute, N. E., 1945, The talc, soapstone, and asbestos deposits of Massachusetts: U.S. Geol. Survey Open-File Rept., 42 p.

Doll, C. G., Cady, W. M., Thompson, J. B., Jr., and Billings, M. P., 1961, Centennial geologic map of Vermont: Montpelier, Vermont Geol. Survey, scale 1:250,000.

Gates, R. M., and Martin, C. W., 1976, Pre-Devonian stratigraphy of the central section of the western Connecticut highlands, *in* Page, L. R., ed., Contributions to the stratigraphy of New England: Geol. Soc. America Mem. 148 (this volume).

Hall, L. M., 1976, Preliminary correlation of rocks in southwestern Connecticut, *in* Page, L. R., ed., Contributions to the stratigraphy of New England: Geol. Soc. America Mem. 148 (this volume).

Harwood, D. S., 1972, Tectonic events in the southwestern part of the Berkshire anticlinorium, Massachusetts and Connecticut: Geol. Soc. America Abs. with Programs, v. 4, p. 19.

Hatch, N. L., Jr., 1967, Redefinition of the Hawley and Goshen Schists in western Massachusetts: U.S. Geol. Survey Bull. 1254-D, 16 p.

——1969a, Geologic map of the Worthington quadrangle, Hampshire and Berkshire counties, Massachusetts: U.S. Geological Survey Geol. Quad. Map GQ-857.

——1969b, Isoclinal folding indicated by primary sedimentary structures in western Massachusetts: U.S. Geol. Survey Prof. Paper 600-D, p. D108-D114.

Hatch, N. L., Jr., and Hartshorn, J. H., 1968, Geologic map of the Heath quadrangle, Massachusetts and Vermont: U.S. Geol. Survey Geol. Quad. Map GQ-735.

Hatch, N. L., Jr., and Norton, S. A., 1969, A talc-serpentine body near Middlefield, Massachusetts: U.S. Geol. Survey Open-File Rept., 7 p.

Hatch, N. L., Jr., Chidester, A. H., Osberg, P. H., and Norton, S. A., 1966, Redefinition of the Rowe Schist in northwestern Massachusetts, *in* Cohee, G. V., and West, W. S., Changes in stratigraphic nomenclature by the U.S. Geological Survey, 1965: U.S. Geol. Survey Bull. 1244-A, p. A33-A35.

Hatch, N. L., Jr., Stanley, R. S., and Clark, S. F., Jr., 1970, The Russell Mountain formation, a new stratigraphic unit in western Massachusetts: U.S. Geol. Survey Bull. 1324-B, p. B1-B10.

Hatch, N. L., Jr., Norton, S. A., and Clark, R. G., 1971, Geologic map of the Chester quadrangle, Massachusetts: U.S. Geol. Survey Geol. Quad. Map GQ-858.

Herz, Norman, 1961, Bedrock geology of the North Adams quadrangle, Massachusetts: U.S. Geol. Survey Geol. Quad. Map GQ-139.

Martin, C. W., 1970, The bedrock geology of the Torrington quadrangle (with map): State Geol. and Nat. Hist. Survey of Connecticut Quad. Report 25, 53 p.

Norton, S. A., 1967, Geology of the Windsor quadrangle, Massachusetts, U.S. Geol. Survey Open-File Rept., 210 p.

——1969, Unconformities at the northern end of the Berkshire Highlands, *in* Bird, J. M., ed., Guidebook for field trips in New York, Massachusetts, and Vermont: New England Intercoll. Geol. Conf., chap. 21, p. 1-20.

——1971, Possible thrust faults between lower Cambrian and Precambrian rocks, east edge of the Berkshire Highlands, western Massachusetts: Geol. Soc. America Abs. with Programs, v. 3, p. 46.

——1975, Chronology of Paleozoic tectonic and thermal metamorphic events in Precambrian, Cambrian and Ordovician rocks at the north end of the Berkshire massif: U.S. Geol. Survey Prof. Paper 888-B, p. 20-31.

Osberg, P. H., 1975, Recumbent folding of the Goshen and Waits River Formations, western Massachusetts: U.S. Geol. Survey Prof. Paper (in press).

Osberg, P. H., Hatch, N. L., Jr., and Norton, S. A., 1971, Geology of the Plainfield quadrangle, Franklin, Hampshire, and Berkshire counties, Massachusetts: U.S. Geol. Survey Geol. Quad. Map GQ-877.

Pumpelly, Raphael, Wolff, J. E., and Dale, T. N., 1894, Geology of the Green Mountains in Massachusetts: U.S. Geol. Survey Mon. 23, 192 p.

Ratcliffe, N. M., 1972, Revised polyphase structural chronology in western Massachusetts and problems of regional correlation: Geol. Soc. America, Abs. with Programs, v. 4, p. 40.

Rodgers, John, Gates, R. M., and Rosenfeld, J. L., 1959, Explanatory text for preliminary geologic map of Connecticut: Connecticut Geol. History Survey Bull. 84, 64 p.

Schnabel, R. W., 1976, Pre-Silurian stratigraphy in south-central Massachusetts and north-central Connecticut, *in* Page, L. R., ed., Contributions to the stratigraphy of New England: Geol. Soc. America Mem. 148 (this volume).

Skehan, J. W., 1961, The Green Mountain anticlinorium in the vicinity of Wilmington and Woodford, Vermont: Vermont Geol. Survey Bull. 17, 159 p.

Stanley, R. S., 1967, Geometry and age relations of some minor folds and their relation to the Woronoco nappe, Blandford and Woronoco quadrangles, Massachusetts, *in* Robinson, Peter, ed., Guidebook for field trips in the Connecticut Valley, Massachusetts: New England Intercoll. Geol. Conf., p. 48-60.

——1968, Bedrock geology of western Connecticut, *in* Orville, P. M., ed., Guidebook for Field Trips in Connecticut: New England Intercoll. Geol. Conf., p. (DO) 1-5.

Thompson, J. B., Jr., and Norton, S. A., 1968, Paleozoic regional metamorphism in New England and adjacent areas, *in* Zen, E-an, White, W. S., Hadley, J. B., and Thompson, J. B., Jr., eds., Studies of Appalachian geology: Northern and maritime: New York, Wiley-Interscience, p. 319-327.

Zartman, R. I., Hurley, P. M., Krueger, H. W., and Giletti, B. J., 1970, A Permian disturbance of K-Ar radiometric ages in New England: Its occurrence and cause: Geol. Soc. America Bull., v. 81, p. 3359-3374.

MANUSCRIPT RECEIVED BY THE SOCIETY JUNE 24, 1974
REVISED MANUSCRIPT RECEIVED FEBRUARY 10, 1975
MANUSCRIPT ACCEPTED FEBRUARY 20, 1975

Printed in U.S.A.

Geological Society of America
Memoir 148

Stratigraphy, Isotopic Ages, and Deformational History of Basement and Cover Rocks of the Berkshire Massif, Southwestern Massachusetts

NICHOLAS M. RATCLIFFE
Department of Earth and Planetary Sciences
City College, City University of New York
138th Street and Convent Avenue
New York, New York 10031
and
U.S. Geological Survey
80 Broad Street
Boston, Massachusetts 02110

AND

ROBERT E. ZARTMAN
U.S. Geological Survey
Denver Federal Center
Denver, Colorado 80225

ABSTRACT

Detailed mapping in the Ashley Falls, East Lee, Stockbridge, Great Barrington, and Monterey quadragles of Massachusetts and Connecticut has established 12 lithostratigraphic paragneiss and metavolcanic units, approximately 1,830 m in total thickness, in the Precambrian gneiss of the Berkshire massif. Coarsely blastoporphyritic, microcline, perthite, biotite, and ferrohastingsite granite gneiss (the Tyringham Gneiss of Emerson, 1898) intruded the foliated paragneiss sequence in the form of broadly concordant granite-granodiorite sills as much as 457 m thick and also in the form of other less well defined intrusions. Fine-grained granodiorite and foliated alaskite locally rich in magnetite form a narrow discontinuous border facies of the granite gneiss. The Tyringham and its country rocks were isoclinally folded along easterly trends, granulated, and foliated prior to deposition of the Dalton Formation of late Precambrian(?) and Early Cambrian age.

U-Pb isotopic data for zircon from the Tyringham and a paragneiss unit, the Washington Gneiss on Beartown Mountain (Great Barrington and Stockbridge quadrangles), are only slightly discordant and give Pb^{207}/Pb^{206} ages of 1,040 to 1,080 m.y. for both units. The zircons from the Tyringham Gneiss have an elongate, euhedral habit compatible with an igneous origin, whereas zircons from the Washington Gneiss commonly are rounded, have overgrowths, and occur as detrital grains in a blue quartz metaconglomerate, which interfingers with metadacitic volcanic rocks. The close agreement in age between the intrusive rock and its country rock may mean either that (1) the age of the source material for the Washington Gneiss and the intrusion of the Tyringham Gneiss are little separated in time, (2) the Tyringham Gneiss was intruded during the same dynamothermal event that reset the Washington Gneiss zircon, or (3) the zircons in both the Tyringham and Washington Gneiss record a metamorphic age. Because of geologic evidence that the Tyringham was intruded syntectonically during a time of deep-seated plutonic dynamothermal metamorphism in Precambrian time, either (2) or (3) is the favored explanation. Some evidence of an inherited component in the zircon from the Washington Gneiss is given by its older (~1,200 m.y.) Th-Pb age.

The Berkshire gneiss sequence resembles closely in total thickness, probable protoliths, character of Precambrian deformation and plutonism, and isotopically determined age the "Grenville" metasedimentary sequence of the Adirondack Mountains described by Buddington and Leonard (1962) and de Waard and Walton (1967).

Upper Precambrian (?) and Lower Cambrian quartzofeldspathic metasedimentary rocks of the Dalton Formation and Cheshire Quartzite overlie with profound angular unconformity the gneissic rocks of the Berkshire massif. Comparison of the metamorphic mineralogy of the Paleozioc and Precambrian rocks indicates that the Precambrian gneiss was affected by dynamothermal metamorphism at least as high as muscovite plus sillimanite grade prior to deposition of the Dalton-Cheshire sequence.

Multiple dynamothermal metamorphism and deformation during Ordovician and Devonian time, however, have greatly affected the configuration and mineralogy of the basement rocks. Large-scale synmetamorphic westward overthrusting of imbricate slices of basement rocks transported the gneiss of the massif at least 21 km across underlying and highly deformed miogeosynclinal and exogeosynclinal metasedimentary rocks of Cambrian to Middle Ordovician age. High-grade (staurolite-kyanite-sillimanite plus muscovite) Barrovian metamophism postdated thrusting during a probable Acadian dynamothermal event. The complex internal structures seen in the Precambrian gneiss of the Berkshire massif result from multiple deep-seated deformations during the "Grenville" orogeny (~1 b.y. ago), Taconic orogeny (~460 m.y. ago), and Acadian orogeny (350 m.y. ago).

INTRODUCTION

The main purpose of this paper is to present new information on the stratigraphy of Precambrian rocks in part of the Berkshire massif (Fig. 1) and to set forth evidence for Precambrian and Paleozoic orogenic activity in this part of the Appalachian orogen. The Berkshire massif lies at the present boundary between an east-facing sequence of Cambrian, Ordovician, Silurian, and Devonian eugeosynclinal rocks to the east and a sequence of highly folded Cambrian and Ordovician miogeosynclinal and exogeosynclinal rocks to the west.

The complexly faulted relationships between the Berkshire gneiss and the rock

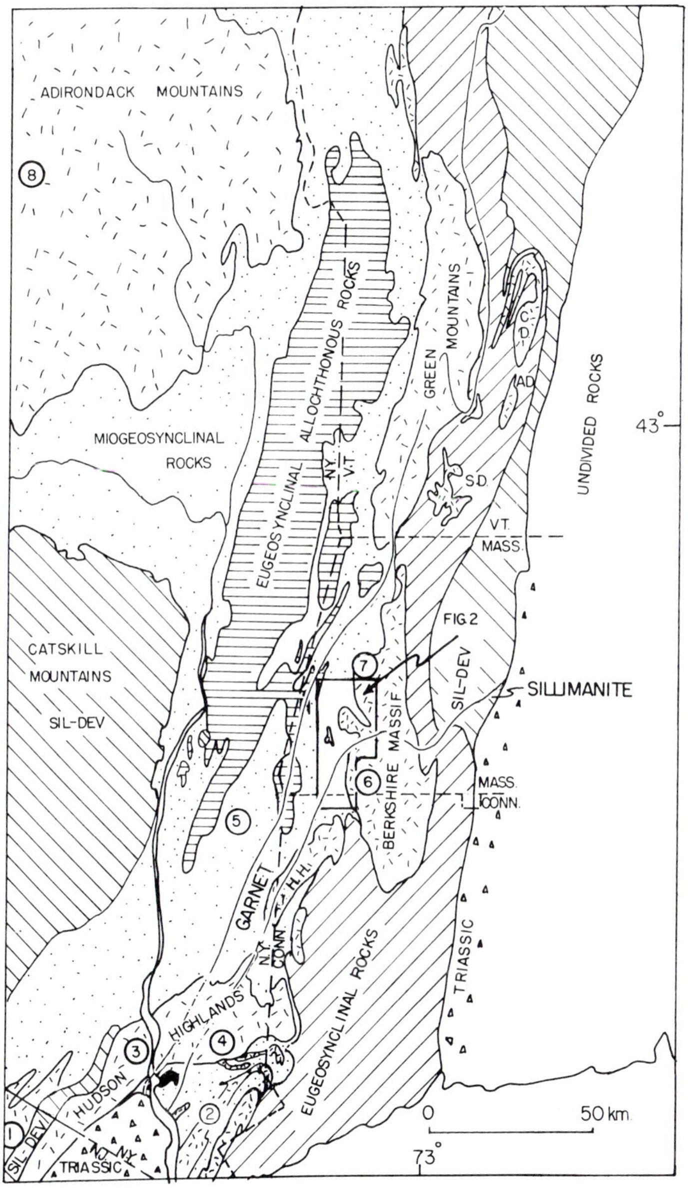

Figure 1. Regional geologic map showing Precambrian massifs and location of area shown in Figure 2. Numbered localities are referred to in text and in Table 2. Key to symbols: C. D., Chester dome; A. D., Athens dome; S. D., Sadawga dome; H. H., Housatonic highlands. Data for Paleozoic garnet and sillimanite plus muscovite isograds from various sources: Hudson Highlands and Manhattan Prong (loc. 2), Ratcliffe (1970); Dutchess County, N.Y., Fisher and others (1961); southwestern Massachusetts, Zen and Hartshorn (1966); Zen and Ratcliffe (1971); and Ratcliffe (1969a, 1969b); information for area north of area shown in Figure 2 from compilation of Thompson and Norton (1968).

sequence to the east are discussed by Norton (this volume). The western front of the Berkshire massif is a series of overlapping low-angle thrust faults, the lowest slice of which overrides the miogeosynclinal rocks (Fig. 2) a minimum of 21 km at lat. 42°12′30″N (Figs. 2, 3, section A-A′).[1] Similar relationships have been shown to the north by Norton (1969, this volume) and to the south by Harwood (see Ratcliffe and Harwood, 1975). Because neither the total nor the relative tectonic foreshortening on basement thrusts is yet known, detailed reconstruction of the Cambrian and Ordovician mioeugeosynclinal transition is not yet possible

[1]Figures 2, 3, 15, and 16 are in pocket inside back cover.

in this area. Moreover, upper Precambrian(?) and lower Paleozoic rocks (Dalton Formation, Cheshire Quartzite, and Stockbridge Formation), deposited as the unconformable sedimentary cover on the Berkshire gneiss, were transported westward on the basement overthrusts and rarely are found in sedimentary contact with the western "autochthonous" basement. A third sequence of rocks, indicated by Xs in Figure 2, is of uncertain structural and stratigraphic position but has eugeosynclinal as well as miogeosynclinal affinities and must eventually be reconstructed into the section.

All the Precambrian rocks shown in Figure 2 were dynamothermally metamorphosed in Precambrian time as well as twice during Paleozoic time and consequently were multiply deformed (Ratcliffe, 1965, 1968, 1969a, 1969b; Zen, 1969). The present mineralogy and texture of the Precambrian rocks are the result of a very complex history of Precambrian prograde metamorphism and of Paleozoic cataclasis, retrogression, and remetamorphism. Despite the severity of the polyphase Paleozoic deformation and metamorphism, the relict Precambrian structural grain has been identified; thus Precambrian and Paleozoic structures may be differentiated. It is beyond the scope of this paper, however, to recount the details of the polyphase deformation. Summaries of regional interpretations may be found in articles by Ratcliffe (1969a, 1969b, 1974a, 1974b, 1974c), Zen (1969, 1972), and Ratcliffe and Harwood (1975).

Precambrian gneissic rocks of the Berkshire massif in the Stockbridge, Great Barrington, Ashley Falls, Monterey, and East Lee quadrangles have been subdivided into 12 lithic units (Fig. 2). Detailed mapping of these lithic subdivisions (1 through 12) suggests that they define a stratigraphic sequence. A thirteenth unit, a foliated granodiorite and quartz monzonite (Tyringham Gneiss), has intrusive contacts with units 1 through 3, 6 through 9, and 12.

Rocks of the Berkshire gneissic basement are unconformably overlain by a quartzofeldspathic schist and metaquartzite unit of late Precambrian(?) and Early Cambrian age (Dalton Formation and Cheshire Quartzite). Within the area shown in Figure 2, Dalton rocks rest on units 1, 6, 7, and 13 of the Precambrian sequence.

The exposed Precambrian stratigraphic sequence of rocks is estimated to be 1,830 to 2,430 m thick, including at least 457 m of metaintrusive rock. The bulk of the rocks are quartz and feldspar-rich graphitic and nongraphitic metasedimentary rocks with smaller amounts of felsic and mafic metavolcanic rock and one distinctive calc-silicate unit (unit 8). In addition to the layered sequence described above, large areas are underlain by granitic gneiss (the Tyringham, unit 13) that is of unknown thickness. The metasedimentary and metavolcanic rocks of the Berkshire massif are similar to paragneiss sequences of the Adirondack Mountains (Buddington and Leonard, 1962; de Waard and Walton, 1967) and of the Reading Prong (Drake, 1969; Offield, 1967; Dodd, 1965; Young, 1971; Prucha and others, 1968) and to the basement rocks of the Manhattan Prong (Fordham Gneiss; Hall, 1968).

Rock types such as anorthosite, charnockitic and fayalitic granite, and syenite, found in the Adirondacks (Buddington, 1939, 1959; de Waard and Walton, 1967) and to a lesser extent in the Reading Prong (Drake, 1969; Young, 1971), have not been recognized in the Berkshire massif.

PRECAMBRIAN STRATIGRAPHY

The Precambrian stratigraphy described here is the result of bedrock mapping in the East Lee, Stockridge, Great Barrington, Monterey, and Ashley Falls quadrangles. Figure 2 shows the generalized geology of this area and the subdivisions of the Precambrian stratigraphy. Stratigraphy of the Dalton Formation and Cheshire Quartzite of Precambrian(?) and Cambrian age will be discussed later in this paper. The relationships of the subdivisions of the younger Stockbridge and Walloomsac Formations and their structural relationships have been discussed by Ratcliffe (1969a, 1969b, 1974a, 1974c), Zen (1969), Zen and Hartshorn (1966), and Zen and Ratcliffe (1971).

The areal distribution of gneissic units shown in Figure 2 shows complexly contorted map patterns for the Precambrian rocks. Because of isoclinal and recumbent folding in both Precambrian and Paleozoic time, the Precambrian succession is inverted in many cases. Unfortunately, definitive indicators of stratigraphic tops have not been found, and the relative ages of the gneiss units (1 through 12) are not known. Moreover, the unconformable Precambrian(?) and Paleozoic cover rocks at different localities rest on Precambrian successions that structurally top toward either unit 1 or unit 12.

Overall, however, the Washington Gneiss (units 1 and 2) and associated intrusive granite gneiss (the Tyringham, unit 13) appear to form the structural base at localities such as Beartown Mountain and Halls Hill. Therefore a stratigraphic succession has been adopted, with unit 1 assumed to be the oldest.

Washington Gneiss: Units 1 and 2

The Washington Gneiss (Emerson, 1898, p. 20; 1899, p. 34) is a distinctive blue quartz-bearing graphitic gneiss and biotite gneiss exposed near Washington, Massachusetts, in the northeastern corner of the East Lee quadrangle; it was mapped in the Beartown Mountain area by Emerson (1899).

Unit 1 is mainly composed of dark, rusty-weathering blue quartz-muscovite-biotite-plagioclase gneiss and muscovite-biotite schist with coarsely ribbed layers, 0.5 to 1 cm thick, of bluish quartz "pebbles" flattened in the plane of foliation. A white, mica-poor, garnetiferous, plagioclase-rich blue quartz-bearing granulite is interlayered on a meter scale and locally takes the place of the more typical rusty-weathering rock. Both the dark two-mica gneiss and the interlayered white granulite have sedimentary features and include quartz-pebble conglomerate beds as much as 0.5 m thick, with graphite flakes as much as 1 cm in diameter. Biolite-rich clots retrograded from garnet (Fig. 4) have elongate tails that extend into the plagioclase-quartz-rich matrix along the foliation. The biotite within the clots is not aligned in this foliation.

Adjacent biotite clots vary markedly within the same thin section and may be composed almost wholly of either green or brown biotite with scattered pink garnet and small grains of zircon. Garnet in various stages of replacement by biotite may be seen in the Washington Gneiss in the October Mountain-Washington area.

Included in unit 1 are layers of dense black garnet-hornblende and hornblende-plagioclase amphibolite too small to be mapped separately and, on Beartown Mountain, a distinctive dark-gray kyanite-muscovite-biotite-quartz-graphite schist.

Unit 2 consists of light-gray to yellowish-gray well-layered muscovite-biotite-plagioclase-quartz gneiss without quartz ribbing. It is interbedded with unit 1 type of Washington Gneiss in the southern part of the Monterey quadrangle and is mapped separately (Fig. 2). Beds of quartzose metagraywacke and metafeldspathic quartzite 3 to 6 cm thick are prominent.

Locally in units 1 and 2 are dark-greenish-gray- to punky yellowish-brown-weathering actinolite-graphite-scapolite-pyrite-bearing calc-silicate rocks and rusty-weathering diopside-calcite marble totaling as much as 3.05 m thick. Although the Washington Gneiss (units 1 and 2) is as much as 915 m thick in adjacent

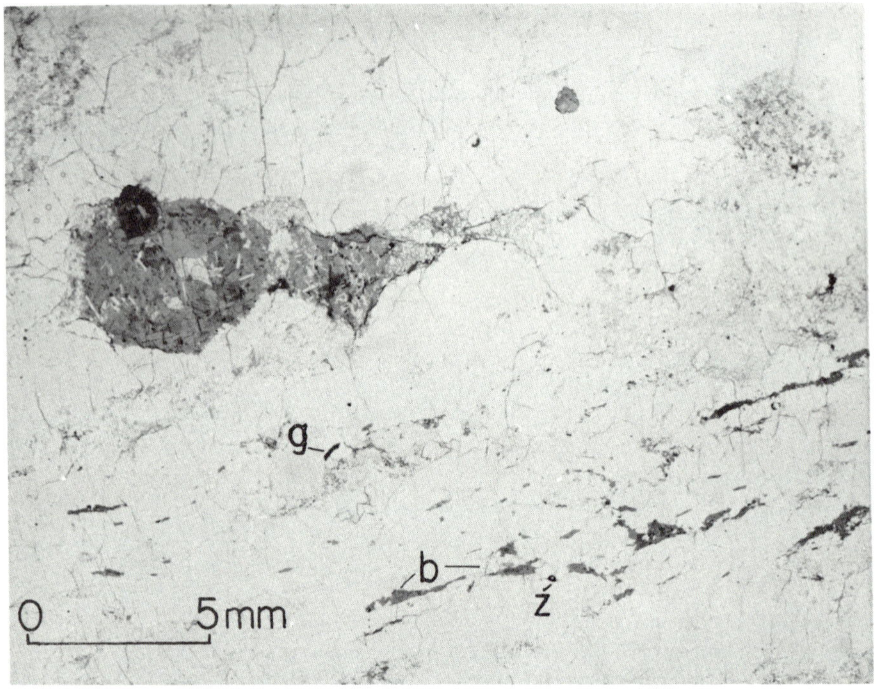

Figure 4. Positive print from thin section (unpolarized transmitted light) of Washington Gneiss (unit 1) showing biotite-rich clot (retrograded from garnet) in blue quartz-rich metaconglomerate matrix. Foliation in gneiss, inclined upward lift to right shown by oriented biotite (b), graphite (g), and lenticular, small, high-relief detrital zircon (z); nonoriented biotite in clots supports replacement origin.

quadrangles (Harwood, 1975), it is locally absent in the area shown in Figure 2. In places it is intruded by the Tyringham Gneiss.

The hornblende-garnet amphibolite layers in units 1 and 2 may represent basaltic volcanic flows intercalated with volcaniclastic deposits (white plagioclase-quartz granulite) derived from reworking of porphyritic dacite. The high percentage of quartz and the presence of the quartz conglomerate, however, suggest a considerable contribution from a nonvolcanic, silicic source area.

Garnet-hornblende Amphibolites

On Beartown Mountain, garnet-hornblende amphibolite and garnet-hornblende-plagioclase amphibolite form two layers separated by a leucocratic gneiss (Fig. 2, units 3, 4, and 5). Together these consititute a tripartite sequence of which unit 3 is laterally equivalent to the Washington Gneiss. The amphibolite of unit 3 intertongues with the Washington Gneiss and is thickest where the Washington is absent or very thin. The relationships of these units are diagrammatically shown in Figure 6. A key to symbols, with probable protoliths and correlation, is given in Figure 7.

Garnet-hornblende Amphibolite: Unit 3. Unit 3 consists mainly of dark-gray, poorly layered to massive hornblende amphibolite and garnet-hornblende amphibolite. Dark-green hornblende constitutes as much as 80 percent of some samples and as little as 30 percent of the more plagioclase-rich varieties. Deep-reddish garnet, present in most samples, may constitute as much as 30 percent of the rock; quartz is commonly absent. Reddish-brown biotite, magnetite, apatite, and sphene are common accessory minerals. Unzoned calcic oligoclase or sodic andesine commonly compose less than 20 percent of the rock.

The massive, nonlayered character of the rocks and the lack of associated calc-silicate rocks suggest that this unit represents basaltic flows or sills interlayered with the Washington Gneiss. In one outcrop on Beartown Mountain, a narrow dike of garnet amphibolite crosscuts 0.5 m of bedded rocks in the Washington Gneiss, indicating an igneous origin for part of this uint (see Fig. 5). Because of its coarse-grained texture and abundant garnets, this unit resembles certain pyroxene-garnet gneiss units found in the Reading Prong in southeastern New York, described by Lowe (1950, p. 142) and Dodd (1965, p. 9, 10).

Leucocratic Biotite Gneiss and Granulite: Unit 4. A light-gray-weathering leucocratic biotite gneiss and granulite (unit 4), 61 to 91 m thick, overlies the Washington Gneiss and garnet-hornblende amphibolite of unit 3 and separates these units from a second amphibolite (unit 5). The leucocratic biotite gneiss and granulite unit contains varying proportions of granoblastic biotite gneiss and interlayered biotite-quartz-plagioclase-microcline paragneiss spotted with small garnet grains. Locally the biotite gneiss predominates; on the eastern slopes and south of the peak of Beartown Mountain it shows gradational contacts with the amphibolites of units 3 and 5. The origin of the more leucocratic granitic rock is uncertain; it may be alaskite intruded lit-par-lit on a scale of meters within the more biotitic gneissic rock.

Garnet-hornblende Amphibolite and Amphibolitic Gneiss: Unit 5. Hornblende-bearing gneiss and massive garnet-hornblende amphibolite, commonly with 5 to 15 percent plagioclase, and massive black- and white-spotted hornblende-plagioclase-microcline-biotite gneiss form the major part of unit 5 in the two outcrop belts on Beartown Mountain. This unit is characterized by the absence of quartz-rich gneiss. Green hornblende is the dominant mafic mineral in these rocks, which were perhaps basaltic flows interlayered with more felsic volcanic material. The unit is 61 to 91 m thick.

Washington Gneiss and Amphibolitic Gneiss Facies Relationships

The facies relationships of the Washington Gneiss (units 1 and 2) and the amphibolitic gneiss (units 3, 4, and 5) are schematically illustrated in Figure 6. Thinning of the amphibolite of unit 3 and thickening of the Washington Gneiss to the east and south, together with coarsening of the grain size of the Washington quartz conglomerate to the west, suggest a shallow-water environment with volcanic land areas and local subsidiary basins. The high quartz content and the aluminous schist suggest a provenance of deeply weathered granitic rocks. The plagioclase-quartz rocks may have been formed from reworking of local dacitic volcanic material, whereas the amphibolite may represent basaltic flows or hypabyssal sills. The persistence of the leucocratic gneiss unit above different units of the Washington Gneiss suggests that the local basins that existed during deposition of the lower part of the section were filled before deposition of the upper amphibolite unit (unit 5).

Biotite-quartz-magnetite-plagioclase-microcline Gneiss: Unit 6

A distinctive, homogeneously fine-grained, magnetite-studded light-gray to pinkish-gray felsic gneiss (unit 6) crops out over broad areas of the Berkshire massif (Fig. 2). Its generally massive nature, local interlayering with amphibolitic gneiss

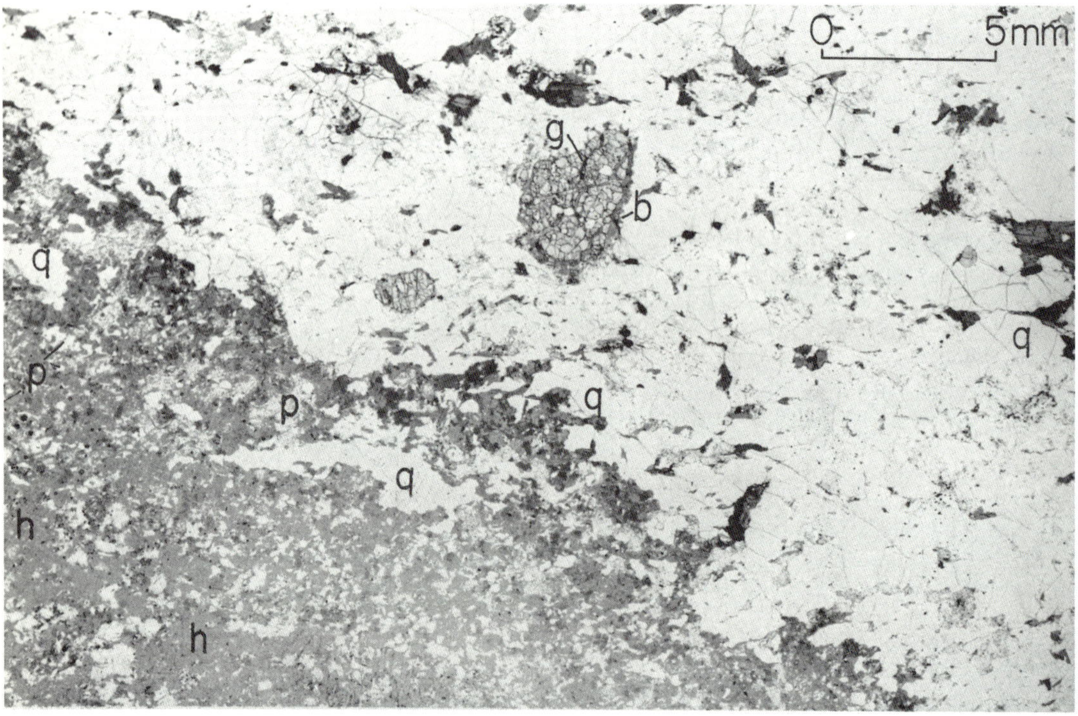

Figure 5. Positive print from thin section (unpolarized transmitted light) showing metaintrusive contact between biotite-hornblende amphibolite of unit 3 (lower left) and blue quartz-pebble metaconglomerate of Washington Gneiss (unit 1) (upper right). At this locality on Beartown Mountain the amphibolite truncates approximately 0.5 m of bedding in the Washington Gneiss, but both units share the foliation in which the Washington quartz is flattened. Key: p = plagioclase; q = quartz; b = biotite; h = hornblende; g = garnet. Note reaction rim of biotite around garnet.

(unit 5), the lack of well-developed bedding and of quartz-rich layers, and the uniformly high potassium feldspar content suggest that the unit is probably a metamorphosed felsic volcanic rock. The unit is 122 to 305 m thick.

Biotite-quartz-plagioclase Paragneiss Sequence: Units 7 through 11

Units 7 through 11 are a metasedimentary sequence with a wide variety of interlayered rock types. The dominant rock is well-layered biotite-quartz-plagioclase paragneiss (unit 7) in which are intercalated distinctive but noncontinuous units including calc-silicate gneiss (unit 8), hornblende-spotted gneiss and layered hornblende-plagioclase gneiss (unit 9), rusty schistose gneiss and granulite (unit 10), and fine-grained amphibolite (unit 11). The rocks replace one another vertically and horizontally but generally maintain a crude stratigraphic succession, with unit 8 or 7 commonly close to unit 6. The base of the rusty schistose gneiss and granulite (unit 10) and associated amphibolite (unit 11) may mark a disconformity, as it appears to occur in different stratigraphic positions. Alternatively, the units may be repeated as separate lenses within the stratigraphic succession.

Rusty Schistose Gneiss and Granulite and Associated Amphibolite: Units 10 and 11. Rusty-weathering, porphyroblastic muscovite-biotite-microcline schistose gneiss and quartz-rich feldspathic granulite with 1- to 3-cm-thick layers of fine-grained rusty quartzite form a discontinuous unit that either rests on the felsic gneiss (unit 6) or is intercalated in the biotite-quartz-plagioclase paragneiss (unit 7). Quartz-pebble-plagioclase-microcline granulite layers alternate with more schistose two-mica schist layers to produce a striking ribbed rock that is interlayered with silvery-gray to yellowish-gray micaceous quartz-plagioclase-microcline granulite. Locally, white-weathering actinolite-plagioclase granulite forms calc-silicate layers as much as 1.5 m thick. Fine-grained garnet-sphene-plagioclase-hornblende amphibolite (unit 11) is locally interbedded in thin, nonpersistent layers. As well as resting on unit 6, units 10 and 11 also are found within the biotite-quartz-plagioclase paragneiss (unit 7), where they are associated with calc-silicate rocks. These separate outcrop belts of units 10 and 11 may not be correlative. However, the low angle of truncation noted beneath unit 10 on Beartown Mountain suggests that the base of the unit transgresses approximately 61 to 91 m of unit 7.

Its irregular stratigraphic position and locally coarse-grained detrital character suggest that unit 10 may mark a disconformity. In this case, potassium-rich aluminous shale and sand could have resulted from the reworking and deposition of chemically weathered soil cover derived from the older felsic rocks of unit 6.

Biotite-quartz-plagioclase Paragneiss: Unit 7. The biotite-quartz-plagioclase paragneiss (unit 7) is markedly different from any of the other map units. Well-layered medium- to dark-gray gneiss predominates. Beds of dark-green, 0.25- to 0.5-m-thick, biotite-rich amphibolite; gray-weathering, fine-grained feldspathic quartz granulite; and gray- and white-banded biotite gneiss are common. Layering (commonly on a centimeter scale) of alternating plagioclase and biotite and (or) hornblende-rich rocks is conspicuous in most exposures.

Pale-brown to greenish-brown biotite is the dominant mafic mineral; pale-green to dark-greenish-brown hornblende occurs in some gneiss, particularly in the vicinity of calc-silicate rocks.

Calc-silicate Rocks: Unit 8. Calcite-chondrodite-zinnwaldite marble, diopside-actinolite-calcite and plagioclase-diopside-hornblende granulite, and, less commonly, dolomite-chondrodite marble, diopside-potassium feldspar granulite, and massive light-green diopside rock form the calc-silicate rocks of unit 8. This unit appears to represent a single horizon in the biotite-quartz-plagioclase paragneiss sequence, although the unit is not traceable continuously because of nondeposition or lack of exposure. Its probable position in areas of poor exposure is shown by the dotted symbol in Figure 2. Marble does not exceed 15 m in thickness in single exposures and commonly occurs in discontinuous pods 2 to 3 m thick. The best exposures of marble are on the south shore of Upper Goose Pond in the East Lee quadrangle. This calc-silicate unit is very well exposed east of Monterey, east of Lee, and at Benton Hill. The Benton Hill belt extends into the South Sandisfield quadrangle (loc. 6, Fig. 1), where it is widespread (Harwood, 1975).

Layered and Spotted Hornblende-plagioclase Gneiss: Unit 9. A white-weathering gneiss (unit 9), either spotted with hornblende or irregularly interlayered with black amphibolite and rusty pyrrhotite containing blue quartz layers, is found locally where the calc-silicate unit is relatively thick. Distinctive hornblende-diopside pegmatite as much as a few meters thick forms concordant pods in the layering. Unit 9 laterally replaces the biotite-quartz-plagioclase paragneiss (unit 7) by intertonguing of the amphibolite and probably represents calcareous shale and graywacke deposited contemporaneously with the calc-silicate unit. The exposures east of Lee belong to the type exposure of rock previously referred to as the Lee Gneiss by Emerson (1899, p. 33), who defined it as a metasedimentary unit. He later referred to this unit as the Lee Quartz Diorite (Emerson, 1917, p. 153) and attributed it to a mafic border facies of the intrusive Tyringham. Emerson's earlier (1899)

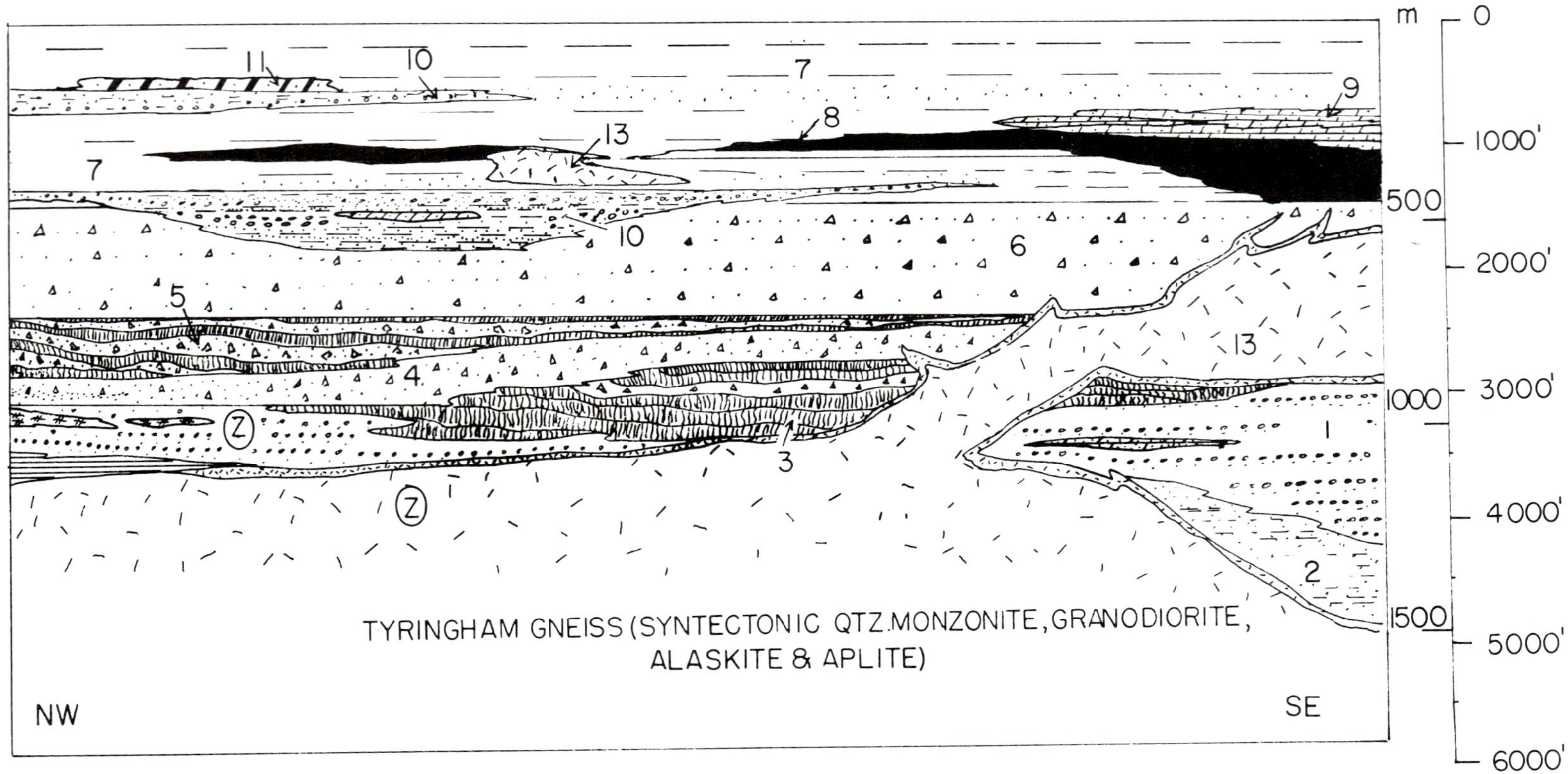

Figure 6. Composite stratigraphic section of the Precambrian rocks in the Stockbridge, Great Barrington, Monterey, and East Lee quadrangles, Massachusetts, showing inferred lateral relationship of lithologic units described in text. Unit 12 not shown, position uncertain. Circled Z shows position of zircon sample sites for U-Th-Pb study. Vertical exaggeration ≃ × 12.

Map unit	Symbol key	Lithology	Probable protolith	Correlation
13		Ferrohastingsite-biotite granite gneiss	Syntectonic granodiorite intrusive rocks	Tyringham Gneiss (Emerson, 1899)
		Alaskite		
11		Garnet-hornblende-plagioclase amphibolite	Basaltic flows or limy shale	Unnamed
10		Rusty-weathering schistose gneiss and granulite	Aluminous and feldspathic sand- stone	Unnamed
		Actinolitic-quartz rock	Iron-rich siliceous dolomite	
9		Layered and spotted hornblende-plagioclase gneiss	Calcareous graywacke and limy shale	Lee Gneiss (Emerson, 1899)
8		Diopside-calc-silicate-chondrodite-calcite or -dolomite marble	Siliceous limestone or dolomite	Hinsdale Limestone (Emerson, 1899)
7		Biotite-quartz-plagioclase gneiss with minor biotite amphibolite	Marine graywacke and shale	Hinsdale Gneiss(?) (Emerson, 1899)
6		Biotite-quartz-magnetite felsic gneiss	Rhyolitic volcanic rocks	Unnamed
5		Garnet-hornblende-plagioclase amphibolite with interlayered biotite gneiss	Basaltic volcanic rocks with interlayered graywacke	Unnamed
4		Leucocratic biotite gneiss and granulite	Rhyolitic volcanic rocks or alaskitic intrusive rocks	Unnamed
3		Garnet-hornblende amphibolite	Basaltic dikes, sills, and flows	Unnamed
2		Rusty-weathering muscovite-biotite gneiss	Sandstone and graywacke	
1		Rusty-weathering blue-quartz gneiss	Quartz-plagioclase conglomerate	Washington Gneiss (Emerson, 1899)
		White garnet-plagioclase-quartz granulite	Dacitic volcanic rocks	
		Pyritiferous diopside-calcite marble	Calcareous sandstone	

Figure 7. Lithologic units, probable protoliths, and correlation of Precambrian rocks shown in Figure 6.

usage is preferred in the light of relationships shown in Figure 2 indicating that the unit does not lie at the border of the Tyringham exclusively, and on the basis of its intertonguing with proven metasedimentary rocks.

Unit 12

Unit 12 consists of light-gray garnet-, biotite-, and hornblende-spotted plagioclase gneiss that is exposed in a fault slice east of Lee (Fig. 2) and has nonfaulted contacts only with the intrusive Tyringham Gneiss (unit 13). The stratigraphic position of unit 12 is therefore, uncertain at present. In lithology, zones with interlayered amphibolite resemble unit 7, and more massive exposures are similar to unit 9.

Tyringham Gneiss: Unit 13

The Tyringham Gneiss (Emerson, 1898, p. 18; 1899, p. 34) was named for exposures in the Lee-Tyringham area; the name specifically referred to the rocks exposed in the core of the Beartown Mountain structure (Emerson, 1899, p. 59). The unit is chiefly light-gray- to pinkish-gray-weathering granite to granodioritic biotite gneiss. The rock has distinctive quartz rodding produced by two or more obliquely intersecting cleavages. Locally the unit contains biotite-granite gneiss with black and pink layers 1 to 20 cm thick that reflect concentrations of microcline and biotite, but commonly it lacks the compositional layering present in other gneiss units. Indistinct color variations result from wispy streaks of brown biotite. Even in the most severely deformed rocks, quartz is uniformly distributed and not concentrated in layers, unlike quartz in other gneiss units.

Individual crystals of perthitic microcline as large as 0.5 cm in diameter are rimmed with or cut by veinlets of granular oligoclase. The rock is characterized by grains of yellow-brown allanite with pale-green to colorless rims of epidote and by abundant scattered zircon grains. The granodiorite variety contains plagioclase grains, zoned regularly from calcic to sodic oligoclase, and perthitic microcline without oligoclase rims or veinlets. The strong granulation along intersecting foliations has nearly destroyed the original texture of the rock (Fig. 8). The mica curves around the larger augen of microcline perthite crystals; the original feldspar shows all degrees of granulation.

Ferrohastingsite and hastingsitic amphibole are common mafic minerals in much of the Tyringham. These amphiboles are known from Precambrian granodiorite gneiss in the western Adirondacks (Buddington, 1957), from late syntectonic syenitic rocks in the Reading Prong (Young, 1971), from rocks of the Pelham dome of central Massachusetts (P. Robinson, 1971, personal commun.), and from the Yonkers Gneiss of the Manhattan Prong (Fig. 1, loc. 2; L. M. Hall, 1971, personal commun.).

Thin, discontinuous, white-weathering, fine-grained, biotite-poor alaskitic (Figs. 9, 10) or aplitic rock is developed locally at the border of the Tyringham, in which accessory magnetite is an important constituent. The unit is too thin to be represented on the map (Fig. 2) but is of great significance because it forms dikes that crosscut the Washington Gneiss. The rock grades imperceptibly into the coarse biotite augen rock by coarsening of grain size of both the groundmass and the microcline and by addition of biotite. This rock is interpreted to be a border facies of the granodiorite magma. Similar alaskite forms the border of the ferrohastingsite granite of the Adirondack highlands (Buddington and Leonard, 1962) and the border of the Storm King Granite of the Hudson Highlands (Lowe, 1950).

Microscopic point count data and staining of rock slabs show that the typical

Figure 8. Photomicrograph of typical Tyringham Gneiss from Beartown Mountain near zircon sample site, showing granulation of a large microcline crystal; microcataclastic foliation slopes from upper right to lower left and consists of granulated microcline, biotite, quartz, and allanite and oriented muscovite and biotite. Section perpendicular to lineation formed by quartz rods and long axis of porphyroclasts of microcline. This granulation and the lineation are Precambrian in age. Crossed polarizers.

Tyringham is a quartz monzonite in composition, with the quartz content ranging between 20 and 40 percent by volume. The composition of the border facies varies greatly from one locality to another and may range from granite to granodiorite. The percentage of ferrohastingsite varies greatly and seems to bear an antipathetic relationship to magnetite and muscovite. Euhedral zircon rarely amounts to more than a few tenths of 1 percent. The zircon forms unzoned crystals with width-to-length ratios ranging from 1:1.5 to 1:5 (Fig. 11A); the crystals do not appear to have inclusions. Abundant elongate zircon (width-to-length ratio of 1:2.4) is found in the alaskitic border facies of the pluton on Beartown Mountain (Fig. 11b).

The Tyringham is found in various structural positions. In the northern part of the Monterey quadrangle and the southern third of the East Lee quadrangle, it forms a sill 457 m thick between unit 6, the biotite-magnetite-felsic gneiss unit, and unit 1, the Washington Gneiss (Fig. 2). Chill zones are preserved at both contacts. These intrusive relationships of the Tyringham are indicated diagrammatically in Figure 6. East of Lee, the Tyringham is in contact with unit 7, the biotite-quartz-plagioclase paragneiss. In this same area, the foliation that produced the strong granulation and augen structure is related to a set of isoclinal folds shared by the adjacent Precambrian country rocks. This foliation and old fold set are unconformably overlain north of Tyringham by the Dalton Formation of late Precambrian(?) and Early Cambrian age (Ratcliffe, 1969a).

At a few localities where the contact is exposed, the Tyringham can be seen to crosscut one foliation and subparallel compositional layering in the country rocks, but both the country rocks and the Tyringham share the same strong

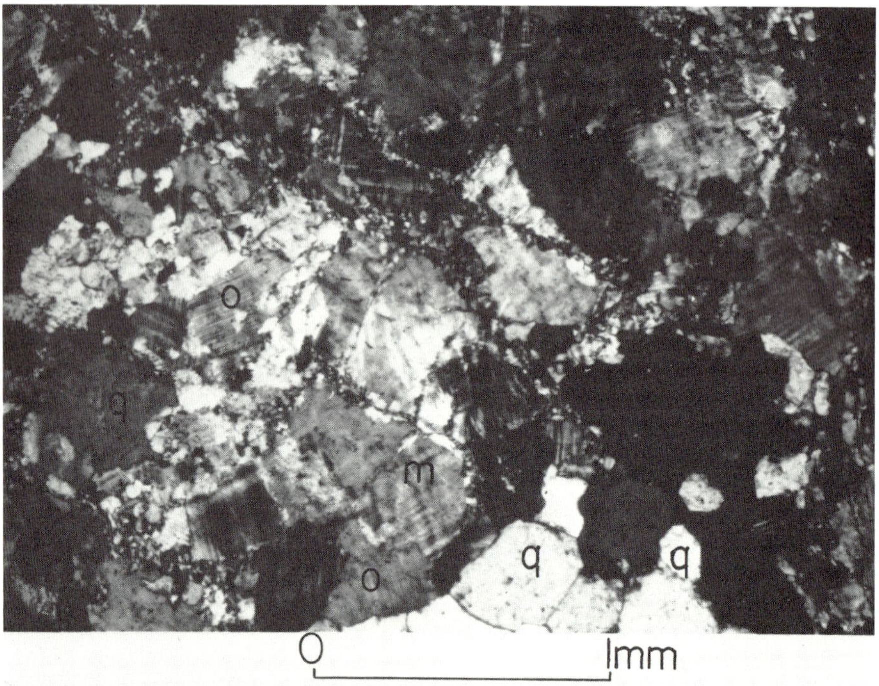

Figure 9. Photomicrograph of fine-grained Tyringham Gneiss taken 61 m from contact with unit 1 of Washington Gneiss in southeastern corner of East Lee quadrangle. Microcataclastic texture is present along granulated zones that slope from upper left to lower right; these zones contain biotite oriented in axial planes of Precambrian folds. Note that texture is finer grained than in Figure 8. Dominant minerals are microcline perthite (m), quartz (q), and oligoclase (o). Crossed polarizers.

postintrusive foliation and granulation. The details of the observed crosscutting relationships, granitic textures, and modal mineralogy indicate that the Tyringham Gneiss is a metaintrusive rock that postdated some deformation of the paragneiss and was deformed by isoclinal folding and penetrative deformation prior to deposition of the Dalton Formation.

Summary of the Precambrian Stratigraphic Relationships of Paragneiss and Orthogneiss

Figure 6 shows the interrelationships of some of the Precambrian units. The composite section incorporates data taken from localities from Beartown Mountain east-southeast of the eastern edge of the Monterey quadrangle. The diagram shows a lower sequence of predominantly coarse, quartz-rich clastic rock and basaltic to dacitic volcanic rock corresponding to the Washington Gneiss (units 1 and 2) and the amphibolite of (unit 3), overlain by mixed felsic and mafic volcanic rock (units 4 and 5). Above this is a rather continuous felsic volcanic unit (unit 6). The lower sequence of units 1 through 6 may represent shallow-water and possibly continental deposits that were succeeded by rhyolitic volcanic activity.

The biotite-quartz-plagioclase paragneiss sequence (units 7 through 11) represents a major break in the sedimentary environment that probably resulted from the development of widespread but shallow-water marine conditions. Calc-silicate rock (unit 8) is an important but nonpersistent unit.

The Tyringham Gneiss is diagrammatically shown as crosscutting all the major units.

U-TH-PB ZIRCON DATA

Zartman made U-Th-Pb isotopic age analyses on zircon from samples of the Tyringham Gneiss and of unit 1 of the Washington Gneiss from Beartown Mountain, both of which contain rather abundant zircon. The Tyringham Gneiss samples were collected from an outcrop at bench mark 1363 along Beartown Mountain Road in the Stockbridge quadrangle. The washington Gneiss samples were collected at the 1870-ft elevation on Wildcat trail, 850 m northwest of Benedict Pond in the Great Barrington quadrangle. Both the locations of the sample sites, in Figure 2, and the approximate stratigraphic positions of the rocks sampled, in Figure 6, are shown by the letter Z.

A morphologic examination of the zircon from the Tyringham Gneiss reveals an elongate, euhedral habit with no evidence of relict cores or later overgrowths

Figure 10. Photomicrograph of alaskitic border of Tyringham Gneiss taken approximately 6.1 m from contact with unit 1 of the Washington Gneiss, same locality as Figure 9. Crossed polarizers. Foliation, defined by muscovite oriented from upper left to lower right, is same as microcataclastic foliation in Figure 9. In outcrop this foliation crosses intrusive contact with Washington Gneiss. In other exposures foliated Tyringham Gneiss is unconformably overlain by Dalton Formation. Dominant minerals are grid-twinned microcline, quartz (q), and muscovite (mc). Decreasing grain size and textural changes commonly found in the Tyringham Gneiss going from the interior to the walls of the pluton are illustrated by the sequence of Figures 8, 9, and 10. The finer grained rocks (Figs. 9, 10) are interpreted as the chilled border.

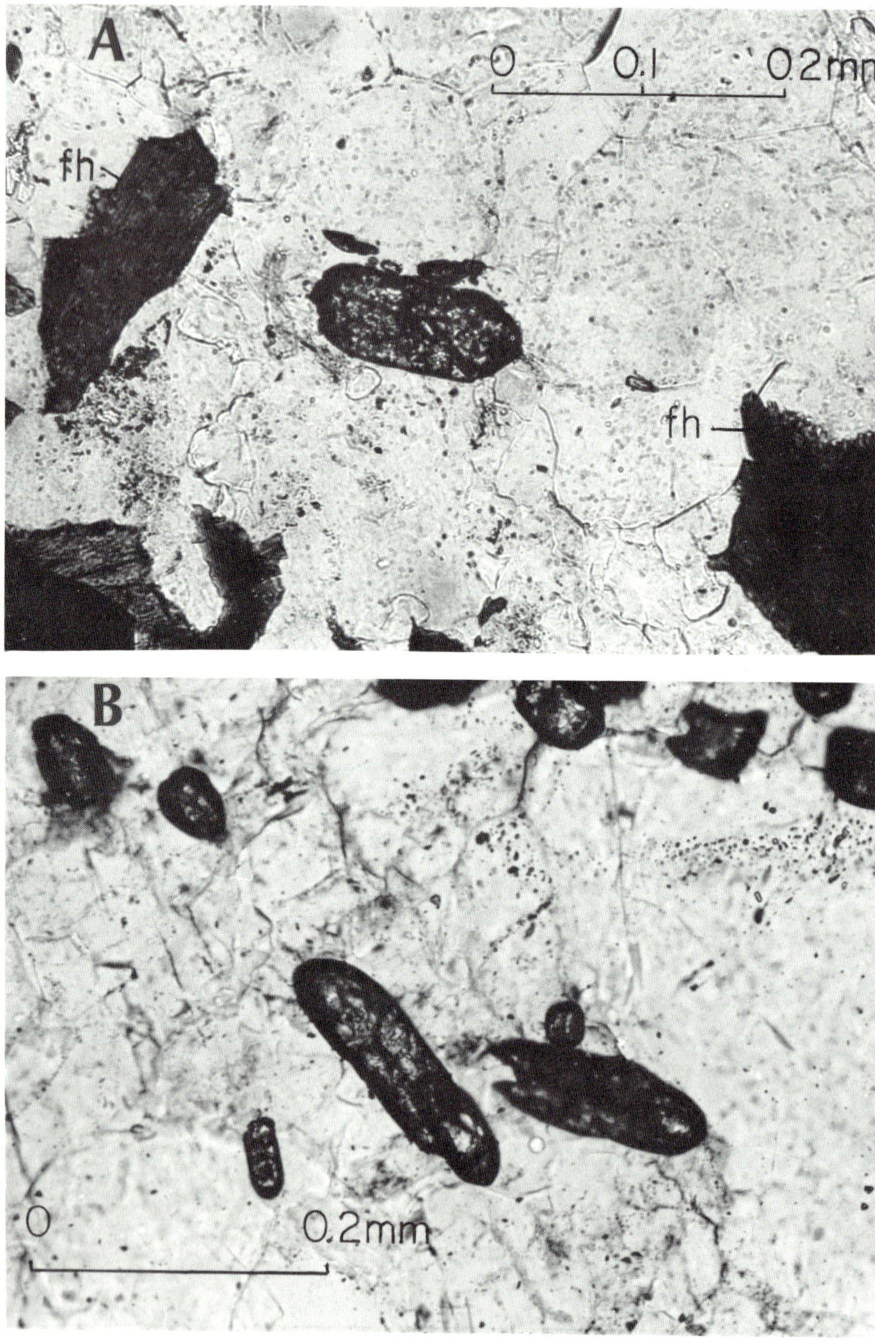

Figure 11. Photomicrographs of zircon from Tyringham Gneiss as seen in thin section, plain light. A, Zircon in fine-grained ferrohastingsite-bearing Tyringham Gneiss, from same locality as Figure 9; fh = ferrohastingsite. B, Zircon from alaskitic border facies on Beartown Mountain. Sample taken approximately 15.1 m from contact with unit 1 of Washington Gneiss. Elongate zircon grains are aligned in microcataclastic foliation that slopes from upper left to lower right, as shown in Figures 8, 9, and 10. Elsewhere in thin section prisms are distorted by zones of granulation.

(Fig. 12). We interpret the mineral to be a primary crystallizing phase of the granite, although there is a possibility that it was affected during subsequent high-grade metamorphism of the rock. The distorted prism faces on some of the grains suggest that the zircon predates at least one period of deformation that affected the rock. By contrast, zircon from the Washington Gneiss shows a much more stubby habit, with many grains appearing to be rounded and with overgrowths faintly developed over the cores (Fig. 13). Heterogeneity in color and clarity may indicate a mixed zircon population that might be expected in a paragneiss of sedimentary and volcanic origin. In thin section, the zircon is seen to occur both in the groundmass of the rock, where it is associated with obvious detrital quartz, and in mafic clots derived from retrogression of garnet metacrysts (Figs. 4, 5).

Several size fractions of zircon were analyzed for each rock, but only minor variations in concentration and parent:daughter ratio were obtained. The zircon was decomposed by sintering with borax, and standard ion-exchange and dithizone extraction techniques were used for chemical purification. All U, Th, and Pb concentrations were obtained by isotope dilution. The analytical data and resultant isotopic ages are given in Table 1. Accuracy limits attached to the ages reflect analytical uncertainty at the 67 percent confidence level, but they do not include possible error in the assumed isotopic composition of the common Pb. Because the Pb in the zircon is highly radiogenic, the ages will not be changed significantly by other reasonable choices of common Pb. The ages generally show only mild internal discordancies, and most diffusion models used to interpret isotopic systematics would give a true age approximately equal to that of the Pb^{207}/Pb^{206} value. When plotted on a concordia diagram (Fig. 14), the U-Pb data are compatible with an age of 1,040 to 1,080 m.y. B.P. for both the Tyringham Gneiss and the Washington Gneiss. This agreement could mean either that (1) little time separates the age of the source material for the Washington Gneiss and the intrusion of the Tyringham Gneiss, (2) the Tyringham Gneiss was intruded during the same dynamothermal event that reset the Washington Gneiss zircon, or (3) the zircons in both the Tyringham and Washington Gneisses mainly record a metamorphic age. Any interpretation of the uniform U-Pb behavior, however, is further complicated by the considerably older Pb^{208}/Th^{232} ages for the Washington Gneiss. Apparently, either the Th-Pb system in this zircon is more controlled by old detrital material than the U-Pb system is, or else it has responded less strongly to later metamorphic disturbance than the U-Pb system has.

If the dacitic volcanic debris in the Washington Gneiss had been derived during a single period of volcanism that culminated in the intrusion of a Tyringham Gneiss, it would not be unreasonable to expect at least some of the zircon in the paragneiss to be of essentially the same age as that in the orthogneiss. This is rather unlikely, however, because the well-rounded character of many of the zircon grains from the Washington Gneiss is suggestive of extreme abrasion, and the geologic data suggest that the Tyringham was a deep-seated syntectonic intrusion rather than shallow-level pluton emplaced into its sedimentary-volcanic cover. Therefore it seems reasonable that the detrital zircon in the Washington may have had its isotopic systems reset during the same Precambrian dynamothermal event in which the Tyringham was intruded. Grauert and Arnold (1968) and Pidgeon and others (1970) have shown in detailed studies that sillimaite-bearing rocks from the Swiss alps contain zircon suites that have been severely isotopically disturbed by metamorphism. They were able to recognize Precambrian apparent ages for detrital zircon in paragneiss even though the isotopic ages had been greatly reduced by intense early Paleozoic metamorphism. Pidgeon (1969) also found detrital zircon grains in the Dalradian rocks of western Ireland that had lost significant amounts of

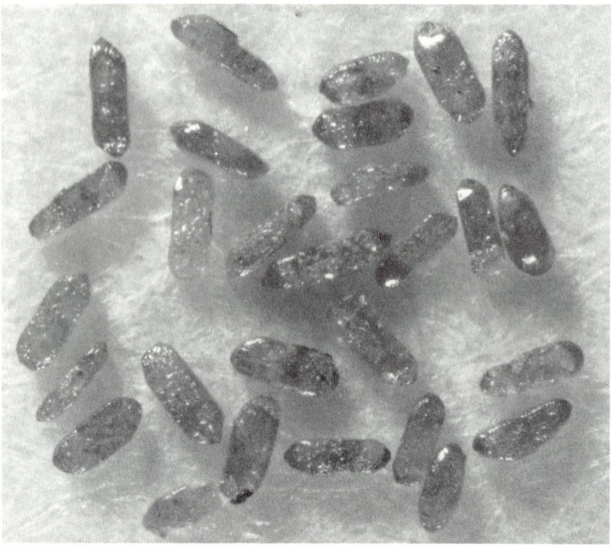

Figure 12. Representative zircon grains from Tyringham Gneiss. Note elongate, euhedral habit. Minor distortion of some prism faces attributed to strain dislocations arising during deformation of the rock. Magnification x50.

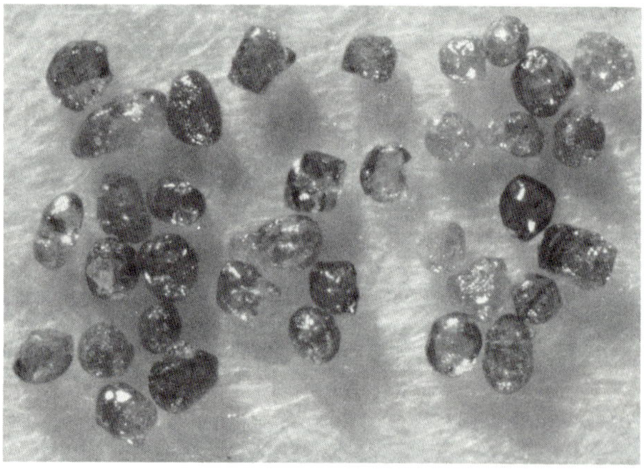

Figure 13. Representative zircon grains from Washington Gneiss. Note generally stubby habit and many rounded grains, indicative of a detrital origin, and faintly developed overgrowths. Heterogeneity in color and clarity are evidence of a mixed population. Magnification x50.

radiogenic Pb during subsequent metamorphism.

Any interpretation of the U-Pb behavior is further complicated by the considerably older Pb^{208}/Th^{232} ages for the Washington Gneiss. These anomalous Pb^{208}/Th^{232} ages provide evidence for an isotopic system inherited from a terrane distinctly older than 1 b.y. B.P. If such a source has also contributed significantly to the U-Pb system, the U-derived radiogenic Pb must have been almost completely expelled at the time of the 1,040- to 1,080-m.y.-old disturbance. It is important to note that neither the geologic relationships nor the U-Th-Pb data support the contention that the Tyringham Gneiss represents pre-Grenville basement rock from which the Washington detritus was derived.[2]

[2]Recent Rb-Sr whole-rock data from the Tyringham Gneiss in the Stockbridge and East Lee quadrangles yield an isochron age of about 1,200 m.y., on the basis of the Rb^{87} decay constant of 1.39×10^{-11} yr^{-1} (D. Mose, 1974, written commun.). This result suggests that the Tyringham and Washington U-Pb zircon ages may both have been affected by metamorphism detectably younger than the time of intrusion of the Tyringham. Discrepancies of such magnitude have been found elsewhere in deeply eroded Precambrian terranes and may represent inconsistencies of retentivity among the dating systems during prolonged orogenic episodes. The minor disagreement, however, does not detract substantially from the main conclusions of this paper.

TABLE 1. U-Th-Pb ISOTOPIC AGES OF ZIRCON FROM THE BERKSHIRE MASSIF

	Concentration (ppm)			Isotopic composition of lead (atom percent)				Age (m.y.)			
Mesh size	U	Th	Pb	Pb^{204}	Pb^{206}	Pb^{207}	Pb^{208}	Pb^{206}/U^{238}	Pb^{207}/U^{235}	Pb^{207}/Pb^{206}	Pb^{208}/Th^{232}
					Tyringham Gneiss						
−150 + 200	1,378	249.5	208.0	0.013	87.71	6.683	5.598	930 ± 10	969 ± 11	1,062 ± 8	959 ± 11
−250 + 325	1,264	217.6	191.2	0.006	88.13	6.570	5.292	937 ± 10	970 ± 10	1,048 ± 6	1,002 ± 10
					Washington Gneiss						
−100 + 150	743.9	54.8	115.8	0.032	89.37	6.997	3.603	970 ± 11	991 ± 12	1,040 ± 11	1,154 ± 15
−150 + 200	757.3	57.0	121.5	0.031	89.33	7.033	3.608	997 ± 11	1,016 ± 12	1,057 ± 11	1,185 ± 15
−250 + 325	824.4	59.2	129.8	0.026	89.49	7.043	3.443	983 ± 11	1,012 ± 12	1,078 ± 9	1,228 ± 13

Note: Constants: U^{238}:λ = 1.54×10^{-10} yr^{-1}; U^{235}:λ = 9.72×10^{-10} yr^{-1}; Th^{232}:λ = 4.88×10^{-11} yr^{-1}; U^{238}/U^{235} = 137.7. Isotopic composition of initial lead assumed to be Pb^{204}:Pb^{206}:Pb^{207}:Pb^{208} = 1:17.0:15.5:36.8.

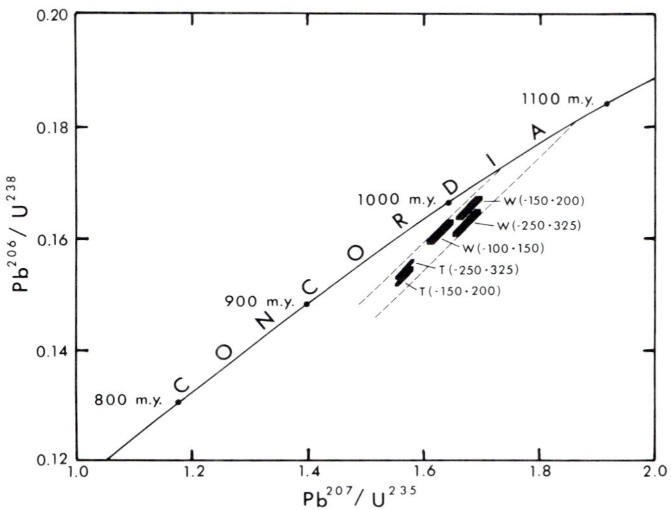

Figure 14. Concordia diagram for zircons from Tyringham (T) and Washington (W) Gneisses from the Berkshire massif of western Massachusetts. Dashed lines delimit Pb^{207}/Pb^{206} age range of all samples.

Despite the problems in explaining details in age of the two units, the radiometric data leave little doubt concerning the Precambrian nature of this terrane and are consistent with the idea of the occurrence approximately 1,050 m.y. ago of a metamorphic and structural event in the Berkshire massif similar to that recognized elsewhere in Precambrian rocks of the Adirondacks (Tilton and others, 1957; Hills and Gast, 1964; Silver, 1964, 1968; Zartman and Wasserburg, 1969), Green Mountains (Faul and others, 1963; Harper, 1968; Cady, 1969), and Reading Prong (Long and others, 1959; Tilton and others, 1960; Long and Kulp, 1962; Ratcliffe and others, 1972). Although the rocks within the Berkshire massif were strongly deformed again during Paleozoic tectonism, the isotopic data for zircon do not show any obvious effects of those events.

BERKSHIRE PRECAMBRIAN ROCKS COMPARED WITH THE PRECAMBRIAN ROCKS OF THE ADIRONDACK MOUNTAINS AND THE READING PRONG

Paragneisses of the Adirondacks and Reading Prong are similar to the rocks of the Berkshires in many important aspects. In Table 2, rocks present in these areas are described and their origins are inferred. Chronostratigraphic or lithostratigraphic correlations are not implied in the table.

There are many similarities in rock types among the Reading Prong sections in Table 2 although the detailed work of Young (1971) in an area of excellent exposure has pointed up the difficulty of correlating units between adjacent structural blocks in the Reading Prong. The predominance of quartzose feldspathic paragneiss and local but distincitive rusty-weathering graphitic, pyritic, and sillimanitic schist and amphibolite is important. Marble and calc-silicate rocks are associated with various other rocks and are scattered throughout the sections. Metavolcanic rocks are represented by oligoclase-quartz-rich gneiss and amphibolite.

Intrusive rocks in the Reading Prong are largely hornblende or biotite granitic gneiss that is thought to be syntectonic (Drake, 1969; Dodd, 1965; Young, 1971) or post-tectonic, as in the case of the Storm King Granite at Bear Mountain (Lowe, 1950). Pretectonic meta-anorthositic gabbro (Young, 1971) and mafic pyroxene diorite (Drake, 1969) are found locally. Post-tectonic Precambrian dioritic to monzonitic plutons with associated spessartitic lamprophyre dikes are found locally along the extension of the Ramapo border fault in New York State (Ratcliffe, 1970, 1971).

If the Pb^{207}/Pb^{206} age of 1,060 m.y. for zircon from the Storm King Granite at Bear Mountain indicates a minimum age for crystallization of that rock (Long and Kulp, 1962, p. 986), then a minimum age for the Precambrian dynamothermal event in the Hudson Highlands is obtained. Concordant U-Th-Pb ages for zircon from the Canada Hill Gneiss (hypersthene-quartz-oligoclase gneiss of Dodd, 1965) of about 1,150 m.y. suggest metamorphism of the country rocks at about that time (Tilton and others, 1960). Likewise, the 1,023 ± 45 m.y. Rb/Sr whole-rock isochron (Rb^{87} decay constant = 1.47×10^{-11} yr.$^{-1}$) for the postmetamorphic Canopus pluton that intrudes Grenville gneiss of the Hudson Highlands confirms an assignment of a 1.0- to 1.1-b.y. age for a dynamothermal event in the Hudson Highlands if the lower Rb^{87} decay constant is also applied. The present data from the Hudson Highlands near the Hudson River suggest a polydeformational Precambrian history (Ratcliffe and others, 1971) involving intense dynamothermal metamorphism to hornblende granulite-grade F_1 folding 1.15 b.y. ago, followed by F_2 folding, strike-slip faulting, and dioritic to granitic plutonism at about 1.0 to 1.1 b.y. ago. We conclude, within the limits of analytical error and theoretical uncertainty, that the Precambrian dynamothermal event in the Hudson Highlands occurred at about 1.0 to 1.1 by. ago. The existence of basement rocks with metamorphic or intrusive ages older than this has not yet been demonstrated, although the 1.5-b.y. age for the Rb/Sr whole-rock isochron (Rb^{87} decay constant = 1.39×10^{-11} yr^{-1}) for granitic gneiss near Peekskill, New York, suggests this possibility (Mose and Hayes, 1974).

The Precambrian sequence of the Adirondacks contains metaigneous granitic, anorthositic, and charnockitic gneiss as well as a sequence of paragneisses (Buddington and Leonard, 1962; de Waard and Walton, 1967). According to de Waard and Walton (1967), the paragneiss sequence (Table 2) in the eastern and south central Adirondacks represents a younger cover sequence deposited unconformably on the older basement rocks. Locally this unconformity is marked by a marble and calc-silicate unit. The paragneiss stratigraphic section includes amphibolite and calc-silicate rocks at several horizons (Table 2), graphitic aluminous schist, graphitic quartzite, massive amphibolite of probable intrusive origin, and probable meta-acidic volcanic rocks. The entire sequence was deformed during the last intense metamorphic phase about 1,100 m.y. ago, according to zircon ages (Silver, 1964, 1968). De Waard and Walton's (1967) interpretation is radically different from that of Buddington (1957, p. 61-62), who held that the Adirondack anorthosite and other igneous rocks, including biotite-ferrohastingsite granitic gneiss with border alaskite, intrude the paragneiss. Although it is not within the scope of this paper to judge the validity of either concept, it is important to note that the Berkshire Precambrian sequence described here resembles in major rock types and approximate total thickness the Grenville paragneiss sequence described by Buddington and by de Waard and Walton. Because the Tyringham Gneiss has crosscutting relationships with the layered gneiss (Fig. 2), the Buddington model for the age of the granite gneiss fits more closely the observations in Massachusetts.

Rusty-weathering graphitic quartzite and quartz conglomerate that might mark unconformities are found in two stratigraphic positions in the Monterey area: they are the Washington Gneiss (units 1 and 2) and the rusty schistose gneiss and granulite (unit 10). Similar rocks in the Grenville metasediment of the eastern Adirondacks (deWaard and Walton, 1967, Fig. 4) and in the southern Green Mountains of Vermont (Skehan, 1961, Fig. 5, Pl. I) are thought to mark important unconformities. Because calc-silicate rocks, marble, and rusty-weathering quartzose metasedimentary rocks are found throughout the eastern Adirondacks (de Waard and Walton, 1967, Fig. 4) as well as in the Berkshires, the presence of such rocks does not everywhere indicate unconformities. No evidence has been found thus far that suggests the calc-silicate rocks (unit 8) in the Berkshires have an angular relationship with other units. None of the calc-silicate rock units in the Berkshires begins to approach the great thickness of such rocks in the Adirondacks.

The late ferrohastingsite-bearing granite and granodiorite from the Adirondacks, described by Buddington (1957) and Buddington and Leonard (1962), and the hornblende granite gneiss of the Reading Prong both resemble the Tyringham Gneiss in mineralogy, texture, and inferred age.

COVER ROCKS: DALTON FORMATION AND CHESHIRE QUARTZITE

Upper Precambrian(?) and Lower Cambrian metaquartzite, meta-arkosic sandstone, biotite-muscovite schist, and biotite-quartz-plagioclase schist and gneiss, as well as quartz-pebble and quartz-cobble conglomerate form the unconformable sedimentary cover of the Berkshire massif in Massachusetts. The basal feldspathic

TABLE 2. LITHOLOGIC SEQUENCES AND PROTOLITHS OF PRECAMBRIAN ROCKS FROM THE READING PRONG, ADIRONDACK MOUNTAINS, AND BERKSHIRE MASSIF

Delaware Valley, N.J.-Pa. (Drake, 1969)	Hopatcong Area, N.J. (Young, 1971) Fig. 1, loc. 1		Popolopen Lake quadrangle Hudson Highlands (Dodd, 1965) Fig. 1, loc. 3
Hornblende granite, microperthite alaskite, and biotite granite (mostly intrusive)	Block C (cumulative thickness, 2,200 m)	Block D (cumulative thickness, ~1,850 m)	Hornblende granite, leucogranite, and biotite granite (syntectonic intrusive)
Microantiperthite alaskite (intrusive)	Clinopyroxene and plagioclase amphibolite (basalt)	Diopside-scapolite-feldspar quartz gneiss (calcitic sandstone)	Plagioclase-quartz leucogneiss (volcanic rocks)
Albite-oligoclase granite and related albite pegmatite (rheomorphic, probably anatectic, with attendant granitization)	Clinopyroxene, mesoperthite granitic and syenitic gneiss with thin amphibolite interlayers (intrusive granite and syenite)	Biotite-feldspar-quartz gneiss (feldspathic sandstone)	Hypersthene-quartz-oligoclase gneiss with thin beds of calcite marble (volcanic and sedimentary rocks)
Quartz diorite (plutonic metamorphism of sodic tuff and mafic volcanic rock)	Biotite-quartz-feldspar gneiss (feldspathic sandstone)	Biotite-plagioclase gneiss (anorthosite sheet)	Gray migmatitic biotite-quartz-feldspar gneiss. Minor beds of garnet-pyroxene gneiss (sedimentary rocks)
Oligoclase-quartz gneiss (volcanic rocks)	Clinopyroxene, mesoperthite granitic and syenitic gneiss (intrusive granite and syenite)	Biotite-feldspar-quartz gneiss with tongues of hornblende granitic gneiss near base (feldspathic sandstone with sills of syntectonic hornblende granite)	Rusty-weathering pyritic biotite-quartz-feldspar gneiss with quartzitic, biotite-rich and graphitic interbeds, sillimanite-biotite schist with lavender garnet and minor beds of calcite marble and calc-silicate rock (sedimentary rocks)
Quartz-feldspathic gneiss, including biotite-quartz-plagioclase gneiss, sillimanite-bearing gneiss, and potassium feldspar gneiss (sedimentary rocks)	Clinopyroxene- and plagioclase-bearing amphibolite (basalt intruded by clinopyroxene syenite)	Hornblende-quartz-oligoclase gneiss (acidic volcanic rocks)	Hornblende-plagioclase-pyroxene amphibolite and hornblende plagioclase gneiss (volcanic rocks)
Marble and its alteration products, including skarn (sedimentary rocks)	Clinopyroxene, mesoperthite granitic and syenitic gneiss (intrusive granite and syenite)	Hornblende granitic gneiss (syntectonic hornblende granite and syenite)	
Amphibolite (igneous, volcanic, and sedimentary rocks)			
Pyroxene gneiss (sedimentary rocks)			
Migmatite: microperthite alaskite neosome in amphibolite paleosome, albite-oligoclase granite neosome in amphibolite paleosome			

Hudson Highlands north of Hudson River (Prucha and others, 1968)	Mountain Lake SW 7-1/2 quadrangle Adirondack paragneiss sequence (de Waard, 1971, written commun.)	Beartown Mountain–Monterey-Tyringham area Berkshire Massif (this paper)	
Quartz-microcline-plagioclase gneiss with clinopyroxene rimmed by hornblende, and foliated pyroxene hornblende amphibolite member (Mahopac Gneiss) (sedimentary?) Hornblende-biotite-quartz-plagioclase gneiss with intercalated marble and calc-silicate rock and sills of pink microcline-perthite-biotite granite (sedimentary rocks impregnated with granite) Rusty-weathering, pyritic biotite-quartz-plagioclase gneiss with minor layers of calc-silicate rock (sedimentary rocks) Biotite-microcline granitic gneiss with minor layers of biotitic and pyroxenic amphibolite (syntectonic to late tectonic intrusive) Pyroxene-quartz-plagioclase gneiss with abundant beds of amphibolite and, locally, quartzite (sedimentary and volcanic rocks) Biotite-hornblende (pyroxene) granitic gneiss with microcline perthite and abundant interlayers of amphibolite (syntectonic granitic intrusive rocks?) Coarse-grained migmatitic pink microcline granite, hornblende-biotite-quartz plagioclase gneiss and local amphibolite (remobilized granitic gneiss?)	Orthopyroxene-bearing quartz-microperthite charnockite and pink hornblende-quartz-microperthite gneiss interlayered with amphibolite, quartzite, calc-silicate rock and one layer of leucocratic K-feldspar-plagioclase-quartz granulite (>915 m thick) Upper marble, coarse-grained calcite marble and layers of amphibolite, calc-silicate rock, and quartzite, with sheet of metagabbro and garnet-bearing leucocratic K-feldspar gneiss at base (273 to 1,210 m thick) Leucocratic gneiss with alternating layers of amphibolite, local calc-silicate rocks and an amphibolite–calc-silicate 46 m thick (total thickness 273 to 820 m) Lower marble, coarse-grained calcitic marble with subordinate layers of amphibolite, and quartzite, plagioclase gneiss, and calc-silicate rocks (91 to 273 m thick) Unconformity Basement complex charnockitic gneiss, meta-anorthosite, and metagabbro	Biotite-microcline microperthite-oligoclase-quartz-ferrohastingsite metagranodiorite, meta-quartz monzonite, and meta-alaskite, Tyringham Gneiss (unit 13) (syntectonic intrusive) Biotite-quartz-plagioclase paragneiss (unit 7) with amphibolite (unit 9), metaquartzite calc-silicate rock, and minor marble (unit 8) (graywacke, limestone, and limy shale) Rusty-weathering muscovite-biotite-feldspar schist and gneiss (unit 10) with minor quartz-pebble conglomerate and actinolitic calc-silicate rock and amphibolite (unit 11) (sandstone, conglomerate, basaltic volcanic rocks) Biotite-microcline-plagioclase-quartz felsic gneiss (unit 6) (acidic volcanic rocks) Hornblende-garnet-plagioclase amphibolite and amphibolitic gneiss (unit 5) (basic volcanic rocks) Leucocratic biotite-poor microcline-plagioclase-quartz gneiss (unit 4) (acidic volcanic rocks)	Hornblende-garnet and hornblende-garnet-plagioclase amphibolite (unit 3) interlayered with rusty graphitic blue quartz gneiss below (flow basalt) Rusty-weathering, pyritic, graphitic blue quartz-pebble conglomerate, schist and granulite (units 1 and 2), with interlayers of blue quartz-garnet-plagioclase granulite and gneiss (dacite), amphibolite (basalt), and minor lenses of pyritic, scapolitic, or actinolitic calc-silicate rock (dolomitic sandstone), Washington Gneiss (shallow-water lagoonal and evaporitic terrestrial deposits)

unit, the Dalton Formation, grades laterally and vertically into clean metaquartzite, the Cheshire Quartzite, that in turn passes upward into relatively quartz-free shelf carbonate, the Stockbridge Formation.

Most of the Dalton-Cheshire sequence in southwestern Massachusetts occurs in thrust slices that override the autochthonous carbonate rocks; only rarely are the Cheshire and Dalton exposed in the autochthonous belt. Therefore, a meaningful analysis of the original facies distribution of those rocks requires palinspastic restoration of the basement slices shown in Figure 2. By "unrolling" the folded thrust sheets and tracing the parautochthonous rocks eastward to their source, the sections have been restored to their original relative positions (Figs. 15, 16; see footnote 1). The original horizontal separation of the sections is difficult to judge and its depiction has not been attempted in the figures. Six stratigraphic sections illustrating lithostratigraphic units in the Dalton-Cheshire sequence are shown in Figures 15 and 16. Sections A, B, and C are roughly aligned east to west, whereas D, E, and F are arranged north to south. Sections A and B are considered to be relatively autochthonous exposures, whereas C through F are from thrust sheets that have moved westward a minimum of 21 km across the autochthon. Section C is a composite taken from exposures of the Dalton-Cheshire on Monument Mountain and on Beartown Mountain. The original site of deposition of these rocks was east of the present position of exposures of section B (Fig. 3, Sec. B-B′).

Autochthonous sections of the cover rock sequence are exposed at Vossburg Hill in the Egremont and Great Barrington quadrangles and at Leffingwell Hill in the northeast corner of the Ashley Falls quadrangle. Since these are the two most complete autochthonous sections yet discovered in southwestern Massachusetts, they are important stratigraphic criteria.

The Vossburg Hill section (Fig. 15, sec A) exposes approximately 91 m of massive and well-bedded vitreous Cheshire Quartzite. At its base the Cheshire appears to be in sharp contact with feldspathic quartzite of the Dalton Formation. The Dalton here is a muscovite feldspathic quartzite spotted with flecks of black biotite and megacrysts of black tourmaline. Although the base of the section is not exposed, approximately 61 m of Dalton is exposed. The Cheshire has a gradational contact at its top with white dolomitic marble (unit a) of the Stockbridge (Zen and Ratcliffe, 1971). Thin vitreous quartzite 3 to 7 cm thick is found interlayered with Stockbridge unit a approximately 16 m above its contact with the Cheshire along strike to the north of Vossburg Hill.

The Leffingwell Hill section (Fig. 15, sec. B) is exposed in an arcuate anticline of autochthonous basement rocks to the east of Vossburg Hill. Local detachment of the Paleozoic rocks from the basement is recognized (Fig. 2; Ratcliffe and Burger, 1975), and therefore the stratigraphic section is a composite one. At the southern, plunging end of this anticline, a normal gradational contact between the Cheshire Quartzite and unit a of the Stockbridge is exposed. The entire Dalton-Cheshire sequence may be less than 150 m thick here. The Cheshire varies from 0 to approximately 6 m in thickness. On the western flank of the anticline, at Umpachene Falls (Fig. 2), the Cheshire Quartzite is approximately 3 m thick and has an apparently gradational contact at its base with a pale, greenish-gray-weathering, biotite-muscovite schistose quartzite and metaconglomerate unit. This unit rests on gneiss units 1 and 13 and at its base consists of dark-gray to silvery-gray muscovite, green biotite schist with white subrounded clasts 1 to 30 mm in diameter. The clasts are granitic gneiss fragments, plagioclase, metamorphic quartz, and microcline. A photomicrograph (Fig. 17) shows detrital texture. Locally the greenish-gray schist contains K-feldspar-rich pseudopegmatitic layers that may represent pods of reconstituted arkosic conglomerate. Similar arkosic beds 0.35 to 1.5 m thick are common in the basal Dalton at other localities such as Day Mountain (Fig. 16, sec. D) and Felton Lake (Fig. 16, sec. E), as well as locally at Tyringham (Fig. 16, sec. F).

At the northern end of Leffingwell Hill, schistose feldspathic quartzite, quartz schist, and fine-grained conglomerate typical of the bulk of the Dalton underlie the Stockbridge (Fig. 15, sec. B). These observations suggest thinning of the Cheshire and Dalton to the east of Vossburg Hill. The local absence of the Cheshire suggests that this area may have been a basement high during Early Cambrian deposition.

The rocks of section C (Fig. 15) now are found in thrust sheets overlying the Stockbridge Formation and have been restored in Figure 15 to a position east of the Leffingwell Hill site. This reconstruction is based on cross-sections A-A′ and B-B′ in Figure 3. The root zone for the Dalton-Cheshire sequence now exposed on East Mountain is identified as the easternmost part of the Lake Buel slice, shown at the east side of cross-section B-B′. The Dalton-Cheshire at East Mountain is approximately 671 m thick; the upper 213 m is largely vitreous quartzite assigned to the Cheshire Quartzite. Aside from the great thickness of this sequence, the presence of a biotite feldspathic schist unit ($€_{DS}$, Fig. 15, sec. C) and of gray plagioclase microcline-rich granulite and schistose quartzite at the base is significant. The schistose facies ($€_{DS}$) closely resembles rocks mapped within the Hoosac Formation to the north (Norton, 1969, this volume) and are thought to represent an eastern deeper water facies of the Dalton. It is significant that these Hoosac-like rocks are thickest in section C (Fig. 15), which is regarded independently, on the basis of structural considerations, to be the most easterly deposited rocks of the Dalton-Cheshire sequence in this area.

Local thickening of the entire sequence eastward from Leffingwell Hill is evident and suggests a basin deepening to the east in Early Cambrian time. Conglomerate

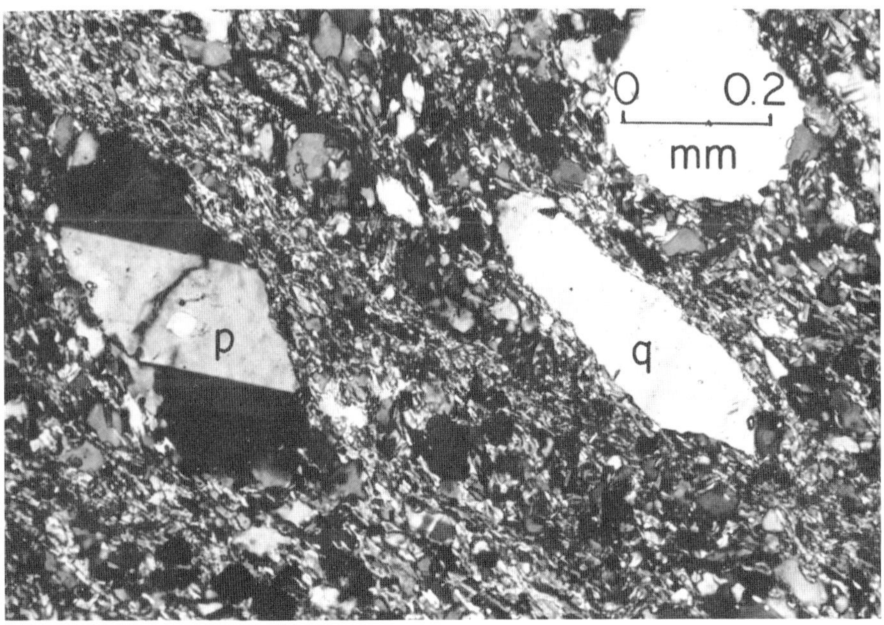

Figure 17. Photomicrograph showing fine-grained pseudodetrital texture of schistose mylonite. Sample taken at base of section at Leffingwell Hill area (Fig. 15, sec. B). Note sub-rounded detrital plagioclase (p) and quartz (q). Crossed polarizers.

is largely absent from section C except close to the base of the sequence directly above the unconformity, where a fine-grained gray metagraywacke with minor quartz-pebble conglomerate is found.

At Day Mountain south of Dalton, Massachusetts, in the Pittsfield East quadrangle, the type Dalton overlies Precambrian gneiss with angular unconformity (Emerson, 1899, 1917). The findings of Emerson have been confirmed by the work of Alavi (1971), and the generalized stratigraphic column for the Day Mountain area (Fig. 16, sec. D) is based on aspects of four separate sections studied in detail by Alavi.

The Dalton is overlain by massive vitreous Cheshire Quartzite approximately 91 m thick. Most of the Dalton is light-tan- to yellowish-brown-weathering schistose feldspathic quartzite and quartzose schist with two prominent coarse-grained conglomerate horizons. The uppermost horizon is a massive, white-weathering, silica-cemented quartz-pebble conglomerate that varies from 0 to 30.5 m in thickness. At or near the base of the Dalton is a second quartz-pebble and quartz-cobble conglomerate of the same variation in thickness. Pebbles of fine-grained white quartz (metachert?) predominate, although rare blue quartz and fine-grained black metachert(?) pebbles are found locally. Although gneissic rock fragments are rare as pebbles, the matrix of the conglomerate is a very feldspathic schistose meta-arkose with small angular fragments of feldspar and gneiss. Locally the basal part of the Dalton is a rusty-weathering muscovite-biotite-potassium feldspar-quartz-tourmaline schist. Throughout the lower 122 m of the Dalton are lenses 0.31 to 1.5 m thick of pink and green biotite-spotted meta-arkose conglomerate (granite wash).

Section E of Figure 16 is taken from a section exposed in the bed of the brook that drains to the southwest from Felton Lake in the northwestern part of the East Lee quadrangle. The conglomerate of the Dalton overlies fine-grained pinkish-gray granitic gneiss (unit 13) with angular unconformity. The contact is exposed at a small waterfall approximately 91 m east of Woodland Street. The base of the Dalton here contains coarse-grained quartz-pebble and quartz-cobble conglomerate similar to the basal unit at Dalton, in beds as much as 1.5 m thick. Thinner (0.15- to 0.6-m) beds of crossbedded pink- and green-mottled meta-arkose and meta-arkose conglomerate are interlayered. The coarse conglomerate is set in well-bedded muscovite-rich feldspathic quartzite typical of the Dalton. The coarse conglomeratic beds extend approximately 45 m up from the base of the unit and are in contact with vitreous quartzite of the Cheshire Quartzite. The total thickness of the exposed section is approximately 106 m. To the north and south along this outcrop belt, the conglomeratic facies is replaced by feldspathic quartzite and locally is absent entirely.

The section at Tyringham (Fig. 16, sec. F) is a composite taken from several excellent exposures. The top of the section shows interbedding with Stockbridge unit a, and at the base it rests with angular unconformity on Precambrian gneiss units 13 (Tyringham Gneiss), 9, 6, and 8. The total thickness of the section is approximately 152 m. A persistent basal unit is a black to silvery-gray schistose muscovite-biotite-rich schist marked by white quartz pebbles. Interlayered with and laterally replacing the basal schistose conglomerate is a light-gray, well-bedded, plagioclase- and microcline-rich biotite-quartz metagraywacke. Coarse conglomerates such as those in sections D and E are not present. The bulk of the Dalton is tan to light-gray biotitic muscovitic feldspathic quartzite with local beds of rusty two-mica tourmaline-rich schist. Approximately 45 to 53 m above the unconformity is a discontinuous, white, quartz-cemented quartz-pebble metaconglomerate that locally grades into beds of vitreous quartzite and shows gradational contacts with flaggy muscovitic quartzite above it. This conglomerate resembles closely the second conglomeratic horizon in the Dalton section (Fig. 16, sec. D). Along strike the typical feldspathic quartzite is replaced by a section of cleaner flaggy quartzite interbedded with vitreous quartzite 2.5 to 15 cm thick. The flaggy quartzite probably represents reworked feldspathic sand of the lower Dalton, with the basal white quartz-pebble conglomerate marking a disconformity in the section. The contact with the Cheshire appears to be sharp.

Marked lateral as well as vertical facies changes are dominant features of the Dalton-Cheshire sequence in southwestern Massachusetts, according to analysis of the exposed sections. Several features of regional significance are noted below.

1. The Dalton-Cheshire sequence apparently thickens eastward and grades by interbedding into schistose rocks attributable to the Hoosac Formation, which crops out east of the Berkshire massif (see Norton, this volume). The sequence of Dalton-Cheshire now found in the thrust slice at Monument Mountain (Fig. 15, sec. C) contains significant amounts of Hoosac-like material.

2. Conglomerate, common near the base of the Dalton Formation, is more abundant and thicker in exposures north of lat 42°22′5″ and is virtually absent south of that latitude. Norton (this volume) also notes that conglomerate is present in the Hoosac north of lat 42°30″ but is absent in exposures south of that latitude along the east side of the Berkshire massif.

3. The quartz pebbles and cobbles in the metaconglomerate of the Dalton Formation are an enigma. Many are white or black, fine-grained, nonmetamorphic chert(?) fragments that are found neither in the local basement rocks nor in the high-grade rocks of the Adirondacks. The source of these pebbles is unknown. Perhaps a younger Precambrian nonmetamorphic source area once existed but is no longer preserved in the geologic record. A detailed study of these and similar pebbles found in the Rensselaer Graywacke of late Precambrian(?) and (or) Early Cambrian(?) age of the Taconic allochthon (Fig. 1) would be highly desirable and might add to the understanding of the pre-proto-Atlantic paleogeography proposed by Bird and Dewey (1970) for the northern Appalachians. The arkosic composition and crossbedding of the matrix of the Dalton conglomerate and the rounding of the pebbles suggest a littoral environment. Although the quartz pebbles are anomalous, the sympathetic coarsening of the matrix with pebble size suggests that a glacial origin is unlikely. However, a glacial-fluviatile environment cannot be ruled out on the basis of currently available data.

Precambrian-Dalton Unconformity

The Dalton Formation is known to overlie the older basement rocks with angular unconformity at Day Mountain (Pittsfield East quadrangle; Fig. 1, loc. 7) on the basis of the work of Emerson (1899), who described the unconformable relationships thus: "Five rods east from the mountain cottage on the high hill above Dalton station the conglomerates, themselves nearly horizontal, can be seen resting upon the vertical beds of the older gneisses which strike N. 80°E. . . ." (p. 70), and "The conglomerate is here so beautifully distinct . . . and the complete unconformity to the underlying so well exposed, that this is the best locality in Massachusetts for the study of all the relations of the two formations." (p. 72)

Emerson (1899) originally described this conglomerate as Becket Gneiss, but in his 1917 compilation he applied the name Dalton Formation to these quartzose, feldspathic, and conglomeratic rocks beneath the Cheshire Quartzite.

Restudy of the Day Mountain area by Alavi (1971) has confirmed Emerson's conclusions. The Dalton Formation overlies with angular unconformity granitic gneiss, Washington Gneiss, and biotite-quartz plagioclase paragneiss within the Day Mountain area.

As shown in Figure 18, the gneissic layering is truncated by the basal conglomerate of the Dalton. For 2 m below the unconformity, the gneiss has been replaced by coarse-grained muscovite and abundant magnetite. Residual quartz remains, and the gneissic layers are well preserved up to the unconformity. Gneiss 3 m below the unconformity is well-layered biotite-hornblende-plagioclase-epidote-quartz gneiss with only minor amounts of muscovite. This peculiar zone has been found at all six of the known exposures of the unconformity from Pittsfield south to Tyringham. At each locality the percentage of replacement muscovite increases toward the unconformity. The enrichment in alumina and magnetite suggests that this is a metamorphosed sapprolite of late Precambrian or early Paleozoic age.

From Day Mountain southward the unconformity is well exposed at several places, some of which are shown in Figure 16. Where the unconformity can be seen or located within a few meters, the older rocks possess fold structures, foliation, and linear features not found in the overlying Dalton Formation. Relationships such as this can be seen at Day Mountain, in the brook that drains Felton Lake in the East Lee quadrangle, at several localities along the ridge northeast of Tyringham, and at Burgoyne Pass in the Stockbridge quadrangle.

These angular relationships are well displayed in the Tyringham area (Fig. 2) along the four-mile-long Precambrian-Dalton contact. Here the basal schistose and conglomeratic unit of the Dalton has an unconformable contact with units 9, 8, and 6, and with the intrusive Tyringham Gneiss (unit 13). Foliations, lineations, minor fold axes, and several major folds trend northeast to east-northeast in the Precambrian rocks, whereas the Paleozoic rocks generally strike N50°W and dip

Figure 18. Angular unconformity between gneiss and basal conglomerate of the Dalton Formation exposed on south slope of Day Mountain, Pittsfield East quadrangle, Massachusetts. Gneissic layering strikes east and is nearly vertical. Unconformity dips gently to northeast. Hammer is 40 cm long; black line on photograph shows trace of unconformity.

gently to steeply to the southwest. Both limbs of the arcuate fold involving units 9, 8, 7, and 13 are unconformably overlain at the northwestern end of the Dalton exposures (Fig. 19). Hornblende lineations, minor fold axes, and a metamorphic axial planar foliation common to the paragneiss and the Tyringham are truncated with an angular unconformity of about 75° in azimuth and about 30° in dip. On the basis of observations at this locality and elsewhere along the unconformity, we conclude that the dominant gneissic fabric, the gneissic structure, and the lineations that lie in that gneissic foliation are Precambrian in age.

It is therefore possible in many cases to differentiate folds and foliation of Precambrian age from numerous structures of Paleozoic age. A simplified tectonic map (Fig. 19) shows the axial traces of folds of Precambrian age. After the Paleozoic deformation, which trends chiefly north-northeast and north-northwest, is "unfolded," the relict Precambrian structures are seen to trend approximately eastward. The Precambrian structures in this part of the Berkshire massif are similar in trend to those in the central part of the Green Mountains of Vermont (Brace, 1953; Doll and others, 1961) and in the Adirondack Mountains (Buddington and Leonard, 1962, Pl. 4).

Precambrian Metamorphism and Possible Metamorphic Discontinuity

Isograds marking the first occurrence of staurolite, diopside, and sillimanite in Paleozoic rocks are shown in Figure 2. The isograds trend approximately N25°E and appear to postdate the low-angle overthrusts of the Berkshire massif. Mineral assemblages within the Precambrian rocks exposed in the central and eastern parts of the map area are broadly consistent with the Paleozoic metamorphic grade. However, exposures of Precambrian rocks on the western edge of Beartown Mountain and north of Lee show retrogressive mineral textures.

Garnet commonly is fractured and rimmed by biotite (Fig. 20) in the Washington Gneiss (unit 1) and in the hornblende-garnet amphibolite (units 3 and 5). Garnets from the garnet amphibolite units exposed east of the Paleozoic diopside isograd are fresh but show evidence of two generations of growth (Fig. 21). Irregular zones of dusty inclusions within the garnets appear to mark old grain boundaries that are overgrown with clear rims (Fig. 21). The secondary growth is interpreted as Acadian and probably represents garnet replacing retrograde biotite of Taconic age.

Precambrian diopside calc-silicate rock is exposed on East Mountain west of the first occurrence of diopside in the Paleozoic rocks. This calc-silicate is extensively altered to tremolite-actinolite. Microperthitic microcline is a widespread metamorphic mineral even in the westernmost exposures of paragneiss units, but it is absent in the Dalton Formation, which contains nonperthitic microcline and plagioclase at all metamorphic grades exposed in the study area.

Relict sillimanite, largely replaced by muscovite, is found in exposures of the Washington Gneiss (unit 1) at the town of Washington, shown in the northeast corner of Figure 2. From Washington south to Monterey, sillimanite is not found in any rocks. Sillimanite is abundant, however, in the Washington Gneiss south of the Paleozoic sillimanite isograd. Norton (1969) also reported relict sillimanite from Precambrian rocks of the Windsor quadrangle in the garnet zone of Paleozoic metamorphism. Retrograde features are widespread in the Precambrian rocks of the Green Mountains (Brace, 1953, p. 62–64). These relict minerals and retrogressive textures, together with the truncation of a Precambrian metamorphic foliation by the upper Precambrian(?) and Lower Cambrian Dalton, support the idea of a metamorphic as well as structural unconformity at the base of the Dalton Formation.

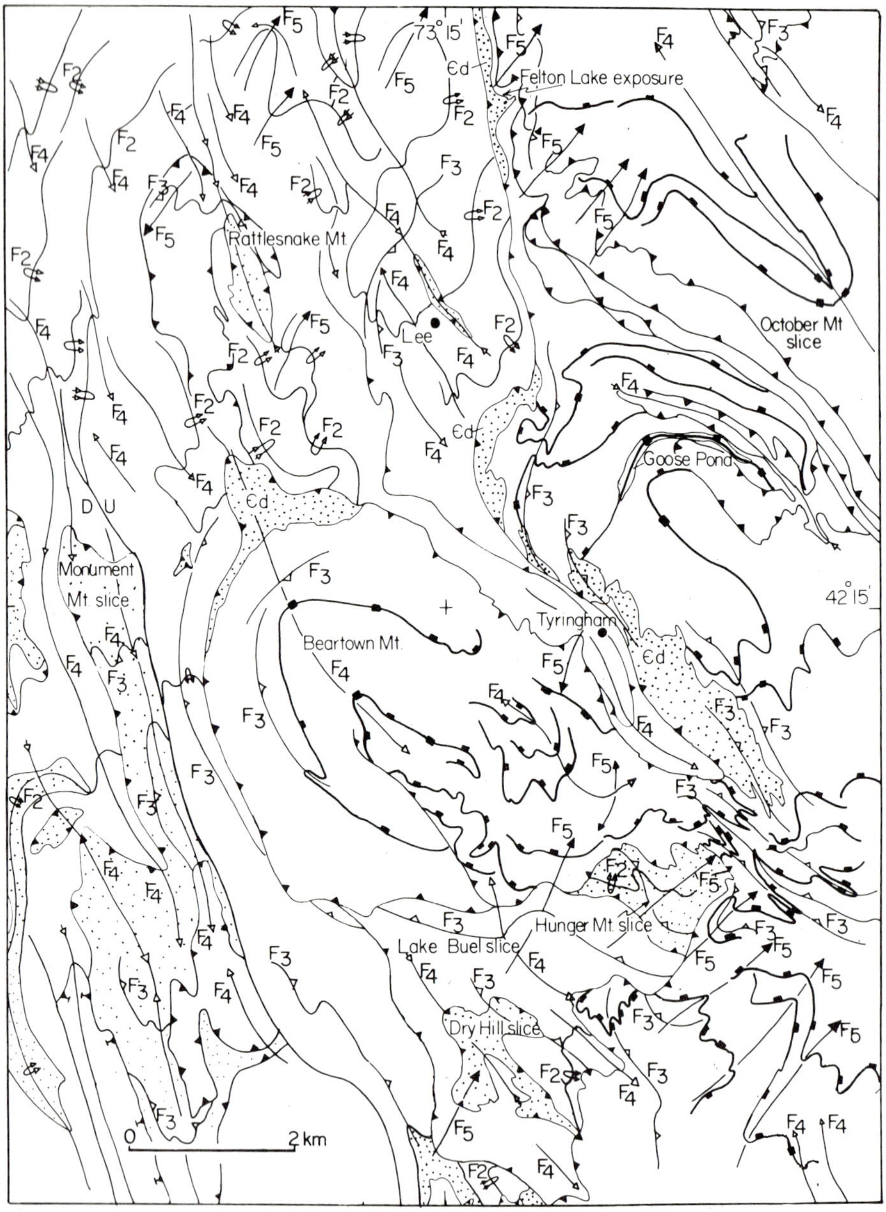

NATURE OF THE PRECAMBRIAN PLUTONISM AND DEFORMATION

On the basis of the geologic and isotopic data now available, it may be concluded that Precambrian deformation in the Berkshire massif involved dynamothermal metamorphism sufficient to produce garnet, diopside, hornblende, and perthitic alkali feldspar and sillimanite. Therefore, regional metamorphism at least as great as sillimanite-plus-muscovite grade affected this area.

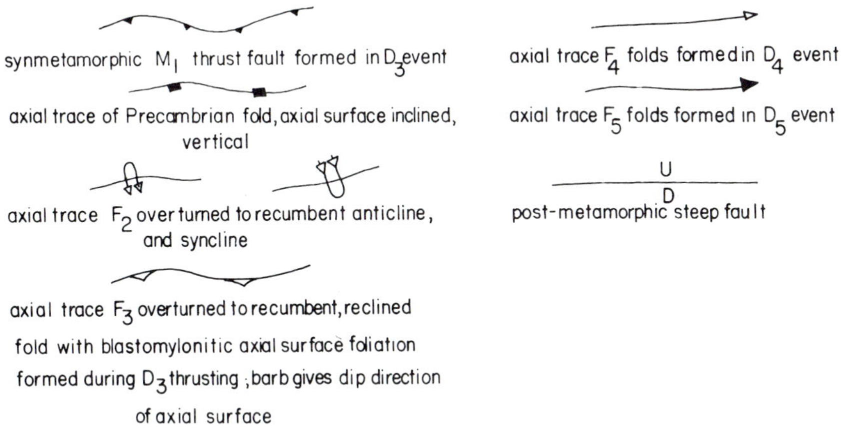

Figure 19. Simplified tectonic map showing fold and fault features formed during Precambrian through Middle Devonian or Triassic(?) orogenesis in Stockbridge, Great Barrington, Monterey, and East Lee quadrangles, Massachusetts. Stippled pattern identifies exposures of upper Precambrian(?) and basal Paleozoic Dalton Formation and Cheshire Quartzite that unconformably overlie the gneiss of the Berkshire massif. Note the general but refolded easterly trend of the Precambrian axial traces. Axial traces of Paleozoic folds are identified and discussed further in Figure 22.

Geologic data indicate that the Tyringham Gneiss intruded foliated rocks and was thoroughly involved in the severe deformation that folded the entire Precambrian sequence along an easterly trend; it can therefore be considered a syntectonic granite. We believe that the U-Pb data for the Tyringham record either the time of its intrusion or the time of severe dynamothermal metamorphism. Therefore, the U-Pb concordia age of 1,040 to 1,080 m.y. can reasonably be expected to approximate the time of the Precambrian dynamothermal event.

The intrusive relationship and inferred syntectonic habit of the Tyringham are remarkably similar to those of the biotite-ferrohastingsite granite in the highlands of the western Adirondack Mountains described by Buddington and Leonard (1962, p. 107).

At present no deformation or metamorphic effects requiring an Avalonian age (600 m.y.) are recognized. However, even a moderately high grade dynamothermal event at that time might not have greatly affected U-Th-Pb isotopic data and could go undetected. Some biotite granite from the eastern edge of the Berkshire massif that recently has been dated as 605 ± 50 m.y. old (Brookins and Norton, 1975) may indicate that Avalonian igneous activity affected the rocks of the massif, although the distribution and significance of these rocks are poorly understood at present.

DEFORMATION OF THE BERKSHIRE MASSIF IN THE PALEOZOIC ERA

In previous sections we have presented evidence for a Precambrian dynamothermal event that greatly affected the gneissic rocks of the Berkshire massif. The most obvious structural trends and structural features now seen in this area, however, were caused by intense polyphase Paleozoic deformation resulting from various phases of the Taconic and Acadian orogenies. The structures and fabrics in the Precambrian rocks are the result of deformations dating from 1 b.y. B.P. to Middle

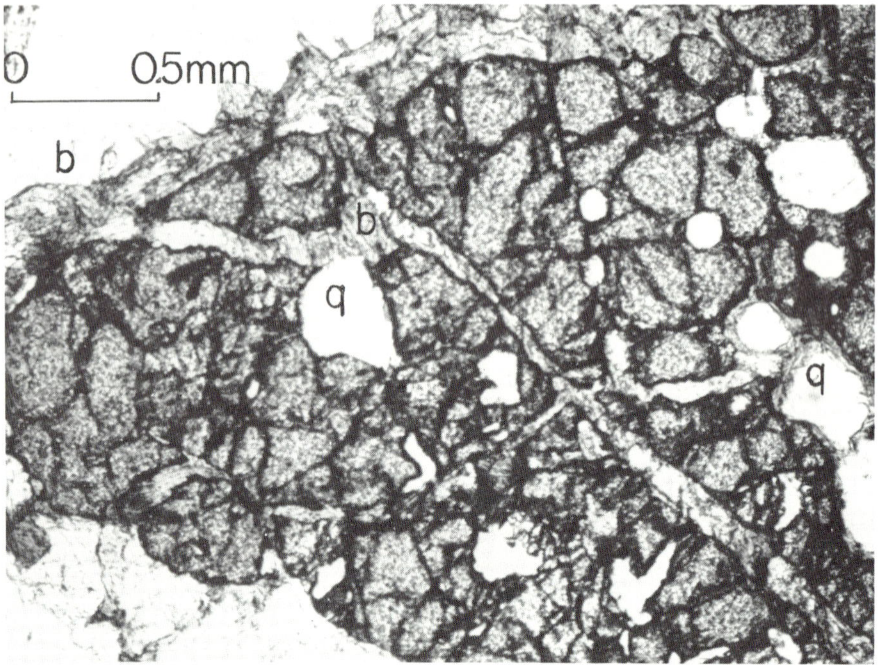

Figure 20. Photomicrograph of poikiloblastic garnet metacryst in Washington Gneiss (unit 1) exposed on west flank of Beartown Mountain in lower staurolite zone of Paleozoic metamorphism, showing alteration to biotite (b) along fractures and the margin; q = quartz. Plain light.

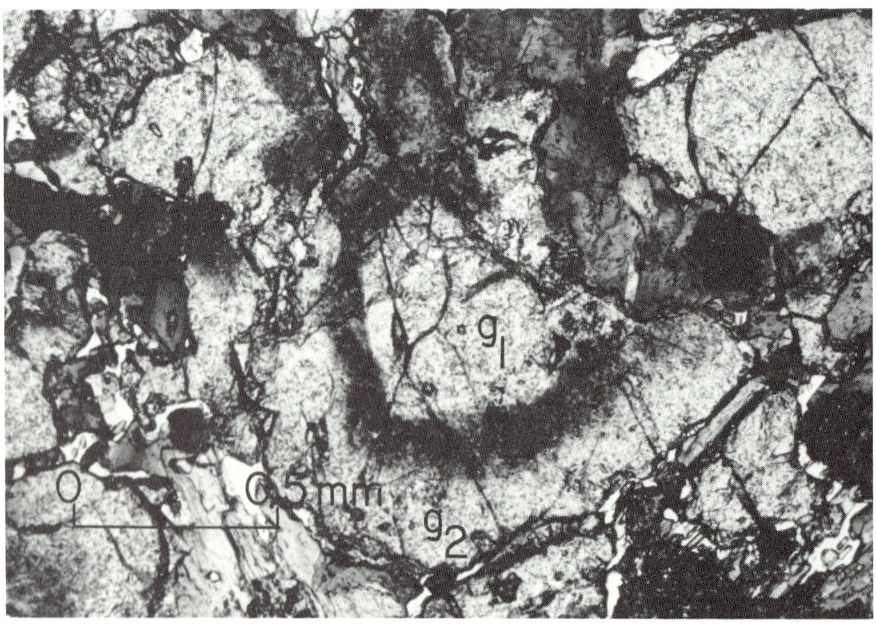

Figure 21. Photomicrograph of garnet in amphibolite (unit 5) in Precambrian gneiss. Sample taken 1.8 km north of Monterey in upper staurolite zone of Paleozoic metamorphism. Clear borders have overgrown previous borders, as shown by irregular zones of dusty opaque inclusions. Overgrowths (g_2) thought to be Paleozoic; interior (g_1), Precambrian. Texture contrasts with that shown in Figure 20. Plain light.

or Late Devonian time. These features are illustrated in Figures 19 and 22. Details of the structural geology resulting from Paleozoic deformation have been described in articles by Ratcliffe (1969a, 1969b, 1974a, 1974b, 1974c) and Ratcliffe and Harwood (1975).

A brief summary of the Paleozoic deformational history as presently understood is given below. The field and petrographic features associated with each deformational episode are described in Figure 22, which is reproduced from Ratcliffe and Harwood (1975).

After the miogeosynclinal shelf sequence (Stockbridge Formation) was deposited in Early Cambrian to Early Ordovician time, crustal warping and possible block faulting (D_0 event) preceded deposition of the exogeosynclinal Walloomsac Formation (Zen, 1967, p. 40–44, 71; Ratcliffe, 1965, 1968, 1969a). This major bathymetric reversal was a precursor to thrusting by gravity sliding of the lower Taconic slices during Middle Ordovician time (D_1 event; Zen, 1967). Emplacement of the higher Taconic slices, the Chatham and Everett slices, involved hard-rock deformation and was accomplished under either nonmetamorphic or extremely low-grade metamorphic conditions (Zen and Ratcliffe, 1966, 1971). Regional metamorphism and deformation along northeasterly structural trends postdated emplacement of allochthonous rocks (D_2 event) and F_2 folds. Either synchronous with or slightly later than this event, intense recumbent folding and low-angle thrusting of the Berkshire massif and its cover rocks took place, also under metamorphic conditions; however, the cover rocks had been metamorphosed prior to thrusting (Ratcliffe, 1972). As a result of this thrust faulting, the gneissic rocks of the Berkshire massif are now exposed in a series of thrust slices that overlap to the northeast (Fig. 2). During this episode of basement reactivation, the pre-existing Precambrian isoclinal fold system in the gneiss was refolded and rotated in a second set of isoclinal, southwest-trending, overturned, and nearly recumbent folds (F_3) whose axes now plunge down the axial surface to the northeast or southwest. Axial surfaces of these recumbent folds and of zones of blastomylonite parallel the exposed thrust faults; the folds and thrusts were formed contemporaneously, giving rise to a distinctive fold-thrust fabric in the fault zones. Analysis of minor drag folds from blastomylonite zones at the soles of the Monument Mountain, Beartown Mountain, and October Mountain slices from Pittsfield south to Connecticut yields fairly consistent azimuths for slip lines. Slip directions commonly vary from N70°E to S70°E on easterly-dipping fault zones. Removal of the effects of F_4 and F_5 folds indicates a general thrust direction from the east. An Ordovician age for this overthrusting is indicated by Zartman's preliminary isotopic results for zircon samples from a granite stock that crosscuts one of the basement overthrusts in the South Sandisfield quadrangle (Harwood, 1972). Deformations D_1 to D_3 are therefore ascribed to various phases of the Taconic orogeny.

The lowermost slice (the Beartown Mountain slice) overrides the Stockbridge miogeosynclinal sequence for 21 km at lat 42°15N. In addition, a higher slice (the Washington Mountain slice) overrides the Beartown Mountain slice for 10 km. The total displacement on all the recognized faults is unknown at present but is large. It is clear that nappe and fault structures comparable to Pennine deformation are found in the Berkshire massif.

Additional deformation along northwest F_4 and northeast F_5 trends postdates the thrusts of basement rock and is in part synchronous with staurolite-kyanite-sillimanite plus muscovite (Barrovian) metamorphism. These late events may be of Acadian age (Ratcliffe, 1968, 1969a, 1969b), whereas the earlier events probably are Middle Ordovician to possibly Silurian (Taconian).

Deformational event	Number of fold system	Types of folds and areal extent	Important tectonic features	Metamorphic event
D_6	F_6	North-south open folds of foliation locally recognized in Stockbridge valley	Northwest- and north-trending normal faults	
D_5	F_5	N. 25°–40° E.-trending upright to northwest-overturned folds of foliation, with axial planar slip or crenulation cleavage. Folds recognized throughout area of figure 1 west to Mount Ida in SW corner of Kinderhook 15-minute quadrangle, N. Y., where Taconic unconformity is folded by N. 40° E. upright folds	Refolds thrust sheets and blastomylonitic foliation	M_2 Thermal maximum Acadian metamorphism
D_4	F_4	Northwest-trending upright to southwest-overturned folds with axial planar slip, crenulation, and flow cleavage. Folds recognized throughout area of figure 1, west to Chatham, N. Y., in center of Kinderhook 15-minute quadrangle	Folds thrust sheets and blastomylonitic foliation resulting in local overturning of thrusts; northwest-trending high-angle reverse faults	
			Granite crosscuts thrust fault and blastomylonitic foliation	Thermal maximum Taconic metamorphism
D_3	F_3	Northwest-trending recumbent to strongly southwest-overturned folds of basement gneiss and large-scale southwestward thrusting of Precambrian rocks of Berkshire massif across autochthon. Fold and thrust style recognized from Windsor quadrangle, Massachusetts (Norton, 1969), south to Norfolk quadrangle, Connecticut (Harwood, unpub. data), along west front of Berkshire massif	Faulted recumbent folds and nappes, mylonite gneiss, blastomylonite associated with major thrusts Thrust sheets at June and Canaan Mountains transported with Berkshire massif	
D_2	F_2	Isoclinal northeast-trending northwest-overturned to nearly recumbent folds with strong axial planar foliation which is dominant foliation in most autochthonous and allochthonous (Taconic) rocks, but not clearly present in Paleozoic rocks attached to Berkshire massif. Folds extend west to Mount Ida where unconformable beneath lowermost Devonian	Folding of Taconic thrust contacts, regional foliation and refolding of slump or soft-rock folds in Taconic allochthonous rocks	M_1
D_2(?)	F_2(?)	Folding and metamorphism of Lower Cambrian metasedimentary rocks attached to Berkshire massif and in independent thrust slices at June and Canaan Mountains	Coarse foliation or schistosity formed	M_1(?)
D_1	F_1	Intrafolial minor folds associated with Taconic thrust contacts. Soft rock or slump folds in Taconic allochthonous rocks; scale of pre-F_2 folds not determined but widespread, area shown in figure 1, west to Mount Ida	Emplacement of upper Taconic slices (here, Chatham and Everett slices) Emplacement of lower Taconic slices to west of area shown in figure 1	No metamorphism recognized
		~~ *Middle Ordovician*		
D_0		Warping of Lower Cambrian to Lower Ordovician carbonate shelf sequence; locally dips near vertical (Ratcliffe, 1969a); possible block faulting		
		~~ *Pre-Dalton*		
$D_{p€}$	$F_{p€}$	Isoclinal east-west-trending folds with generally steeply dipping axial surfaces and strong axial planar foliation; deformation of all Precambrian rocks including granitic intrusions such as Tyringham Gneiss	Gneissosity in Precambrian rocks of Berkshire massif	$M_{p€}$
		Pre-Tyringham foliation		

Figure 22. Chronology of tectonic events recognized in southwestern Massachusetts (from Ratcliff

…event	Important crystalloblastic and other structures	Igneous intrusion	Probable age of rocks in figure 1	Orogeny
	Hematite-cemented breccias		Uncertain (Middle Devonian to Late Triassic)	
…2	Crenulation of sillimanite alined in axial surface of F_4 folds; granulation of garnet and staurolite that includes F_4 foliation		Middle to Late Devonian (Ratcliffe 1969a, b, 1972)	Acadian orogeny
…metamorphism	Muscovite, biotite realined and recrystallized in axial surface foliation; coarse sillimanite crystallized in foliation. Garnet, staurolite include folded F_2 fabric, and blastomylonitic foliation		Middle to Late Devonian (Ratcliffe 1969a, b, 1972)	
	Granite lacks blastomylonitic foliation in country rocks	Granite stock, South Sandisfield quadrangle	Late Ordovician(?) (Harwood, 1972)	Phases of Taconic orogeny
…metamorphism	Alaskite has weakly developed blastomylonitic foliation but intrudes more highly cataclastic rock in fault zones; mylonite gneiss, blastomylonite has muscovite, biotite, hornblende with lepidoblastic texture, cataclasis of F_2 foliation, thrusting synmetamorphic	Alaskite sills in faults and magnetite mineralization	Synchronous with latest movements or thrusts (Late Ordovician?) Thrusting probably late Ordovician based on age of cross-cutting granite	D
…1	Lepidoblastic muscovite, chlorite, biotite, and ilmenite in foliation; chloritoid, albite include foliation but are kinked by F_4 structures		Middle to Late Ordovician(?)	C Taconic orogeny
1(?)	Muscovite, biotite lepidoblastic in schistosity		Time of metamorphism very uncertain depending upon original position of these rocks, and timing of tectonic events at that site (Middle Ordovician to Cambrian?)	(C)?
…recognized	Tectonic breccias with inclusions of Stockbridge Formation along thrusts (Zen and Ratcliffe, 1971)		Uncertain (Middle Ordovician?)	B
	Wild-flysch-like sedimentary rocks along base of thrusts		Middle Ordovician (Zen, 1972b, table 1)	A
unconformity			Late Early to Middle Ordovician (Zen, 1972b, table 1)	Pre-Taconic disturburance
unconformity				
pC	Diopside, sillimanite, hornblende, microcline, perthite formed in dynamothermal event	Granodiorite-quartz monzonite intrusions such as Tyringham Gneiss, syntectonic	Dynamothermal event and granite intrusion approximately 1.04 b.y. (Ratcliffe and Zartman, 1971)	Grenville orogeny

Degree of basement participation and intensity of deformation increasing with time →

…d Harwood, 1975). References and locations referred to are cited in Ratcliffe and Harwood (1975).

SUMMARY AND CONCLUSIONS

1. The Precambrian rocks of the Berkshire massif (Fig. 2) have been subdivided into lithostratigraphic paragneiss units totaling approximately 1,830 m in thickness. Metasedimentary and metavolcaniclastic rocks compose the bulk of the section. Near the inferred top of the section, a minor but rather persistent horizon of calc-silicate gneiss occurs, associated with biotite-quartz-plagioclase paragneiss. The paragneiss units are similar to rocks of the Reading Prong, Adirondack Mountains, and Green Mountains of Vermont, although no specific correlations can be made.

2. The Tyringham Gneiss is a syntectonic ferrohastingsite-bearing granodiorite to quartz monzonite that intrudes and extensively crosscuts the paragneiss sequence. Both the intrusive and the country rocks were foliated and isoclinally folded prior to deposition of the Dalton Formation of late Precambrian(?) and Early Cambrian age.

3. A concordia plot of U-Pb data for zircon from the Tyringham and Washington Gneisses gives ages of approximately 1,040 to 1,080 m.y. B.P. for both units. An igneous origin for the Tyringham zircon grains is suggested by their morphology and occurrence (elongate, euhedral habit), whereas the Washington zircon occurs in rounded, detrital grains. The agreement in age of the zircons may mean either that (1) little time separates the age of the source material for the Washington Gneiss and the intrusion of the Tyringham Gneiss, (2) the zircon from the Tyringham Gneiss gives a true syntectonic igneous age, but the zircon from the Washington Gneiss was reset during the same dynamothermal event, or (3) the zircon in both units mainly reflects a Precambrian metamorphic age. Because of geologic evidence that the Tyringham was intruded during a time of deep-seated plutonic dynamothermal metamorphism in the Precambrian, either (2) or (3) above is the favored explanation.

4. The zircon data and the geologic relationships do not support the contention that the granitic Tyringham Gneiss represents the older basement on which the paragneiss sequence was deposited. At present the source area for these metasedimentary rocks is undetermined.

5. The Precambrian-Dalton contact has been restudied in detail, and abundant evidence indicates the widespread development of an angular unconformity and metamorphic discontinuity, as Emerson stated (1898, 1917). Although the Precambrian rocks were remetamorphosed, deformed, and refoliated during Paleozoic time, the post-Tyringham-pre-Dalton dynamothermal event is responsible for the gneissosity of the Precambrian rocks.

6. Rapid facies changes and variations in thickness characterize the Dalton-Cheshire clastic sequence that overlies the Precambrian rocks. The dominant facies changes involve west-to-east thickening of the sub-Cheshire rocks and a general coarsening and thickening of quartz-pebble conglomerate toward the north end of the massif. These conclusions are based on data cited in this paper and by Norton (1969, this volume) and Gates and Martin (this volume).

7. On the basis of the rapid changes in thickness and complex interfingering of the Dalton-Cheshire lithofacies, we conclude that the basin of deposition was uneven, with local basement highs. It is likely that the Cheshire represents strandline deposits derived from reworking of the previously deposited Dalton, which eroded as a result of local uplift within the sedimentary basin. Although the Dalton-Cheshire lithologies are replaced to the east by metasandstone and schist of the Hoosac Formation (Norton, 1969, this volume), the stratigraphic relationships cited suggest that the easterly-inclined Lower Cambrian paleoslope was marked by basin irregularities and local areas of westward sediment transport.

8. Reconstruction of the lower Paleozoic facies distribution and paleogeography is complicated by strong westward overthrusting of the basement and cover rocks of the Berkshire massif in imbricate thrust slices (Norton, 1969, this volume; Ratcliffe, 1969a; Ratcliffe and Harwood, 1971; Harwood, 1972). The total amount of foreshorteing, as well as the extent of "tectonic bypassing" among individual basement plates, is not known exactly at present but is more than 35 km. Any attempt at detailed reconstructions of the Early Cambrian paleogeography, including rocks of the Taconic klippen, must involve palinspastic reconstruction of the Berkshire basement slices and stratigraphic sections to their original relative positions. Such a study must await the completion of structural investigations of the Berkshire massif that are now in progress.

ACKNOWLEDGMENTS

The mapping of the western side of the Berkshire massif was part of a study begun by Ratcliffe in 1962 with the support of the Penrose Fund of the Geological Society of America and supported from 1966 to the present by the U.S. Geological Survey in cooperation with the Massachusetts Department of Public Works. The isotopic analyses by Zartman were carried out in the laboratories of the U.S. Geological Survey, Denver, Colorado.

Discussions in the field and during informal conversations with Richard Goldsmith, David S. Harwood, Richard H. Jahns, Stephen A. Norton, Lincoln R. Page, Robert W. Schnabel, and E-an Zen have been particularly helpful. We are indebted to Schnabel for initiating and helping with the rock sampling for the zircon studies. The area of study is contiguous on the south with the area mapped by Harwood. The Precambrian stratigraphy proposed here is in substantial agreement with that of Harwood, although minor differences in facies would have to exist between these two areas.

REFERENCES CITED

Alavi, Mehdi, 1971, The bedrock geology of part of the Pittsfield East quadrangle, Massachusetts [M.A. thesis]: New York, City College, 69 p.

Bird, J. M., and Dewey, J. F., 1970, Lithosphere plate-continental margin tectonics and the evolution of the Appalachian orogen: Geol. Soc. America Bull., v. 81, p. 1031-1060.

Brace, W. F., 1953, The geology of the Rutland area, Vermont: Vermont Geol. Survey Bull. no. 6, 120 p.

Brookins, D. G., and Norton, S. A., 1975, Rb/Sr whole-rock ages along the Precambrian-Cambrian contact, east side of the Berkshire massif, Massachusetts [abs.]: Geol. Soc. America Abs. with Programs, v. 7, no. 1, p. 30-31.

Buddington, A. F., 1939, Adirondack igneous rocks and their metamorphism: Geol. Soc. America Mem. 7, 354 p.

——1952, Chemical petrology of some metamorphosed Adirondack gabbroic, syenitic, and quartz syenitic rocks: Am. Jour. Sci., Bowen vol., pt. 1, p. 37-84.

——1957, Interrelated Precambrian granitic rocks, northwest Adirondacks, New York: Geol. Soc. America Bull., v. 68, p. 291-306.

Buddington, A. F., and Leonard, B. F., 1962, Regional geology of the St. Lawrence County magnetite district, northwest Adirondacks, New York: U.S. Geol. Survey Prof. Paper 376, 145 p.

Cady, W. M., 1969, Regional tectonic synthesis of northwestern New England and adjacent Quebec: Geol. Soc. America Mem. 120, 181 p.

deWaard, D., and Walton, M., 1967, Precambrian geology of the Adirondack highlands, a reinterpretation: Geol. Rundschau, Band 56, Heft 2, p. 596-629.

Dodd, R. T., Jr., 1965, Precambrian geology of the Popolopen Lake quadrangle, southeastern New York: New York State Mus. and Sci. Service Map and Chart Ser. no. 6, 39 p.

Doll, C. G., Cady, W. M., Thompson, J. B., Jr., and Billings, M. P., compilers and eds., 1961, Centennial geologic map of Vermont: Vermont Geol. Survey, scale 1:250,000.

Drake, A. A., Jr., 1969, Precambrian and lower Paleozoic geology of the Delaware Valley, New Jersey-Pennsylvania *in* Subitzky, Seymour, ed., Geology of selected areas in New Jersey and eastern Pennsylvania and guidebook of excursions: New Brunswick, New Jersey, Rutgers Univ. Press, p. 51-131.

——1970, The Blue Ridge and Reading Prong—Structural geology of the Reading Prong *in* Fisher, G. W., Pettijohn, F. J., Reed, J. C., Jr., and Weaver, K. N., eds., Studies of Appalachian geology, central and southern: Interscience, New York, p. 271-291.

Emerson, B. K., 1898, The geology of old Hampshire County, Massachusetts, comprising Franklin, Hampshire, and Hampden Counties: U.S. Geol. Survey Mon. 29, 790 p.

——1899, The geology of eastern Berkshire County, Massachusetts: U.S. Geol. Survey Bull. 159, 139 p.

——1917, Geology of Massachusetts and Rhode Island: U.S. Geol. Survey Bull. 597, 289 p.

Faul, Henry, Stern, T. W., Thomas, H. H., and Elmore, P.L.D., 1963, Ages in intrusion and metamorphism in the northern Appalachians: Am. Jour. Sci., v. 261, no. 1, p. 1-19.

Fisher, D. W., Isachsen, Y. W., Rickard, L. V., Broughton, J. G., and Offield, T. W., compilers and eds., 1961, Geologic map of New York: New York State Mus. and Sci. Service, 5 sheets, scale 1:250,000.

Gates, R. M., and Martin, C. W., 1976, Pre-Devonian stratigraphy of the central section of the western Connecticut highlands, *in* Page, L. R., ed., Contributions to the stratigraphy of New England: Geol. Soc. America Mem. 148 (this volume).

Grauert, B., and Arnold, A., 1968, Deutung diskordanter Zirkonalter der Silvrettadecke und des Gotthardmassivs (Schweizer Alpen): Contr. Mineralogy and Petrology, v. 20, no. 1, p. 34-56.

Hall, L. M., 1968, Times of origin and deformation of bedrock in the Manhattan Prong, *in* Zen, E-an, White, W. S., Hadley, J. B., and Thompson, J. B., Jr., eds., Studies of Appalachian geology, northern and maritime: New York, Interscience, p. 117-127.

Harper, C. T., 1968, Isotopic ages from the Appalachians and their tectonic significance: Canadian Jour. Earth Sci., v. 5, no. 1, p. 50-59.

Harwood, D. S., 1972, Tectonic events in the southwestern part of the Berkshire anticlinorium, Massachusetts and Connecticut [abs.]: Geol. Soc. America Abs. with Programs, v. 4, p. 19.

——1975, Fold thrust tectonism in the southern Berkshire massif, *in* Ratcliffe, N. M., ed., Guidebook for field trips in southwestern Massachusetts, Northwestern Connecticut, and adjacent New York State: New England Intercollegiate Geol. Conf., 67th Ann. Mtg., Dept. of Earth and Planetary Sciences, City College of N.Y., N.Y.

Hills, Allan, and Gast, P. W., 1964, Age of pyroxene-hornblende granitic gneiss of the eastern Adirondacks by the rubidium-strontium whole-rock method: Geol. Soc. America Bull., v. 75, p. 759-766.

Long, L. E., and Kulp, J. L., 1962, Isotopic age study of the metamorphic history of the Manhattan and Reading Prongs: Geol. Soc. America Bull., v. 73, p. 969-996.

Long, L. E., Cobb, J. C., and Kulp, J. L., 1959, Isotopic ages on some igneous and metamorphic rocks in the vicinity of New York City: New York Acad. Sci. Annals, v. 80, p. 1140-1147.

Lowe, K. E., 1950, Storm King Granite at Bear Mountain, New York: Geol. Soc. America Bull., v. 61, p. 137-190.

Mose, D. G., and Hayes, John, 1974, Rb/Sr whole-rock age determinations in the Precambrian Reading Prong, New York and New Jersey [abs.]: Geol. Soc. America Abs. with Programs, v. 6, p. 878.

Norton, S. A., 1969, Unconformities at the northern end of the Berkshire Highlands, *in* Bird, J. M., ed., Guidebook for field trips in New York, Massachusetts, and Vermont: New England Intercollegiate Geol. Conf., 61st Ann. Mtg., Albany, New York, 1969, p. 21-1-21-20.

——1976, Hoosac Formation (Early Cambrian or older) on the east limb of the Berkshire massif, western Massachusetts, *in* Page, L. R., Contributions to the stratigraphy of New England: Geol. Soc. America Mem. 148 (this volume).

Offield, T. W., 1967, Bedrock geology of the Goshen-Greenwood Lake area, New York: New York State Mus. and Sci, Service Map and Chart Ser., no. 9, 78 p.

Pidgeon, R. T., 1969, Zircon U-Pb ages from the Galway Granite and the Dalradian, Connemara, western Ireland: Scottish Jour. Geology, v. 5, pt. 4, p. 375-392.

Pidgeon, R. T., Köppel, V., and Grünenfelder, M., 1970, U-Pb isotopic relationships in zircon suites from a para- and ortho-gneiss from the Ceneri Zone, Southern Switzerland: Contr. Mineralogy and Petrology, v. 26, no. 1, p. 1-11.

Prucha, J. J., Scotford, D. M., and Sneider, R. M., 1968, Bedrock geology of parts of Putnam and Westchester Counties, New York, and Fairfield County, Connecticut: New York State Mus. and Sci. Service Map and Chart Ser. no. 11, 26 p.

Ratcliffe, N. M., 1965, Bedrock geology of the Great Barrington area, Massachusetts [Ph.D. thesis]: University Park, Pennsylvania State Univ., 213 p.

——1968, Structural geology and deformational history of the Taconic Range in southwestern Massachusetts near Great Barrington [abs.]: Geol. Soc. America Spec. Paper 101, p. 274.

——1969a, Structural and stratigraphic relations along the Precambrian front in southwestern Massachusetts, *in* Bird, J. M., ed., Guidebook for field trips in New York, Massachusetts, and Vermont, New England Intercollegiate Geol. Conf., 61st Ann. Mtg., Albany, New York, 1969, p. 1-1-1-21.

——1969b, Stratigraphy and deformational history of rocks of the Taconic Range near Great Barrington, Massachusetts, *in* Bird, J. M., ed., Guidebook for field trips in New York, Massachusetts, and Vermont, New England Intercollegiate Geol. Conf., 61st Ann. Mtg., Albany, New York, 1969, p. 2-1-2-23.

——1970, Ancient strike-slip fault tectonics in the Hudson Highlands and Manhattan Prong: New York Acad. Sci. Trans., ser. 2, v. 32, no. 8, p. 1009-1021.

——1971, The Ramapo fault system in New York and adjacent northern New Jersey: A case of tectonic heredity: Geol. Soc. America Bull., v. 82, p. 125-142.

——1972, Revised polyphase structural chronology in western Massachusetts and problems of regional correlation [abs.]: Geol. Soc. America Abs. with Programs, v. 4, p. 40.

——1974a, Bedrock geologic map of the Great Barrington quadrangle, Massachusetts: U.S. Geol. Survey Geol. Quad. Map GQ-1141, scale 1:24,000.

——1974b, Bedrock geologic map of the State Line quadrangle, New York-Massachusetts: U.S. Geol. Survey Geol. Quad. Map GQ-1142, scale 1:24,000.

——1974c, Bedrock geologic map of the Stockbridge quadrangle, Massachusetts: U.S. Geol. Survey Geol. Quad. Map GQ-1143, scale 1:24,000.

——1975, Bedrock geologic map of the Monterey quadrangle, Massachusetts: U.S. Geol. Survey Open-File Rept. 75-126, scale 1:24,000.

Ratcliffe, N. M., and Burger, H. R., 1975, Preliminary bedrock geologic map of the Ashley Falls quadrangle, Massachusetts and Connecticut: U.S. Geol. Survey Open-File Rept. 75-148, scale 1:24,000.

Ratcliffe, N. M., and Harwood, D. S., 1971, Structural geology of the western edge of the Berkshire massif, southwestern Massachusetts and Connecticut [abs.]: Geol. Soc. America Abs. with Programs, v. 3, p. 49-50.

——1975, Blastomylonites associated with recumbent folds and overthrusts at the western edge of the Berkshire massif, Connecticut and Massachusetts: A preliminary report, *in* Tectonic studies of the Berkshire massif, Massachusetts, Connecticut and Vermont: U.S. Geol. Survey Prof. Paper 888, p. 1-19.

Ratcliffe, N. M., Armstrong, R. L., Chai, B. H.-T., and Senechal, R. G., 1972, K-Ar and Rb-Sr geochronology of the Canopus pluton, Hudson Highlands, New York: Geol. Soc. America Bull., v. 83, p. 523-530.

Silver, L. T., 1964, Isotopic investigations of zircons in Precambrian igneous rocks of the Adirondack Mountains, New York [abs.]: Geol. Soc. America Spec. Paper 76, p. 150.

——1968, A geochronologic investigation of the anorthosite complex, Adirondack Mountains, New York, *in* Origin of anorthosite and related rocks: New York State Mus. and Sci, Service Mem. 18, p. 233-251.

Skehan, J. W., 1961, The Green Mountain anticlinorium in the vicinity of Wilmington and

Woodford, Vermont: Vermont Geol. Survey Bull. 17, 159 p.

Thompson, J. B., Jr., and Norton, S. A., 1968, Paleozoic regional metamorphism in New England and adjacent areas, *in* Zen, E-an, White, W. S., Hadley, J. B., and Thompson, J. B., Jr., eds., Studies of Appalachian geology, northern and maritime: New York, Interscience, p. 319-327.

Tilton, G. R., Davis, G. L., Wetherill, G. W., and Aldrich, L. T., 1957, Isotopic ages of zircon from granites and pegmatites: EOS (Am. Geophys. Union Trans.), v. 38, no. 3, p. 360-371.

Tilton, G. R., Wetherill, G. W., Davis, G. L., and Bass, M. N., 1960, 1000-million-year-old minerals from the eastern United States and Canada: Jour. Geophys. Research, v. 65, no. 12, p. 4173-4179.

Young, D. A., 1971, Precambrian rocks of the Lake Hopatcong area, New Jersey: Geol. Soc. America Bull., v. 82, no. 1, p. 143-158.

Zartman, R. E., and Wasserburg, G. J., 1969, The isotopic composition of lead in potassium feldspars from some 1.0-b.y.-old North American igneous rocks: Geochim. et Cosmochim. Acta, v. 33, no. 8, p. 910-942.

Zen, E-an, 1967, Time and space relationships of the Taconic allochthon and autochthon: Geol. Soc. America Spec. Paper 97, 107 p.

——1969, Stratigraphy, structure, and metamorphism of the Taconic allochthon and surrounding autochthon in Bashbish Falls and Egremont quadrangles and adjacent areas, *in* Bird., J. M., ed., Guidebook for field trips in New York, Massachusetts, and Vermont, New England Intercollegiate Geol. Conf., 61st Ann. Mtg. 1969, Albany, New York, p. 3-1-3-41.

——1972, Some revisions in the interpretation of the Taconic allochthon in west-central Vermont: Geol. Soc. America Bull., v. 83, p. 2573-2588.

Zen, E-an, and Hartshorn, J. H., 1966, Geologic map of the Bashbish Falls quadrangle, Massachusetts, Connecticut, and New York: U.S. Geol. Survey Geol. Quad. Map GQ-507, scale 1:24,000.

Zen, E-an, and Ratcliffe, N. M., 1966, A possible breccia in southwestern Massachusetts and adjoining areas, and its bearing on the existence of the Taconic allochthon, *in* Geological Survey Research 1966: U.S. Geol. Survey Prof. Paper 550-D, p. D39-D46.

——1971, Bedrock geologic map of the Egremont quadrangle and adjacent areas, Berkshire County, Massachusetts, and Columbia County, New York: U.S. Geol. Survey Misc. Geol. Inv. Map I-628, scale 1:24,000.

MANUSCRIPT RECEIVED BY THE SOCIETY APRIL 18, 1974
REVISED MANUSCRIPT RECEIVED JANUARY 31, 1975
MANUSCRIPT ACCEPTED FEBRUARY 20, 1975

Printed in U.S.A.

Geological Society of America
Memoir 148

Taconic "Room Problem," North Adams Gap, and Time of Formation of Basement-Related Nappes in Southwestern New England: Discussion of Papers by Harwood, Norton, and Ratcliffe and Zartman

E-AN ZEN
U.S. Geological Survey
National Center
Reston, Virginia 22092

Author's note. This discussion was prepared in 1970 for a specific symposium. Publication of the papers has been delayed, and some of the points discussed in this note are now obsolete. In the context of the publication, however, it seems better not to try to modify it. Surprisingly, some of the possibilities seem to retain current value.

February 1975

New data on the stratigraphic and structural relations of the Dalton Formation, the Hoosac Formation, and the Precambrian gneiss units support the idea that the Paleozoic cover across the North Adams gap in northwestern Massachusetts and southern Vermont was emplaced later than the peeling off of the lower Cambrian to Middle Ordovician rocks that constitute the continuous Giddings Brook slice of the Taconic allochthon, thus admitting the restoration of the allochthonous rocks to this zone. Emplacement of the Taconic allochthon was a protracted event that probably lasted into Late Ordovician or even Early Silurian time; it could have overlapped or even been simultaneous with the development of early thrusting and recumbent folding of the Precambrian gneisses on the west front of the Berkshire massif. The movement on the Precambrian rocks was probably directly responsible for the development of the Monument Mountain thrust slices; could it also be the cause of movement of the high Taconic slices? The relative importance of Taconic and Acadian orogenies in this part of the Appalachian orogen may need new assessment.

Having spent most of my mapping efforts in the Appalachian belt in and around the Taconic allochthon, my comments will naturally be strongly biased. First, I am proceeding on the assumption that the Taconic allochthon exists. Second, I look at the reports by Norton, Harwood, and Ratcliffe and Zartman in this volume through tinted glasses of the possible constraints of Taconic geology on their interpretations and vice versa.

One serious problem that has bothered workers on the Taconic allochthon for years, is, of course, the question of where the allochthonous rocks were originally laid down. The site over the present Green Mountain massif, the Berkshire massif, and possibly over some of the smaller Precambrian massifs farther to the south is a popular candidate (Zen, 1967; Rodgers, 1970, p. 85), but this solution has its own problems. First, even the restored width of the areas of basement rocks is barely room enough for the allochthonous rocks (Zen, 1967). The second problem, potentially disastrous to the allochthonist, is that at North Adams the basement rocks dip under, and Paleozoic rocks, roughly coeval with the rocks of the allochthon, actually bridge the North Adams gap. Thus, if the Paleozoic rocks there are autochthonous, as proposed by Herz (1961), or even if they are involved in nappes locally rooted, we are in serious trouble unless someone can demonstrate that the rocks in the Taconic allochthon are structurally in a stack of discrete imbricated slices.

For rocks of the "high Taconic sequence" (Zen, 1967) this is no problem, because there the rocks are indeed found in discontinuous patches, and, where mapped in detail, so many local thrust faults exist (see, for example, Zen and Ratcliffe, 1971) that it would stretch neither one's conscience nor imagination much to say that there exist many more tectonically discrete pieces of the allochthon than are generally shown. For the "low Taconic sequence," particularly the Giddings Brook slice, however, it is a different story; mapping in critical areas, especially by Potter (Potter and Lane, 1969) and J. M. Bird (1961-1971, personal commun.) west of the Rensselaer Plateau, indicates that a large-scale break probably does not exist.

For this reason Norton's mapping and interpretation of relations in the North Adams, Windsor, and Peru quadrangles are important. Previously I interpreted the presence of a continuous Paleozoic rock cover at the North Adams gap as the result of one or two large thrust faults that formed after the Taconic rocks had peeled off and that now hide the tectonic scar. Norton's (1971) report of a large premetamorphic thrust fault whose trace is now in the wide band of Herz's Dalton Formation, and which apparently continues into the Hoosic thrust fault of Herz (1961), neatly ties up this picture. Previously I (Zen, 1967) placed the southward continuation of the fault on the west side of the north end of the Berkshire massif, but placing it on the east side is just as good, and I will not look a gift horse in the mouth. One does wonder if more large thrust faults might not remain hidden in the eugeosynclinal pile (see Hatch and others, 1968, p. 179), especially if these thrust faults are premetamorphic.

The tectonic position of the Bull Hill Gneiss discussed by Norton may give a clue to the true structural relations. It seems entirely possible that the intercalation of the Bull Hill in the Cavendish is tectonic, analogous with Norton's report of tectonic intercalation of Hoosac and Precambrian rocks in Massachusetts. In fact, I wonder whether the gneisses of the Sadawga and Rayponda domes in the Wilmington quadrangle (Skehan, 1961) might not be allochthonous rocks, resting on the Hoosic fault, that later were updomed as a purely superficial structural phenomenon.

Norton's stratigraphic description of the Hoosac suggests the possibility that the carapace of the lower Hoosac Formation may not have moved far—perhaps not far enough to allow, upon restoration, enough room to accommodate rocks of the Taconic allochthon. This, however, is not fatal to the restoration of the allochthon here, because rocks of the Taconic allochthon probably are all younger, the oldest more comparable with the upper part of the Hoosac Formation and with the Rowe Schist [see also Hatch and others (1968) for description of these units]. Besides, as I have remarked, the high Taconic sequence does not form a tectonically continuous cover.

We can now probably confidently restore the Taconic allochthon across the North Adams gap. Cady's (1969, p. 65) statement that "the Green Mountain-Sutton Mountain anticlinorium cannot possibly be the root of the eugeosynclinal rocks of the Taconic klippe, because it plunges southward and disappears in the southern Green Mountains in southern Vermont and northwestern Massachusetts," is, we hope, a ghost that can finally be put to rest.

Finally, it is nice to see that another reasonably good correlation between rocks of the eugeosynclinal east Vermont sequence and the miogeosynclinal sequence has been achieved. We have a pretty good idea of gross time equivalences of the units through the tracing of the Hawley Formation into the fossiliferous Magog Slate (Currier and Jahns, 1941) and through Osberg's (1965) mapping in southern Quebec; hitherto no direct tie had been made farther south. Norton's interpretation that the Hoosac Formation, or at least the lower part of it, interbeds with the Dalton, if established, would be a most welcome addition.

Turning to the papers by Harwood and Ratcliffe and Zartman, I would like to begin with two remarks. First, both papers report large-scale westward thrusting of the Precambrian basement rocks. These folded Precambrian rocks were refolded along a more nearly north-south axis. Although Ratcliffe and Zartman suggest possibly Devonian age for the thrust movement, I believe Ratcliffe does not rule out an earlier date, and a late Taconic date is quite reasonable. In New England I think we have been conditioned to play up the Acadian deformation and downgrade the Taconic, at least as a deep-seated orogenic process; perhaps this is not so in fact. Various "Taconic" isotopic ages from whole rocks (Harper, 1968) or minerals (Rickard, 1965) in areas to the north demonstrate a late Taconic thermal event; Ratcliffe's (1968) work in the Peekskill, New York, area showed conclusively the importance of deep-seated Taconic disatrophism there. This movement could at least have overlapped in time the development of the large thrust fault in the Hoosac Formation which Norton discussed and which was traced into the Hoosic thrust fault of Herz (1961). If, as Harwood stated, these early folds in the basement rocks were refolded about north-trending, more upright axial surfaces, and if, as seems reasonable, we assign the latter to the Acadian orogeny, then would not it be more consistent to assign the first basement-rock deformation to the Taconic orogeny?

Second, the cause of movement of the nonsoupy high Taconic thrust sheets remains very obscure to me; if we can relate the event to the movement of the Precambrian rocks, we might at least consolidate our miseries. Ratcliffe's mapping in the Monument Mountain-Tom Ball Mountain area, which is directly west of the Precambrian nappes (Ratcliffe, 1965, 1969, p. 1-4; Zen and Ratcliffe, 1971), indicates that that thrusting in this area followed in time the Taconic thrusting; he associated this last movement with the movement of the gneisses. Norton's tectonic and metamorphic chronology now permits us to blame the gneisses for the higher slices of the Taconic allochthon also. The Everett Formation of the high Taconic slices and the Cheshire Quartzite on the Monument Mountain thrust fault are intimately and mutually involved around Tom Ball Mountain (Zen and Ratcliffe, 1971); the geometric relations do not seem to differ drastically from those among other Taconic slices.

If the Precambrian gneisses have been moved relative to the Paleozoic rocks as much as indicated by both Ratcliffe and Harwood's mapping, and if major thrust sheets exist within the Precambrian rocks, we must then ask how we should restore them palinspastically. The Precambrian rocks of the Berkshire massif have been considered more or less parautochthonous; maybe some of these rocks are more allochthonous than hitherto realized, perhaps approaching in tectonic style the Precambrian crystalline rocks of the North Carolina segment of the Blue Ridge anticlinorium (see Rodgers, 1970). How does this loss of crust manifest itself on the east flank of the Berkshire anticlinorium? Is the rearward steepening of the thrust fault that crops out near Woodruff Mountain (described by Harwood) such a root zone, or is it merely a refolded thrust fault? Is the root zone covered by eugeosynclinal rocks that moved later, in the style described by Norton? And how would such a possibility affect the interpretation of Cameron's line and of other geologic realities and speculations for western Connecticut?

It may well be that the late Taconic orogeny is the most important tectonic episode in this part of the Paleozoic world, second to none—not even to the Acadian! We have been conditioned to think of the Acadian orogeny whenever we see basement involvement, but is this valid? Maybe not, and certainly the research of Drake (1970) in the New Jersey highlands and the Reading Prong area indicates otherwise for the area to the southwest. Ratcliffe's work on both sides of the Hudson River in the Peekskill area suggests the same to me (Ratcliffe, 1968, 1970). But in late Middle Ordovician time, the tectonism was near-surface, submarine gravity sliding; a lot must have happened in a very short time, between then and the last part of the Ordovician, when the deformation moved to a deeper seated locus. This span involves perhaps 20 m.y., like between now and the middle Miocene. Viewed in this light, such a chronology does not seem so outrageous after all, if we recall how much had happened to California or the Alps or the Himalayas in this same time span.

REFERENCES CITED

Cady, W. M., 1969, Regional tectonic synthesis of northwestern New England and adjacent Quebec: Geol. Soc. America Mem. 120, 181 p.

Currier, L. W., and Jahns, R. H., 1941, Ordovician stratigraphy of central Vermont: Geol. Soc. America Bull. v. 52, p. 1487-1512.

Drake, A. L., Jr., 1970, Structural geology of the Reading Prong, *in* Fisher, G. W., Pettijohn, F. J, Reed, J. C., Jr., and Weaver, K. N., eds., Studies of Appalachian geology, central and southern: New York, Interscience Pubs., Inc., p. 271-291.

Harper, C. T., 1968, Isotopic ages from the Appalachians and their tectonic significance: Canadian Jour. Earth Sci., v. 5, p. 49-59.

Hatch, N. L., Jr., Schnabel, R. W., and Norton, S. A., 1968, Stratigraphy and correlation of the rocks on the east limb of the Berkshire anticlinorium in western Massachusetts and north-central Connecticut, *in* Zen, E-an, White, W. S., Hadley, J. B., and Thompson, J. B., Jr., eds., Studies of Appalachian geology, northern and maritime: New York, Interscience Pubs., Inc., p. 177-184.

Herz, Norman, 1961, Bedrock geology of the North Adams quadrangle, Massachusetts-Vermont: U.S. Geol. Survey Geol. Quad. Map GQ-139, scale 1:31, 680.

Norton, S. A., 1971, Possible thrust faults between Lower Cambrian and Precambrian rocks, east edge of the Berkshire Highlands, western Massachusetts: Geol. Soc. America Abs. with Programs, v. 3., no. 1, p. 46.

Osberg, P. H., 1965, Structural geology of the Knowlton-Richmond area, Quebec: Geol. Soc. America Bull. v. 76, p. 223-250.

Potter, D. B., and Lane, M. A., 1969, Some major structural features of the Taconic allochthon in the Hoosick Falls area, New York-Vermont, *in* Bird, J. M. ed., Guidebook for field trips in New York, Massachusetts, and Vermont: New England Intercollegiate Geol. Conf., 61st Ann. Mtg., Albany, New York, p. 12/0-12/23.

Ratcliffe, N. M., 1965, Bedrock geology of the Great Barrington area, Massachusetts [Ph.D. thesis]: University Park, Pennsylvania State Univ., 213 p.

——1968, Stratigraphic and structural relations along the western border of the Cortlandt intrusives, *in* Finks, R. M., ed., Guidebook to field excursions: New York State Geol. Assoc., 40th Ann. Mtg., Flushing, New York, p. 197-220.

——1969, Structural and stratigraphic relations along the Precambrian front in southwestern Massachusetts, *in* Bird, J. M., ed., Guidebook for field trips in New York, Massachusetts, and Vermont: New England Intercollegiate Geol. Conf., 61st Ann. Mtg., Albany, New York, p. 1/1-1/21.

——1970, Ancient strike-slip fault tectonics in the Hudson Highlands and Manhattan Prong: New York Acad. Sci. Trans., ser. 2, v. 32, p. 1009-1021.

Rickard, M. J., 1965, Taconic orogeny in the western Appalachians: Experimental application of microtextural studies to isotopic dating: Geol. Soc. America Bull. v. 76, p. 523-536.

Rodgers, John, 1970, The tectonics of the Appalachians: New York, Interscience Pubs., Inc., 271 p.

Skehan, J. W., S. J., 1961, The Green Mountain anticlinorium in the vicinity of Wilmington and Woodford, Vermont: Vermont Geol. Survey Bull. 17, 159 p.

Zen, E-an, 1967, Time and space relationships of the Taconic allochthon and autochthon: Geol. Soc. America Spec. Paper 97, 107 p.

Zen, E-an, and Ratcliffe, N. M., 1971, Bedrock geologic map of the Egremont quadrangle and adjacent areas, Massachusetts and New York: U.S. Geol. Survey Misc. Geol. Inv. Map I-628, with explanatory text, scale 1:24,000.

MANUSCRIPT RECEIVED BY THE SOCIETY JUNE 24, 1974
MANUSCRIPT ACCEPTED DECEMBER 5, 1975

Printed in U.S.A.

Geological Society of America
Memoir 148

Isotopic Dating and New England Stratigraphy

Richard S. Naylor*
Department of Earth and Planetary Sciences
Massachusetts Institute of Technology
Cambridge, Massachusetts 02139

ABSTRACT

Owing to the relative ease with which fossiliferous units can be traced into the highly metamorphosed, highly deformed terrane of southern New England, northern Appalachian orogenic events are subject to much tighter biostratigraphic control than is common in comparable areas. In most instances, for Ordovician through Early Devonian time, the precision of this biostratigraphic dating exceeds that which can be attained by isotopic dating, even for intrusive bodies. For the most part, post-Early Devonian stratigraphic control is lacking, and isotopic methods are commonly the best means of dating younger rocks and events. In spite of the good stratigraphic control and the existence of nearly 800 isotopic analyses in the northern Appalachians, fewer than 10 sets of data meet the rigorous requirements for good time-scale points. The best of these are in the late Precambrian, Middle Ordovician, Early Devonian, and Triassic time periods, where the Appalachian values agree with the commonly accepted time scales.

INTRODUCTION

Radiometric techniques have made the most significant contribution to the understanding of Appalachian geology through the dating of igneous plutons and metamorphic episodes. In keeping with the aims of the symposium on New England stratigraphy, however, this paper is limited to a discussion of the direct application of isotope geology to the dating of stratified units. The major purpose is to discuss data accumulated subsequent to the comprehensive compilation of northern Appalachian geochronologic data by Lyons and Faul (1968). Faul (1966) and Wetherill and Tilton (1967) may be consulted for concise explanations of the dating techniques,

*Present address: Department of Earth Sciences, Northeastern University, Boston, Massachusetts 02115.

the assumptions on which they depend, and the methods by which the data may be interpreted.

There is general agreement on the values of the decay constants (see Wetherill, 1966), except for the Rb-Sr method for which two values differing by 6 percent are in common use. The Rb-Sr ages in this paper are calculated using Rb decay constant of 1.39×10^{-11} yr^{-1}.

DATING OF BASEMENT ROCKS

Most Appalachian rocks, even in structurally complex areas, show a characteristic stratigraphic continuity that facilitates lithologic and fossil correlation. Even the meager fossil finds in highly metamorphosed rocks serve to correlate large stratigraphic sections where this continuity prevails. In most areas these well-stratified rocks are underlain by structurally complex gneissic basement rock in which even lithologic correlation is difficult. Long-range correlation is further hampered by the fact that most exposures of the different basement-rock complexes are not connected at the present erosion surface. Radiometric methods are the best means available for dating these rocks.

Two major findings have emerged from isotopic studies of the basement rocks in New England: (1) The age of the basement rocks differs on the two flanks of the Appalachian geosyncline. (2) In the center of the geosyncline, most of the gneiss in the local basement rocks is Ordovician and possibly Cambrian in age rather than Precambrian. Basement rocks on the west margin were subjected to high-temperature, high-pressure metamorphism about 1,000 to 1,200 m.y. ago and may include units that formed significantly earlier. On the eastern margin, the basement is chiefly greenstone that is cut by intrusive bodies of granite, 500 to 600 m.y. old, in which the presence of significantly older rocks has not been verified.

Radiometric dating confirms the lithologic and structural correlation of the basement rocks of the Green Mountain massif with the "Grenville" basement of the Adirondacks and farther west. Dating is complicated by disturbance of the mineral-isotopic systems that occurred during various Paleozoic orogenies. Pegmatite at Buttermilk Falls (lat 43°26.11′N, long 72°43.89′W) in the Green Mountains is at least 1,050 m.y. old, as measured by Rb-Sr on the central portions of large deformed muscovite books (Naylor, 1967, p. 110). Fine-grained muscovite from the pegmatite and the outer parts of the large books yield ages as young as 800 m.y. The samples have probably exchanged Rb and Sr with the country rocks to yield anomalously young ages. This effect is least in the center of the largest book, which yields the least-disturbed age. Small granite bodies along the eastern margin of the massif have apparent ages as young as 850 m.y. (Naylor, 1967, p. 114), but the data scatter due to the uneven response of the samples to subsequent disturbance. A similar range of apparent ages is reported by Faul and others (1963). Ratcliffe and Zartman (this volume) discuss U-Th-Pb isotopic ages in the billion-year range for samples from the Berkshire massif. Long (1969, p. 2087) reported Rb-Sr whole-rock ages as young as 570 m.y. for the Yonkers Gneiss north of New York City, but the interpretation of these ages is difficult. The samples with low Rb/Sr ratios, which are probably the more common rocks, yield ages approaching 1,000 m.y. when calculated using an initial Sr^{87}/Sr^{86} ratio of 0.708. The young ages come from samples with unusually high Rb/Sr ratios that could have been reset quite readily during the intense Paleozoic metamorphism to which the area has been subjected. Some rocks in the area could be as young as 570 m.y., but further work is needed before this conclusion should be extended to the bulk of the Yonkers Gneiss.

The Chester dome and other gneiss domes in southeastern Vermont are lithologically and structurally correlated with the basement rocks of the Green Mountain massif (Doll and others, 1961). Faul and others (1963, p. 7) reported a Pb-alpha age of 900 m.y. from the core-gneiss of the Chester dome, but further geochronologic study employing U-Pb isotopic data is needed to confirm the correlation of these rocks. Faul and others (1963, p. 7) reported a Pb-alpha age of 1,100 m.y. for zircons from their sample Vt6 on the east flank of the Green Mountain massif. Pinpointing the sample locality in the field demonstrates that sample Vt6 is from the basal Hossac Formation rather than from the basement rock as reported. The data indicate that the apparent age of detrital zircons was not significantly affected by later disturbance. More data are needed to evaluate the possibility that the zircons from the Chester dome core-gneiss might also be detrital.

The next basement exposures to the east are gneissic rocks in the cores of domes along the Bronson Hill anticlinorium in western New Hampshire. Naylor (1969) showed that the oldest rocks exposed in the Mascoma dome (a representative member of this group of gneiss domes) are Ordovician dacitic metavolcanic rocks (U-Pb isotopic zircon age 450 ± 25 m.y.) cut by slightly younger quartz-monzonite rocks. Whole-rock Rb-Sr data on the same samples confirm these results, but with less precision. Naylor (1968) concluded that the quartz monzonite in the core of the Mascoma dome was intruded at about the same time as the Highlandcroft Plutonic Series of Billings (1956). Plutonic rocks similar to the Highlandcroft are associated with Ordovician volcanic rocks as far north as the Presque Isle area in Maine. I interpret this volcanic belt extending from northern Maine to near New Haven, Connecticut, as the remnant of a chain of Ordovician volcanic islands, possibly an island arc.

Stratigraphic and structural arguments have been advanced to suggest that Precambrian rocks are exposed at several localities along this belt (Boucot, 1969, p. 55; Boudett, 1970, p. C-7; Robinson, 1967, p. 37). This has been confirmed by Naylor and others (1973), who reported zircon ages of 575 ± 15 m.y. for the Dry Hill Gneiss of the Pelham dome in Massachusetts and 1,500 ± 20 m.y. for rocks of the Chain Lakes massif in western Maine. Previously, the oldest ages reported were a U-Pb isotopic age of about 470 m.y. for zircons from the Attean Quartz Monzonite in western Maine (by L. T. Silver as reported by Boone and others, 1970, p. 19) and a Rb-Sr whole-rock age of 480 ± 15 m.y. for the Monson Gneiss in central Connecticut (Brookins and Methot, 1971, p. 20). Older ages are reported from gneiss along the southeast coast of Connecticut, but, as discussed below, I do not consider these to be a continuation of the Bronson Hill belt.

The next exposures of basement rocks to the east occur in a belt that extends from Dover, New Hampshire, southwestward through Fitchburg, Massachusetts. These rocks were mapped by Billings (1956) as the Fitchburg pluton. Sriramadas (1966) showed that the Fitchburg pluton in the Manchester area of New Hampshire consists of two components: cross-cutting muscovite-biotite granite (Concord Granite) and highly contorted, sillimanite-bearing gneiss (Massabesic Gneiss). I have obtained a Rb-Sr age of 350 ± 20 m.y. on coarse muscovite from Concord Granite from the Swenson Quarry in Concord, New Hampshire. Reconnaissance Rb-Sr whole-rock ages of 600 ± 200 m.y. (Naylor and others, 1969) suggest that the Massabesic Gneiss is an older basement complex. Along strike in York County, Maine, Hoefs (1967) reported a Rb-Sr whole-rock age of 460 ± 20 m.y. for the Webhannet and Lyman plutons. Much more work is needed in this very complex area, but the present isotopic data strongly suggest that rocks significantly older than Silurian and Devonian age (as shown on current maps) exist in this belt.

Along the southeast coast of New England and adjacent New Brunswick, a thick sequence of predominantly mafic volcanic rocks with intercalated metasedimentary rocks and granitic to gabbroic plutons lies underneath fossiliferous Lower and Middle Cambrian strata. I refer to these rocks as the "eastern-block" basement. Fairbairn and others (1967) cited the following Rb-Sr whole-rock ages for granite that is part of this basement in the Boston, Massachusetts, area: Hoppin Hill Granite, 514 ± 17 m.y.; Northbridge and Milford Granites, 569 ± 4 m.y.; and Dedham Granodiorite, 591 ± 28 m.y. Fairbairn and others (1966) reported the following Rb-Sr whole-rock ages from the basement volcanic rocks: Coldbrook (New Brunswick), 495 ± 48 m.y.; Fourchu (Cape Breton Island, Nova Scotia), 509 ± 40 m.y. Lowdon and others (1963) and Leech and others (1963) reported K-Ar ages of 500 m.y. to 640 m.y. for muscovite from a clast in conglomerate intercalated in the Coldbrook Group. The Hoppin Hill age may be anomalously low due to alteration, and the apparent ages of the volcanic units may be low for reasons discussed in the next section.

I consider gneisses that occur along the southeastern coast of Connecticut to be a part of this "eastern-block" basement and not an eastward extension of the Bronson Hill sequence. Hills and Dasch (1968, p. 136) reported a Rb-Sr age of 650 ± 50 m.y. on whole-rock samples of gneiss from the Stony Creek dome east of New Haven, Connecticut. Lithologically, these gneisses could be assigned to the Bronson Hill sequence (Dixon and Lundgren, 1968) or to the "eastern-block" basement. Geochronologic data support the latter correlation. Two sequences of units that resemble each other in lithology but differ in age by about 200 m.y. have apparently been structurally juxtaposed in the vicinity of New Haven.

DATING OF EARLY PALEOZOIC STRATIFIED ROCKS

Although it should be possible to date stratified volcanic rocks by radiometric means, the results in New England have been disappointing and indicate the great need for caution in interpreting the data. Mafic volcanic rocks seldom have Rb/Sr ratios high enough for precise dating of rocks younger than about 500 m.y. Intermediate and felsic volcanic rocks more commonly have suitable (higher) Rb/Sr ratios, but very commonly they yield anomalously low Rb-Sr whole-rock ages. Zartman (1968, oral commun.) suggested that the apparent sensitivity of these rocks to subsequent disturbance may be related to their content of primary glass. Ions of Rb and Sr readily migrate in and out of glass, and Zartman suggested that volcanic rocks may not obey the closed-system assumption until their primary glass had devitrified. In some cases, these rocks may behave as open systems (not reliably recording the passage of time) until they have been recrystallized by some later metamorphic event. Rocks containing calcite also fail to behave as closed systems under some conditions. Dating is seriously hampered by the lack of reliable petrographic criteria for rejecting "bad" samples prior to analysis. Fossils from volcanoclastic rocks continue to provide the most reliable means of dating early Paleozoic volcanic rocks in New England.

The direct dating of sedimentary rocks is hampered by the inherent difficulty of finding samples that satisfy the assumptions required for isotopic dating. Detrital grains contribute radiogenic isotopes derived from the parent rock, indicating apparent ages older than the time of sedimentation. The requirement for Rb-Sr whole-rock dating is that all the samples must have been formed at the same time and must have begun with the same initial ratio of Sr^{87}/Sr^{86}. The coarser detrital fraction of a sediment seldom satisfies this constraint because of inherited radiogenic Sr^{87} (Whitney and Hurley, 1964). Authigenic or fine-grained detrital components of sediment might satisfy this requirement if they were isotopically equilibrated with seawater at the time of deposition, but these fine-grained components seldom satisfy the closed-system assumption subsequent to deposition. Reliable petrographic criteria for selecting appropriate samples have not been developed. Fossils remain the best method for dating Paleozoic sedimentary and metasedimentary rocks.

DATING OF POST-DEVONIAN ROCKS

Rock- and biostratigraphic control is meager in the northern Appalachians for post-Lower Devonian strata; hence, radiometric dating provides much of the available data in this time range. Several pertinent studies have been published since the compilation of Lyons and Faul (1968).

Armstrong and Besancon (1970) reported K-Ar ages clustering at 200 m.y. for Upper Triassic diabase dikes and sills and basalt flows in the Connecticut valley. They concluded that these ages are slightly young due to loss of radiogenic argon.

Foland and others (1971) and Armstrong and Stump (1971) have made new K-Ar and Rb-Sr age determinations on plutons of the White Mountain Plutonic-Volcanic Series. Their most significant conclusion is that plutons of this series appear to have been emplaced over a span of 70 m.y.

ACKNOWLEDGMENTS

This work was supported by National Science Foundation Grant GA-25721.

REFERENCES CITED

Armstrong, R. L., and Besancon, James, 1970, A Triassic time scale dilemma—K-Ar dating of Upper Triassic mafic igneous rocks, eastern U.S.A. and Canada, and post-Upper Triassic plutons, western Idaho, U.S.A.: Eclogae Geol. Helvetiae, v. 63, no. 1, p. 15-28.

Armstrong, R. L., and Stump, Edmund, 1971, Additional K-Ar dates, White Mountain magma series, New England: Am. Jour. Sci., v. 270, p. 331-333.

Billings, M. P., 1956, The geology of New Hampshire, Pt. 2—Bedrock geology: New Hampshire Div. Econ. Devel. Mineral Resources Survey, Concord, N.H.

Boone, G. M., Boudette, E. L., and Moench, R. H., 1970, Bedrock geology of the Rangeley lakes-Dead River basin region, *in* Boone, G. M., ed., Guidebook for field trips in the Rangeley lakes-Dead River basin region, western Maine: New England Intercoll. Geol. Conf., 62d Ann. Mtg., Rangeley, Maine, 1970, p. 1-24.

Boucot, A. J., 1969, Geology of the Moose River and Roach River synclinoria, northwestern Maine: Maine Geol. Survey Bull., v. 21, 117 p.

Boudette, E. L., 1970, Pre-Silurian rocks in the Boundary Mountains anticlinorium, northwestern Maine, *in* Boone, G. M., ed., Guidebook for field trips in the Rangeley lakes-Dead River basin region, western Maine: New England Intercoll. Geol. Conf., 62d Ann. Mtg., Rangeley, Maine, 1970, p. C1-C21.

Brookins, D. G., and Methot, R. L., 1971, Geochronologic investigations in south-central Connecticut. Pt. 1—Pre-Triassic basement rocks: Geol. Soc. America Abs. with Programs, v. 3, p. 20.

Dixon, H. R., and Lundgren, L. W., Jr., 1968, Structure of eastern Connecticut, *in* Zen, E-an, White, W. S., Hadley, J. B., and Thompson, J. B., Jr., eds., Studies of Appalachian geology—Northern and maritime: New York, Interscience Pubs., p. 219-229.

Doll, C. G., Cady, W. M., Thompson, J. B., and Billings, M. P., 1961, Centennial geologic map of Vermont: Vermont Geol. Survey.

Fairbairn, H. W., Bottino, M. L., Pinson, W. H., Jr., and Hurley, P. M., 1966, Whole-rock age and initial $^{87}Sr/^{86}Sr$ of volcanics underlying fossiliferous Lower Cambrian in the Atlantic provinces of Canada: Canadian Jour. Earth Sci., v. 3, p. 509-521.

Fairbairn, H. W., Moorbath, Stephen, Ramo, A. O., Pinson, W. H., Jr., and Hurley, P. M., 1967, Rb-Sr age of granitic rocks of southeastern Massachusetts and the age of the Lower Cambrian at Hoppin Hill: Earth and Planetary Sci. Letters, v. 2, p. 321-328.

Faul, Henry, 1966, Ages of Rocks, planets, and stars: New York, McGraw-Hill Book Co., 109 p.

Faul, Henry, Stern, T. W., Thomas, H. H., and Elmore, P.L.D., 1963, Ages of intrusion and metamorphism in the northern Appalachians: Am. Jour. Sci., v. 261, no. 1, p. 1-19.

Foland, K. A., Quinn, A. W., and Giletti, B. J., 1971, K-Ar and Rb-Sr Jurassic and Cretaceous ages for intrusives of the White Mountain magma series, northern New England: Am. Jour. Sci., v. 270, p. 321-330.

Hills, F. A., and Dasch, E. J., 1968, Rb-Sr evidence for metamorphic remobilization of the Stony Creek Granite, southeastern Connecticut [abs.]: Geol. Soc. America Spec. Paper 121, p. 136-137.

Hoefs, J., 1967, A Rb-Sr investigation in the southern York County area, Maine: M.I.T. 15th Ann. Prog. Rept. to U.S. Atomic Energy Comm., 1381-15, p. 127-129.

Leech, G. B., Lowdon, J. A., Stockwell, C. H., and Wanless, R. K., 1963, Age determinations and geological studies (Including isotopic ages—Report 4): Canada Geol. Survey Paper 63-17, 140 p.

Long, L. E., 1969, Whole-Rock Rb-Sr age of the Yonkers Gneiss, Manhattan Prong: Geol. Soc. America Bull., v. 80, p. 2087-2090.

Lowdon, J. A., Stockwell, C. H., Tipper, H. W., and Wanless, R. K., 1963, Age determinations and geological studies (Including isotopic ages—Report 3): Canada Geol. Survey Paper 62-17, 140 p.

Lyons, J. B., and Faul, H., 1968, Isotope geochronology of the northern Appalachians, *in* Zen, E-an, White, W. S., Hadley, J. B., and Thompson, J. B., Jr., eds., Studies of Appalachian geology—Northern and maritime, New York, Interscience Pubs., p. 305-318.

Naylor, R. S., 1967, A field and geochronologic study of mantled gneiss domes in central New England [Ph.D. thesis]: Pasadena, California Inst. Technology.

——1968, Origin and regional relationships of the core-rocks of the Oliverian Domes, *in* Zen, E-an, White, W. S., Hadley, J. B., and Thompson, J. B., Jr., eds., Studies of Appalachian geology—Northern and maritime: New York, Interscience Pubs., p. 231-240.

——1969, Age and origin of the Oliverian Domes, central-western New Hampshire: Geol. Soc. America Bull., v. 80, p. 405-428.

Naylor, R. S., Gates, T. M., and Fryer, B. J., 1969, Basement rocks in the northern Appalachians—Preliminary data on the Massabesic Gneiss, Manchester area, New Hampshire: M.I.T. 17th Ann. Prog. Rept. to U.S. Atomic Energy Comm., 1381-17, p. 21-22.

Naylor, R. S., Boone, G. M., Boudette, E. L., Ashenden, D. D., and Robinson, Peter, 1973, Pre-Ordovician rocks in the Bronson Hill and Boundary Mountain anticlinoria, New England, U.S.A. [abs.]: Am. Geophys. Union Trans., v. 54, p. 495.

Ratcliffe, N. M., and Zartman, R. E., 1976, Stratigraphy, isotopic ages, and deformational history of basement and cover rocks of the Berkshire massif, southwestern Massachusetts, *in* Page, L. R., ed., Contributions to the stratigraphy of New England: Geol. Soc. America Mem. 148 (this volume).

Robinson, Peter, 1967, Progress of bedrock geologic mapping in west central Massachusetts, *in* Farquhar, O. C., ed., Economic geology in Massachusetts, Proc. Conf., Amherst, 1966: Amherst, Univ. Massachusetts, p. 29-44.

Sriramadas, Aluru, 1966, Geology of the Manchester quadrangle, New Hampshire: New Hampshire Dept. Resources and Econ. Devel. Bull. 2, 78 p.

Wetherill, G. W., 1966, Radioactive decay constants and energies, *in* Clark, S. P., Jr., ed., Handbook of physical constants: Geol. Soc. America Mem. 97, p. 513-520.

Wetherill, G. W., and Tilton, G. R., 1967, Geochronology, *in* Abelson, P. H., ed., Researches in geochemistry: New York, John Wiley & Sons, v. 2, p. 1-28.

Whitney, P. R., and Hurley, P. M., 1964, The problem of inherited radiogenic strontium in sedimentary age determinations: Geochim. et Cosmochim. Acta, v. 28, no. 4, p. 425-436.

MANUSCRIPT RECEIVED BY THE SOCIETY APRIL 18, 1974
REVISED MANUSCRIPT RECEIVED OCTOBER 22, 1974
MANUSCRIPT ACCEPTED NOVEMBER 12, 1974

Printed in U.S.A.

Index